Proceedings of the Conference on

Construction of Power Generation Facilities

Experience with the Implementation of Construction
Practices, Codes, Standards and Regulations

Pennsylvania State University
University Park, Pennsylvania

September 16-18, 1981

Edited by
Jack H. Willenbrock
Department of Civil Engineering,
Pennsylvania State University

Published by the
American Society of Civil Engineers
345 East 47th Street
New York, New York 10017

TH
4581
.C64
1981

The Society is not responsible for any statements made or opinions expressed in its publications.

Copyright © 1982 by the American Society of Civil Engineers,
All Rights Reserved.
Library of Congress Catalog Card Number 82-70491
ISBN 0-87262-306-8
Manufactured in the United States of America.

EXPERIENCE WITH THE IMPLEMENTATION OF CONSTRUCTION
PRACTICES, CODES, STANDARDS AND REGULATIONS IN THE
CONSTRUCTION OF POWER GENERATION FACILITIES

Sponsored By

A.S.C.E. Construction Division: Committee on
Construction of Nuclear Facilities

Conference Co-Chairman:

1. Jack H. Willenbrock*
   Professor of Civil Engineering
   The Pennsylvania State University

2. Aldo Palmeri*
   Supervisor, Civil Engineering
   Ebasco Services, Inc.
   (Chairman: Committee on Construction of Nuclear Facilities)

Conference Committee:

1. Eugene J. Gallagher* - U.S.N.R.C.
2. John C. Archer* - Burns & Roe, Inc.
3. William F. Mercurio* - Ebasco Service, Inc.
4. James H. Olyniec* - T.V.A.
5. Joseph F. Artuso - Construction Engineering Consultants, Inc.
6. Stephen R. Toth* - Northeast Utilities Service Company
7. Clyde B. Tatum* - Ebasco Services, Inc.
8. John M. Fisher, Jr.* - Gilbert Commonwealth Companies
9. Glen A. Chauvin* - Sargent & Lundy Engineers
10. Alice Gannon - Gilbert Commonwealth Companies

*Members: Committee on Construction of Nuclear Facilities

FOREWORD

This Speciality Conference was developed by the A.S.C.E. Construction Division: Committee on Construction of Nuclear Facilities and presented at the University Park Campus of The Pennsylvania State University in September 1981. A total of 12 session chairman, 54 authors of papers and 108 attendees interacted during the conference in a keynote session, 8 concurrent sessions and a wrap-up session.

The conference was both timely and important because over the next twenty-five years, if the critical energy needs of the United States are to be met, utility companies must continue to make a large capital investment in the construction of both fossil fuel and nuclear electrical-power generation facilities. The industry is, however, at a critical crossroad as it attempts to balance factors such as complexity of design and construction, length of time required for completion and magnitude of financial investment required against the ever increasing industry codes and standards, and governmental regulatory and quality assurance requirements which are being imposed.

The conference is designed to focus on this critical issue by not only documenting what has occurred on completed projects (or on projects which are in progress) but also by serving as a forum which will establish what improvements can be made in the future as a result of the lessons learned in the past. In order to meet this objective each individual session was organized to include the following elements:

- A review of the current industry codes, standards and regulations which are applicable to the topic area and an analysis of any conflicts, similarities and problem areas which exist.

- Examples of actual construction practices and experiences related to the topic area which indicate how the requirements of the applicable industry codes, standards and regulations have been implemented on actual construction projects. The effect which these practices and experiences have had on quality, cost and schedule will be indicated.

- A panel-audience discussion which will identify the four or five most pressing improvements which must be made in the topic area in the future.

This proceeding provides the reader with the complete text of the papers in each session, as well as individual "Session Summaries," written by each session chairman, which capture not only the central theme of the papers in each session but also document the results of the panel-audience discussion which followed the presentation of the papers.

It should be noted that this conference represents a natural extension of a liaison between the electric power plant industry and the Department of Civil Engineering and the College of Engineering of The Pennsylvania State University which began in 1975 when a Power Plant Construction Advisory Group was formed. This group (which consisted of: William A. Frederick, Pennsylvania Power and Light Company; Glenn J. Davidson and Alice Gannon, Gilbert Commonwealth Companies; Thomas P. Gotzis, Philadelphis Electric Company; Louis R. Larsen, Stone & Webster Engineering Corporation; Saxon B. Palmeter, GPU Service Corporation; Edmund F. Raspa, United Engineers & Constructors, Inc.; and John P. Theriault, Bechtel Power Corporation) has assisted faculty members since then with regard to the following accomplishments:

   a. Presentation of two Seminars at Penn State on "The State of the Art of Power Plant Construction" - August 1-13, 1976 and May 22-26, 1978.

   b. Presentation of a "Power Plant Construction Educational and Research Needs Workshop" at Penn State - May 10, 11, 1977.

   c. Participation in a U.S. D.O.E. sponsored research project entitled, "A Comparative Analysis of Structural Concrete Quality Assurance Practices on Nine Nuclear and Three Fossil Fuel Power Plant Construction Projects" - Completed December 1978.

   d. Development of a textbook entitled, "Planning, Engineering and Construction of Electric Power Generation Facilities," edited by J. H. Willenbrock and H. Randolph Thomas, published by Wiley Interscience, May 1980.

It will be noted that a number of the speakers at this Speciality Conference are employed by firms which are members of the above mentioned Advisory Group. Hopefully, such interaction between industry and engineering educators, at both individual universities and under the auspices of the American Society of Civil Engineers, will help to solve some of the problems and take advantage of some of the opportunities which the electric power plant industry will face in the years ahead.

This particular Speciality Conference, with its somewhat narrow focus of interest will be considered successful if the information is instrumental in causing fundamental changes to be made to industry codes and standards and governmental regulatory and quality assurance requirements <u>where they are required</u>. Those changes will certainly influence how effectively electric power generation facilities are built from both a time and a cost standpoint during and beyond the 1980's.

Jack H. Willenbrock
University Park, PA
December 1981

TABLE OF CONTENTS

OPENING KEYNOTE SESSION

    Session Objectives/Session Chairman Summary 1
    -Aldo Palmeri, Jack H. Willenbrock

1. A Utilities Perspective 6
   -William G. Counsil

2. An Architect-Engineer Perspective 25
   -Russell J. Christesen

3. Regulations Affecting Power Plant Construction 41
   -Howard W. McCall

4. Nuclear Plant Construction—Government Perspective 51
   -James G. Keppler

5. Concrete Industry Perspective 58
   -Robert E. Philleo

SESSION I: NUCLEAR STANDARDS, LICENSING, ENFORCEMENT, AND CONSTRUCTION INSPECTION PROGRAMS

    Session Objectives/Session Chairman Summary 68
    -Eugene J. Gallagher

6. Nuclear Plant Civil-Structural Design Standards 71
   -John D. Stevenson

7. The Nuclear License Picture—Overview and Outlook 107
   -Richard H. Vollmer

8. Nuclear Regulatory Commission Enforcement Policy 112
   -Charles E. Norelius

9. USNRC Inspection of Reactors Under Construction 120
   -Charles F. Heishman

10. Have We Effectively Adjusted to Nuclear Regulation? 128
    -Robert J. Washabaugh

SESSION II - CONCRETE MATERIALS: REQUIREMENTS AND PERFORMANCE

    Session Objectives/Session Chairman Summary 137
    -John C. Archer

11. Concrete Aggregate Requirements and Performance 140
    -Ward R. Malisch

12. Portland Cements: Requirements vs. Performance 150
    -William F. Perenchio

13. Fly Ash, Admixtures, and Water Quality 162
    -Rupert E. Bullock

14. Mix Design Requirements for Nuclear Power Plants 170
    -Gerald R. Murphy, Chaman L. Grover

15. Concrete Materials: Requirements vs. Performance 185
    -Richard A. Bradshaw, Jr.

SESSION III - QUALITY ASSURANCE/QUALITY CONTROL PRACTICES

Session Objectives/Session Chairman Summary 200
-William F. Mercurio, James H. Olyniec

16. Organizing for QA/QC Functions 205
    -Clyde L. Hawn

17. Qualification and Training of Inspectors 225
    -Donald R. Johnson, Dennis L. Vanderpol

18. QA/QC Training, Certification/Recertification 234
    -Kevin T. Kimmel

19. Civil Engineering Unit Documentation Practices 242
    -James H. Olyniec

20. Information Processing Methods and Applications 260
    -Dennis D. Millican

SESSION IV: CONCRETE TEST REQUIREMENTS AND PROCESS CONTROL PRACTICES

Session Objectives/Session Chairman Summary 276
-Joseph J. Artuso

21. Control of Concrete Mixing and Testing 279
    -J. R. Wells

22. Sampling of Concrete and Concrete Constituents 288
    -Douglas J. Haavik

23. Strength and Special Properties of Concrete 299
    -James P. Allen, III

24. Relevance of ASTM C-342 Volume Change Test 309
    -Ashok J. Desai, Allen J. Hulshizer

25. Utilizing Test Data for Greater Quality Control 325
    -Reginald Coupland

26. Creep/Shrinkage Studies: Containment Structures 335
    -Mauro J. Scali, Donald W. Pfeifer

SESSION V: STEEL FABRICATION PRACTICES, CONCRETE ANCHORAGE SYSTEMS, PLANT MODIFICATION AND RETROFIT EXPERIENCE

    Session Objectives/Session Chairman Summary 357
    -Stephen R. Toth

27. Field Installation of Concrete Anchorage Systems 361
    -James A. Flaherty, Louis J. DiLuna

28. Replacement of Steam Generators: PWR Nuclear Plant 372
    -Wallace G. Sanborn

29. Millstone III Containment Dome Lift 389
    -James A. Galinsky

30. Steel Fabrication in the Nuclear Power Industry 396
    -Denis Mason

31. Nuclear Construction Concrete Anchor Experiences 403
    -Thomas W. Deshefy

SESSION VI: CONCRETE BATCH PLANT PRODUCTION TRANSPORTATION AND FIELD ADJUSTMENT PRACTICES

    Session Objectives/Session Chairman Summary 415
    -C. B. Tatum, John M. Fisher

32. Concrete Plant and Production Operation Standards 417
    -Clyde B. Tatum, Albert P. Demers

33. Standards Impact on Mix Verification/Adjustment 435
    -Bruce C. Bennett, Robert V. Potter, John R. McCutchen

34. From Batch Plant to Form 447
    -William H. Brown

35. Intent of Concrete Adjustment Criteria 453
    -Timothy L. Moore

36. Is Criteria as Written Enough? 465
    -Robert M. Eshbach

SESSION VII: CONCRETE FORMWORK, REINFORCEMENT AND PLACEMENT TOLERANCES AND PRACTICES

Session Objectives/Session Chairman Summary 472
-Glen A. Chauvin, Aldo Palmeri

37. ACI Standard 117: Concrete Tolerances 474
    -J. Doug Sykes

38. Relevance of Enforced Construction Tolerances 488
    -Allen J. Hulshizer

39. Engineering Tolerances: A Construction Viewpoint 497
    -Frank J. Freiseis

40. Nuclear Tolerance Requirement Impact on Supplier 508
    -Robert A. Yockin

41. Modeling of Complex Reinforcing Steel Placements 522
    -Alan J. Boos

SESSION VIII: CONCRETE PLACEMENT AND POST PLACEMENT PRACTICES

Session Objectives/Session Chairman Summary 529
-Alice Gannon, Jack H. Willenbrock

42. Concrete Practice Changes at the Limerick Project 532
    -Alex C. McLean

43. Hot Weather Temperature Placement of Mass Concrete 542
    -James P. Lonergan

44. Cold Weather Placement and Post Placement Control 554
    -Donald E. Dixon

45. Quality Evaluation of Nuclear Concrete Structures 566
    -Joseph J. Artuso

FINAL WRAP—UP SESSION: RESULTS OBTAINED ON THE U.S.D.O.E. SPONSORED RESEARCH PROJECT

Session Objectives/Session Chairman Summary 587
-Jack H. Willenbrock

46. Concrete Q. A. Practices on Nuclear Power Projects 588
    -Jack H. Willenbrock, H. Randolph Thomas

PROGRAM 605

SUBJECT INDEX 611

AUTHOR INDEX 613

OPENING KEYNOTE SESSION

SESSION OBJECTIVES/SESSION CHAIRMAN SUMMARY

by

Aldo Palmeri[1], M. ASCE and Jack H. Willenbrock[2], M. ASCE

Objective of Session

The keynote session presents the viewpoints of five prominent leaders in the Power Industry on the Theme of the Conference. In this session representatives of the utilities, architect/engineers, constructors, U. S. Nuclear Regulatory Commission and the concrete industry analyze the impact of the present construction practices, codes, standards and regulations on the construction of power generation facilities. From their respective points of view they examine the major problems related to codes and regulations, the difficulty of implementing some of the requirements, and the possible improvement recommendations that can be made.

Construction for power generation is a large portion of the construction industry, therefore, improvements in the codes, standards and regulations for power facilities represent improvements to the construction industry as a whole.

Session Chairman Summary

A. UTILITY VIEWPOINT

The licensing process for the construction of power plants, and especially nuclear plants, is very complex since the power industry is the most regulated industry in the nation. The time from the decision to build a nuclear power plant to completion is about 14 years.

The basic laws that regulate the design and construction of nuclear plants are contained in Title 10 of the Code of Federal Regulation (10 CFR). These regulations are "clarified" by an enormous number of Regulatory Guides. Currently, there are over four hundred in effect. Since both the regulations and the guides are subject to constant updating, the industry is plagued by the "backfit problem." The effect of all those regulations is an escalation in costs, and long delays. The delays are also derived from an NRC inability to issue licenses in a timely fashion. In reference to the NRC's recently adopted value impact analysis, it was felt that the evaluation of alternative courses of action is generally done incorrectly. The costs are not properly weighed against benefits and the schedules are grossly under-estimated. To improve the overall licensing process, the following suggestions were offered:

---

[1] Supervisor, Civil Engineering, Ebasco Services, Inc., Princeton, NJ.

[2] Professor, Department of Civil Engineering, The Pennsylvania State University, University Park, PA 16802

1. Prioritization of NRC Rulemaking Proceedings.
2. Stabilization of NRC Regulations
3. Streamlining of NRC Licensing Proceedings.

Some of the specific steps proposed were:

1. A review of all rules, regulations, policies, positions, etc., and the performance of cost/benefit analysis and comparison of each result to established safety goals in order to justify their use.
2. Consideration of the one-stage licensing process.
3. Consideration of the elimination of public hearings and the hearing boards.
4. Certification of plant designs as adequate analogous to FAA Airworthiness Certification.

As for the industry, its primary responsibility is for codes and standards. Simplifying or improving these documents is therefore an industry chore. Some of the problems encountered by the utilities in the construction of nuclear plants are related to code dates and to the overlapping and conflicting code requirements.

Since some of the codes specify that fabrication shall meet the requirements in force at the time of the contract, and since contracts are issued at different times, it may occur that the same piping system built in different time spans will be fabricated to different code requirements. An example of overlapping a requirement in the design of fire protection for electrical equipment was noted.

Another example of conflicting requirements was related to the inspection of piping systems. These systems are inspected during fabrication by radiographic techniques. They must then be reinspected by ultrasonic methods, with often conflicting results occurring.

It was felt, in conclusion, that the NRC, with all the rules and regulations imposed on the utilities, has almost removed the nuclear option from further consideration.

B. ARCHITECT/ENGINEER'S VIEWPOINT

It was noted that the electric power industry is the largest user of the engineering/construction industry resource with huge capital expenditures and substantial employment of construction workers and engineers. The growth of the generation of electricity has been at an average of 7.6 percent per year up to 1970, but since 1973, the growth has slumped to a rate less than 1.1 percent per year. The electric utility in recent years has been beset by several problems. Proliferation of government regulations, high interest rates on construction capital, escalation of labor and materials, and construction delays have increased the cost of new facilities by a factor of more than seven since the mid-sixties. Regulatory excesses have had a major negative impact on the ability of the electric power industry to plan and finance new facilities. These regulations have increased costs dramatically and have stretched schedules considerably. The major regulations for the design and construction of coal-fired power plants are the National Environmental Policy Act, the Clean Air Act, and the Water Pollution Control Act. For nuclear plants the NRC regulation is Title 10 of the Code of Federal Regulations together with 540 regulatory guides. The impact on direct craft manhours and costs of these regulations has been staggering. For instance, for nuclear plants, the new regulatory requirements have increased costs to four times the rate of inflation. Streamlining and making the regulatory process realistic, responsive and cost effective is one of the critical national agenda items of the 1980's.

To the engineering/construction industry, the following three recommendations are offered:

    Reduce conservatism in design.

    Resolve the ambiguities and conflicts between codes.

    Improve acceptance criteria and tolerances.

Conservatism in design has resulted from the inability to accurately define the loads and from the fact that conventional codes, standards and practices are not compatible with the actual requirements.

To remedy this situation, the industry should initiate a test program to provide reliable information on the loads and update the codes, making them simpler.

Codes today interact, overlap and conflict with other codes, engineering standards and client requirements. Therefore, the industry should establish a group to provide an overall management of code-related problems. As for tolerances, the engineer should examine the requirement of his specific design before adapting tolerances from a code. The intended usage of structure or system should dictate the tolerances required. As for acceptance criteria, more "hard numbers" should be provided for "fitness for purpose" acceptance.

In conclusion, it was felt that the nuclear policy of the new administration in Washington offers significant opportunities for making regulations of the design and construction of nuclear facilities more cost effective while still adequately protecting the health and safety of the public.

C. CONSTRUCTOR'S VIEWPOINT

It was felt that regulations have increased dramatically every year, especially in the last 10 years, and that requirements have been added to the construction of power generation facilities without adequately assessing their overall impact. Many of these requirements stem from codes, others from design. The various Code committees and the engineers must take a realistic approach to establishing these requirements. The design must be innovative and not rely on "good old" standards just because they have been used before. The design must also be constructible.

Another problem also arises from the difference between the intent of the requirements as they appear in the code and the way they are applied or interpreted during inspection. To alleviate the problems, a joint team should be established between constructors, designers, code committee, and regulatory bodies. To avoid potential problems, constructors with the participation of the designer should do more "up-front" work. A well-planned program of construction engineering and quality organization should be established. As for the recently proposed third-party inspection by the Federal Energy Regulatory Commission, it was felt that this inspection would not be beneficial since it would create more divisions between the various parties. Inspections in the field should be performed by the constructor who should police himself.

It was noted that the constructor quality organization should do more than inspection and verification. It should also perform quality engineering. It should insure that the project requirements are correctly identified, understood, and properly communicated to the inspectors. On the other hand, the construction engineer should inform the craftsman and provide him with the materials to perform his job well and thus increase his productivity. Due to the complexity of the projects of today, a constant flow of communication should be established between the designer, construction engineer, and the craftsman. This communication should be in both directions - from the designer to the craftsman and vice versa. A major role of the constructor is to make sure that all the information required to perform a job is available at the level of implementation.

In summary, there should be better communication in all areas, between code committee, designers, and constructors.

D. GOVERNMENT PERSPECTIVE

Since the Three Mile Island accident, the NRC has been placing, and continues to place, many new requirements on the construction and the operation of nuclear power plants. Specifically, the NRC has increased the inspection program, has toughened its enforcement posture, and has established an annual review of the performance of the licensees.

Past experiences in the construction of nuclear plants have indicated failures in the QA programs both by the licensee as well as by the NRC. Primarily, these failures have occurred because most, if not all, of the QA programs have been handled by the contractors at the site. To improve this situation, the NRC in 1978 initiated a Resident Inspection Program. Generally, two inspectors are assigned full-time at the construction site. In addition, there are periodic visits by specialists from the NRC regional office. NRC has also toughened the enforcement policy by substantially increasing fines for licensee violations.

Recently, the NRC introduced a new program - SALP (Systematic Assignment of Licensee Program). With this program the NRC reviews the entire yearly performance of the licensee and compares it to the performance of other licensee The SALP appraisal has been found to be very useful. It has revealed patterns of deficiency that a day-by-day review did not reveal.

In conclusion, it was noted that a recent congressional report states that the NRC has been too tolerant in the past. Therefore, a stronger enforcement posture must be taken. It must be made clear to the industry that a strong licensee performance and a demanding regulatory overview are essential for the public acceptance of nuclear power.

E. CONCRETE INDUSTRY PERSPECTIVE

The construction industry is a big user of energy. Part of this energy is used for the production of cement. In view of the high cost of energy, the concrete industry has taken several conservation measures. In the production of cement, long rotary kilns with transportation of material by water slurry have been replaced by smaller, single-temperature kilns with mechanical handling of the material. The kilns are fired with high sulfur coal instead of gas or oil. The use of this coal made necessary a change in the cement specifications by increasing the amount of permissible SO_3.

Fly ash or blast furnace slag have also replaced part of the cement in the production of concrete. These two materials are readily available since they are waste materials of the power-generating or steel-producing industries. Fly ash is found useful in mass concrete since it provides less heat generation during hydration.

As for concrete standards, the ACI has written them mostly in the form of recommended practice. It was noted that ambiguities are introduced when they are adopted directly into the engineer's specification. The ACI is now responding to this problem by writing more of its standards in specification language. In reference to nuclear standards, a comparison was made between standards issued for high dams where the consequences of accidents is also very high. Where the Code for Concrete for Reactor Vessels and Containment had 280 pages on construction and inspection, the dam manual has only 84 pages. Poor construction is not attributable to simply written standards. A simple set of standards is perfectly adequate if coupled with a good enforcement program. Inspection is the key to safe and successful construction.

Consensus Recommendations

There was insufficient time at the end of the session for a Panel of Speakers/Audience discussion period.

A UTILITIES PERSPECTIVE

By

William G. Counsil[1]

Abstract

Northeast Utilities is the principal supplier of electric power in the state of Connecticut and in the western third of the state of Massachusetts. Currently, nuclear power provides about 70 percent of the electrical needs of our service area. In 1980, nuclear power supplied 50% of our electricity. This large percent of nuclear electric generation is directly attributable to an ambitious nuclear power plant construction program carried out over the past fifteen years. Currently, Northeast Utilities operates three nuclear power plants, Millstone Units 1 and 2 and Connecticut Yankee. We have a fourth nuclear power plant under construction, Millstone Unit 3.

The Millstone Unit 3 project was started in 1970 with an initial completion date of 1978. Since then, due to a series of difficulties, primarily inadequate rate relief from the State Regulators, the Project has been stretched out to the current completion date of 1986. This significant extension of the original completion schedule has exposed Millstone Unit 3 to a frustrating array of ever increasing licensing and regulatory requirements.

Today, I plan to share with you, from a utility point of view, some of the difficulties we encounter in licensing the design, construction, and operation of nuclear power plants. We believe that some of these difficulties stem from poorly planned, and in some cases, excessive requirements.

As a starting point, I will describe briefly the licensing process for the design and construction of a nuclear power plant, including the array of statutes controlling the process. Next, since the Nuclear Regulatory Commission (NRC) has the overall responsibility for regulating the nuclear industry, I will address the many regulations they impose and the proliferation of these requirements over the past decade. I will discuss some of the unnecessary cost the NRC has imposed on the industry because of their delay in issuing licenses in a timely fashion; and difficulties they seem to have in performing value-impact analysis of their proposed regulations. I will then address several areas in NRC licensing arena which need improvements and make specific recommendations. Next, since we in the industry also have problems that need correcting, I will point out some specific examples of troubles we have had with the current codes and standards.

---

[1] Senior Vice President, Nuclear Engineering & Operations Group
Northeast Utilities Service Company

THE REGULATORY PROCESS

The licensing of a nuclear power plant is one of the most complex legal processes that a corporation can undertake. It has become a gamble taken with significant financial risk.

The utility industry in general, and particularly those organizations which construct or operate nuclear power plants are the most regulated industry in the nation. Over the years, the United States Congress and our Connecticut Legislature has enacted a great number of laws and statutes applicable to nuclear power facilities. The most important of these are the Atomic Energy Act, the National Environmental Policy Act, the Clean Water Act and the Clean Air Act.

To give you a flavor of the extent of these laws and statutes, I have prepared a tabulation (Figure 1) of the Federal Laws that confront a utility desiring to construct a nuclear power plant. In addition to these 17 Federal Laws, we, in Connecticut, must conform to 4 State Laws.

(1) Connecticut Environmental Policy and Protection Acts
(2) Connecticut Air Pollution Control Laws
(3) Connecticut Clean Water Act
(4) Connecticut Public Utility Environmental Standards Act

The Nuclear Regulatory Commission (NRC), in conformance with the authority invested in it by the Atomic Energy Act, as amended, is responsible for regulating the design, construction and operation of nuclear power plants to ensure that they present no undue hazard to public health and safety. The NRC also acts as the lead agency in the licensing process and is charged with assessing the environmental impact of the nuclear facility.

The Environmental Protection Agency is responsible for non-radiological aspects of plant design, construction and operation. The Army Corps of Engineers must approve the design and construction of structures, and dredging in navigable waters. Two State agencies, the Department of Environmental Protection and the Power Facility Evaluation Council, have important roles in Connecticut to ensure that state, as well as federal statutory mandates are met.

The state and federal agencies, in carrying out their mandate, produce voluminous, complex, and many times overlapping regulations. Further, these regulations are issued with a multitude of guidelines, standards and codes to implement the regulatory policy.

To obtain permission to build a nuclear station, a utility files a Construction Permit Application with the NRC. The application consists of four separate documents.

TABLE 1

FEDERAL LAWS APPLICABLE TO NUCLEAR FACILITIES

Atomic Energy Act of 1954

National Environmental Policy Act of 1969 (NEPA)

Clean Air Act of 1977 (CAA)

FWPCA (Clean Water Act of 1977) (CWA)

Marine Protection, Research and Sanctuaries Act of 1972 (MPRSA)

Coastal Zone Management Act of 1972 (78 Amendment) (CZMA)

Noise Control Act of 1972 (NCA)

Federal Environmental Pesticide Control Act of 1972 (FEPCA)

Ports and Waterways Safety Act of 1972 (PWSA)

Marine Mammal Protection Act of 1972 (MMPA)

Endangered Species Act of 1973

Deepwater Port Act of 1972 (DPA)

Safe Drinking Water Act of 1974 (SDWA)

Resource Conservation and Recovery Act of 1976 (RCRA)

Toxic Substances Control Act of 1976 (TSCA)

Federal Land Policy Management Act of 1976

Power Plant and Industrial Fuel Use Act of 1978 (PPIFUA)

(1) General and Financial Information
(2) Antitrust Information
(3) Applicants' Environmental Report (ER)
(4) Preliminary Safety Analysis Report (PSAR)

The first document is intended to show the financial arrangements of applicants and their financial ability to build and operate the plant. The second document is reviewed by the U.S. Justice Department to assure that all antitrust laws will be met. The third document, the applicants' ER, is a multi-volume, detailed study of the anticipated environmental affects of plant construction and operation on the surrounding areas and population. The final document, the PSAR, is a voluminous report which specifies the engineering and construction bases for all structures and systems in the plant. This includes detailed specifications of the codes and standards to be utilized.

At the same time or closely following the NRC application, a separate application must be filed with the Connecticut Power Facility Evaluation Council (PFEC) to obtain a Certificate of Public Need and Environmental Compatibility. PFEC ensures that the interests of Connecticut are met by conducting its own review of the utility's proposal.

While this activity is on-going, the utility also must seek as many as 30 other federal, state and local permits and licenses (See Table 2). Some of these require separate public hearings and rejection of any permit may delay the overall licensing schedule.

About three years before the completion of the construction, the utility must submit a Final Safety Analysis Report (FSAR) and a Operating License Stage Environmental Report (EROLS). These documents are updated versions of the PSAR and ERCPS. They reflect approved revisions during construction and the final design of the plant. Also submitted are the Safety and Environmental Technical Specifications for plant operation. These specifications detail operating procedures and monitoring requirements, reporting responsibilities and limits under which the plant can operate.

The result of having to satisfy all these requirements is a time span of approximately fourteen years from the decision to build the plant to the production of electricity.

NUCLEAR REGULATIONS

The Nuclear Regulatory Commission (NRC) is the governmental agency with overall responsibility for the regulation of nuclear power. The NRC carries out this responsibility through the issuance and enforcement of regulations contained in Title 10 of the Code of Federal Regulations (10 CFR) titled "Energy". The regulations are changed or expanded continually. The process involves publication of the proposed regulations in the Federal

TABLE 2
TYPICAL PERMITS REQUIRED

FEDERAL

	NUCLEAR PLANT
Nuclear Regulatory Commission	Early Notification of Proposed Facility
	Construction Permit (CP)/Limited Work Authorization (LWA)
	Special Nuclear Material License
Environmental Protection Agency	316(a) & (b) Determination/NPDES Permit*
Corps of Engineers	Dredging Permit
	Structure Permit
	Fill Permit (Rip Rap)
U.S. Coast Guard	Intake Structure/Cofferdam Lighting
Federal Aviation Administration	Obstruction & Lighting of Containment and/or Cooling Towers
National Marine Fisheries Service (Marine)	Endangered Species Consultant
Fish and wildlife Service (Inland)	
Advisory Council on Historic Preservation	Archeological & Historic Consultation
Department of Energy	None

STATE

	NUCLEAR PLANT
Power Facilities Evaluation (CT)	Certificate of Need and Environmental Compatibility
Energy Facilities Siting Council (MA)	Similar to PFEC
Department of Enviromental Protection	(1) Aux. Boiler (2) Diesel Generators (3) NPDES Permit (4) Water Quality Certification (5) Dredging Permit (6) Structures Permit (7) Fill Permit

LOCAL

Building Inspector	Building Permit
Conservation Commission	Inland Wetlands Permit

---

*Only if located in State without approved NPDES Implementation Plan.

Register and the solicitation of comments from interested persons
and organizations. The NRC staff then reviews the comments and a
final version of the regulation is prepared. If approved by the
Commissioners, it is incorporated into 10 CFR. If the regulation
is of great concern, controversial or if information and testimony
are required, it then goes into Rulemaking before finally being
incorporated into the Regulations. Rulemaking is a quasi-judicial
process presided over by a hearing board where testimony is given
by experts and concerned persons. The resulting hearing record
is summarized by the Board and presented to the Commissioners.
If the Commissioners approve, the final rule is incorporated into
the regulations.

The regulations, in many cases, tend to be vague or subject to
interpretation. To provide clarification as to acceptable means
for meeting the regulations, the NRC publishes several type of
documents. The three most significant ones are:

(1) Regulatory Guides
(2) NUREG Reports
(3) Standard Review Plan

These three documents generally state explicit requirements
and/or reference specific industry codes and standards, which
adds further to the proliferation of requirements.

Although the above three documents do not have the force of the
laws as does Title 10 CFR, they are considered to provide guidance
for conforming to the requirements. Failure to follow the
guidance is generally looked upon as failure to meet the regulations.
Alternative approaches, although legal, are frowned upon and
seldom found acceptable.

This introduces one of the problems that deeply concerns the
utility industry. That is over-regulation of the nuclear industry
through the proliferation of regulatory requirements and the
adverse consequences of this over-regulation.

Proliferation of Requirements
In order to illustrate the gravity of this situation, I have
tabulated the number of Regulatory guides and their revisions
(Division 1, 4, 5, 7 and 8 only) over the past decade. These are
shown in Table 3. Notice that in 1970, we had only a handful,
but by the mid 1970's the number of Regulatory Guides and their
revisions had grown to approximately three hundred. Currently,
there are over four hundred in effect.

The Backfit Problem
Initially, many of these Regulatory Guides were issued to provide
guidance only, or for implementation at a later date on future
nuclear plants. However, as time passed, the requirements were
made applicable to plants currently under construction. This
concept of imposing new requirements on plant designs already
approved for construction is referred to as "backfitting". Many

TABLE 3

PROLIFERATION OF REGULATORY GUIDES

Year	Number of Original & Revisions Issued	Cumulative* Total
1970	4	4
1971	17	21
1972	20	41
1973	71	112
1974	67	179
1975	62	241
1976	54	295
1977	59	354
1978	54	408
1979	18	428
1980	24	450
1981	7	457

*Does not account for those cancelled.

TABLE 4

NRC REPORTED NUCLEAR PLANT LICENSING DELAYS
(January 1981)

Nuclear Plant	Construction Complete	Operating License Issuance	Months Delay
Summer	10/81	6/82	8
Diablo Canyon-1	3/81	3/82	12
Diablo Canyon-2	10/81	3/82	5
San Onofre	7/81	4/82	9
Zimmer	11/81	7/81	8
McGuire	2/81	3/81	13
Enrico Fermi-2	11/82	6/83	7
Susquehanna-1	3/82	11/82	8
Waterford-3	10/82	4/83	6
Shoreham	9/82	10/82	1
Comanche Peak-1	12/82	2/83	2
Salem-2*	4/80	6/81	14
Farley-2*	3/81	3/81	0
		Total	93 mos.

*Plants with fuel-loading/zero-power licenses which are not listed as impacted plants by NRC.

requirements were also "backfitted" on operating plants. In fact, the NRC has a formal program to review and evaluate 11 of the oldest nuclear plants in the country relative to meeting current requirements. This program has imposed as many of the current requirements as can possibly be met with little regard for cost or benefit.

Effect of Excessive Nuclear Regulation
What is the effect of this myriad of nuclear regulations? It is causing nuclear plant licensing and design to be accomplished in a crisis mode. It is causing delays in the issuance of operating licenses for new plants that are ready to start up. It is placing requirements that have limited or no definite safety value-impact or cost-benefit assessment. And it is requiring nuclear plant modifications for safety enhancement on schedules that are unrealistically short and for assumed costs that are often-times ridiculously low.

Nuclear Plant Licensing Delays. In January 1981, the NRC published a schedule for the issuance of operating licenses for the 13 nuclear power plants that will complete construction during the 1981-1982 time period. This schedule is shown in Table 4. Notice that several plants are currently ready to operate but cannot for lack of a license. The delay in licensing, after construction is completed, varies from 1 to 14 months. The current total delay for all 13 plants is 93 plant-months. Now these delays have little to do with plant safety. They are caused because the NRC is unable to do its job in a timely fashion.

These delays are expensive. They cost the applicants, the stockholders and the electric customers. An American Nuclear Energy Council report estimates the cost of delay at Diablo Canyon - 1 and - 2 at about $1 billion a year, or $83 million per month. For San Onofre - 2 and 3 it estimates the delay cost at $3 million a day or $90 million per month. The above figures include the cost of interest plus the cost of replacement power. For each of the 13 affected plants, the average cost incurred from the delay are "...in the range of $30 to $40 million per plant per month..." the report states. For an accumulated delay of 93 months, the total cost for all 13 plants "...would be between $2.8 and $3.7 billion..." the report states.

The above comparison reflects only the current delays in nuclear plant licensing. Some of these nuclear plants have also experienced extensive delays in the past. For example, the Shoreham plant was originally scheduled for completion at about the same time as Northeast Utilities Millstone Unit 2. In 1975, Millstone Unit 2 went into commercial operation, however, Shoreham is not scheduled to receive an operating license until late 1982. Much of this past delay is directly attributable to the regulatory process.

Faulty Value-Impact Analysis. The NRC Value-Impact guidelines for evaluating alternate courses of action were adopted in 1978 and state:

> "The policy of the Nuclear Regulatory Commission is that value-impact analysis be conducted for any proposed regulatory actions that might impose a significant burden on the public (where the term public is defined in the broadest sense). Such policy is not to be construed to mean that cost considerations take precedence over considerations of health, safety, environment, or national security. These factors remain paramount. However, where there are alternative means of realizing equivalent benefits in regulatory matters, cost should be a prime consideration."

The stated purpose of a value-impact analysis is to assure that the expenditure of funds and manpower by both the licensee and the NRC will result in significant increased plant safety. In a value-impact analysis, there is always a trade-off between cost and benefit. The value-impact analyses form relevant input in determining the appropriateness of new requirements by the NRC. As with any decision making process, accurate input is a prerequisite for making informed decisions.

Experience has demonstrated that NRC estimates of industry costs required to achieve compliance have been consistently low, sometimes by an order of magnitude or more. I believe that had the NRC performed realistic value-impact analyses on recent TMI Action Plan requirements and properly weighed cost against benefit, a number of the requirements would not be justified in terms of increased plant safety.

Table 5 provides a listing of some recent NRC requirements and a comparison of the NRC estimated costs with actual costs incurred to date for the Connecticut Yankee and Millstone Units 1 and 2. It should be noted that the costs listed in the Attachment are not the final costs in all cases, but only represent dollars expended to date. In addition, these figures do not include replacement power costs incurred by plant down-time. I am not aware of any quantification of the benefits to justify the promulgation of these requirements.

Effect of Unrealistic Schedules. It is not uncommon for a utility to receive a NRC mandate that an operating plant be modified within a certain time period after which it cannot be operated. As I noted earlier, the NRC is not particularly adept at estimating costs of plant modifications; they are also not very good at estimating the time to accomplish the modification.

Upon notification of a deadline for either complying with a regulatory directive or shutting down a nuclear plant, the utility, in effect, pushes the "panic button". Engineering is forced to seek a prompt solution that will allow the plant to continue to operate or at least minimize the length of the plant

TABLE 5

NRC COST ESTIMATES VS. ACTUAL CAPITAL COSTS FOR TMI ACTION PLAN REQUIREMENTS

TMI Action Plan No.	Item Description	NRC Cost* Estimate	Haddam Neck Plant	Millstone Unit No. 1	Millstone Unit No. 2
11.B.1	Reactor Vessel Head Vent	100,000	1,102,000	NA	988,000
11.B.2	Plant Shielding Review	50,000	6,000	252,000	155,000
11.B.3	Post Accident	100,000	652,000	651,000	400,000
11.E.1.2	AFWS Initiation and Flow Indication	20,000**	305,000	NA	714,000
11.F.1	Accident Monitoring	250,000	1,245,000	1,746,000	832,000
111.A.1.2	Emergency operations Center	4.54 million***	6,237,000		5,251,000
11.A.2	Emergency Preparedness				
111.D.1.1	Systems Integrity	5,000	1,149,000	125,000	6,000

*Obtained from NUREG-0660
**The NRC has admitted this item does not increase plant safety, only reliability
***Estimate from 1979 drafts of NUREG-0660, later estimates more realistic

shutdown. At the time, that engineering is working at solving the problem, construction is preparing to build and install the modifications engineering specifies. Often, while engineering is still identifying the scope of the work to be accomplished, construction is mobilized.

Since the replacement power costs associated with a nuclear plant shutdown are so great, typically $1 million a day for a 1000 MW plant, the grossly inefficient approach in engineering and construction that the utility may be forced to take, becomes justified. Inevitably, due to the "panic atmosphere", the solution to the problem is an overkill. I am convinced that it is in the best interest of the NRC and our customers that sufficient time be allowed for efficient engineering, material and equipment procurement, and construction activities.

An NRC Problem. It is not only the nuclear industry which is critical of the NRC's activities at overregulation. Some employees within the NRC also share this concern. In fact, some feel very strongly about the adverse effect the overregulation is having on the industry.

I would like to read a portion of a letter sent to President Reagan last December by a NRC engineer.

December 16, 1980

Office of the President Elect
1726 M. Street NW
Washington, D.C.   20270

Gentlemen:

To introduce myself, I am Donald Gene Anderson, an inspector in the Office of Inspection and Enforcement (OIE) of the U.S. Nuclear Regulatory Commission (USNRC). Coincident with this letter to you, I am resigning my position with the Commission effective December 26, 1980, because I have deep concerns that as presently structured, the Commission is not meeting its mandate of "Protecting the Health and Safety of the Public." Instead, by overregulating the nuclear industry, the Commission has only succeeded in eliminating the nuclear alternative..."

Mr. Donald Anderson, teaches nuclear engineering at the University of Texas. I believe he expressed the sense of frustration that all of us in the nuclear business feel.

Suggestions for Regulatory Improvement
A review of the current NRC licensing arena reveals three areas where improvements are needed over the near-term.

(1) Prioritization of NRC Rulemaking Proceedings
(2) Stabilization of NRC Regulations
(3) Streamlining of NRC Licensing Proceedings

Prioritization of NRC Rulemaking Proceedings. There are currently over 150 proposed NRC rulemaking activities in progress. Priorities in addressing each issue are crucial; the industry has limited resources available to actively participate in the rulemakings. The excessive number of issues the NRC considers of safety significance is diverting industry attention from potentially truly important safety issues.

I believe that the development of quantitative safety goals should be the *first* order of business. To accomplish this, it may be necessary for Congress to pass legislation directing the NRC to quantify safety goals, issue regulations and implement the regulations. The issuance of safety goal regulations will then form the basis for the selection and development of other rules.

Specifically, the following steps should be taken:

(1) NRC activities on current rulemaking should be halted and further assessment made.
(2) Rulemaking activities on safety goals should be initiated on an expedited schedule.
(3) The halted rulemaking activities should be subjected to critical review and prioritized relative to safety significance.
(4) At the conclusion of the safety goal rulemaking, the results should be incorporated, as appropriate, into the remaining proposed rulemakings.
(5) After appropriate review and comment, and incorporation of safety goals, the previously halted rulemaking proceedings should be restarted on the priority basis established.
(6) Rules which cannot be justified on the basis of the developed safety goals should be dropped from further consideration.

Stabilization of NRC Regulations. The stabilization of NRC regulations is an area that has needed attention for a number of years. It has been a major factor in utility decisions to forego nuclear plant expansion.

The problem has grown to epidemic proportions since TMI and must be corrected if the nuclear option is to remain viable. The stabilization must occur in both the backfit requirements on older operating plants and on the imposition of new requirements for plants at the construction permit or operating license stage.

Specifically, the following steps should be taken:

(1) The issuance of new criteria should be held in abeyance until the appropriate rulemaking activities, including those on safety goals, are completed.
(2) Procedures should be instituted whereby new requirements are issued only after the performance of a risk assessment and a

comparison of the results to established safety goals. For
new plants, a cost/benefit analysis would also be performed
to determine if the change is in the public interest. For
operating plants, meeting established safety goals is
sufficient.
(3) The current "Value/Impact Statement" that accompanies new
regulations is worthless and does not balance costs vs.
benefits as was the original intent. The Staff should be
directed to utilize cost/benefit data, i.e., $/man-rem,
developed as part of the safety goal rulemaking, as justification for each new regulation.
(4) All new regulations should be limited to the specification
of criteria, not prescriptive action.
(5) The NRC should implement procedures that assure industry and
other comments are considered and factored into the regulations,
as appropriate.

Streamlining of NRC Licensing Process. The successful
pursuit of the items in the above two paragraphs would be effective
in streamlining future NRC rules and regulations. However, a
major effort should be undertaken to undo the current web of
regulations that are strangling the nuclear industry.

Recently, construction was completed on a nuclear plant in Japan
in a time span from docketing to licensing to operate, of four
years. In Taiwan, the Kuosheng plant, which is the first BWR-6
Mark III, was completed in 62 months. France regularly schedules
and completes nuclear plants on a 6 year schedule. Why is it
that plants in the United States require about 13 years to
complete. Much of this is due to a regulatory system that has
become self-serving and is no longer capable of simply fulfilling
its mandate of "---protecting the health and safety of the
public---". Instead, by overregulating the nuclear industry, the
NRC is slowly succeeding in eliminating the nuclear option.
Unfortunately, many of these regulations have not been critically
examined to determine whether they improve safety or whether they
are needed at all. I believe it is time that the entire NRC
nuclear power plant licensing process be critically examined.
This would include:

(1) A review of all rules, regulations, policies, positions,
etc., and the performance of cost/benefit analysis and
comparison of each result to established safety goals
to justify their continued use.
(2) Consideration of the one-stage licensing process.
(3) Consideration of elimination of the public hearings and
the hearing boards.
(4) Certification of plant designs as adequate analogous to
FAA Airworthyness Certification.

Significant improvement in the three broad problem areas discussed
above would go a long ways towards reducing the licensing difficulties currently encountered. It would also permit, indeed
force, attention to the more significant problems rather than

diluting this attention as is done now. This alone would represent an improvement in safety.

INDUSTRY CODE AND STANDARDS PROBLEMS

Until now, I have been addressing the difficulties caused by NRC overregulation. Now I would like to focus on some of the problems caused by industry.

As was discussed earlier, during the planning and early design phase of a nuclear power station, the applicant (the utility) is required to prepare a Preliminary Safety Analysis Report (PSAR) which specifies the applicable codes, standards and NRC Regulatory Guides that will be implemented during the final design and construction of the plant. The NRC Regulatory Guides frequently invoke additional industry codes and standards by reference. In addition, other NRC documents such as Bulletins, Circulars, and Information Notices impose more codes and standards. The requirements then become very complex, difficult and frequently overlapping when applying all the various codes; conflicts and interferences seem inevitible.

Since these codes and standards are developed and written by the industry, it would seem to be our responsibility to correct their shortcomings. A sample tabulation of some of the currently required codes is given in Table 6.

Some specific problems Northeast Utilities has experienced during the construction of Millstone Unit 3 are described in the following sections.

Code Date Requirements

Applicable code dates (revisions) are initially specified in the PSAR. Extended design and construction schedules have caused considerable confusion at various stages as to the applicability of newer code revisions. For example, the construction phase for Millstone Unit 3 will be approximately twelve (12) years under the current schedule. The ASME (American Society of Mechanical Engineers) code requirements are determined by the contract issue date of the specification.

The original pipe supports for Millstone Unit 3 were contracted to meet the code requirements of the 1971 ASME Section III through the Summer of 1973 Addendum. Requirements for additional or revised pipe supports identified and contracted for under a newer specification were required to meet a newer code date (revision). This situation results in having supports on the same piping system that are fabricated to different code date requirements.

Table 7 presents seven examples of equipment and pipe supports on Millstone Unit 3 that are required to meet a code date other than the original 1971 ASME Section III through the Summer of 1973

TABLE 6

SAMPLE LIST OF VARIOUS CODES
APPLIED TO NUCLEAR PLANT
DESIGN AND CONSTRUCTION

Code	Title
ASME	American Society of Mechanical Engineers-Boiler & Pressure Vessel Code
IEEE	Institute of Electrical & Electronics Engineers
ANSI	American National Standards Institute
SMACNA	Sheet Metal and Air Conditioning Contractors National Association
AWS	American Welding Society
AISC	American Institute of Steel Contractors
ASTM	American Society for Testing and Materials
ASNT	American Society of Nondestructive Testing
NEMA	National Electrical Manufacturers Association
SPCC	Steel Structures Painting Council
ACI	American Concrete Institute
IPCEA	Insulated Power Cable Engineers Association
NFPA	National Fire Protection Association

TABLE 7

EXAMPLES OF VARIOUS APPLICABLE CODE DATES

EQUIPMENT NAME	SPECIFICATION NO.	APPLICABLE CODE
Main Steam Valve Building Pipe Support	2280.000-515	ASME III Sum '77 NF Third Party Insp/Stamped
Reactor Vessel Support	2213.100-029	ASME III Sum '74 NF Third Party Insp/Unstamped
Hydraulic Snubbers	2221.189-127	ASME III Sum '73 NF No Third Party Insp/Unstamped
Reactor Vessel Leveling Device	2211.110-128	ASME III Sum '74 NF No Third Party Insp/Unstamped
Pressurizer Support	2221.311-129	ASME III Sum '74 NF Third Party Insp/Unstamped
Reactor Coolant Pump Supports	2221.180-130	ASME III Win '74 NF Third Party Insp/Stamped
Residual Heat Removal Pump Supports	2214.322-132	ASME III Sum '77 NR Third Party Insp/Stamped

UTILITIES PERSPECTIVE

Addendum. Also note the differing inspection and stamping requirements.

The varied code requirements identified do not substantially increase the quality of the system, but only cause increased costs due to varied inspections and recordkeeping requirements.

Overlapping Requirements
The following regulatory documents have been utilized in the design of fire protection measures for electrical equipment at Millstone Unit 3.

(1) Regulatory guide 1.75, "Physical Independence of Electrical Systems".
(2) IEEE Standard #384, "Criteria for Independence of Class 1.E. Equipment and Circuits".
(3) NUREG-0050, "Recommendations Relating to the Brown's Ferry Fire".

Recently, 10CFR50 Appendix R was issued and imposed by the NRC. It is more restrictive in the electrical fire protection area than the preceding documents which were previously approved as acceptable to use. It is considered that many manhours will be required for design review which may result in structure or equipment changes in the plant design due to the imposition of these new requirements.

Generally, Regulatory Guide 1.75 addresses the physical separation requirements of electrical equipment as stated in IEEE Std. 384 and the recommendations identified in NUREG-0050. IEEE Std. 384 identifies minimum separation distances of cables and raceways of one foot between raceways separated horizontally and three feet between raceways separated vertically. Also, IEEE Std. 384 will allow lesser separation distances by using enclosed raceways that serve as barriers.

In comparison, 10CFR50 Appendix R, Section III, Paragraph G.2 addresses minimum separation of cables and equipment as follows:

(1) Separation of cables and equipment by a fire barrier having a three hour rating or;
(2) Separation of cables and equipment by a horizontal distance of twenty feet with no intervening combustibles or fire hazards. In addition, fire detectors and an automatic fire suppression system shall be installed in the fire area, or;
(3) Enclosure of cable and equipment to be separated from other circuits in a fire barrier having one hour rating. In addition, fire detectors and an automatic fire suppression system shall be installed in the fire area.

In some areas of the plant, the equipment will be tested to assure that there will be no flame propagation. Therefore, the requirements of 10CFR50 Appendix R would obviously be overrestrictive, particularly the twenty foot separation requirement. Also, the above documents apply the separation requirements for Safety Related Class IE Circuits and Equipment in different terms.

The continuing review of these requirements as they change in scope, applicability, and inter-relationship creates a cost burden on the utility and the Engineer/Constructor for both the review and the implementation.

Another case of codes overlapping which create unnecessary costs is in the hydrostatic pressure testing of piping systems. Piping systems are designed and installed in accordance with the applicable ASME or ANSI Codes. These codes specify the hydrostatic pressure testing requirements. When the piping system is encased in structural concrete, the ACI Code also specifies hydrostatic pressure test on the encased piping. This hydrostatic pressure testing is different than the ASME or ANSI, even though its purpose is to insure the integrity of the piping. The only difference is that the ACI Code is interested in ensuring the pipe's integrity to protect the structural concrete, while the ASME or ANSI Code's intent is to ensure piping system reliability.

Since the testing requirements are different, two tests must be conducted. These duplicating tests do not result in higher quality but do increase costs and extend construction schedules. I believe the code committees should examine overlapping jurisdictions. When formulating code requirements, consideration should be given to the other codes so that unnecessary requirements can be avoided.

Conflicting Code Requirements
Components which are manufactured in accordance with ASME Section III are required to pass a volumetric inspection of welds and adjacent base metal. The inspection is performed using a specified radiography technique. The radiographic acceptance/rejection criteria are provided by the code.

Components that meet all the requirements of Section III are deemed acceptable and turned over to the owner. At this point, ASME Section XI becomes the governing code for all future tests and inspections. ASME Section XI specifies that a complete "base line" and periodic volumetric inspections be performed using an ultrasonic inspection technique. The ultrasonic acceptance/rejection criteria are provided by the code.

The two different nondestructive inspection methods should compliment each other. However, a component which was accepted by radiography can be rejected by ultrasonic inspection due to different sensitivity and acceptance requirements. In fact, this has occurred in several instances and has led to extensive rework.

It appears that the code committees should provide consistency between the different Section requirements of the same code in order to eliminate any conflicting acceptance criteria.

Redundant Qualification Requirements
Utility companies that are constructing and/or operating nuclear power plants must meet the requirements of ASME Section IX which governs qualification of welders and welding procedures.

The requirements set forth in Article 3 QW-301 of ASME Section IX states, "Each manufacturer or contractor shall be responsible for conducting tests to qualify the performance of welders and welding operators". Performance qualification basically determines the ability of the welder to deposit sound weld metal. An individual welder could take ten or more qualification tests over a period of time.

Article 1 of the code states, "It is not the purpose of the code to cause extensive retesting of previously employed welding or brazing procedures, welders, brazers or welding or brazing operators". However, the requirements of Section IX, specify that a welder must re-qualify whenever he changes employers, even though the new employer is a contractor for the same utility company. Thus, every time a change in employer is made, the tests are repeated.

It is not unusual for welders to change employers frequently to take advantage of overtime, but still work on nuclear plants for the same utility. One pipefitter/welder estimates that, in the past fifteen years, he has taken between 60 and 90 qualification tests for different employers contracted by Northeast Utilities. Each welder qualification test costs approximately $500.00, so the above tests cost between $30 and $45 thousand. These costs are billed to the utility.

Much of the same repetition is required when qualifying weld procedures. Article 2 QW-201 states, "Each manufacturer or contractor shall qualify the weld procedure by the welding of test coupons and the testing of specimens as required by the Code". Many welding procedures used today are generic, the only difference is the contractor name on the letterhead.

The American Welding Society (AWS) has eliminated some of the repetitious procedure qualifications by publishing "prequalified procedures" which do not require qualification by each contractor. There is a real need for the various ASME codes to streamline and standardize welding procedures and welder qualifications.

SUMMARY

In summary, I have pointed out the heavy burden of statutes, permits and regulations the utility industry must bear. In the case of the NRC, this has led to the current regulatory morass which has effectually removed the "nuclear option" as a viable means for planning our future electrical generating requirements. The utility industry is not alone in this view. As I have earlier pointed out, there are even people within regulatory agencies who share these thoughts. I have strived to keep my criticism constructive, and have offered suggestions which address the major issues.

I have also indicated that the industry needs to do some house cleaning. It appears to me that our own regulations, that is the codes and standards that we develop and invoke, have serious flaws. I urge that those of you who are on code committees or other bodies that review and develop codes and standards pay particular attention to vaguely defined, overlapping or unnecessary requirements that are costly to the industry.

And, finally, I would like to praise all of you who are involved in the preparation of codes and standards for our industry. It is generally a voluntary effort, often without thanks and frequently with criticism. I think you are doing a fine job.

An Architect-Engineer Perspective

by

Russell J Christesen [1]/ Member ASCE

Abstract

The impact of regulatory excesses and reduced construction effectiveness have caused a dramatic increase in power plant costs. Therefore, streamlining and making the regulatory process realistic and cost effective is one of the critical national agenda items for the 1980's. The engineering-construction industry, for its part, must improve its performance in developing and implementing codes, standards and practices. Industry should sponsor a truly industy-wide seminar to aggressively seek out and address problems in these areas. We need to consider and support the President's energy program which offers significant opportunities for betterment in regulation of the design and construction of power generation facilities. The aim is cost-effectiveness in accord with public health and safety, and a revitalized nuclear industry.

Introduction

The first reaction of most of us to the title "Gulliver's Travels" is that it is a book for kids. This is because of the imagery it shows. Yet, "Gulliver's Travels" is recognized as a highly sarcastic satire of the workings of England in the 1700's.

---
[1] Executive Vice President - Operations Ebasco Services Incorporated, New York, NY

Today, in the design and construction of power generation facilities with our ever increasing and entangled web of regulations and restrictions, it is an exercise in self-control to avoid the sarcastic, the satirical outlook. Surprisingly though, the electric utility industry--its planners, engineers and operators; and the architect-engineers, suppliers and contractors who serve them--have tried to meet these regulations and restrictions head on...in a positive and supportive way.

Gulliver, you will recall, started out tied down to the point of immobility. He wound up showing the Lilliputians that he could provide certain of their necessaries.

I believe this conference provides a unique forum for identifying some of the necessaries for improving the performance of our industry. I am especially pleased, therefore, to be one of your keynote speakers.

For the good of the consumers, electric utility companies, and the economic well-being of the nation, we must strive for a better balance between quality, productivity and regulation in the construction of power generation facilities. This morning I shall share with you, concerns about the impact of excesses by regulatory statutes and by the administrative regulations of governmental agencies. I will give an overview of the development and implementation of codes and standards, and conclude with recommendations for coming to grips with some of the problems.

But first, a brief look at the industry we serve may help put the importance of these issues into perspective.

The electric utilities are the largest user of the engineering-construction industry. As you see, the annual addition of generation capacity reached a peak in 1973. Annual capital expenditures have continued to rise since the early 1960's reaching a total of $27.4 billion in 1980. Last year, with 235,000 MW of generating capacity underway or planned, workers on power plant construction totalled over 200,000. The manpower involved with the engineering and design of this generating capacity numbered over 25,000. (Figure 1)

The electric power industry is the most capital-intensive of all major U.S. industries. With spending of over $27 billion electric utility companies absorbed about 10 percent of vital U.S. capital investment in 1980. According to the Electric Power Research Institute, the average electric utility needs $5 in plant, property and equipment to produce $1 in revenue. In contrast, steel needs $3 for each $1, while the auto industry requires $1 per dollar. (Figure 2)

FIGURE 1
ELECTRIC POWER INDUSTRY....
LARGEST INDUSTRIAL USER OF CONSTRUCTION

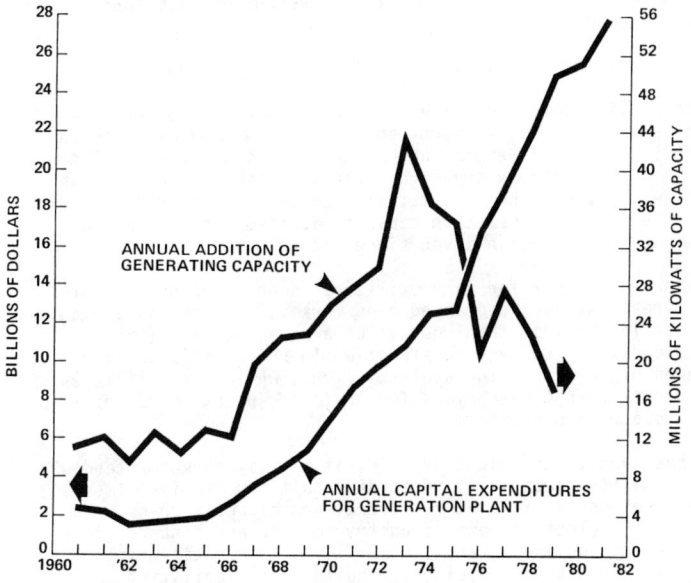

FIGURE 2
U.S. NON-FARM INDUSTRY CAPITAL INVESTMENT

INDUSTRY	CAPITAL SPENDING BILLION DOLLARS 1980	PERCENT OF TOTAL NON-FARM INDUSTRY
ELECTRIC UTILITIES	$ 27.4	9.3%
CHEMICAL	12.8	4.3
PETROLEUM	20.3	6.9
IRON AND STEEL	3.4	1.2
NON-FERROUS METALS	3.1	1.1
AUTOMOTIVE	9.0	3.1
TOTAL NON-FARM INDUSTRY INVESTMENT	$294.3	

For 50 years, from 1920 to 1970, generation of electricity grew continuously at a rate of more than 7.6 percent per year. Actually greater than the average growth rate of the Gross National Product! During the period from the 1973 oil embargo to 1980, growth of generation of electricity slumped to an average of less than 3.0 percent per year.

These are facts of life in today's electric power industry.

Electric utility companies have been beset by a host of problems. Some self-inflicted, some energy-source related, and many caused by a proliferation of Federal and State regulations and regulatory excesses. Compounded by sky-high interest rates on construction capital, escalating labor and material costs and construction delays, the cost of generating capacity additions has increased by a factor of more than seven since the 1960's.

The financial outlook for many utilities is anything but bright. High interest rates, a depressed bond market, and utility common stocks selling at less than book value are making it difficult to raise capital. Delayed construction in progress is pushing costs even higher, draining available funds while most utilities are unable to charge ratepayers for these new plant costs until the plants are in operation.

Given these additional facts of life, it is easy to understand why electric utility companies find it difficult to make investment decisions responsive to long-run energy and capacity needs. The ability of the electric power industry to plan and finance generating capacity additions, and the ability of the engineering-construction industry to design and build these facilities has been further affected by socio-political trends of the last decade. Evidence of this influence is seen in the proliferation of government regulations, the unwillingness of the public to endorse new power plants and in the litigious climate in our society today, which substantially increases liability risks--even in the most routine dealings with clients.

The tendency to overprotect people, places and things in our society today has resulted in rampant regulatory excesses. The resulting surfeit of regulations in turn causes waste of financial, physical and human resources, inefficiency, and increased costs.

A look at some of the principal legislation which relates to the design and construction of power generation facilities, clearly demonstrates that regulatory excesses by statute are pervasive because of overlapping, conflicting laws. Regulatory agencies created by such statutes promulgate administrative regulations which are subject to different interpretations by Federal, State and local administrators, thereby creating further excesses. (Figure 3)

FIGURE 3
REGULATORY EXCESSES BY STATUTE

- NATIONAL ENVIRONMENTAL POLICY ACT OF 1969
- ATOMIC ENERGY ACT, AS AMENDED
- CLEAN AIR ACT OF 1977
- CLEAN WATER ACT OF 1977
- NOISE CONTROL ACT OF 1972
- SAFE DRINKING WATER ACT OF 1974
- TOXIC SUBSTANCE CONTROL ACT OF 1976
- RESOURCE CONSERVATION AND RECOVERY ACT OF 1976
- NATIONAL ENERGY ACT
- SURFACE MINING AND CONTROL ACT
- ENDANGERED SPECIES ACT OF 1973
- NUCLEAR NON-PROLIFERATION ACT OF 1978
- WETLANDS AND FLOOD PLAIN EXECUTIVE ORDERS
- ENVIRONMENTAL EFFECTS ABROAD EXECUTIVE ORDER
- FIFTY STATE LEGISLATIVE ACTS
- FIFTY STATE EXECUTIVE ORDERS

FIGURE 4
REGULATORY EXCESSES BY GOVERNMENTAL AGENCIES

- 14 COMMITTEES OF U.S. CONGRESS
- U.S. DEPARTMENT OF ENERGY
- U.S. ENVIRONMENTAL PROTECTION AGENCY
- U.S. NUCLEAR REGULATORY COMMISSION
- U.S. DOMESTIC POLICY COUNCIL
- FEDERAL ENERGY RELIABILITY COUNCIL
- U.S. COUNCIL ON ENVIRONMENTAL QUALITY
- U.S. DEPARTMENT OF INTERIOR
- U.S. ARMY CORPS OF ENGINEERS
- U.S. COURTS
- FIFTY STATE COURT SYSTEMS
- FIFTY STATE PUBLIC UTILITY REGULATORY BODIES
- FIFTY STATE ENVIRONMENTAL AGENCIES

These are but a few of the principal government agencies involved with administration of regulations relating to electric power installations. Many of these agencies are frequently at odds with one another. Some have "special interest constituencies". Most of them appear to be oblivious to the collective impact of their actions and inactions. They make it nearly impossible for the electric utility industry to produce and distribute the electric energy so vitally needed. (Figure 4)

The significant statutory requirements imposed since 1969 on the design and construction of coal-fired power plants are primarily environmental in nature. They have their origin in the National Environmental Policy Act (NEPA) of 1969, the Clean Air Act Amendments of 1977, the Water Pollution Control Act of 1972, and various interpretations that were placed on the Legislation by the courts, By EPA, by other Federal agencies, and in many cases by State agencies.

By comparison, the most significant impact on nuclear plant design is the need to comply with Nuclear Regulatory Commission (NRC) regulations and their many revisions, together with the 540 plus regulatory guides and other regulatory position documents. These regulatory guides, which are in effect de facto regulations in themselves, are still being issued at the average rate of 5 or 6 per month.

It is plain to see from the increasing numbers of regulatory criteria from 1969 to 1980 that the regulatory guides issued to date cover almost every facet of plant and systems engineering design criteria and implementation. Some of these required redesign of plants already in design or construction and some apply to future plants. Still others require backfitting modifications to operating plants. Accomodating regulatory guides often has a significant impact on both schedule and cost. In developing and applying new regulatory criteria, the NRC appears to have a single goal, to make things safer than they are, pursuing safety improvements on a fragmented basis. An item of concern is identified and criteria are then formulated to deal with the issue. Resolution of the issue becomes an end to itself, and the resultant criteria, while solving the item of concern, frequently has a negative impact on safety in other areas. There does not seem to be any concerted effort to counter balance such negative safety benefits brought on by new criteria. (Figure 5)

The impact of industry codes, standards and practices under the nuclear regulatory syndrome on the complexity of design, magnitude of material quantities, and the resultant time and cost of placing generating capacity on the line are best illustrated by the installation requirements for plant structures and equipment. Look at the craft man-hour resources required on typical current coal-fired and nuclear plant construction programs! (Figure 6)

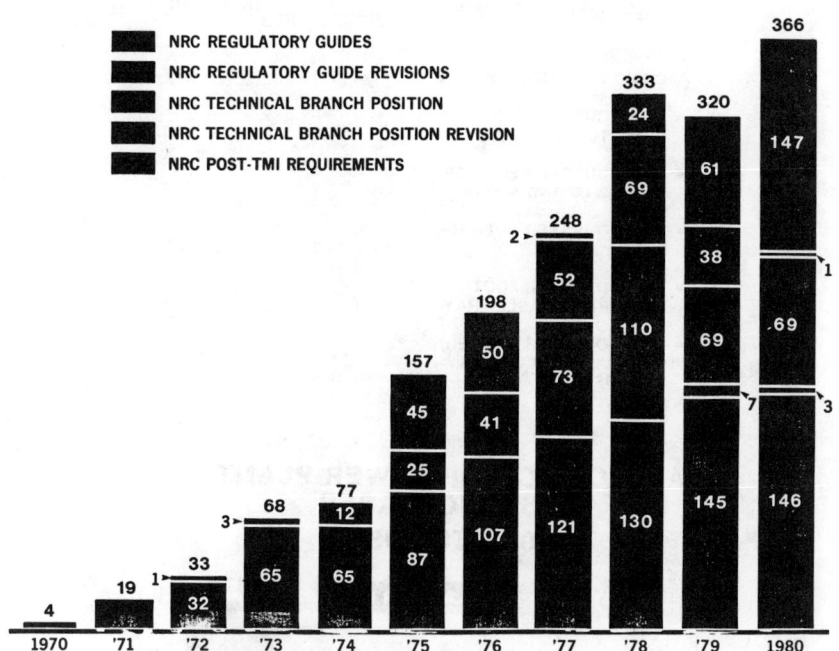

FIGURE 5
ADDED STATUTORY & REGULATORY REQUIREMENTS 1969 - 1980
(CUMULATIVE NE NUCLEAR PLANT GUIDES)

FIGURE 6

REDUCED EFFECTIVENESS ....
Nuclear Power Plant Construction

TYPICAL MANPOWER RESOURCES REQUIRED	COAL-FIRED PLANT	NUCLEAR PLANT
• MAN-HOURS PER CUBIC YARD OF CONCRETE	7.8	18.7
• MAN-HOURS PER TON STRUCTURAL STEEL	15.0	28.0
• MAN-HOURS PER FOOT OF PIPE — 2-1/2 AND ABOVE	3.3	9.7
• MAN-HOURS PER AVERAGE WELD	14.0	30.0
• MAN-HOURS PER FOOT ELECTRICAL CABLE TRAY	1.6	3.1
• MAN-HOURS PER HANGER/ RESTRAINT, INCLUDING EMBEDS	15.0	70.0

FIGURE 7

ALLOCATION OF POWER PLANT COST INCREASES 1969 TO 1980

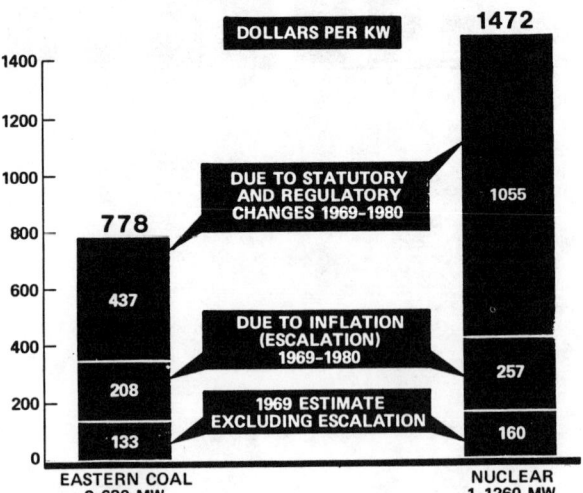

ARCHITECT/ENGINEER PERSPECTIVE

- The direct craft man-hours required per cubic yard of nuclear concrete in place are about three times that for a coal-fired plant. This decreasing effectiveness is due primarily to the complex reinforcing systems and tolerances imposed by conventional codes and standards (practices) which did not anticipate the structural requirements and magnitude of loads imposed on nuclear plant structures.

- The direct craft man-hours required per foot of pipe installed in nuclear plant systems, exclusive of hangers and restraints, are about four times that for a coal-fired plant. This decreasing effectiveness is due to more rigid installation and welding procedures, non-destructive testing requirements, and often times by unreasonable inspection and documentation requirements.

- The higher unit man-hour expenditures also reflect the greater congestion, as well as the changes and additions that often occur in partially or totally completed systems because of additional regulatory requirements and interferences resulting from design complexity.

- The direct craft man-hours required per foot of installed electrical cable tray, conduit and cable is about 2 times that for a coal-fired plant. As with the piping installations, this decreased effectiveness is due to greater complexity and congestions, and the changes and additions which occur in completed or partially completed systems.

- The direct craft man-hours required for installation of hangers and supports for piping, electrical and HVAC systems and the embeds to which they are attached, are about five times that for such support systems in a coal-fired plant. Hangers and restraints have become so numerous that they are major structural and space considerations. For example, on a typical nuclear plant, large bore piping systems require upwards of 16,000 supports, equating to 2000 or more tons of structural steel.

Obviously, congestion, complexity and the degree of required modifications to completed installations are major causes of reduced effectiveness on nuclear plant construction.

The first reaction most people have to reducing the complexity and congestion would be to provide more volume within the structures. Added building volume, however adds cost, both in terms of the structures themselves, and in terms of the additional bulk quantities of piping and electrical raceway and cable, and associated supporting systems.

There is an optimum volume for a given plant design that permits optimum construction, equipment installation, and plant maintenance. Space allocations must be a key consideration in nuclear plant design as we strive to optimize plant volume, reduce congestion and facilitate construction.

Again, the impact of regulatory excesses and reduced construction effectiveness has caused a dramatic increase in power plant costs. Our estimate of the allocation of plant investment cost increases between 1969 cost and 1980 level estimates, excluding escalation beyond 1980 illustrates this. For coal-fired plants, starting with 1969 level average two-unit cost for $133/KW, escalation has added $208/KW. It is not suprising that statutory and regulatory requirements have added almost twice that, or $437/KW. For a nuclear plant, starting with the one-unit average cost of $160/KW in 1969, escalation to 1980 has added $257/KW. Statutory and regulatory requirements including the impact of TMI, have added $1055/KW, almost four times that of inflation. (Figure 7)

To put these costs into perspective, for a 1200 MW nuclear plant planned for 1990 service (based on today's known regulatory requirements), utilities are facing costs of over $2500/KW, representing capital investment of over $3 billion. For a 1200 MW coal-fired capacity installation, costs of over $1400/KW representing capital investment of over $1.75 billion are expected.

It should be noted that for the 1200 MW coal-fired plant, approximately $230/KW or one-sixth of the capital investment will be spent on interest during construction (IDC). On the 1200 MW nuclear installation, approximately $720/KW, or almost one-third of the capital investment is spent on IDC. In fact, stretched-out schedules due to regulatory excesses and accompanying litigation have a staggering impact on electrical utility financing requirements.

Clearly, we must drastically reduce regulatory impediments to nuclear plant design and construction, and we can do so without in any way jeopardizing the public health and safety.

Streamlining and making the regulatory process realistic, responsive and cost effective is one of the critical national agenda items for the 1980's.

It would be easy to spend the rest of my allotted time on that subject. Instead, I will address those challenges that confront the engineering-construction industry in getting its own house in order.

Implicit in the theme of this conference is the need to improve industry performance in the development and implementation of codes, standards and practices. I commend those who have had the foresight to see the need for this meeting. Subjects included on the agenda for the next three days cover a broad spectrum of areas where improvement is necessary and achievable. Let me suggest three areas:

- Reducing conservatism in design
- Resolution of ambiguities and conflicts in codes and standards and
- Improving acceptance criteria and tolerances

Before proceeding, a brief look at how codes and standards are developed.

A code is a group of administrative and technical rules and standards covering any combination of materials, design, construction, installations, inspection and operation of equipment which is prepared for ready adoption into law of a legal jurisdiction.

Standards or "Standard Practices" are generally intended to present the recommended and acceptable methods and materials to be used in design, planning, execution and inspection of construction and in preparing specifications. Some standards are written in obligatory language and can be incorporated into regulations.

Increasingly, codes and standards are subject to a self-imposed consensus process under the management of the American National Standards Institute (ANSI). ANSI establishes a structure and criteria for assuring that before designation as an ANSI standard, it meets the minimum due process requirement and can be considered industry consensus on the subject. It is through this consensus process that opportunity is provided for incorporation of industry-wide expertise into the standards.

I have a high regard for the many dedicated people who devote time and effort to code and standard committee work. However, I do not believe that these working groups receive enough broad-based support and direction from industry. And often times, a lack of direct involvement of experienced hands-on construction representation has resulted in issues of constructability arising after standards have been adopted. I further suggest that in view of conflicts and ambiguities in existing codes and standards, that a greater sense of urgency must prevail in committee activities.

Both industry and government are guilty of ultra conservatism. In the design of nuclear plant structures and supports, large complex loads, coupled with dynamic effects, require large structural members to comply with code stress levels. Problems in plant design arise from conservatism in applied loads. Reinforced concrete and structural steel designs allow stresses approaching yield strength for extremely remote accident cases. The calculations of many of the loads, though, are less than sophisticated and often are unable to be verified; hence the so-called "conservative simplifying assumption" is applied, which in essence applies the worst possible case, the limiting condition, to the load. Even if not this extreme, as you would expect, conservative assumptions are always consciously made when assumptions are necessary. The BWR hydrodynamic loads are a prime example of loads being adjusted by the assumptions and conservatism of the analysis. When the maximum loads got so large as to be virtually impossible to handle, the assumptions and analytical techniques were refined and changed so as to reduce the load.

One of the factors in the complexities of nuclear plant design, particularly structures and support systems, is that conventional codes, standards and practices are not compatible with the perceived or actual requirements. Most seismic hangers, pipe anchors, supports and restraints are overly bulky and expensive because our analytical capabilities, lacking specific test data, often cause us to make very conservative assumptions, particularly as regard to dynamic conditions.

In view of this, the AIF Task Force on building cost-effective nuclear power plants has recommended an industry supported test program to provide the basis for updating codes and practices, achieving some significant simplifications in certain areas of nuclear plant design, and reducing construction costs.

The designs tested would use standard components, practical tolerances, with ease of construction as a primary consideration for establishing test configuration. The test results would be factored into code revisions and new standard practices by the industry.

An example of the type of programs envisioned is that for electrical cable tray and support systems. This program would involve:

First Identification of classes of electrical cable tray and their support systems by their role in nuclear plants--high seismic, low seismic, and non-safety related.

Second Performance of full scale static and dynamic tests of alternate systems for each class--using standard manufactured components.

And Then Based on the tests, selection of standard design or designs, and acceptance of these standards as references for design and manufacturing.

Other testing programs would include seismic design criteria for pipe hanger and support design; and seismic, dynamic and static capabilities of concrete structures, with more simplified criteria for reinforcing steel design including tolerances

Implementation of the AIF Task Force recommendation would be one means of reducing the complexity of nuclear plant designs. Such a program deserves broad industry support; including a commitment by the NRC that the results of such a program will be accepted references for licensing.

The dimensions of the task of resolving ambiguities and conflicts may be illustrated by examination of implementation of codes and standards in the welding segment of power plant construction.

The majority of the joining requirements governing the fabrication of various power plant components are specified in Section IX of the ASME Boiler & Pressure Vessel Code. This Section represents the dominant control document for welding since it relates to the qualification of personnel and the procedures they employ in welding or brazing according to the code. Moreover, section IX establishes basic criteria for welding which are observed in the preparation of welding requirements that affect procedure and performance. Section IX, as such: Quote "is an active document subject to constant review, interpretation and improvement", unquote.

This statement underscores the grim task of implementation--a critical control document which is subject to <u>constant review and interpretation.</u> The fact that over 100 inquiries requesting interpretation were submitted for code review, and over 200 page changes were issued last year highlights the need for a more stabilized document.

Taking matters a step further, as the primary or "core" welding specifications, Section IX interacts with and relates to other requirements which may supplement or supersede Section IX. These include: architect-engineer welding requirements; requirements of codes and standards applicable to equipment; and client and regulatory requirments. Therefore, welding procedures and the welding criteria that affect performance may be influenced by a number of codes and project specifications rather than Section IX alone, a common misconception.

Confronted with this complex framework of interacting and interdependent specifications, it is no wonder that we experience unproductive redundancy and conflict and encounter significant impacts on cost and schedule.

To improve the implementation of codes and standards, I believe industry should sponsor a truly industry-wide seminar to agressively seek out problems. Ambiguous, inconsistent and unjustified regulatory requirements; code and standard requirments which have redundant and conflicting requirments, and which conflict with state-of-the-art construction practices, all should be identified analyzed and prioritized for action. Once this is accomplished, the concerns can be addressed through inquiries, code cases, committee work, and technical data accumulation. To carry out such an approach, we should establish a permanent representative industry supported group or body to provide overall management of code related problems encountered in the design and construction of power generation facilities. The group could act as a "watchdog" to assess the impact of proposed changes on a timely basis and recommend appropriate action by established code and standards committees. Success of the recommended program would demand broader industry commitment to committees and full support and direction by the industry.

There is a general recognition by industry that presently imposed acceptance standards may be overly conservative. And there are some involved in building power generation facilities who say that some tolerances specified are rarely met.

I believe we need to examine much more critically the philosophical, theoretical, and practical aspects of specified tolerances. Careful planning and analysis of requirements must be accomplished before broad general requirements of particular codes and standards are referenced in specifications. Intended usage of a structure or system should dictate the tolerances required. The engineer must establish a level of quality and determine that the cost to achieve it is consistent with service requirements. That is engineering!

Inspection criteria for steel and concrete structures should be established on an industry-wide basis. A standard specification defining how inspections should be made, as well as acceptance criteria should be developed by the industry. Present guides do little more than identify what is to be inspected. We need to put hard numbers to inspection criteria, and provide for "Fitness-For-Purpose" acceptance.

That is also engineering!

Because I believe it is compatible with the basic theme of this conference, I will take the last few minutes to ask you to consider the attractive opportunity the nation has to establish the kind of nuclear industry visualized by the President in his proposed initiative. In keeping with his pledge for regulatory reform, he has spelled out his draft nuclear policy aimed at wider use of nuclear energy in electric power generation. To remove obstacles from construction of the current generation of nuclear reactors, he has directed the Task Force on Regulatory Relief to give immediate priority attention to improving licensing.

The Administration's planned program offers positive opportunities for making regulation of the design and construction of power generation facilities more cost-effective, while at the same time adequately protecting the public health and safety.

- We must continue to encourage "one-stop" construction permits, as the President has outlined.

- We should be entitled to and request true cost-benefit analysis of proposed regulations, pursuant to the President's regulatory concept.

- We should endorse a game-plan which allows industry to determine the means of compliance with "performance" regulations, based on societal goals, rather than "prescriptive" regulations.

- The following specific NRC actions are recommended:

    a. Establish quantitative safety goals for regulation and integrate them with the development of any new regulations, as well as with assessment of TMI related changes.

    b. Support and expand the present effort to develop probabilistic risk assessment (PRA) methodology, including a down-to-earth dialogue with industry on its practical implementation.

    c. Require that each new regulatory requirement be accompanied by a value-impact assessment that relates the new rule to the safety goals.

. Finally, we should insist that the purpose, scope and effectiveness of Federal Regulations and Regulatory Agencies be periodically reviewed by the Congress.

The secret to improving performance in construction of power generation facilities is both complex and simple. It is complex because we have so entangled ourselves in a web of restrictions, constraints, standards, guidelines and controls that like Gulliver, it will be hard to shake ourselves loose. It is simple because it requires only the unleashing of the powerful, innovative, ingenious hard working spirit inherent in the industry. This conference provides a unique opportunity for bellwether action to improving industry performance.

I extend my best wishes for a meaningful and productive conference here at Penn State.

REGULATIONS AFFECTING POWER PLANT CONSTRUCTION

by

Howard W. McCall[1]
Fellow, ASCE

ABSTRACT

The problems of establishing, design-engineering, and implementing realistic regulations for the power plant construction industry and the impact of these regulations on cost and value to the customer are presented. The observations present current practices and how they affect the construction and quality of the project. Productivity, lack of innovation, and ineffective conservatism are items that must be considered in establishing and implementing the regulations affecting power plant construction. It is suggested that improvement of communications between designer, constructor, and code committees would allow reasonable construction practices to be followed. By working together, everyone -- the utilities, the engineers, the suppliers, the constructors, and the regulators -- can deliver a project that gives good value to the ultimate consumer -- us.

INTRODUCTION

The ultimate test of our experience with the implementation of construction practices, codes, standards, and regulations is the grade we all would get from the consumer who pays for his or her electricity. All of us get paid by that consumer with scarcely any test of the cost benefit we deliver. We at Daniel believe it is time to take a hard look at practices, codes, standards, and regulations from the standpoint of the consumer. Is the consumer getting good value for his or her money?

The constructors and the manufacturers, more so than other groups in the industry, are in the best position to see the impact of practices, codes, standards, and regulations and to evaluate the benefit resulting from that impact. Daniel was constructing uranium and coal-fueled power plants during the 1970's when practices, codes, standards, and regulations were undergoing their biggest change; and we are convinced that the consumers are not getting their money's worth.

---

[1]President, Power Group, Daniel Construction Company, a Division of Daniel International Corporation, Daniel Building, Greenville, South Carolina 29602

The industry could deliver better value for the ultimate customers. I would like to focus on three of the major reasons for this failure: productivity, lack of innovation and ineffective conservatism.

Before anyone jumps to conclusions about productivity, we believe that workers have the same abilities and attitudes that they have always had. However, we as managers could provide these workers with better information, materials, training, coordination, and direction to allow them to perform more effectively and efficiently. Industry's approach to practices, codes, standards, and regulations has contributed to inhibiting our managers' ability to deliver the necessary support to permit workers to perform more effectively and efficiently.

Lack of innovation and ineffective conservatism have also been significantly effected by our approach to practices, codes, standards, and regulations; and I believe that we can address all three reasons at the same time because what we do is common to all three.

BACKGROUND

The fact that the electric power industry and its supporting contractors and suppliers must operate in a controlled, regulated environment is a given fact, and we should spend no time discussing ways to achieve an uncontrolled and unregulated environment. The question is not whether the industry will be regulated but how it will be regulated. The industry has done a reasonably good job to date, better than most other regulated industries, but not good enough. We at Daniel believe that we -- the utilities, the engineers, the suppliers, the constructors, and the regulators -- can, by working together, deliver a much better controlled and regulated environment than we are delivering today. These comments apply to all types of generation: fossil, hydro, and nuclear.

One other point concerns reference to "they." When we are asked why we are in this or that strange situation, we always reply, "they...." Ladies and gentlemen, I can tell you that I have met "they," and "they" are us. "They" are us as a group and as individuals, and we all can and must make improvements as individuals, as groups, and as a whole.

You might ask what is the incentive for us to make it better. I have also met the consumer who is paying the bills, and the consumer is us.

THE PROBLEMS

The number of regulations and codes applicable to power plant construction is increasing each year, and previously established requirements are constantly being revised. In the last 10 to 12 years, the changes have been dramatic and have occurred at such a rate that it is a major task to maintain awareness of them.

With the increasing number of standards and revisions, there is also a greater likelihood of inadequacies and problems associated with them. Unquestionably, there are many regulations that apply restrictions for which very little benefit is obtained when compared to the associated impact of their implementation on cost and schedule. In some cases, inadequate consideration is given to the impact of the regulations on the user. Many of these requirements come from our own committees. Each organization in the industry needs to look at how we are supporting this committee work. If we do not support committee work from all phases of the industry, and with our best people, our time and our corporate commitment, we will greatly contribute to our own burdens.

There are numerous examples of undue problems and unnecessary repetition induced by codes, standards, and regulations that are experienced by the constructor on a daily basis. One illustrative example is welder qualification requirements.

The two most commonly used standards for qualifying welders are AWS D1.1 and ASME Section IX. There are enough differences between the two that welders qualified to one of these standards would not necessarily be qualified to weld to the other. Due to these differences, ridiculous situations can exist. For example, a welder may take a test to ASME Section IX, which will allow him to weld a main steam-pipe joint. This same qualification will allow him to fillet weld attachments to pressure piping. However, he would not be qualified to use the same welding procedure to weld a similar fillet weld on structural steel requiring qualification to AWS D1.1.

Some of the problems induced by codes, standards, and regulations are compounded by additional designer-imposed requirements. One significant example is the problem imposed by interrelating regulations controlling design and spacing tolerance of concrete-reinforcing steel in a complex structure such as a reactor shield wall or auxiliary building wall.

In a typical reactor shield wall or auxiliary building wall, the amount of reinforcing steel and the number of embedments, penetrations, and weldments have proliferated to the extent that filling the void (forms) with the specified concrete mix has become a major construction concern. Regulatory requirements that affect design criteria have dramatically increased the amount of reinforcing steel specified for typical wall sections; and the subsequent congestion has compounded the problem of meeting installation tolerances for the rebar, formwork, embedments, and penetrations. The number of systems that penetrate a typical wall have also increased as have the size and number of weld plates needed to support the additional systems. It becomes evident that the extreme congestion caused by these structural concrete components creates a nightmare for the constructor when tolerances must be met for location of each individual component. Each time a reinforcing bar is moved to make room for a weld plate or penetration, the potential is great that it will create a new interference with some other component or that the rebar-placing tolerance will be violated. As congestion increases due to compounding tolerances, new problems are created for the concrete placement personnel.

This difficulty is illustrated by the following example. We were asked to construct 1-foot-thick structural concrete walls with an 8-inch-flange deep web beam embedded at the upper grade elevation of the wall and a layer of #6 rebar on both faces of the wall. In this situation, we were required to maintain a 1-inch clear cover on each face of the wall while the concrete specification required a 3/4-inch nominal maximum aggregate mix design. The addition of the minimum clear cover distance, the width of the beam, and the rebar diameters totalled $11\frac{1}{2}$ inches of the 12-inch-thick wall. Obviously, concrete placement from the top of the wall would have been impossible, so unconventional and costly placing methods had to be employed. We accomplished these concrete placements, but not without substantial increases in man-hours and cost of formwork in the process.

Problems of this type should be viewed as a challenge by the designer. New ideas and approaches should be considered and developed to create a more constructible design. Where the potential for problems of the type previously discussed exists, the designer should attempt to foresee the problem and to minimize changes for occurrence in his design. For example, for walls requiring numerous penetrations for pipe, HVAC, cable tray, etc., consideration may be given to consolidating the penetrations. Where this could be done, fewer penetrations would be required, thus reducing the congestion in the walls and reducing the cost of concrete placement.

To minimize installation of surface-mounted plates, consideration could be given to the feasibility of using large sections of steel plate (with steel-welded shear connectors) as permanent forming. This could serve the purpose of increasing the weldable area on the wall face while negating the cost of form removal for this portion of the wall and considerably reducing the quantity of reinforcing steel.

Designers can provide greater assistance to constructors by making realistic constructibility evaluations of their designs. For example, seemingly cost-effective designs that reduce the size of concrete members or that concentrate several systems in a smaller structure may appear to save space and money, when, in fact, the cost of constructing the member/system more than offsets any savings in the design stage. Structural details should be considered from the standpoint of how they integrate with the plant as a whole. For example, it appears that designers in the nuclear industry view a concrete wall or slab as a solid entity and then proceed to punch holes in the wall as each new system is added into the design without regard for previous or future penetrations or for the total development of reinforcing. Major problems have developed on some projects due to the need to post-apply piping, mechanical and/or electrical systems. Designer and constructor omissions, design revisions, and stricter regulatory requirements have led to a virtual explosion in the installation of concrete expansion anchor bolts and supports. Very specific tolerances are invoked for the installation of anchor bolts and supports; however, the burden is on the constructor to ensure that installation and tolerances mesh with existing rebar, penetrations, and supports.

Many potential problems can be prevented by imaginative and innovative design. However, there are problems in our approach to the design of power plants that have minimized the ability or willingness to meet this challenge. To minimize liability, the design engineer relies upon standards because they have been used before. Sometimes these are drawn from more restrictive or totally different applications, thus resulting in gross overcommitments. In effect, the degree of innovation allowed by the codes, standards, and regulations is not effectively utilized. This seems more prevalent in the design of safety-related items where innovation is viewed as risky and risk is associated with liability. As a result, designers become slaves to the system, which appears to demand nothing short of absolute or zero risk at any cost.

The constructor's problems in working to codes and standards are not always as obvious as those previously described. Many of the constructor's problems arise from the difference in the original intent of a standard or requirement, the manner in which the requirement is applied at the point of usage, or the manner in which the requirement is interpreted during inspection. This type of situation can occur when those who develop the standards or requirements lack a full understanding of state-of-the-art construction practices and do not foresee the various possible results that may be fully adequate but that do not conform completely with a strict interpretation of the standard.

A fillet weld may be used to illustrate this situation. A fillet weld, when deposited, fluctuates in size, as can be determined by measuring in several locations along its length. This is normal for a manually deposited fillet weld. The designer may be satisfied with a fillet weld with an average size conforming to his specified size. However, a strict interpretation by an inspector may require the rejection of the weld if any single point on the weld is slightly less than the size called for by the designer. The constructor is put in the position of intentionally overwelding to ensure that no isolated locations along the fillet weld are less than the size called for by the designer. No relief is provided to the constructor until the instructions provided by those preparing the standards make provision for acceptance of this condition. For this situation, the

American Welding Society standards make provision for acceptance, but the American Society of Mechanical Engineers standards do not. Something as seemingly insignificant as this can place considerable hardship upon the constructor and expense to the owner.

In some instances, specification writers unconsciously create construction discrepancies by not delineating specific requirements within codes or standard manuals, e.g., ACI, ASME. For example, a typical concrete construction specification may go into great detail in defining the requirements for the type of finish to be placed on various slabs, walkways, pads, etc.; but this same specification will make only a broad general reference to controlling concrete during hot weather in accordance with ACI 305, Recommended Practice for Hot Weather Concreting. By making a general reference, the specification writer has minimized his liability and has certainly increased the options of the construction engineer; however, he has also left too much to interpretation of the referenced standard and has made it extremely difficult for engineering or quality control to enforce any of the recommendations within the referenced standard. Unfortunately, the specification writer may later find that his general reference has been misapplied and a particular portion of the work did not get the attention he intended. The burden should always be on the designer to furnish directions that are not subject to wide degrees of interpretation.

ALTERNATE APPROACHES

The types of conflicts previously discussed exist in almost every phase of power plant construction. Greater insight is needed by the designer to eliminate these problems through awareness of the constructor. An increased spirit of teamwork between the constructor, designer, code committees, and regulatory bodies is needed to identify and resolve these problems as they occur, and before they occur, when possible. When those who establish the standards are not attuned to the constructor who must implement the requirements, problems with the enforcement of these requirements are bound to occur.

The problem of eliminating design-related problems often rests with the constructor. Many potential design-related construction problems can be eliminated before construction with adequate up-front attention by the constructor. This may be done by a thorough review of the design requirements by the constructor to identify and resolve with the designer potential construction problems before construction commences. Various types of problems lend themselves to early detection by these constructibility reviews, especially problems associated with implementation of various technical and regulatory requirements.

To make the constructibility review effort effective, the designer must accept the constructor's expertise and should use the constructor's input in a positive fashion. The most effective means of problem-solving occurs when effective, honest communication from both the design organization and the construction organization interface with team spirit and a common goal of true accomplishment. There should be no attempt to pressure the designer to accept changes with which he does not agree.

The voluminous amount of regulations, codes, and standards imposes on the constructor a compliance task of gigantic magnitude. Compliance can only be accomplished effectively with a well-planned and well-organized program in the constructor's engineering, quality, and construction organizations. The normal approach to ensure that the work conforms to codes, standards, and regulations is through verification by the constructor's quality organization. This organization should have unrestricted latitude to identify deviations from codes, standards, and specifications.

Presently, the Federal Energy Regulatory Commission (FERC) is considering the establishment of a requirement that the inspection currently being performed by the constructors be performed by third-party inspection organizations on construction projects under FERC jurisdiction. The purpose of this requirement would be to establish independence between the constructor and the inspector. While there is question that some degree of independence is necessary, total disassociation from the constructor, as this proposal would require, would have a negative total effect when applied.

The constructor is responsible for complying with specified requirements, and therefore he needs the assurance that he is doing so. To assign external responsibility for inspection still does not relieve the constructor from complying with the requirements; it merely gives him less control over ensuring that he has complied unless he duplicates the inspection effort.

This degree of governmental policing could be expected to make all parties more defensive; and rather than achieving the needed increase in teamwork, communication, and innovation, this approach in the nuclear power construction industry would be expected to have a disastrous negative effect. It would provide a hasty, misdirected attempt at resolving an industry problem of implementation of regulatory requirements based on a lack of complete understanding of the problem. In an industry where a greater degree of teamwork and communication among the owner, designer, and constructor is vital, replacement of the constructor's inspection organization with third-party inspectors would only create more division.

For the inspection and verification program to be performed effectively, the program must be performed by the constructor, who must police himself through a well-planned and -implemented full-service quality organization. This is most effectively accomplished by a construction project's quality management reporting above and outside the project's organization, thus providing the necessary independence and an effective approach to establishing a total quality program.

A weak quality organization providing only an inspection and verification function will create obstacles to communication ties that are absolutely necessary for implementation of codes, standards, and regulatory requirements.

The constructor's quality organization needs to make a greater contribution than simply providing an inspection and verification function. More attention needs to be focused on quality engineering to convert the codes and standards into workable procedures and systems, which will allow accurate methods and means for ensuring that the work conforms to the requirements. The quality engineer should ensure that the requirements are correctly identified, understood, and communicated to the inspectors. Quality engineering, while ensuring that requirements are fully met, also ensures that realistic judgments are made when comparing the finished product to the engineers' requirements. The construction engineers, design engineers, and the quality organization all need to direct their efforts to achieving the same end result. The quality engineer's role is to ensure that the constructor's inspection organization is in tune with design engineering. The quality engineer must also maintain the interface with construction engineering and the project construction organization to ensure that these groups have a common understanding of the requirements and a means for complying with them. The quality engineer provides a means for ensuring that accurate and clearly understood information is received by each inspector.

With the nearly endless quantity of inspection that must be performed and documented, the resulting quantity of records and inspection documents is staggering. Well-designed quality systems that minimize the total number of documents and finalize the documents for record storage as soon as possible without endless reviews are an important responsibility of the quality organization.

The construction engineer is to the craftsman what the quality engineer is to the inspector. Without knowledge of existing requirements, the construction side of the project organization is not likely to comply with the requirements. This information must be provided by the construction engineer in such a manner that it becomes meaningful and usable by the craftsman. Standardized methods and systems need to be used whenever possible to present meaningful information to construction when and where it is needed. Innovative construction engineering that will allow the craftsman to efficiently utilize his skills with as few disruptions as possible while still complying with all project requirements will improve the attitude and productivity of the worker.

Systems that do not provide the worker with the information, materials, and means of performing his job efficiently create bad attitudes and decrease productivity beyond the decrease built into a poorly designed system. At the present state of complexity of codes, standards, and regulatory requirements, construction engineering must keep pace to provide a smooth flow of information from the designer to the craftsman and also from the craftsman back to the designer. Providing the required information to construction is just as important as providing this information in a manner that will allow it to be effectively received. At present, the mechanism for transferring information from the designer to the craftsman is through time-consuming, manually prepared work packages. Mechanized systems can relieve the construction engineer and allow efforts to be directed toward improving the quality of instruction provided to the craftsman.

The constructor has the job of changing seasoned construction workers to deal with new requirements, and any improvement in the construction engineer/craftsman communication can be expected to have a positive effect on the craftsman's attitude toward and reception of the imposed requirements and restrictions.

Obviously, the quality engineer and construction engineer must have a common understanding and awareness of project requirements. Also important to the project is the need for information to be transmitted to the craftsman and the inspector through organized training programs. Too often the information is available, but has not been disseminated to the level of implementation. The constructor's job is to make sure that the information required to perform the job is available at the level of implementation.

The size of our organizations limits our communications and information dissemination. In the past, the chief engineer personally reviewed many of the details that his junior engineers and designers were developing. The sheer size of our organization today, whether manager, designer, constructor, or operator, makes this kind of personal attention from the senior, seasoned people physically impractical, if not impossible. We need to find ways to re-create that kind of leadership and input. The quality circle concept so widely used in Japan may be one such way.

SUMMARY

The need for improvement in the implementation of codes, standards, and regulatory requirements is evident; but to effect the improvement, attention must be addressed to all sources of the problem. Greater attention needs to be given to the establishment of more realistic requirements that fit with reasonable construction practice. Better communication should be established between designer, constructor, and code committees so that the real intent of the standards is effectively achieved. Refined systems are needed in the constructor's organization to effectively provide information at the point of implementation.

These problems are not new ones; they are basic and have always been an underlying problem of the industry. With the virtual explosion of requirements in recent years, these basic problems have become more visible and significant. The need is for dedicated, honest communicators to work together and to direct their efforts to the end product. The resolution lies with people who can admit their mistakes, analyze them and take positive steps; this attitude should be rewarded. Such an approach should apply to all organizations impacting on construction, including code committees, regulatory bodies, the designer, the constructor, and the owner.

Nuclear Plant Construction--Government Perspective

by

James G. Keppler[1]

Abstract

Major observations, from a regulator's perspective, are presented on experiences at nuclear power plants under construction. Changes in the NRC's regulatory program and insights on regulatory philosophy related to regulation of nuclear plant construction are also discussed.

Introduction

Over two years have passed now since the Three Mile Island Accident; yet that event continues to have a dramatic impact on the nuclear industry and the NRC's regulatory program. Licensing of plants for operation or for construction slowed to a standstill while the accident was analyzed again and again by the NRC, the Presidential Kemeny Commission, the NRC-sponsored Rogovin Commission, Congressional Committees, industry, and others. Consequently, the NRC began placing and has continued to place many new requirements on utilities operating and building nuclear plants in order to upgrade the safety of nuclear activities.....requirements involving equipment modifications, staffing improvements, procedural changes, and upgraded emergency preparedness.

At the same time, in response to a clear mandate from the Congress, from the states, from the public and the Commissioners themselves, the NRC staff began focusing on identifying and dealing with substandard licensee regulatory performance as a means of shoring up the nation's confidence in nuclear safety and NRC's regulatory program. In this regard, three areas are worth noting. First, there has been a beefing up of the NRC's inspection program to the point that each operating site now has at least two resident NRC inspectors and we expect to have at least one NRC inspector at most construction sites by the end of this year. This effort has been directed toward providing more independent verification that licensee activities are being conducted safely and in compliance with regulatory requirements.

---

[1]Director, Region III, U.S. Nuclear Regulatory Commission, 799 Roosevelt Road, Glen Ellyn, IL 60137

Second, there has been a toughening of the NRC's enforcement posture with an eye on dealing more decisively with safety related problems and with chronic poor regulatory performance. Here, the NRC requested and received from Congress authority to issue higher civil penalties and late last year implemented a new Interim Enforcement Policy that better focused on the significance of violations. I might add that this policy placed increased emphasis on enforcement in our construction inspection program. And third, the agency has established and implemented a program of annual reviews of the regulatory performance of nuclear power plant licensees. This program has as its main objective the upgrading of the regulatory performance of the weak licensees through comparison of licensee performance on a national level.

In focusing on licensee regulatory performance we have become increasingly concerned with the number and seriousness of quality related problems being encountered in the construction of nuclear power plants --- problems having safety importance which resulted from a major breakdown in the overall quality assurance program. Perhaps just as important are the related questions that have been raised concerning industry's ability to assure proper construction and the effectiveness of the NRC's inspection program for monitoring site construction activities. The dominant issue, recognizing that some problems are going to occur, centers around whether too many problems are occurring and why these problems are not identified on a more timely basis.

Let's reflect on some experiences that have and continue to generate widespread public interest. Within my region alone, there have been serious quality assurance breakdowns with broad repercussions at the Marble Hill, Midland, and Zimmer construction sites.

At the Marble Hill Nuclear Plant construction site in southern Indiana, our inspection program in 1979 identified weaknesses in the concrete program and its related quality assurance system. We were dealing with these problems, and had met with the utility management to review them, when a local environmental group brought forward a concrete worker who claimed that honeycombing, voids and surface defects were being improperly patched.

These allegations, which were substantiated, led to a more detailed, timely inspection and investigation that broadened into other areas of work at the site. About the same time, code compliance problems were identified by the Indiana Boiler Code Inspector and the National Board of Boiler and Pressure Vessel Inspectors.

This overall effort led to a halting of all safety-related work at the site in August, 1979 --- a move taken by the utility and confirmed by an NRC order. Work was not permitted by the NRC to resume until December, 1980, some 16 months later, when the utility's quality assurance program --- and that of its contractors --- had been substantially upgraded and the adequacy of completed construction work had been verified.

In the case of the Midland facility in Michigan, excessive settlement of the diesel generator building was observed in 1978. The settling was subsequently attributed to inadequate and poorly compacted soil under the building. Further investigation by the licensee revealed that other safety-related systems and structures were affected. All of these systems and structures were well along toward completion at the time the problem was discovered. The NRC's investigation determined that design and construction specifications had not been followed during placement of the soil fill materials and that there was a lack of control and supervision of the soil placement activities by the utility and its contractors.

The NRC issued an Order in December 1979 suspending remedial work on the substandard soil and fill --- an Order which the licensee contested. Over three years have passed since the problem was first identified. The matter has still not been resolved and the issues are currently being litigated before an NRC Hearing Board with intervenor participation.

As for the Zimmer facility in southern Ohio, that is the most recent problem case and, in my view, offers the potential to be the most serious. Since January of this year we have been investigating allegations of construction deficiencies. This investigation effort, which is still ongoing, started out with allegations from a couple of sources, but soon broadened to many workers and ex-workers. To date, we have interviewed over 100 individuals and expended over 200 man-days onsite pursuing these allegations.

The investigation, thus far, has identified a number of quality related problems at the Zimmer site. Although some actual construction deficiencies have been identified, the majority of these problems focus on the ineffectiveness of controls implemented by the licensee and its contractors for assuring the quality of work performed. The total impact of these quality assurance deficiencies on the actual quality of construction has yet to be determined; however, from the public perspective, it is an indictment against both the utility and the NRC to have a nuclear power plant better than 95 percent complete and have serious outstanding questions as to the adequacy of construction.

Before the plant can be licensed a comprehensive quality confirmation program will have to be conducted and identified problem areas resolved. The impact of this quality assurance breakdown is far-reaching. By itself, without factoring in any rework, the quality confirmation program will be both costly and time consuming. Whether or not this delays operations remains to be determined. But, it is quite likely that the Operating License Hearings will be reopened and interest conveyed by some congressional committees suggests the distinct possibility that congressional hearings will be held on this case.

Problems with construction of nuclear power plants are not limited to Region III. Similar experiences at South Texas Project, Washington Public Power, Wolf Creek and Watts Bar, just to mention a few, have also served to fuel the fire in this regard.

Assessment of Problems

During the early 1970's much attention was given by the then Atomic Eneergy Commission and the nuclear industry to the importance of quality assurance in connection with nuclear power plants. The Commission's regulations concerning quality assurance, Appendix B to 10 CFR 50, were dissected and restated, cussed and discussed. By the middle part of the decade, the role of quality assurance had been literally "milked to death" and there were really not many original thoughts left on the subject. No one disputed the need for quality assurance in the nuclear reactor industry. Yet, despite the elequent rhetoric spoken in support of quality assurance, substantive deficiencies in the implementation of licensee QA programs continue to occur and the problems are not being identified and resolved in a timely manner.

In looking at the Marble Hill, Midland, and Zimmer problems, questions have been raised as to why the licensee's Quality Assurance Program and the NRC inspection program had not identified the problems sooner. Clearly, in each case, there was an overreliance by the utility on its contractors for maintaining a thorough quality assurance program. The utility's own QA staff was small, too small in retrospect, to maintain sufficient surveillance over the work of contractors. In two of the cases we saw instances where the construction management dominated or controlled the quality assurance program and personnel. And, in each of the cases where problems had been identified, the corrective action taken was not sufficiently broad. Too frequently, the response was one of treating the symptom, rather than finding the basic cause and correcting it.

Without doubt, there have been shortcoming in our inspection program at construction sites.

There have been cases where we, like licensees, have failed to see the breadth or depth of a problem. We identified specific items of noncompliance without requiring that the basic cause of the problem be corrected.

We interviewed QC personnel and workers only infrequently -- relying instead on direct observation, document review and management discussions. But we are finding working level personnel can have valuable observations, too.

We also --- through our inspection program sample --- often look at an area once and, if it is satisfactory, may not recheck for some time, if at all. As construction schedules stretch out, however, the possibility of changes in performance in a particular area increases. There may be deterioration in work quality and we might miss it.

Related Regulatory Actions

We are continuing to review our construction inspection program to see where improvements can be made. In addition to augmentation of our resident inspection program we have: (1) increased our interface with the construction workers and QC inspectors; (2) broadened our capability to perform independent nondestructive examinations; and (3) implemented a trial team inspection approach whereby several NRC inspectors go to a construction site for two or three weeks to do a broad, intensive inspection of the quality assurance program for the work underway. With respect to this latter effort, several team inspections have been completed around the country and we are presently evaluating the overall results from a cost benefit consideration.

Another major area of regulatory change that I previously mentioned is the NRC's enforcement program which was put out for public comment and adopted for interim use in October, 1980.

The most obvious difference in the new enforcement program is the significant increase in possible fines, reflecting the higher fine authority granted by Congress last year. Single violations may now be assessed a fine of up to $100,000, instead of the former $5,000 limitation.

The two measures I've discussed -- both boradened inspections and stricter enforcement -- have been strongly endorsed by the NRC's Congressional overseers. In an October 1980 report, the House Committee on Government Operations urged the NRC on to even greater vigilance -- "The NRC," according the Committee "should be even more aggressive in its inspection of nuclear power plants, more stringent in its requirement for technical competence and management performance by licensees and more resolute in the enforcement response toward licensees with protracted problems."

This Congressional Committee also strongly supported the third element which I see growing in importance in the coming years -- the NRC's new program of stepping back once a year to assess licensee performance, called the Systematic Assessment of Licensee Performance, or SALP.

This review is done by the inspectors and supervisors who have been involved with the facility during the year with additional input from the other major NRC Offices.

The SALP review includes an examination of the licensee's enforcement record, the information required to be reported to the NRC, and a somewhat subjective appraisal of the management performance and attitudes in dealing with problems. Licensee performance is rated in various functional areas as above average, average, or below average compared to other licensees throughout the country. The first round of SALP appraisals for the period July 1979 through June 1980 is pending issuance.

I feel the SALP appraisals have been useful from the NRC point of view to assess where we need to place our regulatory priorities -- and, hopefully, the review process has assisted licensees in measuring their own performance against other licensees. Looking at the entire year's performance gives us a better regulatory perspective, and quite frankly it has revealed a pattern of less than satisfactory performance that might not have been evident in day-to-day activities.

With the Reagan Administration leaning more favorably toward the uses of nuclear energy than the previous administration, it has been suggested that we can expect a softening in NRC regulation. Clearly, the NRC has been given direction to accelerate the licensing process associated with nuclear power plants and the agency is moving to respond to this direction. Furthermore, I see several other changes that may tend to reduce the regulatory burden placed on licensees.... perhaps most significant is the awareness by NRC management that all the requirements imposed upon industry by the NRC since Three Mile Island might actually be producing a negative safety impact rather than a positive safety impact. And, while I do not necessarily look for a significant backing down in the number of requirements with the present Commission, I see indications that this awareness will at least result in NRC making a better effort to prioritize these requirements. That, in turn, should enable industry to plan its activities more effectively and achieve better utilization of available resources.

While the NRC's efforts to accelerate the licensing process and to improve its coordination of regulatory requirements may be a welcome sign to the industry, I think it would be a mistake to view such actions as a softening in our approach to monitoring the safety and compliance of licensee activities. I'm convinced our Congressional Oversight Committees will continue to demand a strict inspection and enforcement posture. The House Committee on Government Operations Hearings last year, that I referred to earlier, took the NRC to task for tolerating the sub-par utility efforts for excessive periods of time, stressed the premise that the public had the right to expect the NRC to demand nothing less than excellent utility safety performance, and demanded the NRC take strong actions in dealing with sustained poor regulatory performance. There is also indication that the

GOVERNMENT PERSPECTIVE

present administration believes strong nuclear regulation is in order, since NRC's Inspection and Enforcement budget and personnel ceilings have survived to date the regulatory ax wielded by the Office of Management and Budget on many other regulatory agencies.

Lastly, I'm personally convinced that strong licensee perfromance and a demanding regulatory posture are essential ingredients to public acceptance of nuclear power.

In summary, while I see new initiatives being taken by the NRC to shorten the Licensing process, the public and the government are going to expect top performance from the nuclear industry in the construction and operation of nuclear power plants. Our job at NRC will be to make sure that such high level performance is achieved.

CONCRETE INDUSTRY PERSPECTIVE

by

Robert E. Philleo[1]

Abstract

The developments within the concrete industry which currently have the biggest impact on standards are those associated with energy conservation. These include changes in cement manufacturing technology, burning of coal which cannot be conveniently used in power plants, and the use of waste materials such as fly ash and blast furnace slag as a partial replacement for portland cement. The standards system for reactors and containment structures in nuclear plants has been complicated by the overlapping jurisdictions of ACI and ASME. The problem has been resolved by a joint ACI-ASME committee. Quality construction in critical structures can be achieved by rather simple standards if they are rigidly enforced.

INTRODUCTION

Power generating facilites produce usable energy. The most significant developments which are currently taking place in the concrete industry are those intended to reduce energy consumption in concrete construction. Thus, in a nation concerned about energy stewardship there is a double benefit if energy producing facilities are built with energy-saving methods. But in addition to this general favorable correlation between energy creation and energy conservation, there are in the case of concrete construction, some unique interfaces between power-generating facilities and energy conservation. Waste from power plants, for example, is used directly as an ingredient of concrete, and fuels no longer permitted to be used in thermal power plants are now used in cement plants. Both phenomena are having a direct and continuing effect on standards for concrete construction.

BACKGROUND

The construction industry has long been a target for energy conservationists. The most recent figures I have seen state that 11 percent of all the energy produced in the United States is consumed by the construction industry. It is interesting to note that in Great Britain the figure is 15 percent. One might conclude

---

[1] Chief, Structures Branch, Directorate of Civil Works, Office of the Chief of Engineers, Washington, D.C. 20314

that our industry is somewhat more efficient than the British, but the truth probably is that in this country more energy is wasted in non-construction activities. Now that waste in such uses is decreasing, the fraction of total energy used in construction will increase unless similar conservation measures are adopted by the construction industry. At any rate, the construction industry is the biggest industrial user of energy, and it presents a big opportunity for energy conservation.

When the Federal Energy Administration, one of several forerunners of the Department of Energy, came into being a few years ago, one of its first efforts was to compile statistics on how energy was being used. The figures were presented in many different ways. In one presentation they displayed the cost of energy of manufactured materials as a percent of the value of the materials. When applied to principal construction materials, it yielded the results in Table 1.

Table 1. Cost of Energy Expressed as a percent of the Value of the Material Produced

Portland Cement	15.6%
Steel	8.5%
Structural Clay Products	7.5%
Concrete	2.7%
Cut Stone	2.2%

One is immediately impressed by the spectacular value for portland cement. In fact, all of materials manufactured in the United States portland cement leads the list. Proponents of concrete were quick to point out that portland cement itself is not a construction material. It is an ingredient of concrete; and when it is put in concrete, its energy content is greatly diluted. This is demonstrated in Table 1 by the very low value for concrete. Such proponents prefer other ways of displaying the data. In Table 2, which shows the tons of oil required to produce a ton of materials, even portland cement looks good.

Table 2. Tons of Oil Required to Produce a Ton of Material

Aluminum	5.6
Copper	1.2
Steel	1.0
Cement	0.18
Concrete	0.02

But the FEA continued to be fascinated by the figure for portland cement in Table 1. Its fascination increased when it discovered that in Japan and western Europe cement was produced with 30 percent less energy than in the United States. The immediate conclusion was that Americans don't know how to design cement plants. The truth is that, when American plants which existed at that time were designed, the designers gave labor conservation a higher priority than energy conservation. It is impossible to hold the world championship for both labor efficiency and energy efficiency. American designers went for the former. The two striking features of an American cement plant are the lack of people and the long rotary kiln. It is only necessary to transport raw materials from the quarries to the upper end of the kiln. From there they are untouched by human hands as they roll and tumble downhill through the various temperature zones, undergoing a particular part of their pyroprocessing in each, until they fall out of the hot end as cement clinker. But a long kiln is inherently energy inefficient. The creation of the various temperature zones depends on the loss of heat through the walls of the kiln. It would be counterproductive to insultate kilns to retain the heat because the necessary zones would not exist. The problem was further compounded by the fact that the least labor intensive method for getting raw materials from the quarries to the kiln was to grind the materials and transport them in a water slurry. One man at a control panel could throttle valves and accurately control the proportioning of materials. Unfortunately all the water that entered the kiln had to be evaporated at the rate of 144 BTUs per pound. Europe and Japan were confronted with rising energy costs earlier than the United States. They abandoned the wet process and long rotary kilns in favor of smaller single-temperature thermal processing units between which the material had to be handled. The FEA exacted from the cement industry a promise to reduce energy consumption to the level which had been demonstrated as feasible, and there began a trek of American cement plant designers across the Atlantic and Pacific. The most inefficient American Plants were shut down, and the new plants look much like their counterparts in western Europe and Japan. The revolution that has taken place would probably have occurred naturally as a result of rising energy costs but possibly not as quickly. It was convenient for the United States that the technology had already been developed.

CURRENT DEVELOPMENTS

Two current developments interface with the power-generating industry directly. They concern the availability of fuel and the disposal of waste products.

High Sulfur Coal
The cement industry has gone through the same fuel cycle as many other industries. Fifty years ago most cement kilns were fired by coal. Then there was a trend away from coal until there was almost complete conversion to gas and oil. Now conversion back to coal is nearly complete. Cement kilns can make use of coal with a sulfur content too high to be used in power plant boilers by current environmental restrictions. Thus, the cement industry

offers some relief to those segments of the coal industry which
have lost electric utilities as customers as a result of standards
outside the construction industry.

The use of high sulfur coal, however, has not been without its
problems in the cement industry. Those who follow the ASTM
portland cement specification closely will have noted changes in
one specification limit in three consecutive years. While portland
cement consists almost entirely of ground clinker, it is necessary
to add a few percent of gypsum to control setting and strength gain
characteristics. However, excess gypsum produces adverse effects.
Thus, the specification contains an upper limit on SO_3 content.
But high sulfur coal introduces sulfur into the clinker with a
resultant increase in the calculated SO_3 content in the chemical
analysis with no particular effect, either favorable or
unfavorable, on performance. Last year the specification was
modified to permit an SO_3 content half a percent over the
allowable maximum value when test data were supplied to demonstrate
that the higher amount was needed to achieve optimum performance
and there were no adverse effects. This year the half-percent
ceiling was removed so tht any amount of SO_3 is permissible if
supported by test data. Next year, if a revision currently
underway is successfully completed, the waiver may be implemented
whenever the optimum SO_3 value is within half a percent of the
specification limit. This relaxation accomodates the inherent
testing error in determination of optimum SO_3.

Waste Materials

Since portland cement is by orders of magnitude the most energy
intensive material in concrete, it is natural to search for less
energy-intensive materials which may be used to replace at least
part of the cement in concrete. Fortunately there are waste
materials from both the power-generating and steel-producing
industries which fill the bill: fly ash and blast furnace slag.
While it takes a great deal of energy to produce both of these
materials, our system of energy bookeeping charges the BTUs to the
utility and steel industries so that they come to the concrete
industry unencumbered by an energy charge.

Fly ash. This residue collected in he stacks of thermal power
plants which burn powered coal is part of the general class of
materials known as pozzolans. A pozzolan is defined by ASTM as "a
siliceous or siliceous and aluminous material, which in itself
possesses little or no cementitious value but which will, in finely
divided form and in the presence of moisture, chemically react with
calcium hydroxide at ordinary temperatures to form compounds
possessing cementitious properties."

Pozzolans were first used not because they saved energy but
because they reduced the cost and improved certain properties of

concrete. In my own organization, which builds dams many of which include hydroelectric generating facilities, we were interested in the fact that in mass concrete pozzolans used as a replacement for about a third of the portland cement produce concrete with adequate strength but with less heat generation during hydration than concrete containing portland cement only. The entire technology of mass concrete construction revolves around the prevention of thermal cracking attributable to heat generation resulting from the hydration of cement. We began using pozzolans in the late 1950s and have used them in virtually all the mass concrete we have placed since then so that we are the largest user of pozzolans in the country. Initially we used both fly ashes and natural pozzolans. Natural pozzolans, which are produced by calcining certain clays or shales, are cheaper than cement but more expensive than fly ash. However, twenty years ago in the western part of the country it was frequently cheaper to produce natural pozzolans near the job site than to ship fly ash from the middle west. Now just about the only pozzolan used is fly ash since it is available in most parts of the country and the production of natural pozzolans is energy intensive. While a recent shift in our national priorities has reduced the number of water resource projects under construction, in the years of maximum pozzolan usage we regularly used 150,000 tons of fly ash annually. In fact in those days we were concerned about the trend away from coal to oil and gas in the utility industry. We were particularly concerned about those plants equipped to burn coal which burned economical "dump" gas during the summer since summer was the principal construction season when the need fly ash was greatest. These concerns seem to have subsided.

Fly ash is also widely used in the ready-mix concrete industry. It is by no means universally used. The extent of its use appears to be related more to the sucess of promotional efforts than to the availability of material. The most recent use is as a blending material in the production of portland pozzolan (IP) cement. Although this cement has been in the ASTM specification for years, the cement industry never felt much incentive to produce it. Now, however, it not only offers an opportunity for energy saving but it has provided a means for extending the clinker supply during the period following the closing of inefficient cement plants prior to the opening of new more efficient plants. Such cement commonly has about 20 percent of the clinker replaced with fly ash. It still does not comprise more than a minor part of total cement production, but several companies are producing it. If there is a marked upswing in construction activity, a threatened cement shortage will probably stimulate increased interest in the product.

Another use for fly as is as a raw material providing a source of silica and alumina for production of clinker. Statistics on the collection and use of fly ash are given in Table 3.

Table 3. Use of Fly Ash in Concrete
       1979

Total Collected	57,500,000 Tons
Total Used	10,000,000
Use in Concrete	
Clinker Production	620,000
Type I-P Cement	210,000
Added at Mixer	1,900,000

The amount of fly ash used in concrete may be drastically affected by deliberations currently underway at the Environmental Protection Agency. EPA administers the Resource Conservation and Recovery Act which is concerned with the "establishment of environmentally sound disposal practices for all wastes." Section 6002, Federal Procurement, directs all procuring agencies which use federal funds to procure items containing the highest percentage of recovered materials practicable. The agency decided that its first attempt at implementation would be a directive requiring that in federal and federal-funded construction all concretes will contain fly ash. A public hearing was held which may modify the position, but presumably there will be an order either manadating the use of fly ash or prohibiting specifications which prevent its use.

Blast-furnace slag. In most parts of the world granulated blast furnace slag is a much more common cement extender than fly ash. Slags tend to be more hydraulic than pozzolanic in that they do not depend on the presence of calcium hydroxide produced in the hydration of portland cement in order to be cementitious. Like fly ash, slag may be either an ingredient of a blended cement (Type 1S) or a material added separately at the mixer. Reasons given for the lack of use in this country have included the difficulty of grinding and the lack of interest in the steel industry in granulating its slag. The slag must be rapidly chilled in order to have acceptable hydraulic properties. Prior to World War II two companies in steel producing areas produced a portland-blast-furnace-slag cement, and during the first cement shortage following the war several companies temporarily produced the material. Granulated blast furnace slag has never been commercially available in the United States as a separate concrete ingredient, although for several years there has been a source in Canada. Now interest is escalating in the United States. A large facility in Baltimore will produce a very satisfactory material which will be marketed as a separate concrete ingredient by a major cement company. A new ASTM standard specification for slag is currently being processed.

FUTURE PROSPECTS

Practically all portions of the earth's crust contain suitable raw materials for cement as well as aggregates for concrete. As other materials become scarce or objectionably energy intensive, the prospects for concrete appear excellent. Statistics on world-wide consumption of portland cement are instructive. In 1968 Canada ranked 29th in the world in percapita use of cement and the United States ranked 31st. Last year, with the addition of many petroleum-producing countries which were unranked in 1968, neither ranked in the top 50. It is clear that Canada and the United States are the only countries in the world with a real choice in construction materials. Other countries have found lumber scarce, steel expensive, and plastics and asphalt too dependent on petroleum feed stocks so that they build predominately of concrete. There are indications that we are headed in the same direction. The skyscraper used to be the domain solely of the steel frame. Yet since the Sears Tower was completed a decade ago all the buildings in the world over 700 feet in height, and half of them are in the United States, have with a single exception reinforced concrete frames (possibly one-and-a-half exceptions, depending how one classifies the compound frame of the recently-topped out Texas Commerce Tower in Houston).

Yet in spite of what seems to be a bright future for concrete there are impediments to the expansion of its use. We continue to hear that "we are running out of good aggregate." What has happened is that good developed aggregate deposits near cities are surrounded by housing tracts which preclude expansion of the development, and opening of new deposits is prevented by zoning or environmental restrictions. Since the cities are the major markets for concrete, the only alternatives in some cases are to import aggregates from great distances at great expense or to rework portions of existing deposits which have previously been discarded as inferior material. The impact on a power generating facility depends on its location. Most of our water resource projects are located in remote areas where we open virgin aggregate deposits and generally find excellent material. Those building power plants well removed from cities should have the same favorable experience. Those building in or near cities may have a more difficult time.

One of the challenges of the future which may help the situation in cities is the recycling of concrete. Whenever large redevelopment projects are undertaken in cities, it is not uncommon to see one fleet of trucks hauling broken concrete out of the area while another hauls aggregate into the area. Experimental work to date indicates that recycled concrete can be a good source of aggregate. Another challenge is the utilization of the lignitic and sub-bituminous fly ash which form an ever-increasing fraction of the total production of fly ash. The present technolgy is based primarily on experience with bituminous coal fly ash. The newer forms differ from bituminous in that they contain significant quantities of lime, a constituent that might be either an advantage or a disadvantage.

STANDARDS

Traditionally concrete construction has been controlled by standards of the American Concrete Institute and the ASTM. The system has served well. The difficulty experienced most frequently is attributable to the fact that many ACI standards, such as those for hot weather and cold weather construction, are in the form of recommended practices. While they provide guidance to specification writers, they are not of such a nature that they can be cited by reference only in a project specification. Specification writers continue to do so, however, not realizing that they are introducing an array of alternative provisions without specifying which are applicable and are, thereby, introducing ambiguity into their project specifications. ACI is responding to the problem by writing more of its standards in specification language.

The advent of nuclear power plant construction indroduced a major problem in standards jurisdiction. Pressure vessels of concrete were suddenly being designed. ASME has jurisdiction over pressure vessels. ACI has the know-how for concrete design. The bureaucratic solution which was worked out seemed ideal on paper. There was formed joint ACI-ASME Committee 359 on Concrete Components for Nuclear Reactors. It produced the first "Standard Code for Concrete Reactor Vessels and Containment" in 1974, and the document has been revised regularly since. It forms a part of the Boiler and Pressure Vessel Code.

The concrete code is a lengthy one. While the authors have stuck mostly to references to the accepted standards, they have taken the most conservative option where there was a choice, added new conservatism of their own, established very high testing rates, and introduced ambiguity by referencing standard practices. To evaluate the effectiveness of the code it is interesting to compare it with the guidance provide by may own organization for the contruction of high-head dams. In comparing nuclear and water resources concrete it might be well to start with a comparison of the hazard sensitivity. This group does not need to be reminded of the gravity of nuclear accidents. But in our field we have constructed concrete dams to heights exceeding 700 feet, and a 700-foot wall of water moving down a narrow valley creates all the devastation and adverse publicity of a nuclear accident, as has been amply demonstrated during the past 5 years by the failure of non-concrete dams.

Testing Philosophy

To compare the actual practices of the nuclear industry and the Corps of Engieners it is instructive to note the length of the guidance documents. The latest issue of the Code for Concrete Reactor Vessels and Containments available to me contains about 280 printed pages on construction and inspection. Our guidance is a manual consisting of 84 typewritten pages and a double-spaced guide specification 114 pages long. In our guidance we make rather frequent use of numerical values to control or accept the work of the contractor. While it is true that numbers can lead to irritating confrontations, the confrontations which arise which cannot be adjudicated by objective criteria are equally irritating and usually result in claims. The objective is to devise a system of numbers which can be enforced without holding up the project.

The first decision is to decide on the quantity of numbers that is needed. Ideally we might desire complete testing of the concrete in critical structures. But concrete tested is concrete that does not wind up in the structure; hence, complete testing is a contradicition in terms. We must satisfy ourselves with the reality that even if we are willing to pay very large sums for testing, the test specimens will constitute only a very small representative sample of concrete similar to that in the structure. Suppose, for example, that in a placement requiring 100 batches 5 are defective. If the sampling plan calls for testing 3 batches there is an 86% probability that none of the defective batches will be discovered. If the sampling is increased to 6 batches, a rather high rate, the probability is reduced only to 74%. Thus, doubling the cost of testing only reduces the contractor's probability of getting away with his or her non-compliance by one-seventh, and the cost of testing quickly reaches the point of diminishing returns for detecting infractions. Clearly the test is not a police device. A collection of test data characterizes the population of concrete batches from which the structure is built so that it may be determined whether the concrete is significantly different from that assumed in design, but some other form of inspection must be used for a continuous check of compliance with the specifications.

We use a testing frequency based on time rather than on volume of concrete. It is our feeling that normal variations as well as blunders are related to time rather than to volumne. On our very largest projects in relatively noncritical areas cylinders may be taken not oftener than once ever 2,000 cubic yards and mandatory slump tests every 1000 cubic yards. On the other hand, we require continuous recording of batch plant quantities.

Another feature of Corps specifications is a mandatory contractor quality control section. While quality-control testing does not replace acceptance testing, since to do so would place the contractor in an introlerable conflict of interest situation, the contractor is required to maintain certain quality control charts which give warning of impending specification violations while there is still time to react.

CONCLUSION

There have been several documented cases of poor performance by contractors in the construction of nuclear power plants. I have participated in the correction of some of them. The problems have not been attributable to poor standards but to the failure to comply with standards. A simple set of standards will suffice even for critical work if they are rigidly enfored. Complicated standards do not insure quality construction if they are not enforced. Inspection, backed up by appropriate testing, is the key to acceptable and safe construction.

Standards in the concrete industry are generally well organized, but they are undergoing continual change in response to the impacts of energy shortage and environmental constraints.

SESSION I - NUCLEAR STANDARDS, LICENSING, ENFORCEMENT
AND CONSTRUCTION INSPECTION PROGRAMS

SESSION OBJECTIVES/SESSION CHAIRMAN SUMMARY

by

Eugene J. Gallagher[1] M, ASCE

Objective of Session

The objectives of this session are to provide information on the current development of nuclear standards, codes and regulations, the Nuclear Regulatory Commission's licensing, enforcement and construction inspection programs and the adjustments which the nuclear industry must consider in order to deal effectively with nuclear regulatory requirements.

Session Chairman Summary

The significant points presented by the session speakers were:

Development of Codes, Standards and Regulations

- The nuclear power industry is probably the most regulated industry in the world today.

- There now exists a tremendous volume of codes, standards, regulations and guidelines which are intended for use in the regulation and control of nuclear power plant design, construction and operation and which are still being continuously augmented and revised.

- There has been a significant increase in engineering manpower in the design of a nuclear power plant as a result of the development of codes, standards and regulations.

Nuclear Licensing Outlook

- The lessons learned from Three Mile Island can be accommodated by current generation plants without the need for significant redesign.

- Additional construction permit requirements dealing with containment integrity are to be included in near term construction permits.

---

[1] Civil Engineer, U.S. Nuclear Regulatory Commission, Washington, DC 20555

- Whether new construction permit applications for nuclear power plants will come forward is not clear in view of the financial, political and social constraints.

- The long-term outlook for nuclear power construction would seem to depend more on the real and perceived need for power than on the ability to meet design challenges.

NRC Enforcement Program Objectives

- Establish criteria for utilizing increased civil penalty authority.

- Establish tougher enforcement sanctions.

- Establish more uniform licensee treatment.

- Establish enforcement capabilities to all NRC licensed activities.

- Focus escalated enforcement actions on significant events.

NRC Construction Inspection Program

- The licensee is considered responsible for the proper design, construction, testing and operation of nuclear power plants.

- The construction program consists of routine and reactive elements.

- The NRC has utilized the resident inspection program to provide better communication with licensee's and more involvement in day-to-day site activities.

- The NRC inspection program attempts to factor in better ways of assessing the quality of construction work.

- Although management supports the inspection program, the NRC does not have adequate resources to fully conduct the program.

Industry Adjustments to Nuclear Regulatory Requirements

- The goals of the nuclear power industry are to engineer and construct a facility that conforms to the regulatory requirements within cost and schedule restraints.

- Major factors responsible for the increase in manhours and resultant increase in schedule time are the increase in complexity of plant design and changes to management control resulting from 10 CFR 50, Appendix B, Quality Assurance requirements.

- For construction to reach acceptable levels of productivity the complete scope of the engineering and design effort must be established; an engineering/design schedule which peaks prior to ground breaking must be developed and schedule milestones for engineering/design production must be established.

Consensus Recommendations*

1. The engineering and planning of a nuclear power plant should be completed to a much greater degree than is current practice before initiating construction.

2. A reasonable stabilization of the licensing process and regulatory requirements for design and construction must be established.

3. Standardized designs can be instrumental in reducing nuclear plant lead-times and licensing reviews. Standardization is one method of achieving efficient application of engineering resources and should be encouraged to meet increasing demands for design and construction of nuclear plants.

4. In view of manpower limitations a method of prioritizing inspections must be established to provide adequate inspection of those areas important to safety.

5. Corrective measures for identifying deficiencies should be taken in order to avoid enforcement actions by the NRC which can result in project interruptions and increase project schedule and costs.

6. The power plant construction industry should establish an experimental program which will provide a basis for updating codes and standard practices and for identifying effective constructibility methods.

---

*Based upon the Panel of Speakers/Audience discussion period at the end of the session.

NUCLEAR PLANT CIVIL-STRUCTURAL
DESIGN STANDARDS
by
John D. Stevenson[1]

ABSTRACT

The nuclear power industry is perhaps the most regulated industry in the world today. The manpower statistics of individuals and organizations active in nuclear standards development and the number of approved standards and those under development are summarized in Tables 1 and 2.[1,2] This paper is essentially an update summary of a survey paper given at the Civil Engineering and Nuclear Power Mini-Conference held in Boston, April 2-3, 1979.[3]

INTRODUCTION

Any discussion of standards should include a definition of standards as they are applied in the nuclear industry. In Figure 1 is presented a suggested hierarchy of reference material used in the design and construction of engineering work which distinguish standards from other reference materials.

Before proceeding to the discussion of particular standards it is well to describe the types of standards used in the nuclear industry. These "standards" generally fall into one of three categories.

The first category is that of a Code. A Code is a document prepared to regulate the methods, procedures, assumptions, loads and permissible behavior used to design and construct engineering works. It is prepared assuming it will be adopted by a legally constituted regulatory body and as such will have the backing of the force of law. All provisions are usually mandatory and contain the "Shall" as opposed to "Should" or "Recommended" terminology. Explanation of requirements given are normally contained in a Commentary which is legally not part of the Code. Provision is usually made by the organization originating the Code to periodically update the Code and to formally answer inquiries.

The second category is that of a Specification. A Specification is a document prepared to define the methods, procedures, assumptions, loads and permissible behavior used to design and construct engineering works. It is prepared assuming it will form part of a legally binding contract between two civil parties and as such will have the backing of the law governing civil contracts. However, commercial terms and conditions or contract responsibilities as to liabilities or warranties or lack thereof should not be part of such a technical document. All provisions are usually mandatory and contain the "Shall" terminology. Commentaries to explain requirements may or may not be provided and provision for scheduled periodic update and formal interpretation are usually not provided.

---

(1)Vice President and General Manager, Structural Mechanics Associates, Cleveland, Ohio.

The third category is that of Guidelines, Criteria or Recommended Practice. Such documents are prepared to give guidance as to generally acceptable methods, procedures, assumptions, loads and permissible behavior used to design and construct engineering works. It is prepared assuming it will be used as guidance in the design and construction of engineering works and may contain permissive as well as manatory requirements. Terminology such as "May", "Should", "Recommended" as well as, "Shall" are typically used. The document is not prepared assuming it will be adopted in total as a "Code" or a "Specification" but instead should be used as a reference document. In general, there is no separate Commentary since Commentary type material is included in the body of the Guideline, etc. Provision for scheduled periodic update or interpretation are not normally made.

The dividing line between "standards" and other documents used in design and construction is the establishment of a required, recommended or preferred position by the issuing organization. Such preference usually implies some liability or responsibility by such an organization.

Nuclear standards are being developed by both governmental agencies and by voluntary industry groups sponsored by technical societies or by trade associations. In general, voluntary industrial standards are used supplemented by specific government agency requirements.

INTERNATIONAL NUCLEAR STANDARDS

The governmental group currently developing international standards is the International Atomic Energy Agency, IAEA, which is headquartered in Vienna, Austria. It is an organ of the United Nations and has the responsibility for the development of nuclear safety standards which are applicable to developing or third world countries. The nuclear power plant standards under preparation by the IAEA are grouped in five categories; regulation or governmental licensing organization, siting, design, operation and quality assurance. The primary document in these five areas are the Codes of Practice which establish the objectives and minimum requirements which should be fulfilled to provide adequate safety for the Nuclear Power Plant. Safety Guides are also prepared which serve to supplement the Codes of Practice in that they recommend procedures that might be followed to implement the relevant Codes of Practice. In Table 3 can be found a summary of the standards currently under development by the IAEA in the siting and design categories of particular interest to structural engineers.

Actual preparation of the IAEA Standards is done in three phases. First, the text of the draft standard is prepared by a group of technical experts. It is then passed on to a Technical Review Committee, TRC, for review, comment and approval. Finally, it is sent to a Senior Advisory Group, SAG, representing member nations of the IAEA where it is reviewed by the member nations and finally approved. The procedure of review, comment and approval may undergo several interactions involving the technical experts committee, TRC and SAG.

NUCLEAR PLANT DESIGN

The international industrial counterpart of the IAEA is the International Organization for Standards, ISO. Nuclear standards development by this group of particular interest to the structural engineer is under Technical Committee 85, Nuclear Energy, Sub-Committee 3, Power Reactor Technology. This effort is performed by individual volunteers which are named to the various committees by their national standards groups. In the U.S., this is the American National Standards Institute, ANSI. In Table 4 can be found a listing of international standards of particular interest to civil engineers which currently are under development.

Standards developed by ISO tend to be a compromise between the various national industrial standards developed in the leading industrial nations. As such they tend to be used in smaller countries which do not have a well developed standards activity of their own and are particularly useful in providing a common bid basis for suppliers who would otherwise use their own national standards as a bid basis.

Prior to 1980 most of the detailed nuclear standards in use worldwide were developed in the U.S. In the past two years national governments and industrial standards organizations outside the U.S. have made important strides in developing their own detailed body of national standards. In many cases these new standards were based on an original U.S. standard but have been modified and contain significantly more detail than is contained in the original U.S. Standard. An example of such additional detail is shown in Table 5 extracted from the German GRS Safety Codes and Guides.[4]

NATIONAL NUCLEAR STANDARDS

GOVERNMENT STANDARDS

In the U.S., the prime source of nuclear standards from government is the U.S. Nuclear Regulatory Commission, NRC. This independent federal commission was created in 1975 by separating the regulatory function of the old Atomic Energy Commission from the promotional function of the application of atomic energy.

The Commission as a whole formulates and has published Federal Regulations which regulate the design and construction of nuclear power plant facilities. There are three regulations which are of particular interest to the structural engineer, 10CFR50 Appendix A, 10CFR50 Appendix B and 10CFR100 Appendix A. In addition there is a Federal Regulation 10CFR21 Reporting of Defects and Noncompliance which should be known to all engineers involved in nuclear power.

Federal Regulations

10CFR50 Appendix A

In this document are set forth the General Design Criteria to be used in the design of safety class nuclear plant facilities. It contains the following specific criteria which are of particular interest to the civil-structural engineer.

POWER GENERATION FACILITIES

Criterion 1 - Quality Standards and Records

Structures, systems and components important to safety shall be designed, fabricated, erected, and tested to quality standards commensurate with the importance of the safety functions to be performed.

Where generally recognized codes and standards are used, they shall be identified and evaluated to determine their applicability, adequacy, and sufficiency and shall be supplemented or modified as necessary to assure a quality product in keeping with the required safety function. A quality assurance program shall be established and implemented in order to provide adequate assurance that these structures, systems, and components will satisfactorily perform their safety functions. Appropriate records of the design, fabrication, erection, and testing of structures, systems, and components important to safety shall be maintained by or under the control of the nuclear power unit licensee throughout the life of the unit.

Criterion 2 - Design Bases for Protection Against Natural Phenomena

Structures, systems, and components important to safety shall be designed to withstand the effects of natural phenomena such as earthquakes, tornadoes, hurricanes, floods, tsunami, and seiches without loss of capability to perform their safety functions. The design bases for these structures, systems, and components shall reflect: 1) Appropriate consideration of the most severe of the natural phenomena that have been historically reported for the site and surrounding area, with sufficient margin for the limited accuracy, quantity, and period of time in which the historical data have been accumulated 2) appropriate combinations of the effects of normal and accident conditions with the effects of the natural phenomena, and 3) the importance of the safety functions to be performed.

Criterion 3 - Fire Protection

Structures, systems, and components important to safety shall be designed and located to minimize, consistent with other safety requirements, the probability and effect of fires and explosions. Non-combustible and heat resistant materials shall be used wherever practical throughout the unit, particularly in locations such as the containment and control room. Fire detection and fighting systems of appropriate capacity and capability shall be provided and designed to minimize the adverse effects of fires and structures, systems, and components important to safety. Firefighting systems shall be designed to assure that their rupture or inadvertent operation does not significantly impair the safety capability of these structures, systems, and components.

Criterion 4 - Environmental and Missile Design Bases

Structures, systems and components important to safety shall be designed to accomodate the effects of and to be compatible with the environmental conditions associated with normal operation, maintenance, testing and postulated accidents, including loss of coolant accidents. These structures, systems, and components shall be appropriately protected against dynamic effects, including the effects of missiles, pipe whipping, and discharging fluids, that may result from equipment failures and from events and conditions outside the nuclear power unit.

Criterion 16 - Containment Design

Reactor containment and associated systems shall be provided to establish an essentially leaktight barrier against the uncontrolled release of radioactivity to the environment and to assure that the containment design conditions important to safety are not exceeded for as long as postulated accident conditions require.

Criterion 50 - Containment Design Basis

The reactor containment structure, including access openings, penetrations and the containment heat removal system shall be designed so that the containment structure and its internal compartments can accommodate, without exceeding the design leakage rate and, with sufficient margin, the calculated pressure and temperature conditions resulting from any loss of coolant accident. This margin shall reflect consideration of 1) the effects of potential energy sources which have not been included in the determination of the peak conditions, such as energy in steam generators and energy from metal-water and other chemical reactions that may result from degraded emergency core cooling functioning, 2) the limited experience and experimental data available for defining accident phenomena and containment responses, and 3) the conservatism of the calculation model and input parameters.

Criterion 51 - Fracture Prevention of Containment Pressure Boundary

The reactor containment boundary shall be designed with sufficient margin to assure that under operating, maintenance, testing, and postulated accident conditions 1) its ferritic materials behave in a nonbrittle manner, and 2) the probability of rapidly propagating fracture is minimized. The design shall reflect consideration of service temperatures and other conditions of the containment boundary material during operation, maintenance, testing and postulated accident conditions, and the uncertainties in determining 1) materials properties, 2) residual, steady-state, and transient stresses, and 3) size of flaws.

Criterion 52 - Capability for Containment Leakage Rate Testing

The reactor containment and other equipment which may be subjected to containment test conditions shall be designed so that periodic integrated leakage rate testing can be conducted at containment design pressure.

Criterion 53 - Provisions for Containment Testing and Inspection

The reactor containment shall be designed to permit 1) appropriate periodic inspection of all important areas, such as penetrations, 2) an appropriate surveillance program, and 3) periodic testing at containment design pressure of the leaktightness of penetrations which have resilient seals and expansion bellows.
10CFR50 Appendix B

This Federal Regulation presents the quality assurance criteria for nuclear power plants and fuel reprocessing plants. It is divided into 18 sections which cover the following subject areas:

1. Organization
2. Quality Assurance Program
3. Design Control
4. Procurement Document Control
5. Instructions, Procedures and Drawings
6. Document Control
7. Control of Purchased Material, Equipment and Services
8. Identification and Control of Materials Parts and Components
9. Control of Special Processes
10. Inspection
11. Test Control
12. Control of Measuring and Test Equipment
13. Handling, Storage and Shipping
14. Inspection, Test and Operating Status
15. Nonconforming Materials, Parts or Components
16. Corrective Action
17. Quality Assurance Records
18. Audits

10CFR100 Appendix A

    This Federal Regulation defines the Operational Basis and Safe Shutdown Earthquakes and the manner in which their intensities are determined. It also sets the locations where the earthquake motion is defined and sets with width of fault zones as a function of fault length and defines capable and non-capable faults.

10CFR21 Reporting of Defects and Noncompliance

    All individuals having any responsibility regarding design, construction or operation of nuclear power plants should be familiar with contents of this regulation. In summary, this regulation establishes procedures and requires any defect or noncompliance in a basic component which could create a substantial safety hazard in a nuclear power station be reported to the NRC within 2 days by any individual director or responsible officer of a firm constructing, owning, operating or supplying components to a nuclear power station. Any director or responsible officer who knowingly and consciously fails to notify the NRC as required shall be subject to a civil penalty not to exceed $5000.00 for each failure to notify.

NRC Regulatory Guides

Regulatory Guides are prepared by the Office of Standards Development and are issued to describe and make available to the public, methods acceptable to the NRC Regulatory staff of implementing specific parts of the Commission's regulations, to delineate techniques used by the staff in evaluating specific problems or postulated accidents or to provide guidance to applicants. Regulatory Guides are not substitutes for regulations and compliance with them is not required. Methods and solutions different from those set out in the guides will be acceptable if they provide a basis for the findings requisite to the issuance or continuance of a permit or license by the Commission.

The guides are issued in the following ten broad divisions:

1. Power Reactors
2. Research and Test Reactors
3. Fuels and Materials Facilities
4. Environmental and Siting
5. Materials and Plant Protection
6. Products
7. Transportation
8. Occupational Health
9. Antitrust Review
10. General

In Table 6 can be found a listing of all existing and contemplated Regulatory Guides for Division 1. The following is a brief description of those Regulatory Guides of particular interest to the Civil-Structural Engineer.

R.G. 1.12 - Instrumentation for Earthquake

This guide provides for a least two strong motion triaxial accelerographs be installed on the containment base mat and shell above the mat. Other instrumentation including that recording free field motion may also be required.

R.G. 1.26 - Quality Group Classifications and Standards

This guide provides for the establishment of four groups of quality standards, A through D, and describes a method for determining acceptable quality standards for safety class components.

R.G. 1.28 - Quality Assurance Program Requirements (Design and Construction)

This standard references the N-45.2 (now NQA-1) Quality Assurance Program Requirements for Nuclear Power Plants, ANSI standard as being generally acceptable and provides an adequate basis for complying with the program requirements of Federal Regulation 10CFR Part 50 Appendix B.

R.G. 1.29 - Seismic Design Classification

This guide defines those systems and components which require verification of seismic design adequacy. Such seismic design adequacy is required to assure (1) the integrity of the reactor coolant pressure boundary, (2) the capacity to shutdown the reactor and maintain it in a safe shutdown condition, (3) prevent or mitigate consequencies of accidents which could result in radiation release in excess of 10CRF Part 100 guideline exposures.

R.G. 1.136 - Materials, Construction and Testing of Concrete Containments

This guide makes reference to the ACI-ASME Code for Concrete Reactor Vessel and Containments. It lists specific exceptions and augmentations of the Code by the NRC.

R.G. 1.46 - Protection Against Pipe Whip Inside Containment

The guide specifies what, where and how high energy piping systems break inside containment.

R.G. 1.48 - Design Limits and Loading Combinations for Seismic Category I
Fluid System Components

This guide defines some specific load combinations including Safe Shutdown Earthquake, the Operational Basis Earthquake and faulted plant conditions and the ASME Operating Conditions to which the load combination apply. It also introduces the concept of active as well as passive components.

R.G. 1.57 - Design Limits and Loading Combinations for Metal Containment
Systems

This guide defines specific load combinations to be used in metal containment design. It also includes specific limitations to be used with rigorous shell instability analysis.

R.G. 1.59 - Design Basis Flood for Nuclear Power Plants

This guide describes an acceptable method of determining for sites along streams or rivers the design basis floods that nuclear power plants must be designed to withstand without loss of safety related functions. It also discusses the phenomena producing comparable design basis floods for coastal estuary, and Great Lake Sites.

R.G. 1.60 - Design Response Spectra for Seismic Design of Nuclear Power Plants

This guide establishes the ground response spectra for various levels of assumed damping which can be used to seismically qualify structures located at ground and to qualify time history input motion which are used to develop amplified spectra for use in design of components located within building structure. Horizontal as well as vertical spectra are specified.

R.G. 1.61 - Damping Values for Seismic Design of Nuclear Power Plants

   This guide presents interim modal damping values which are currently acceptable to the NRC Regulatory staff. Values are tabulated for both the OBE and SSE seismic events.

R.G. 1.70 - Standard Format and Content of Safety Analysis Reports for Nuclear
            Power Plants - LWR Edition

   This document identifies the principal information that should be provided by applicants in Safety Analysis Reports for light-water cooled reactors and provides a uniform format for presenting the information. The Standard Format was developed in parallel with and is keyed to the Standard Review Plans prepared by the NRC's Office of Nuclear Reactor Regulation.

R.G. 1.76 - Design Basis Tornado for Nuclear Power Plants

   This guide describes a design basis tornado acceptable to the NRC Staff that a nuclear plant should be designed to accomodate for each of three geographical regions in the U.S.

R.G. 1.91 - Evaluation of Explosions Postulated to Occur on Transportation
            Routes Near Nuclear Power Plant Sites

   This guide describes a method acceptable to the NRC Regulatory staff for determining safe distances from a nuclear power plant to a transportation route over which explosive material may be carried. It also provides data for the development of peak positive over pressure associated with accidental explosions as a function of distance and equivalent TNT charge.

R.G. 1.92 - Combination of Modes and Spatial Components in Seismic
            Response Analysis

   This guide provides a suggested method for combining seismic response effects between modes including closely spaced modes as well as spatial combinations considering independence of directional components of the earthquake.

R.G. 1.100 - Seismic Qualification of Electrical Equipment for Nuclear
             Power Plants

   This guide in general endorses the industry test standard IEEE 344-1975 developed to seismically qualify electrical components and has been widely used to qualify mechanical components by test as well. It also states where it takes exception to provisions of the IEEE-344-NS Standard.

R.G. 1.103 - Post-Tension Prestressing Systems for Concrete Reactor Vessels
             and Containments

   This guide covers the generic qualifications of post-tensioned prestressing systems used in concrete reactors vessels and containments. It also gives the status (1976) of the types of tendon systems which have been submitted to the NRC for use in containments and reactor vessels.

R.G. 1.115 - Protection Against Low-Trajectory Turbine Missiles

This guide gives guidance as the types of turbine missiles which should be considered in design, exclusion criteria and the design procedures acceptable for design of concrete and steel missile barriers.

R.G. 1.117 - Tornado Design Classification

This guide describes a method acceptable to the NRC Staff for identifying those structures, systems and components of light water reactors that should be designed to withstand the effects of the Design Basis Tornado (R.G. 1.76), and remain functional.

R.G. 1.122 - Development of Floor Design Response Spectra for Seismic Design of Floor-Supported Equipment of Components

This guide describes methods acceptable to the NRC staff for developing two horizontal and one vertical floor design response spectra at various elevations in building structures from the time-history motions resulting from the dynamic analysis of the supporting structure.

R.G. 1.130 - Design Limits and Loading Combinations for Class 1 Plate-and-Shell-Type Component Supports

This guide delineates acceptable design limits and appropriate combinations of loadings associated with normal operation, postulated accidents, and specified seismic events for the design of Class 1 plate-and-shell-type component supports as defined in Subsection NF of Section III of the American Society of Mechanical Engineers (ASME) Boiler and Pressure Vessel Code. This guide applies to light-water-cooled reactors.

R.G. 1.132 - Site Investigations for Foundations of Nuclear Power Plants

This guide describes programs of site investigations that would normally meet the needs for evaluating the safety of the site from the standpoint of the performance of foundations and earth works under most anticipated loading conditions, including earthquakes. It also describes site investigations required to evaluate geotechnical parameters needed for engineering analysis and design.

R.G. 1.138 - Laboratory Investigations of Soils for Engineering Analysis and Design of Nuclear Power Plants

This guide gives acceptable laboratory test methods to determine soil or site foundation properties.

R.G. 1.143 - Safety-Related Concrete Structures for Nuclear Power Plants (Other Than Reactor Vessels and Containments)

This guide in general endorses industry standard ACI-349 used in the design of safety class concrete structures with exceptions as listed in the guide.

NUCLEAR PLANT DESIGN

NRC Standard Review Plants, SRP

Regulatory Standard Review Plans are prepared for the guidance of the NRC Office of Nuclear Reactor Regulation staff responsible for the review of applications to construct and operate nuclear power plants. These documents are made available to the public as part of the Commission's policy to inform the nuclear industry and the general public of Regulatory staff scope and policies relative to the safety analysis reports. Standard Review Plans are not substitutes for Regulatory Guides or Federal Regulations and compliance with them is not required. The Standard Review Plan sections are keyed to the Standard Format and Content of Safety Analysis Reports for Nuclear Power Plants, R.G. 1.70. Not all sections of the Standard Format have a corresponding review plan.

The Standard Review Plans in general incorporate requirements of "Branch Position Papers" which in the past were informally issued for the guidance of the applicant to assist preparation of Safety Analysis Reports. Those sections of the Standard Review Plans of particular interest to Civil-Structural Engineers are Section 2.5 covering Seismically and Geology and all of Chapter 3.0 as summarized in Table 7.

INDUSTRY STANDARDS

The principal coordinating organization for industry development of nuclear standards is the American National Standards Institute. ANSI, Nuclear Standards Management Board, NSMB. This organization develops priorities, coordinates and enlists the aid of technical societies and other interest organizations in the development of consensus, voluntary industry standards. The principal organizations actively preparing nuclear standards of particular interest to the civil-structural engineer are as follows:

1. American Nuclear Society, ANS
2. American Society of Mechanical Engineers, ASME
3. American Society of Civil Engineers, ASCE
4. American Concrete Institute, ACI
5. American Institute of Steel Constructing, AISC
6. Institute of Electrical and Electronic Engineers, IEEE
7. American Society of Testing and Materials, ASTM

American Nuclear Society, ANS(5)

The American Nuclear Society (ANS) typically prepares systems or functionality standards related to nuclear plant design while ASME, ACI, ASCE etc., typically prepare the hardware standards which relate to the detailed design and stress or behavior analysis of specific components. In Table 8 can be found a listing of ANS Standards of particular interest to civil-structural engineers. The three major ANS committees of interest are ANS-2, ANS-4 and the ANS-51-54 series.

ANS-2 - Site Criteria

The ANS-2 management committee covers those criteria which affect the plant which have their basis in natural phenomena and relate to siting of the nuclear plant facility.

ANS-4 - General Criteria

This ANS management committee has responsibility for development of the generic standards on accident missiles and pipe rupture and probabilistic risk assessment.

ANS-51-54 series - Reactor System Safety Criteria

The third major ANS Committee activity is that which develops the safety criteria for the design of the four major nuclear reactor types used in the U.S., i.e. Pressurized Water Reactors, Boiling Water Reactors, Gas Cooled Reactors and Sodium Cooled Fast Breeder Reactors. In addition to the four major ANS-50 subgroups dealing with each reactor type this committee is concerned with the plant process conditions of design and their load combinations including interaction with natural phenomena and the associated operating or design condition criteria and the safety classification of components.

American Society of Mechanical Engineers, ASME

In Table 9 can be found a listing of those standards being prepared by ASME of interest to civil-structural engineers.

Boiler Pressure Vessel Nuclear Components Code, Section III, Division 1

This code covers the material, design, fabrication and inspection of metal pressure retaining components used in nuclear service including their metal supports. Three levels of safety classes are generally recognized and are designated Classes 1, 2 and 3. The actual classification of components is the responsibility of the owner of the component using the guidelines as established by ANS or NRC criteria.

Boiler and Pressure Vessel Code Section III Concrete Components in Nuclear Service Division 2

This code covers concrete pressure retaining components in nuclear service. Specifically, the prestressed concrete HTGR reactor vessel and the reinforced concrete containment both deformed bar and prestressed types are covered.

Committee on Nuclear Quality Assurance

This committee has prepared a large number of specific standards related to Quality Assurance requirements for mechanical components and structures.

American Concrete Institute, ACI

This organization jointly with ASME shares the responsibility of developing the concrete pressure retaining components in nulcear service code through its committee ACI-359. In addition to ACI-359, Committee ACI-349 has developed a standard for design of safety (seismic) Category I concrete building structures other than those which are pressure restrained.

American Institute of Steel Construction, AISC

This organization is developing a standard for design of safety (seismic) Category I structural steel building structures (ANSI N690). As such it will parallel in structural steel the efforts of ACI-349 in concrete.

Institute of Electrical and Electronic Engineers, IEEE

IEEE-308 - Criteria for Class IE Electrical Systems for Nuclear Power Stations

This standard develops the rationale for seismic classifications of electrical equipment.

IEEE-322 - General Trial Use Guide for Qualifying Class I Electrical Equipment for Nuclear Power Stations

This standard covers the requirements for environmental testing of electrical components to determine their functionality particularly under extreme environmental conditions. Included are the requirements on aging of equipment to simulate end of life conditions plus the evaluation of statistical variation of test results.

IEEE-344 - Guide for Seismic Qualification of Class I Electrical Equipment for Nuclear Power Generator Stations

This standard covers the procedures for seismic testing or analysis of electrical components to demonstrate the seismic design adequacy of the component. Both fragility testing as well as proof testing are discussed as well as the requirements for multi-direction as well as multi-frequency testing are covered.

American Society of Civil Engineers, ASCE

The ASCE has developed a Manual of Engineering Practice, No. 58, Structural Analysis and Design of Nuclear Plant Facilities, published in August, 1980, which gives practicing structural engineers guidance in nuclear plant structural design and analysis. It also is developing a number of standards which interface with ANS and ASME Standards. The ANS Standards develop the physical phenomena and the ASCE Standards reduces the physical phenomena to loadings and develop applicable stress criteria where such criteria has not already been developed by other societies such as ACI and ASME as defined in Table 10.

American Society of Testing and Materials, ASTM

This society has developed a large number of standards which have been either directly adopted by other standards groups or referenced by them. These standards are designed to control the mechanical, dimensional and chemical properties of structural materials and their fabrication or provide procedures for testing the various properties of structural materials.

SUMMARY AND CONCLUSION

As can be seen bythis paper there now exists a tremendous volume of codes, specifications and guidelines which are continuously being augmented and revised which are intended for use in the regulation and control of nuclear power plant design, construction and operation which effect the civil-structural engineer. This may be one reason why the civil-structural engineering manpower required to design a nuclear power plant has increased from approximately 100,000 manhours in 1965 to approximately 900,000 manhours in 1980. Not withstanding the relative merit or lack of merit for such an increase in engineering effort, this paper is an attempt to summarize and identify those standards which have a particular impact on the civil-structural engineer.

REFERENCES

(1) ANSI-NTAB Status Report Section 1, "Nuclear Projects Status Report," NTAB-SB-6 American National Standards Institute, Inc., February, 1975.

(2) Personnel Involved in the Development of Nuclear Standards in the United States 1976, ORNLYENG-4 Special, RDT STANDARDS office, Oak Ridge National Laboratory.

(3) Stevenson, J.D., "Current Development of Nuclear Standards," Volume 1, Civil Engineering and Nuclear Power, Preprint 3594 ASCE National Convention, Boston, Mass., April, 1979.

(4) Gesellschaft fur Reaktorsicherheit "RSK Guidelines for Pressurized Water Reactors," 2nd Edition, January, 1979.

(5) Status Report of Projects Under the Nuclear Standards Management Board NSMB SR-24 American National Standards Institute, November, 1980.

TABLE 1 - Numbers of Individuals Involved in Nuclear Standards Development in the U.S. - 1970 - 1976

	1970	1976
Number of Personnel[1]	1465	7685
Number of Program Members[1]	1950	13141
Number of Organizations (Societies)	-	49
Number of Committees	515	906
Number of Employers	1015	3225
Number of Standards	500	700

[1]Member is one person per position; that is a person with four committee memberships is counted as four members.

TABLE 2 - Nuclear Standards Status - 9/78

Standards sponsor	Stds. appr.	Stds. in prep.	Total projects
ACI	2	0	2
AIA	11	14	25
AIChE	14	18	32
AISC	-	1	1
ANS	55	130	185
ASCE	0	3	3
ASME	19	17	36
ASTM	151	79	230
HPS	16	26	42
IEEE	42	60	102
INMM	17	16	33
NBS	7	6	13
NFPA	1	1	3
Totals	335	371	706
% approved	49.9		

TABLE 3 - IAEA Siting and Design Standards of
Particular Interest to Structural Engineers

SAFETY CODE OF PRACTICE ON NUCLEAR POWER PLANT SITING

SG-S1	Earthquakes and Associated Topics for Nuclear Power Plant Siting
SG-S2	Aseismic Analysis and Testing of Nuclear Power Plants
SG-S3B	Meteorology - Extreme Meteorological Conditions in Nuclear Power Plant Siting
SG-S5	Man-Induced Events Related to Nuclear Power Plant Siting
SG-S6B	Phenomena due to Extreme Hydrological Conditions (floods, etc.)

CODE OF PRACTICE ON DESIGN FOR SAFETY OF NUCLEAR POWER PLANTS

SG-D1	Safety Guide on Safety Functions and Component Classification for BWR, PWR and PTR
SG-D2	Safety Guide on Fire Protection in Nuclear Power Plants
SG-D4	Safety Guide on Protection Against Internally Generated Missiles and Their Secondary Effects in Nuclear Power Plants
SG-D5	Safety Guide on Man-induced Events
SG-D6	Safety Guide on Ultimate Heat Sink and Directly Associated Heat Transport System(s)
SG-D10	Safety Guide on Fuel Handling and Storage Systems in Nuclear Power Plants
SG-D11	Safety Guide on Containment

TABLE 4 - List of Standards Under Preparation by
the International Standards Organization

ISO Technical Committee 85/Subcommittee 3

1. Working Group 3 - Containment Structures

    a. Steel Containment
    b. Concrete Containment

2. Working Group 6 - Pressure Boundaries of Primary Circuits

3. Working Group 7 - Seismic Considerations

4. Working Group 9 - Reliability Data

5. Working Group 10 - Pressure Vessels

    a. Steel Vessels
    b. Concrete Vessels

TABLE 5 GENERAL SPECIFICATION "BASIC SAFETY"-
Pipe Design Breaks and Loads to be Considered for a PWR

	Systems Important to Safety	Design Breaks				Loads to be Considered	
		Longitudinal Location of Break	Br.Area	Circumferential Location of Break	Br.Area		
Ferritic Systems / Components	Main Steam System	Any	Subcritical Crack 1)	High-Duty Weld	2F-u	Reactivity Behavior	
					2F-b	Buildings	(differential pressure) (jet force)
					2F-b	Pressure-Retaining Components	(reaction force)
				Outside Valve Housing	2F-b	Support Structures	(reaction force)
					2F-u	Internal Loads	(internal pressure surge)
	Feedwater System, Steam Generator Blowdown System *	Any	Subcritical Crack 1)	High-Duty Weld	2F-u	Buildings	(differential pressure) (jet force)
					2F-b	Pressure-Retaining Components	(reaction force)
				Outside Valve Housing	2F-b	Support Structures	(reaction force)
					2F-u	Internal Loads	(internal pressure surge)
Austenitic Systems	Volume Control System* Volume Control Surge System** (Injection leg, extraction leg up to HP reducing station) p > 20 bar	Any	Subcritical Crack 1)	High-Duty Weld	2F-u	Buildings	(differential pressure) (jet force)
					2F-b	Pressure-Retaining Components	(reaction force)
					2F-b	Support Structures	(reaction force)
					2F-u	Internal Loads	(internal pressure surge)
	Nuclear Residual Heat Removal System (Accumulator injection pipe*, core flooding pipe**)	Any	Subcritical Crack 1)	High-Duty Weld	2F-u	Buildings	(jet force)
					2F-b	Pressure-Retaining Components	(reaction force)
					2F-b	Support Structures	(reaction force)
					2F-u	Internal Loads	(internal pressure surge)
Low-Energy Systems Subject to Brief Loads	Emergency Feed System* Emergency Feedwater System** Single Passive Fault	Any	Subcritical Crack 1)	Any	Subcritical Crack	Buildings	(flooding)
	Nuclear Residual Heat Removal System, Residual Heat Removal Case; Single Fault					Buildings System Analysis Emergency Core Colling	(differential pressure) (flooding) (leg failure)
Low-Energy Systems (p≤20 bar; ϑ≤100°C to nominal bore DN≤50mm)	Nuclear Component Cooling System	Any	Subcritical Crack 1)	Any	Subcritical Crack	Buildings	(flooding)
	Secondary Cooling Water System					Buildings	(flooding)
	Makeup Boration System * Safety Boration System **					System Analysis	(leg failure)

Legend: -u=unimpeded; -b=impeded; -on items without stop - 2F-u; -on items with stop - 2F-b; the stationary substitute load is used for subcritical cracks

*   - KWU-specific
** - BBC/BBR-specific
1) - Subcritical cracks are determined on a fracture mechanics basis and limited to a maximum of 0.1 F

TABLE 6

U.S. NUCLEAR REGULATORY COMMISSION REGULATORY GUIDE SERIES

DIVISION 1 - POWER REACTORS

TABLE OF CONTENTS

This table of contents lists every revision of each regulatory guide in Division 1 with the date it was issued. If the latest version of a guide (as of the date of this table of contents) was issued for early comment, this is so noted.

Division 1, one of ten broad divisions in which regulatory guides are issued, contains those guides that were developed for power reactors. There may also be some guides issued in other divisions that would be of interest to those whose primary concern is in the area of power reactors. Accordingly, this issue of the table of contents includes, for the first time, a listing of regulatory guides issued in the other divisions that the NRC staff has identified as possibly of interest to recipients of Division 1 guides. This listing will be updated from time to time, and suggestions for additions to it are encouraged.

Most regulatory guides contain a section headed "Implementation" that is intended to provide information to applicants and licensees regarding the NRC staff's plans for using the guide. If a guide does not contain such a section or if detailed information is needed on the staff's plans for using a regulatory guide with respect to a specific permit or license or application therefor, requests for such information should be addressed to the appropriate licensing project manager in the Office of Nuclear Reactor Regulation or Office of Nuclear Material Safety and Safeguards.

At an appropriate point in the development of a new regulatory guide or a proposed revision to an existing guide, the guide and the associated value/impact statement are issued in draft form to involve the public in the early stages of the development of a regulatory position. These drafts have not received complete staff review and do not represent an official NRC staff position. They are temporarily identified by their task number and issued to the same distribution list that is used for published guides in each division. Draft guides issued in this division are listed in this issue of the table of contents.

All regulatory guides, including draft guides, proposed revisions, and all published revisions, may be examined at the Commission's Public Document Room at 1717 H Street NW., Washington, D.C. Regulatory guides are not copyrighted and Commission approval is not required to reproduce them. Requests for single copies of draft guides and proposed revisions should be made in writing to the U.S. Nuclear Regulatory Commission, Washington, D.C. 20555, Attention: Director, Division of Technical Information and Document Control. Copies of active guides may be purchased at the current Government Printing Office (GPO) price. A subscription service for future guides in specific divisions is available through the Government Printing Office. Information on subscription service and current GPO prices may be obtained by writing to the U.S. Nuclear Regulatory Commission, Washington, D.C. 20555, Attention: Publications Sales Manager.

Regulatory Guides

Number	Title	Rev.	Issued Year/Month
1.1	Net Positive Suction Head for Emergency Core Cooling and Containment Heat Removal System Pumps (Safety Guide 1)	---	70/11
1.2	Thermal Shock to Reactor Pressure Vessels (Safety Guide 2)	---	70/11
1.3	Assumptions Used for Evaluating the Potential Radiological Consequences of a Loss-of-Coolant Accident for Boiling Water Reactors	--- 1 2	70/11 73/06 74/06
1.4	Assumptions Used for Evaluating the Potential Radiological Consequences of a Loss-of-Coolant Accident for Pressurized Water Reactors	--- 1 2	70/11 73/06 74/06

Number	Title	Rev.	Issued Year/Month
1.5	Assumptions Used for Evaluating the Potential Radiological Consequences of a Steam Line Break Accident for Boiling Water Reactors (Safety Guide 5)	---	71/03
1.6	Independence Between Redundant Standby (Onsite) Power Sources and Between Their Distribution Systems (Safety Guide 6)	---	71/03
1.7	Control of Combustible Gas Concentrations in Containment Following a Loss-of-Coolant Accident	--- 1 2	71/03 76/09 78/11
1.8	Personnel Selection and Training	--- 1 1-R	71/03 75/09 77/05
1.9	Selection, Design, and Qualification of Diesel-Generator Units Used as Onsite Electric Power Systems at Nuclear Power Plants (For Comment)	--- 1	71/03 78/11
1.10	Mechanical (Cadweld) Splices in Reinforcing Bars of Category I Concrete Structures	--- 1	71/03 73/01
1.11	Instrument Lines Penetrating Primary Reactor Containment (Safety Guide 11) Supplement to Safety Guide 11, Backfitting Considerations	---	71/03 72/02
1.12	Instrumentation for Earthquakes	--- 1	71/03 74/04
1.13	Spent Fuel Storage Facility Design Basis (For Comment)	--- 1	71/03 75/12
1.14	Reactor Coolant Pump Flywheel Integrity (For Comment)	--- 1	71/10 75/08
1.15	Testing of Reinforcing Bars for Category I Concrete Structures	--- 1	71/10 72/12
1.16	Reporting of Operating Information--Appendix A Technical Specifications (For Comment)	--- 1 2 3 4	71/10 73/10 74/09 75/01 75/08
1.17	Protection of Nuclear Power Plants Against Industrial Sabotage	--- 1	71/10 73/06
1.18	Structural Acceptance Test for Concrete Primary Reactor Containments	--- 1	71/10 72/12
1.19	Nondestructive Examination of Primary Containment Liner Welds (Safety Guide 19)	--- 1	71/12 72/08
1.20	Comprehensive Vibration Assessment Program for Reactor Internals During Preoperational and Initial Startup Testing	--- 1 2	71/12 75/06 76/05
1.21	Measuring, Evaluating, and Reporting Radioactivity in Solid Wastes and Releases of Radioactive Materials in Liquid and Gaseous Effluents from Light-Water-Cooled Nuclear Power Plants	--- 1	71/12 74/06
1.22	Periodic Testing of Protection System Actuation Functions (Safety Guide 22)	---	72/02
1.23	Onsite Meteorological Programs (Safety Guide 23)	---	72/02
1.24	Assumptions Used for Evaluating the Potential Radiological Consequences of a Pressurized Water Reactor Radioactive Gas Storage Tank Failure (Safety Guide 24)	---	72/03

POWER GENERATION FACILITIES

Number	Title	Rev.	Issued Year/Month
1.25	Assumptions Used for Evaluating the Potential Radiological Consequences of a Fuel Handling Accident in the Fuel Handling and Storage Facility for Boiling and Pressurized Water Reactors (Safety Guide 25)	---	72/03
1.26	Quality Group Classifications and Standards for Water-, Steam-, and Radioactive-Waste-Containing Components of Nuclear Power Plants (For Comment)	--- 1 2 3	72/03 74/09 75/06 76/02
1.27	Ultimate Heat Sink for Nuclear Power Plants (For Comment)	--- 1 2	72/03 74/03 76/01
1.28	Quality Assurance Program Requirements (Design and Construction)	--- 1 2	72/06 78/03 79/02
1.29	Seismic Design Classification	--- 1 2 3	72/06 73/08 76/02 78/09
1.30	Quality Assurance Requirements for the Installation, Inspection, and Testing of Instrumentation and Electric Equipment (Safety Guide 30)	---	72/08
1.31	Control of Ferrite Content in Stainless Steel Weld Metal	--- 1 2 3	72/08 73/06 77/05 78/04
1.32	Criteria for Safety-Related Electric Power Systems for Nuclear Power Plants	--- 1 2	72/08 76/03 77/02
1.33	Quality Assurance Program Requirements (Operation)	--- 1 2	72/11 77/02 78/02
1.34	Control of Electroslag Weld Properties	---	72/12
1.35	Inservice Inspection of Ungrouted Tendons in Prestressed Concrete Containment Structures	--- 1 2	73/02 74/06 76/01
1.36	Nonmetallic Thermal Insulation for Austenitic Stainless Steel	---	73/02
1.37	Quality Assurance Requirements for Cleaning of Fluid Systems and Associated Components of Water-Cooled Nuclear Power Plants	---	73/03
1.38	Quality Assurance Requirements for Packaging, Shipping, Receiving, Storage, and Handling of Items for Water-Cooled Nuclear Power Plants	--- 1 2	73/03 76/10 77/05
1.39	Housekeeping Requirements for Water-Cooled Nuclear Power Plants	--- 1 2	73/03 76/10 77/09
1.40	Qualification Tests of Continuous-Duty Motors Installed Inside the Containment of Water-Cooled Nuclear Power Plants	---	73/03
1.41	Preoperational Testing of Redundant On-Site Electric Power Systems to Verify Proper Load Group Assignments	---	73/03
1.42	(Withdrawn--See 41 FR 11891, 3/22/76)	---	---
1.43	Control of Stainless Steel Weld Cladding of Low-Alloy Steel Components	---	73/05
1.44	Control of the Use of Sensitized Stainless Steel	---	73/05
1.45	Reactor Coolant Pressure Boundary Leakage Detection Systems	---	73/05

NUCLEAR PLANT DESIGN

Number	Title	Rev.	Issued Year/Month
1.46	Protection Against Pipe Whip Inside Containment	---	73/05
1.47	Bypassed and Inoperable Status Indication for Nuclear Power Plant Safety Systems	---	73/05
1.48	Design Limits and Loading Combinations for Seismic Category I Fluid System Components	---	73/05
1.49	Power Levels of Nuclear Power Plants	--- 1	73/05 73/12
1.50	Control of Preheat Temperature for Welding of Low-Alloy Steel	---	73/05
1.51	(Withdrawn--See 40 FR 30510, 7/21/75)	---	---
1.52	Design, Testing, and Maintenance Criteria for Post Accident Engineered-Safety-Feature Atmosphere Cleanup System Air Filtration and Adsorption Units of Light-Water-Cooled Nulear Power Plants	--- 1 2	73/06 76/07 78/03
1.53	Application of the Single-Failure Criterion to Nuclear Power Plant Protection Systems	---	73/06
1.54	Quality Assurance Requirements for Protective Coatings Applied to Water-Cooled Nuclear Power Plants	---	73/06
1.55	Concrete Placement in Category I Structures	---	73/06
1.56	Maintenance of Water Purity in Boiling Water Reactors (For Comment)	--- 1	73/06 78/07
1.57	Design Limits and Loading Combinations for Metal Primary Reactor Containment System Components	---	73/06
1.58	Qualification of Nuclear Power Plant Inspection, Examination, and Testing Personnel	---	73/08
1.59	Design Basis Floods for Nuclear Power Plants	--- 1 2	73/08 76/04 77/08
1.60	Design Response Spectra for Seismic Design of Nuclear Power Plants	--- 1	73/10 73/12
1.61	Damping Values for Seismic Design of Nuclear Power Plants	---	73/10
1.62	Manual Initiation of Protective Actions	---	73/10
1.63	Electric Penetration Assemblies in Containment Structures for Light-Water-Cooled Nuclear Power Plants	--- 1 2	73/10 77/05 78/07
1.64	Quality Assurance Requirements for the Design of Nuclear Power Plants	--- 1 2	73/10 75/02 76/06
1.65	Materials and Inspections for Reactor Vessel Closure Studs	---	73/10
1.66	(Withdrawn--See 42 FR 54478, 10/06/77)	---	---
1.67	Installation of Overpressure Protection Devices	---	73/10
1.68	Initial Test Programs for Water-Cooled Reactor Power Plants	--- 1 2	73/11 77/01 78/08
1.68.1	Preoperational and Initial Startup Testing of Feedwater and Condensate Systems for Boiling Water Reactor Power Plants	--- 1	75/12 77/01

Number	Title	Rev.	Issued Year/Month
1.68.2	Initial Startup Test Program to Demonstrate Remote Shutdown Capability for Water-Cooled Nuclear Power Plants	--- 1	77/01 78/07
1.69	Concrete Radiation Shields for Nuclear Power Plants	---	73/12
1.70	Standard Format and Content of Safety Analysis Reports for Nuclear Power Plants	--- 1 2 3	72/02 72/10 75/09 78/11
1.71	Welder Qualification for Areas of Limited Accessibility	---	73/12
1.72	Spray Pond Piping Made from Fiberglass-Reinforced Thermosetting Resin	--- 1 2	73/12 78/01 78/11
1.73	Qualification Tests of Electric Valve Operators Installed Inside the Containment of Nuclear Power Plants	---	74/01
1.74	Quality Assurance Terms and Definitions	---	74/02
1.75	Physical Independence of Electric Systems	--- 1 2	74/02 75/01 78/09
1.76	Design Basis Tornado for Nuclear Power Plants	---	74/04
1.77	Assumptions Used for Evaluating a Control Rod Ejection Accident for Pressurized Water Reactors	---	74/05
1.78	Assumptions for Evaluating the Habitability of a Nuclear Power Plant Control Room During a Postulated Hazardous Chemical Release	---	74/06
1.79	Preoperational Testing of Emergency Core Cooling Systems for Pressurized Water Reactors	--- 1	74/06 75/09
1.80	Preoperational Testing of Instrument Air Systems	---	74/06
1.81	Shared Emergency and Shutdown Electric Systems for Multi-Unit Nuclear Power Plants	--- 1	74/06 75/01
1.82	Sumps for Emergency Core Cooling and Containment Spray Systems	---	74/06
1.83	Inservice Inspection of Pressurized Water Reactor Steam Generator Tubes	--- 1	74/06 75/07
1.84	Design and Fabrication Code Case Acceptability--ASME Section III, Division 1	--- 1 2 3 4 5 6 7 8 9 10 11 12 13 14 15	74/06 75/04 75/06 75/09 75/11 76/02 76/05 76/08 76/11 77/03 77/08 77/11 78/03 78/07 78/11 79/05
1.85	Materials Code Case Acceptability--ASME Section III, Division 1	--- 1 2 3	74/06 75/04 75/06 75/09

NUCLEAR PLANT DESIGN

Number	Title	Rev.	Issued Year/Month
		4	75/11
		5	76/02
		6	76/05
		7	76/08
		8	76/11
		9	77/03
		10	77/08
		11	77/11
		12	78/03
		13	78/07
		14	78/11
		15	79/05
1.86	Termination of Operating Licenses for Nuclear Reactors	---	74/06
1.87	Guidance for Construction of Class 1 Components in Elevated-Temperature Reactors (Supplement to ASME Section III Code Cases 1592, 1593, 1594, 1595, and 1596)	--- 1	74/06 75/06
1.88	Collection, Storage, and Maintenance of Nuclear Power Plant Quality Assurance Records	--- 1 2	74/08 75/12 76/10
1.89	Qualification of Class 1E Equipment for Nuclear Power Plants	---	74/11
1.90	Inservice Inspection of Prestressed Concrete Containment Structures with Grouted Tendons	--- 1	74/11 77/08
1.91	Evaluations of Explosions Postulated to Occur on Transportation Routes Near Nuclear Power Plants (For Comment)	--- 1	75/01 78/02
1.92	Combining Modal Responses and Spatial Components in Seismic Response Analysis	--- 1	74/12 76/02
1.93	Availability of Electric Power Sources	---	74/12
1.94	Quality Assurance Requirements for Installation, Inspection, and Testing of Structural Concrete and Structural Steel During the Construction Phase of Nuclear Power Plants	--- 1	75/04 76/04
1.95	Protection of Nuclear Power Plant Control Room Operators Against an Accidental Chlorine Release	--- 1	75/02 77/01
1.96	Design of Main Steam Isolation Valve Leakage Control Systems for Boiling Water Reactor Nuclear Power Plants	--- 1	75/05 76/06
1.97	Instrumentation for Light-Water-Cooled Nuclear Power Plants To Assess Plant Conditions During and Following an Accident	--- 1	75/12 77/08
1.98	Assumptions Used for Evaluating the Potential Radiological Consequences of a Radioactive Offgas System Failure in a Boiling Water Reactor (For Comment)	---	76/03
1.99	Effects of Residual Elements on Predicted Radiation Damage to Reactor Vessel Materials	--- 1	75/07 77/04
1.100	Seismic Qualification of Electric Equipment for Nuclear Power Plants	--- 1	76/03 77/08
1.101	Emergency Planning for Nuclear Power Plants	--- 1	75/11 77/03
1.102	Flood Protection for Nuclear Power Plants	--- 1	75/10 76/09
1.103	Post-Tensioned Prestressing Systems for Concrete Reactor Vessels and Containments	--- 1	75/11 76/10

POWER GENERATION FACILITIES

Number	Title	Rev.	Issued Year/Month
1.104	(Withdrawn--See 44 FR 49321, 8/22/79)	---	---
1.105	Instrument Setpoints	--- 1	75/11 76/11
1.106	Thermal Overload Protection for Electric Motors on Motor-Operated Valves	--- 1	75/11 77/03
1.107	Qualifications for Cement Grouting for Prestressing Tendons in Containment Structures	--- 1	75/11 77/02
1.108	Periodic Testing of Diesel Generator Units Used as Onsite Electric Power Systems at Nuclear Power Plants	--- 1	76/08 77/08
1.109	Calculation of Annual Doses to Man from Routine Releases of Reactor Effluents for the Purpose of Evaluating Compliance with 10 CFR Part 50, Appendix I.	--- 1	76/03 77/10
1.110	Cost-Benefit Analysis for Radwaste Systems for Light-Water-Cooled Nuclear Power Reactors (For Comment)	---	76/03
1.111	Methods for Estimating Atmospheric Transport and Dispersion of Gaseous Effluents in Routine Releases from Light-Water-Cooled Reactors	--- 1	76/03 77/07
1.112	Calculation of Releases of Radioactive Materials in Gaseous and Liquid Effluents from Light-Water-Cooled Power Reactors	--- O-R	76/04 77/05
1.113	Estimating Aquatic Dispersion of Effluents from Accidental and Routine Reactor Releases for the Purpose of Implementing Appendix I	--- 1	76/05 77/04
1.114	Guidance on Being Operator at the Controls of a Nuclear Power Plant	--- 1	76/02 76/11
1.115	Protection Against Low-Trajectory Turbine Missiles	--- 1	76/03 77/07
1.116	Quality Assurance Requirements for Installation, Inspection, and Testing of Mechanical Equipment and Systems	--- O-R	76/06 77/05
1.117	Tornado Design Classification	--- 1	76/06 78/04
1.118	Periodic Testing of Electric Power and Protection Systems	--- 1 2	76/06 77/11 78/06
1.119	(Withdrawn--See 42 FR 33387, 6/30/77)	---	---
1.120	Fire Protection Guidelines for Nuclear Power Plants (For Comment)	--- 1	76/06 77/11
1.121	Bases for Plugging Degraded PWR Steam Generator Tubes (For Comment)	---	76/08
1.122	Development of Floor Design Response Spectra for Seismic Design of Floor-Supported Equipment or Components	--- 1	76/09 78/02
1.123	Quality Assurance Requirements for Control of Procurement of Items and Services for Nuclear Power Plants	--- 1	76/10 77/07
1.124	Service Limits and Loading Combinations for Class 1 Linear-Type Component Supports	--- 1	76/11 78/01
1.125	Physical Models for Design and Operation of Hydraulic Structures and Systems for Nuclear Power Plants	--- 1	77/03 78/10
1.126	An Acceptable Model and Related Statistical Methods for the Analysis of Fuel Densification	--- 1	77/03 78/03

NUCLEAR PLANT DESIGN

Number	Title	Rev.	Issued Year/Month
1.127	Inspection of Water-Control Structures Associated with Nuclear Power Plants	--- 1	77/04 78/03
1.128	Installation Design and Installation of Large Lead Storage Batteries for Nuclear Power Plants	--- 1	77/04 78/10
1.129	Maintenance, Testing, and Replacement of Large Lead Storage Batteries for Nuclear Power Plants	--- 1	77/04 78/02
1.130	Service Limits and Loading Combinations for Class 1 Plate-and-Shell-Type Component Supports	--- 1	77/07 78/10
1.131	Qualification Tests of Electric Cables, Field Splices, and Connections for Light-Water-Cooled Nuclear Power Plants (For Comment)	---	77/08
1.132	Site Investigations for Foundations of Nuclear Power Plants	--- 1	77/09 79/03
1.133	Loose-Part Detection Program for the Primary System of Light-Water-Cooled Reactors (For Comment)	---	77/09
1.134	Medical Evaluation of Nuclear Power Plant Personnel Requiring Operator Licenses	--- 1	77/09 79/03
1.135	Normal Water Level and Discharge at Nuclear Power Plants (For Comment)	---	77/09
1.136	Material for Concrete Containments	--- 1	77/11 78/10
1.137	Fuel-Oil Systems for Standby Diesel Generators (For Comment)	---	78/01
1.138	Laboratory Investigations of Soils for Engineering Analysis and Design of Nuclear Power Plants (For Comment)	---	78/04
1.139	Guidance for Residual Heat Removal (For Comment)	---	78/05
1.140	Design, Testing, and Maintenance Criteria for Normal Ventilation Exhaust System Air Filtration and Adsorption Units of Light-Water-Cooled Nuclear Power Plants (For Comment)	---	78/03
1.141	Containment Isolation Provisions for Fluid Systems (For Comment)	---	78/04
1.142	Safety-Related Concrete Structures for Nuclear Power Plants (Other Than Reactor Vessels and Containments) (For Comment)	---	78/04
1.143	Design Guidance for Radioactive Waste Management Systems, Structures, and Components Installed in Light-Water-Cooled Nuclear Power Plants (For Comment)	---	78/07
1.144	Auditing of Quality Assurance Programs for Nuclear Power Plants (For Comment)	---	79/01
1.145	Atmospheric Dispersion Models for Potential Accident Consequence Assessments at Nuclear Power Plants (For Comment)	---	79/08

Draft Regulatory Guides

Task Number	Title	Issued Year/Month
EM 805-5	Nuclear Analysis and Design of Concrete Radiation Shielding for Nuclear Power Plants	79/02
RS 810-5	Qualification of Quality Assurance Program Audit Personnel for Nuclear Power Plants	79/02
SC 704-5	Functional Specification for Safety-Related Valve Assemblies in Nuclear Power Plants	79/02
SC 807-4 (Proposed R.G. 1.35.1)	Determining Prestressing Forces for Inspection of Prestressed Concrete Containments	79/04
SC 705-4	Ultrasonic Testing of Reactor Vessel Welds During Inservice Examination	79/05
RS 809-5	Qualification Test for Cable Penetration Fire Stops for Use in Nuclear Power Plants	79/07
SC 721-4	Inservice Inspection Code Case Acceptability--ASME Section XI, Division 1	79/08
RS 705-4	Lightning Protection for Nuclear Power Plants	79/08
FP 811-4	Safety-Related Permanent Dewatering Systems for Nuclear Power Plants	79/09
SC 521-4	LWR Core Reloads; Guidance on Applications for Amendments to Operating Licenses and on Refueling and Startup Tests	79/09

Proposed Revisions to Regulatory Guides

Task and R.G. Numbers	Title	Proposed Revision	Issued Year/Month
RS 807-5 1.8	Personnel Selection and Training	2	79/02
SC 810-4 1.35	Inservice Inspection of Ungrouted Tendons in Prestressed Concrete Containments	3	79/04
RS 901-5 1.58	Qualification of Nuclear Power Plant Inspection, Examination, and Testing Personnel	1	79/07
RS 050-2 1.131	Qualification Tests of Electric Cables and Field Splices for Light-Water-Cooled Nuclear Power Plants	1	79/08
RS 902-4 1.33	Quality Assurance Program Requirements (Operation)	3	79/08
RS 908-5 1.94	Quality Assurance Requirements for Installation, Inspection, and Testing of Structural Concrete, Structural Steel, Soils, and Foundations During the Construction Phase of Nuclear Power Plants	2	79/09

NUCLEAR PLANT DESIGN

Regulatory Guides Under Development

* Interim Guide on Tornado Missiles
* Single-Failure Criteria for Light-Water Reactor Plants Fluid Systems
* Earthquake Instrumentation Data Handling for Nuclear Power Plants
* Pressurized Water Reactor and Boiling Water Reactor Containment Spray Design Criteria
* Criteria for Electric, Instrumentation, and Control Portions of Safety Systems
* Design and Construction Deficiency Reporting Requirements
* Qualification of Electric Modules for Nuclear Power Plants
* Quality Assurance Requirements for Packaging, Shipping, Receiving, Storage, and Handling of Items for Nuclear Power Plants
* Meteorological Extreme Air Temperatures for Design and Operation of Nuclear Power Plants
* Extreme Windspeeds in Coastal Areas for Design and Operation of Nuclear Power Plants
* Geochronologic Techniques Applied to Nuclear Power Plant Siting
* Procedures and Criteria for Assessing Soil Liquefaction Potential at Nuclear Facility Sites
* Foundation and Earthwork Construction for Nuclear Power Plants
* Snow and Ice Accumulations for the Design and Operation of Nuclear Power Plants
* Geological Mapping of Excavations for Nuclear Power Plants
* Fracture Analysis of Flaws at Structural Discontinuities
* Inservice Monitoring of Core and Core Support Structure Motion Via Neutron-Flux Measurement
* Requirements for Qualification Tests and Production Tests for Piping and Equipment Snubbers
* Ultrasonic Testing (UT) of ASME Code Class 1 and 2 Austenitic Piping Systems
* Methods of Analysis and Design of Reinforced Concrete Containment Structures
* Recommendations for Inservice Testing of Valves Required to Perform a Safety Function in Light-Water Reactors

POWER GENERATION FACILITIES

Regulatory Guides Being Revised

* Revision 1 to Regulatory Guide 1.9, "Selection, Design, and Qualification of Diesel-Generator Units Used as Standby (Onsite) Electric Power Systems at Nuclear Power Plants"

* Revision 2 to Regulatory Guide 1.12, "Instrumentation for Earthquakes"

* Revision 2 to Regulatory Guide 1.14, "Reactor Coolant Pump Flywheel Integrity"

* Revision 5 to Regulatory Guide 1.16, "Reporting of Operation Information--Appendix A Technical Specifications"

* Revision 1 to Regulatory Guide 1.25, "Assumptions Used for Evaluating the Potential Radiological Consequences of a Fuel Handling Accident in the Fuel Handling and Storage Facility for Boiling and Pressurized Water Reactors"

* Revision 1 to Regulatory Guide 1.50, "Control of Preheat Temperature for Welding of Low-Alloy Steel"

* Revision 2 to Regulatory Guide 1.56, "Maintenance of Water Purity in Boiling Water Reactors"

* Update of Revision 3 to Regulatory Guide 1.70, "Standard Format and Content of Safety Analysis Reports for Nuclear Power Plants"

* Revision 1 to Regulatory Guide 1.71, "Welder Qualification for Areas of Limited Accessibility"

* Revision 1 to Regulatory Guide 1.80, "Preoperational Testing of Instrument and Control Air Systems"

* Revision 16 to Regulatory Guide 1.84, "Design and Fabrication Code Case Acceptability--ASME Section III, Division 1"

* Revision 16 to Regulatory Guide 1.85, "Materials Code Case Acceptability--ASME Section III, Division 1"

* Revision 1 to Regulatory Guide 1.89, "Qualification of 1E Equipment for Nuclear Power Plants"

* Revision 2 to Regulatory Guide 1.101, "Emergency Planning for Nuclear Power Plants"

* Revision 1 to Regulatory Guide 1.133, "Loose-Part Detection Program for the Primary System of Light-Water-Cooled Nuclear Power Plants"

* Revision 1 to Regulatory Guide 1.135, "Normal Water Level Discharge at Nuclear Power Plants"

* Revision 2 to Regulatory Guide 1.136, "Material for Concrete Containments (Article CC-2000 of the 'Code for Concrete Reactor Vessels and Containments')"

* Revision 1 to Regulatory Guide 1.137, "Fuel Oil Systems for Standby Diesel Generators"

* Revision 1 to Regulatory Guide 1.139, "Guidance for Residual Heat Removal"

* Revision 1 to Regulatory Guide 1.140, "Design, Testing, and Maintenance Criteria for Normal Ventilation Exhaust System Air Filtration and Adsorption Units of Light-Water-Cooled Nuclear Power Plants"

* Revision 1 to Regulatory Guide 1.141, "Containment Isolation Provisions for Fluid Systems"

* Revision 1 to Regulatory Guide 1.142, "Safety-Related Concrete Structures for Nuclear Power Plants (Other Than Reactor Vessels and Containments)"

* Revision 1 to Regulatory Guide 1.143, "Design Guidance for Radwaste Management Structures, Systems, and Components Installed in Light-Water-Cooled Nuclear Power Plants"

Other Regulatory Guides of Possible Interest to Division 1 Recipients

Number	Title	Rev.	Issued Year/Month
4.1	Programs for Monitoring Radioactivity in the Environs of Nuclear Power Plants (For Comment)	1	75/04
4.2	Preparation of Environmental Reports for Nuclear Power Stations	2	76/07
4.4	Reporting Procedure for Mathematical Models Selected to Predict Heated Effluent Dispersion in Natural Water Bodies	---	74/05
4.6	Measurements of Radionuclides in the Environment--Strontium-89 and Strontium-90 Analyses	---	74/05
4.7	General Site Suitability Criteria for Nuclear Power Stations	1	75/11
4.8	Environmental Technical Specifications for Nuclear Power Plants (For Comment)	---	75/12
4.11	Terrestrial Environmental Studies for Nuclear Power Stations	1	77/08
4.13	Performance, Testing, and Procedural Specifications for Thermoluminescence Dosimetry: Environmental Applications	1	77/07
4.15	Quality Assurance for Radiological Monitoring Programs (Normal Operations)--Effluent Streams and the Environment (For Comment)	---	77/12
5.1	Serial Numbering of Fuel Assemblies for Light-Water-Cooled Nuclear Power Reactors	---	72/12
5.7	Control of Personnel Access to Protected Areas, Vital Areas, and Material Access Areas	---	73/06
5.12	General Use of Locks in the Protection and Control of Facilities and Special Nuclear Materials	---	73/11
5.29	Nuclear Material Control Systems for Nuclear Power Plants	1	75/06
5.43	Plant Security Force Duties	---	75/01
5.44	Perimeter Intrusion Alarm Systems	1	76/06
5.54	Standard Format and Content of Safeguards Contingency Plans for Nuclear Power Plants (For Comment)	---	78/03
7.1	Administrative Guide for Packaging and Transporting Radioactive Material	---	74/06
7.2	Packaging and Transportation of Radioactively Contaminated Biological Materials	---	74/06
7.3	Procedures for Picking Up and Receiving Packages of Radioactive Material (For Comment)	---	75/05
7.4	Leakage Tests on Packages for Shipment of Radioactive Materials (For Comment)	---	75/06
7.5	Administrative Guide for Obtaining Exemptions From Certain NRC Requirements Over Radioactive Material Shipments	O-R	77/05
7.6	Design Criteria for the Structural Analysis of Shipping Cask Containment Vessels	1	78/03
7.7	Administrative Guide for Verifying Compliance With Packaging Requirements for Shipments of Radioactive Materials (For Comment)	---	77/08
7.8	Load Combinations for the Structural Analysis of Shipping Casks (For Comment)	---	77/05

POWER GENERATION FACILITIES

Number	Title	Rev.	Issued Year/Month
7.9	Standard Format and Content of Part 71 Applications for Approval of Packaging of Type B, Large Quantity, and Fissile Radioactive Material (For Comment)	---	79/03
8.1	Radiation Symbol	---	73/02
8.2	Guide for Administrative Practices in Radiation Monitoring	---	73/02
8.3	Film Badge Performance Criteria	---	73/02
8.4	Direct-Reading and Indirect-Reading Pocket Dosimeters	---	73/02
8.6	Standard Test Procedure for Geiger-Müller Counters	---	73/05
8.7	Occupational Radiation Exposure Records Systems	---	73/05
8.8	Information Relevant to Ensuring That Occupational Radiation Exposures at Nuclear Power Stations Will Be As Low As Is Reasonably Achievable	3	78/06
8.9	Acceptable Concepts, Models, Equations, and Assumptions for a Bioassay Program	---	73/09
8.10	Operating Philosophy for Maintaining Occupational Radiation Exposures As Low As Is Reasonably Achievable	1-R	77/05
8.13	Instruction Concerning Prenatal Radiation Exposure	1	75/11
8.14	Personnel Neutron Dosimeters	1	77/08
8.15	Acceptable Programs for Respiratory Protection	---	76/10
8.19	Occupational Radiation Dose Assessment in Light-Water Reactor Power Plants--Design Stage Man-Rem Estimates	1	79/06
8.20	Applications of Bioassay for I-125 and I-131	1	79/09
10.1	Compilation of Reporting Requirements for Persons Subject to NRC Regulations	3	77/05

TABLE 7 - List of NRC Standard Review Plans of Particular Interest to Structural Engineers

CHAPTER 2 SITE CHARACTERISTICS

- 2.5.1 Basic Geologic and Seismic Information
- 2.5.2 Vibratory Ground Motion
- 2.5.3 Surface Faulting
- 2.5.4 Stability of Subsurface Materials and Foundations
- 2.5.5 Stability of Slopes

CHAPTER 3 DESIGN OF STRUCTURES, COMPONENTS, EQUIPMENT AND SYSTEMS

- 3.2.1 Seismic Classification
- 3.2.2 System Quality Group Classification
- 3.3.1 Wind Loadings
- 3.3.2 Tornado Loadings
- 3.4.1 Floor Protection
- 3.4.2 Analysis Procedures
- 3.5.1.1 Internally Generated Missiles (Outside Containment)
- 3.5.1.2 Internally Generated Missiles (Inside Containment)
- 3.5.1.3 Turbine Missiles
- 3.5.1.4 Missiles Generated by Natural Phenomena
- 3.5.1.5 Site Proximity Missiles (Except Aircraft)

TABLE 7 - List of NRC Standard Review Plans of Particular Interest to Structural Engineers - cont.

CHAPTER 3 DESIGN OF STRUCTURES, COMPONENTS, EQUIPMENT AND SYSTEMS

- 3.5.1.6 Aircraft Hazards
- 3.5.2 Structures, Systems, and Components to be Protected from Externally Generated Missiles
- 3.5.3 Barrier Design Procedures
- 3.6.1 Plant Design for Protection Against Postulated Piping Failures in Fluid Systems Outside Containment
- 3.6.2 Determination of Break Locations and Dynamic Effects Associated with the Postulated Rupture of Piping
- 3.7.1 Seismic Input
- 3.7.2 Seismic System Analysis
- 3.7.3 Seismic Subsystem Analysis
- 3.7.4 Seismic Instrumentation
- 3.8.1 Concrete Containment
- 3.8.2 Steel Containment
- 3.8.3 Concrete and Steel Internal Structures of Steel of Concrete Containments
- 3.8.4 Other Seismic Category I Structures
- 3.8.5 Foundations
- 3.9.1 Special Topics for Mechanical Components
- 3.9.2 Dynamic Testing and Analysis of Mechanical Systems and Components
- 3.9.3 ASME Code Class 1, 2 and 3 Components, Component Supports, and Core Support Structures

TABLE 8 – American Nuclear Society Nuclear Standards Projects of Prime Interest to Structural Engineers

ANS 2	SITE EVALUATION
ANS 2.2	Earthquake Instrumentation Criteria for Nuclear Power Plants
ANS 2.3	Guidelines for Estimating Tornadoes, Hurricane and Other Extreme Wind Parameters at Power Reactor Sites
ANS 2.4	Guidelines for Determining Tsunami Criteria for Power Reactor Sites
ANS 2.7	Guidelines for Assessing Capability for Surface Faulting at Nuclear Power Reactor Sites
ANS 2.8	Standards for Determining Desing Basis Flooding at Power Reactor Sites
ANS 2.10	Guidelines for Retrieval, Review, Processing and Evaluation of Records Obtained from Seismic Instrumentaton
ANS 2.11	Guidelines for Evaluating Site-Related Geotechnical Parameters at Power Reactor Sites
ANS 2.12	Guidelines for Combining Natural and External Man-Made Hazards at Power Reactor Sites
ANS 2.19	Guidelines for Evaluating Site Related Parameters for Independent Spent Fuel Storage Facilities
ANS 4	CRITERIA, CONTROL AND DYNAMICS
ANS 4.1	Design Basis Criteria for Safety Systems in Nuclear Power Generating Stations
ANS 58.1	Plant Design Against Missiles
ANS 58.2	Design Basis for Protection of Nuclear Power Plants Against Effects of Postulated Pipe Rupture
ANS 58.5	Probabilistic Risk Assessment
ANS 51	PRESSURIZED WATER REACTOR
ANS 51.5	Nuclear Safety Criteria for the Design of Stationary Pressurized Water Reactor Plants
ANS 51.7	Single Failure Criteria for PWR Fluid Systems

POWER GENERATION FACILITIES

ANS 52	BOILING WATER REACTOR
ANS 52.1	Nuclear Safety Criteria for the Design of Stationary Boiling Water Reactor Plants
ANS 58.9	LWR Single Failure Criteria
ANS 53	HIGH COOLED GAS-COOLED REACTOR MANAGEMENT COMMITTEE
ANS 53.1	Nuclear Safety Criteria for the Design of Stationary Gas Cooled Reactor Plants
ANS 53.6	Gas-Cooled Reactor Plant Containment System
ANS 53.21	Gas-Cooled Reactor Plant Secondary Coolant Systems
ANS 54	LIQUID METAL FAST BREEDER REACTOR
ANS 54.1	LMFBR General Design Criteria
ANS 54.3	Principal Design Criteria for LMFBR Containments
ANS 54.6	LMFBR Safety Classification and Related Requirements
ANS 54.10	Risk Limit Guidelines for LMFBR Design
ANS 54.11	Application of Risk Limit Guidelines for LMFBR Design
ANS 54.12	Event Categorization Guidelines for LMFBR Design
ANS 55	FUEL AND RADWASTE
ANS 57.7	Away from Reactor Spent Fuel Storage Facilities
ANS 56	CONTAINMENT
ANS 56.3	Overpressure Protection of Low Pressure Systems Connected to the Reactor Coolant Pressure Boundary
ANS 56.4	Pressure/Temperature Transient Analysis for LWR Containments
ANS 56.8	Reactor Containment Leakage Testing Requirements
ANS 56.9	Environmental Envelopes to be Considered in Safety Related Equipment

TABLE 9 - American Society of Mechanical Engineers Nuclear Standards Projects of Particular Interest to Structural Engineers

Subsections NCA - NG, Boiler and Pressure Vessel Code Section III, Division 1 - Nuclear Power Plant Components

Boiler and Pressure Vessel Code Section III, Division 2 -

Subsections CB and CC Concrete Reactor Vessels and Containments

Section XI Boiler and Pressure Vessel Code - Fuels for Inservice Inspection of Nuclear Power Plant Components

N626.0 - 1974	Qualifications and Duties of Authorized Nuclear Inspection
N626.1 - 1975	Qualifications and Duties of Authorized Inservice Inspection
N626.2 - 1976	Qualifications and Duties for Authorized Nuclear Inspection (Concrete)
N626.3 - 1978	Qualifications and Duties of Personnel Engaged in ASME Boiler and Pressure Vessel Code Section III Division 1 and 2 Certifying Activities
ANSI/ASME NQA-1 - 1979	Quality Assurance Program Requirements for Nuclear Power Plants
N45.2.20	Supplementary Quality Assurance Requirements for Subsurface Investigations Prior to Construction Phase of Nuclear Power Plants
P/NC-77-3	Proposed Standard on Overhead and Gantry Cranes

TABLE 10 American Society of Civil Engineers Nuclear Standards Projects of Particular Interest to Civil Engineers

P/N167	Guideline for Seismic Analysis of Safety Class Structures
P/N173	Water Borne Load Design Criteria
P/N175	Foundation Design Criteria
P/N725	Guideline for Design and Analysis of Safety Class Earth Structures

STANDARDS

ASCE TECHNICAL COUNCIL ON CODES & STANDARDS

    *1. CODES

    *2. SPECIFICATIONS

    *3. GUIDELINES, CRITERIA, RECOMMENDED PRACTICE

DOCUMENTS OTHER THAN STANDARDS

ASCE TECHNICAL DIVISIONS

    1. MANUALS OF PRACTICE

    2. COMMITTEE REPORTS

    3. PUBLISHED PAPERS IN ASCE JOURNALS BY INDIVIDUAL OR MULTIPLE AUTHORS

    4. PREPRINTS AND PAPERS AT ASCE CONVENTIONS AND SPECIALTY CONFERENCES

OTHER

    1. PUBLISHED PAPERS SUBJECT TO PEER REVIEW BY INDIVIDUAL OR MULTIPLE AUTHORS

    2. PUBLISHED PAPERS NOT SUBJECT TO PEER REVIEW BY INDIVIDUAL OR MULTIPLE AUTHORS

    3. REPORTS PREPARED FOR CLIENTS

    4. LETTER COMMUNICATIONS

    5. INTERNAL MEMORANDUM

    6. ORAL OPINIONS

    7. OFF HAND COMMENTS

* THE DEVELOPMENT OF REFERENCE MATERIAL WHICH ESTABLISHES RECOMMENDATIONS OR PREFERRED POSITIONS HENCE POTENTIAL LIABILITY FOR ASCE ARE IDENTIFIED AS ASCE STANDARDS AND ARE THE RESPONSIBILITY OF THE ASCE TECHNICAL COUNCIL ON CODES AND STANDARDS. STANDARDS DEVELOPED BY OTHER INDUSTRIAL GROUPS, PROFESSIONAL SOCIETIES AND GOVERNMENT BODIES CARRYING RESPONSIBILITY AND LIABILITY AS DEFINED BY THE ISSUING ORGANIZATION AND AS OTHERWISE PRESCRIBED BY LAW.

FIGURE 1 A SUGGESTED HIERARCHY OF REFERENCE MATERIAL USED TO DESIGN AND CONSTRUCT ENGINEERING WORKS BY CIVIL ENGINEERS

THE NUCLEAR LICENSE PICTURE--OVERVIEW AND OUTLOOK

by

Richard H. Vollmer[1]

ABSTRACT

This paper provides the U.S. Nuclear Regulatory Commission's outlook on licensing by discussing the impact of the Three Mile Island accident and other projects of interest to the Civil Engineer.

It is certainly a pleasure to speak before the ASCE on nuclear issues here at Penn State. Of course, as you all know, Penn State has taken on an extra significance for us at the NRC since the recent appointment of Nunzio Palladino as Chairman of the Commission. However, even before that we always had a certain relationship with Penn State because of his participation on the ACRS and the benefit we have had of many of your graduates.

My talk today is entitled "The Nuclear License Picture--Overview and Outlook." What I would like to do is to relate to you some of the problems and success we are having in the licensing of current nuclear power plants and discuss the near-term outlook for getting plants already constructed on the line and long-term outlook for licensing new facilities.

IMPACT OF TMI ON LICENSING

Regarding the licensing of plants that have been constructed, you are probably all aware of the year and a half or so moritorium on licensing that occurred post-TMI. This resulted from a number of factors including the need for us in the licensing business to reassess our mission and priorities and to alter or affirm them depending on the lesson learned from TMI. This process was very lengthy and sometimes painfull but resulted in some new licensing initiatives which I believe will enhance the safety of plants and public confidence in them. As you are all aware, TMI was more the result of people and systems problems than of individual components or structures. So most of the lessons learned did not have a direct bearing on nuclear work that most civil engineers are engaged in. However, it is safe to say that the fallout to the civil areas and associated activities has been significant and requires in many cases an enhanced level of evaluation and analyses.

First, I would like to discuss the most direct impact of TMI on structures for nuclear power plants. As you may know, in the hours and days following the accident, there was substantial concern about the generated hydrogen and its potential for exploding, burning, or otherwise threatening the integrity of containment. As it turns out, the

---
[1]Director, Division of Engineering, Office of Nuclear Reactor Regulation, U.S. Nuclear Regulatory Commission, Washington, D.C. 20555

TMI structure was sturdy indeed, and in retrospect would not likely be challenged by a hydrogen event. But not all containments were constructed to the pressure capability of TMI and not many have the capability to withstand the impact of a 727 jet. So as a matter of prudent regulation, since we did have an example of significant hydrogen which far exceeded that which was part of the design basis for licensed plants, we started considering what capability plants being licensed should have to cope with hydrogen generation and subsequent burns or detonations.

For BWRs with small containments, hydrogen was not a problem because most of them were inerted and the remaining ones could be required to inert. With respect to the large dry containments, hydrogen was less of a problem because they generally have a significant pressure capability and a large heat capacity, the combination of which mitigates the danger of hydrogen even from a 100% metal-water reaction. The intermediates, however, for example, the BWR Mark IIIs and the ice condenser containments, are generally of a low pressure design as well as a low volume so that in the event of substantial hydrogen generation and burn the design pressure capability could be exceeded by a significant amount. Therefore, in the licensing of ice condenser plants last Fall, we needed to look retroactively to their capability for coping with the hydrogen event. We had some analyses performed which tried to establish a lower, expected, and upper bounds of containment pressure capability for these type containments. We also looked at ways to reduce the impact of hydrogen burning on containments by such things as igniters which deliberately sequence the burn, and post-accident inerting. As a result of all of this, we found that for typical steel and concrete containments, a margin of 2 1/2 to 3 was available between the design pressure and the pressure at which these containments would be expected to leak significantly in excess of their design leak rate. This, coupled with systems designed to burn the hydrogen a bit at a time as it was being generated and released from the core following an accident, gives high confidence that containment, and therefore the public health and safety, would not be threatened by this type event. It should go without saying that this specific sequence at TMI was a special subset of low probability reactor accidents that had been considered for some time in safety analyses. The actions taken by the Commission over the past couple of years, including additional facility changes, procedural upgrading, and extensive training and technical augmentation of the operation staff, should make such a recurrence an even more unlikely event.

Another item which should be of substantial interest to the civil engineers' concerns is the potential for release of highly radioactive liquids to the ground water following not only a TMI-type but other types of nuclear accidents. As you may know, and most of you from Pennsylvania in particular will recall, the subject of 700,000 gallons of highly radioactive water in the reactor building sump was of considerable concern. We have had for some time a generic study of the potential consequences of release of radioactivity to ground and surface waters in the form of the Liquid Pathway Study. TMI, however, provided a real life example of potentials for such releases. It was of particular note because there are drinking water intakes below the plant on the Susquehanna and significant fishery activity in the Susquehanna-Chesapeake Bay region. So it was no small concern to begin with and was considerably enhanced when very low but detectable amounts of radioactivity were found in the ground months after the accident. Fortunately, however, an analysis showed a period of time on the order of months between release from the buildings and entry into the body of the Susquehanna and even then with great reduction in radioactive intensity because of cleanup by the soils. It was, however, enough of a lesson so that we now take particular note of the potential for onsite interdiction. In this case, preventing by physical means any radioactive water from getting into potable or ground waters. This could be done by drilling wells and pumping to lower the ground water such that the gradient would be toward the plant rather than away or by setting up curtains such as with grout to effectively isolate the radioactivity from the environment. These things are now considered in the licensing review and a plant which was sitting on a major aquifer and had ready access to same would likely have a tough time in the environmental review process without special means to protect the waters. Of course, as with TMI, contamination of waters is more of a social and economic impact than a human health impact because in the final analysis, if drinking water and fish were to get contaminated then they would have to be interdicted and taken out of the food chain. The social and economical cost of that type of process, however, could be quite significant depending on the location.

CURRENT LICENSING ISSUES

I would like now to relate a couple of areas of immediate licensing concern to us, and are of interest to your discipline. At several plants there has been a significant settling problem of safety-related structures or components. Since we license these plants for a long period of time, the soils problem and integrity of structures is an important one. In most of these cases not enough attention was given to soil engineering and/or quality of process of upgrading the soil properties where this was found to be necessary. We have also had a number of cases where liquefaction in the event of an earthquake could have threatened safety-related components. Again, of course, the design basis for such structures should be that they have an adequate base to withstand any normal or natural event loadings. At a recent event at a West Coast facility, a severe but not terribly abnormal winter storm wiped out a good part of the breakwater system which was designed to protect the cooling water intake from that very type of storm. Thirty-five-ton dolos were moved and scattered by the storm, rendering the breakwater much less effective. These are but a couple of examples where the civil engineer is challenged to provide unique and highly certain protection to nuclear systems from the wear and tear of nature both in its slow processes and stormy ones. Unique protection is required because in the event of natural disasters the plants must not only keep its own integrity so that radioactive releases will not take place, but would undoubtedly be counted on as a highly reliable source of power.

And finally, I'd like to relate how one plant design overcame some significant soils problems. This plant, scheduled for operation in 1983, was designed and constructed using the "floating foundation" principle. This principle prescribes that the weight of soil removed from the plant excavation be approximately equal to the weight of the plant placed in the excavation; therefore, there is no change in soil stress at the base of the plant. The foundation of the plant can be conceived as "floating" in the soil in the same manner as a boat floats in water. This concept is sometimes described as a "fully compensated" foundation.

This plant, on the bank of the Mississippi River, is underlain by more than 2,000 ft. of loosely consolidated soils. Any structural loads applied to the foundation soils would have caused consolidation of the underlying sediments, resulting in settlement and differential settlement of the structure; similarly, any significant unloading of the foundation soils during excavation would have caused heave of the underlying sediments, resulting in re-compression settlements as structural loads were applied.

Potential settlement problems at the site have been avoided by containing all the safety-related structures and components within a large, reinforced concrete, boxlike structure and adopting the floating foundation principle. The structural assemblage, described as the nuclear plant island structure (NPIS) is 270 ft. wide and 380 ft. long. The NPIS foundation mat is 12 ft. thick and it is located 65 ft. below ground surface. This depth of embedment was necessary to provide sufficient effective weight of excavated soil to compensate for the weight of the plant.

During the excavation phases, the groundwater level was lowered by pumping from 250 wells located around the perimeter of the NPIS; during the NPIS construction stages, the ground water level was raised by pumping into recharge wells and wetting the backfill around the NPIS. The construction operations required careful monitoring of ground water levels to assure compliance with the design criteria throughout construction. The movement of the soil beneath the NPIS was also monitored by continually recording the vertical position of anchors set a few feet below the NPIS mat.

The well planned and executed engineering approach took advantage of what is regarded as adverse site and foundation support conditions. The entire construction operation was performed without the use of sheet piling or slurry walls, and yet the excavation was maintained in a dry condition. The complexity and cost of a deep foundation system, e.g., piling, was avoided, and the potential problem of long-term settlement associated with conventionally-constructed, soil-supported foundations has been eliminated.

Although some of the issues discussed above have given reactor designers as well as regulators some new and interesting challenges, they have not held up the licensing process. In fact, if you look at the schedules for review and licensing of those plants that are now completing construction, you will find in many cases that they are being accomplished in less time than pre-TMI. The NRC has responded to wishes of Congress and the needs of the consumers and industry in reprogramming its efforts significantly in areas of licensing. The moritorium had set us back significantly but within a year or so we will have pretty much caught up to a point where neither the staff nor the anticipated hearing schedules will delay start-up of the plant.

LICENSING OUTLOOK

Regarding the outlook, it seems that all of the lessons from TMI can be accommodated by current generation nuclear plants without the need for significant redesign. Certainly new systems have been added, procedural aspects have been upgraded, and the technical qualifications of the operating staff has been impacted. But none of the basic features of the plant, as long as they have met prior design requirements, would likely be impacted at all. It is just perhaps that we would be looking closer at some of these features in the regulatory process. It is sort of hard to say what is in the cards for coming construction permit applications. We have new near-term CP requirements which cover extensively, among other things, the issue of containment integrity following an accident.

For example, the pending CP applications will be required to show that even for a 100% fuel clad metal-water reaction and subsequent combustion of the hydrogen generated, the containment will meet Service Level C requirements. If the facility design were to include a system to inert the containment following the accident, then Service Level C would need to be met in consideration of the LOCA pressure, the addition of the hydrogen, plus the post-accident inerting agent, for example, carbon dioxide. In any event, we are looking for a Service Level C pressure limit of at least 45 psig. If the option chosen is one of post-accident inerting, we are also requiring that the containment be designed to withstand inadvertent actuation of the post-accident inerting system and demonstrate by pressure test that it can be accommodated.

Obviously, these and other measures will set forth a firm basis upon which future generation of plants could be designed. It is not yet clear if and when new applicants will come forth for a host of reasons, including financial, political, and social. However, it is the intent of the Commission to provide a licensing basis upon which a utility could order new plants. As such the long-term outlook for nuclear power would seem to depend more on the real and perceived need than the ability to meet the design challenges.

NUCLEAR REGULATORY COMMISSION ENFORCEMENT POLICY

Charles E. Norelius[1]

ABSTRACT

The Nuclear Regulatory Commission has recently revised its enforcement policies to provide for increased civil penalties and generally to develop a firm and fair program which covers all aspects of reactor construction and operations. An interim enforcement policy, developed as part of the revision process and which was approved for use in September 1980, is described in detail. Potential changes, based on its use and comments received, are also discussed.

BACKGROUND

First, I will provide general background information. Prior to 1969, the NRC's enforcement program did not include civil penalties. The enforcement actions in that era were primarily Notices of Violation, supplemented by the occasional use of Orders for the more serious safety and chronic noncompliance cases. In 1969, Congress granted the then AEC authority to levy civil penalties for items of noncompliance. Civil penalties of up to $5,000 per item of noncompliance with a maximum civil penalty of $25,000 for all violations occurring within a monthly period were permitted. In August 1971, a Rule was published to implement the statute and in October 1972, the Commission first published its enforcement policy in the Federal Register.

The next important milestone was December 31, 1974, when the staff provided all licensees an update and further clarification of its enforcement criteria.

Another key milestone occurred in early 1978, when the Commission, recognizing that a $5,000 civil penalty did not represent a serious financial disincentive to larger licensees, submitted a request to Congress to increase the maximum civil penalty from $5,000 to $100,000 per item of noncompliance. Congress enacted such legislation and it was signed into law on June 30, 1980. This law provided for civil penalties up to $100,000 per item of noncompliance, but did not place any limit on the total amount of a civil penalty which could be issued.

---

[1]Director, Division of Engineering and Technical Inspection
U. S. Nuclear Regulatory Commission, Region III, Glen Ellyn, IL

NRC ENFORCEMENT POLICY

While civil penalties and other escalated enforcement actions were used cautiously during the early and middle seventies, there has been increasing emphasis on enforcement actions over the past few years with a significant increase in the number and severity of enforcement actions since Three Mile Island. This increase has come about because of a mandate given to the NRC by the Congress to be strong regulators.

It will be of interest to this group to note that the issuance of civil penalties to facilities under construction was almost nonexistent until recently. This was true in large part to the fact that no problem at a construction site represented an immediate threat to the public health and safety. Since there was always time to take corrective action, problem areas identified at construction facilities did not receive the same degree of regulatory attention as those which occurred at operating facilities.

However, considering the general increase in regulatory action following the Three Mile Island accident and considering findings of significant construction deficiencies at certain facilities, the Commission directed the staff to review the need for escalated enforcement action at construction sites.

In late 1979, Mr. Victor Stello, Director of the Office of Inspection and Enforcement, appointed a Task Force to revise the enforcement policy in consideration of the general mandate to be tougher regulators and the pending legislation request to provide for increased civil penalty authority. The Task Force was headed by James G. Keppler, Director, NRC Region III; I also served as a member.

The Task Force established six specific objectives in revising the enforcement policy.

- We wanted to establish criteria for utilizing the increased civil penalty authority.

- We wanted to make the enforcement program tough, but fair.

- We wanted to achieve greater uniformity in the treatment of licensees by taking equivalent actions against similar licensees having similar problems.

- We wanted to better define our enforcement capabilities with respect to NRC licensed activities at other than operating reactors.

- We wanted to focus escalated enforcement actions on the specific event or problems which lead to the decision to take escalated enforcement rather than focus on the total number of noncompliance items identified.

- We wanted to clearly articulate our enforcement policy and define more clearly the criteria for taking various enforcement actions.

On September 4, 1980, the Commission approved an interim enforcement policy. This was published for public comment in the <u>Federal Register</u> on October 7, 1980. The Commission also approved the use of this enforcement policy by the staff at that time.

INTERIM ENFORCEMENT POLICY

I would like to now discuss in some detail the major provisions of the interim enforcement policy.

Severity Levels

We have had for the past several years three categories of noncompliances termed violations, infractions, and deficiencies. While we have found that having different severity categories is beneficial in judging the significance of noncompliances, our experience has shown that more categories were needed to capture the differing thresholds of noncompliance. In defining severity categories, we wanted to relate them to the fundamental problem or event involved. In determining the proper number of severity categories, we looked at actual experiences at operating reactors as a starting point. From this, we determined that the significance of various events which had occurred and the various types of noncompliance which were identified lent themselves toward distributing the items of noncompliance into six severity levels. After going through this exercise for operating reactors, we next applied the same general principles to other licensed activities including developing guidance for six separate severity levels for reactors under construction. Generally, we believe the severity levels I, II, and III are serious violations that should occur infrequently if appropriate attention is being given to NRC requirements. We believe the severity level IV violations also should not occur often and we view the severity level V violations to be equivalent to most of the infractions that we have had in the past. The different severity levels are defined separately for each of the seven different program areas which we regulate. These program areas are shown below.

DIFFERENT NRC LICENSED ACTIVITIES

Reactor operations

Reactor construction

Safeguards

Health physics - 10 CFR 20

Transportation

Fuel cycle operations

Materials operations

NRC ENFORCEMENT POLICY

While the severity levels show the relative importance of violations within the same program area, it is important to recognize that severity levels are not equatable in terms of safety importance from one program area to another. Said another way, the severity level I is the most significant violation in each of the seven different program areas shown, but a severity level I violation in the area of reactor operations obviously does not have the same immediate safety significance as a severity level I in facility construction, for example.

As I mentioned earlier, the determination of severity categories is event oriented. To help understand this, it may be well to look at the specific guidance given for the different severity levels for facility construction. In this area, severity I violations are those involving all or part of a structure or system that is completed in such a manner that it would not have satisfied its intended safety related purpose. Severity II violations are those involving a significant deficiency in quality assurance program implementation related to more than one work activity (i.e., structural, piping, electrical, foundations) as shown by multiple program implementation violations that were not identified and corrected by more than one quality assurance/quality control checkpoint relied upon to identify such violations. Severity level III violations are those involving a lack of quality assurance program implementation related to a single work activity as shown by multiple program implementation violations that were not identified and corrected by more than one quality assurance/quality control checkpoint relied upon to identify such violations. The purpose of these violation descriptions was to focus on the seriousness of a situation rather than on the particular number of violations.

It must be recognized that the guidance in the policy is general in nature. Not all identified violations will fit the policy guidance. In such cases, it is recognized that judgment will have to be exercised in selecting the proper severity category.

Notices of Violation

It is expected that Notices of Violation will continue to be sufficient enforcement action for greater than 90% of the violations which are identified during our inspections. One difference under the interim enforcement policy is that responses to Notices of Violation must be submitted under oath or affirmation as provided for in Section 182 of the Atomic Energy Act.

Civil Penalties

Let me now turn to a discussion of civil penalties. There are four general areas that are likely to lead to assessment of a civil penalty. The first is for severity level I, II, or III violations which have occurred. Secondly, it is possible to assess civil penalties for recurring severity level IV and V violations. Thirdly, the knowing and conscious failure to report a defect by a responsible official of a licensee or vendor organization may result in the assessment of a civil penalty against that particular individual as provided for in Section 206 of the Energy Reorganization Act. Fourthly, willful violations may result in civil penalties.

I want to go back and make some additional comments on the first two items mentioned above. We recognize that some technical judgment will enter into the categorization of severity levels I, II, and III and whether they warrant a civil penalty. Normally, however, if it has been determined that a severity level I, II, or III violation existed, it is the Commission's intent to issue a civil penalty.

Civil penalties will generally be assessed for recurring severity level IV violations that are similar in nature to those which were the subject of an enforcement conference and which occurred within two years following the enforcement conference. An enforcement conference is a meeting specifically designated as such between NRC and licensee management for the purpose of discussing specific violations, the planned corrective action and the enforcement options available to the NRC. If similar violations occur after such an enforcement conference and it is concluded that their occurrence resulted from ineffective licensee action, a civil penalty will generally be assessed.

Set forth below is a Table of Base Civil Penalties for different types of licensed programs and for different severity levels of noncompliance. You will note that for the construction of power reactors, the first line in the table will apply.

BASE CIVIL PENALTIES

	Severity Levels of Violations				
	I	II	III	IV	V
Power reactors Other SNM licensees associated with Category I material for safeguard purposes only	$80,000	$80,000	$40,000	$15,000	$5,000
Test reactors Fuel facilities. Other SNM licensees for safeguard purposes only	40,000	40,000	20,000	7,500	2,500
Research reactors Critical facilities	16,000	16,000	8,000	3,000	1,000
All other licensees and persons subject to civil penalties	8,000	8,000	4,000	1,500	500

You will note that from the Table the Base Civil Penalty values for severity level I and II violations are the same. You will find, however, later in the discussion, that the actual sanction against a licensee is greater for a severity level I violation in that such a violation normally results in issuance of some type of Order in addition to a civil penalty.

It is also noteworthy that while the law provides that a civil penalty of $100,000 may be assessed for each violation, the policy provides that for severity level I, II, and III violations, the civil penalty will be assessed for each event, irrespective of the number of violations associated with event. For example, if several violations were identified at a reactor construction site which lead to the conclusion that a breakdown in quality assurance occurred in multiple phases of construction, this would be considered as a severity level II event. However, the civil penalty would be assessed for the event; that is, a cumulative civil penalty of $80,000 would be assessed for all the violations which constituted the event regardless of their number. We believe such an approach will help to focus licensee and public attention on the significance of overall events as opposed to the individual violations which may be identified. The actual mechanics for assessing civil penalties will remain the same; that is, a Notice of Proposed Imposition of Civil Penalties and Notice of Violation must clearly state which violation occurred and which violation civil penalties are being assessed for. For example, if eight violations constitute a severity level II event, the $80,000 base civil penalty may be equally assessed for all eight items which make up the event or the entire civil penalty may be assessed against only one violation. The actual distribution will be determined on a case-by-case basis.

The civil penalties shown in the Table are Base civil penalties. The policy identifies various other modifying factors which may enter into increasing or decreasing the actual civil penalty from the base amount. I do not plan to discuss the mitigating factors in detail. Generally, however, the final civil penalty amount is based on the gravity of the violation involved, the duration of the noncompliance, how the problem was identified, the financial impact on the licensee, the good faith of the licensee, a licensee's prior enforcement history, and whether the violation was willful.

<u>Orders</u>

We will next look at the types of Orders which may be issued by the Commission. There are Orders to Modify, Suspend, or Revoke Licenses or Permits, and Orders to Cease and Desist any particular operation. These Orders may affect all or part of a licensed activity. Normally, Orders for modification, suspension, or revocation will be issued with the Show Cause provision; that is, they will require a licensee to show cause why such action as proposed should not be taken. Such orders always provide a licensee opportunity for a hearing on the issues. However, if a determination is made by the Director of the Office of Inspection and Enforcement, that the public health and safety, common defense and security, or public interest so demands, the Order may be made effective immediately. It is possible for

Orders to be issued which combine these provisions; that is, an Order may require the immediate suspension of a particular operation and may, at the same time, include a Show Cause provision as to why the license should not be revoked.

Combined Sanctions

Table 2 shows a progression of escalated enforcement action which may be taken for repetitive serious violations. This Table is not intended to prohibit the NRC from taking a different action if the case warrants. The basic intent of this table is to show that the Commission will not accept repetitive escalated problems. At some point, more stringent enforcement actions will be taken until such time as the problem is corrected or, if necessary, the licensed activity stopped.

TABLE 2

Examples of Progression of Escalated Enforcement Actions For Violations in the Same Activity Area Under the Same License

Severity of Violation	Number of similar violations from the date of the last inspection or within the previous two years (whichever is greater)		
	1st	2nd	3rd
I	a+b	a+b+c	d
II	a	a+b	a+b+c
III	a	a	a+b

a - Civil penalty.

b - Suspension of affected operations until the Office Director is satisfied that there is reasonable assurance that the licensee can operate in compliance with the applicable requirements; or modification of the license, as appropriate.

c - Show cause for modification or revocation of the license, as appropriate.

d - Further action, as appropriate.

Let me run through an example of how this table might be applied. If a severity level II violation occurred, its first occurrence would result in a civil penalty. A second similar violation within the two year period would result in a civil penalty and an Order to either suspend affected operations until the Office Director is satisfied that there is reasonable assurance that the licensee can operate in compliance or to modify the license or permit to impose additional requirements to provide equivalent assurance. If a third similar violation occurred within a two year period, then in addition to the actions taken the previous time, additional action to show cause for further modification or revocation would be the next step.

PENDING CHANGES TO THE ENFORCEMENT POLICY

As I indicated early in our discussion, the enforcement policy was published in the <u>Federal Register</u> on October 7, 1980, and was provided for the staff's use on an interim basis. As a result of the publication in the <u>Federal Register</u>, 160 separate written comments were received on the policy. Additionally, five public meetings were held at major cities across the country and comments were received from members of the general public who attended. These comments will be reviewed and experience will be considered. On this basis, the policy will be revised accordingly. While we expect some modifications, based on these comments and our experience, at this point, it appears that the revisions will not be major in terms of the concept and basic format of the policy.

Upon final approval of a policy by the Commission, we intend to publish the policy in the <u>Federal Register</u>. The advantage that I see of its publication is that the general public, licensees, and the NRC staff will all have a common basis to work from. This will help everyone to know the "rules of the game."

In conclusion, let me note again that the NRC is giving more regulatory attention to the construction phase of nuclear power plants. This increased attention is shown in the enforcement policy, in that it provides definitive situations where plants under construction may be subject to civil penalties. Of course we would hope that the result would not be the issuance of more civil penalties, but rather that more attention would be given by licensees and their contractors to building the facilities right such that noncompliant activities do not occur. Said another way, we hope that the increased emphasis on enforcement would have a deterrent action which helps improve facility construction, and thereby the public health and safety.

USNRC INSPECTION OF REACTORS UNDER CONSTRUCTION

by

Robert F. Heishman[1]

ABSTRACT

This paper provides the reader with an understanding of the inspection program that was developed and is utilized by the Nuclear Regulatory Commission (NRC) for its inspection of nuclear reactors under construction. The Office of Inspection and Enforcement in the Nuclear Regulatory Commission is responsible for the development and administration of programs and policies for inspecting licensees to ascertain whether they are complying with NRC regulations, rules, orders and license provisions, and to determine whether these licensees are taking appropriate actions to protect the health and safety of the public and also as a basis for recommending issuance of construction permits and operating licenses.

INSPECTION PROGRAM CONCEPTS

One of the major philosophies adopted by the Nuclear Regulatory Commission is that the licensee is responsible for the proper design, construction, testing and operation of a nuclear plant. The NRC's role is one of evaluating the licensees' efforts to meet this responsibility and to assure that corrective action is taken whenever the proper controls are not in place to carry out construction in accordance with commitments to the NRC. The principal basis used by the NRC in assuring itself that the plant has been constructed in accordance with the commitments is that the nuclear requirements governing construction are mandatory and enforceable by law.

ORGANIZATION

Enclosed as attachment 1 and 2 are two simplified organizational charts that will aid in understanding the lines of authority and responsibility of those NRC staff groups directly associated with the inspection and review of nuclear project construction and design activities. Looking at the NRC headquarters staff, it should be noted that the office of Nuclear Reactor Regulation (NRR) has the principle responsibility for the review of reactor designs and licensing. This includes technical evaluations as well as acceptability of the plant design.

---

[1] Chief, Performance Appraisal Section, Office of Inspection and Enforcement, U.S. Nuclear Regulatory Commission, Washington, D.C. 20555

FIGURE 1
NUCLEAR REGULATORY COMMISSION

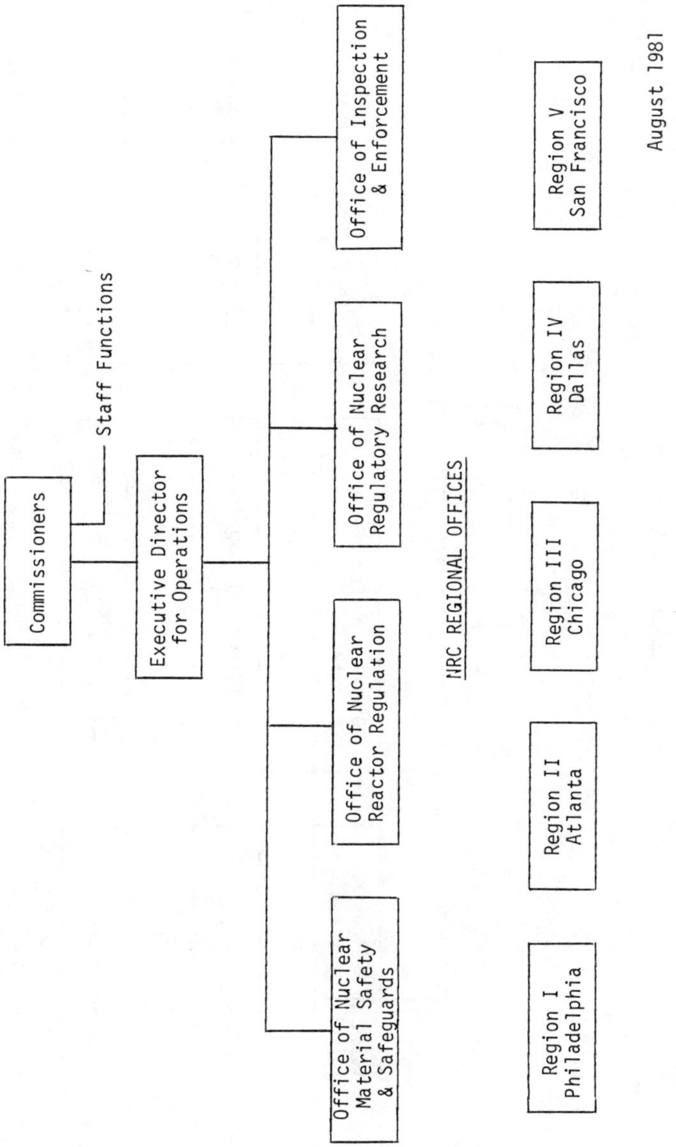

NRC REGIONAL OFFICES

August 1981

FIGURE 2
NRC REGIONAL OFFICE

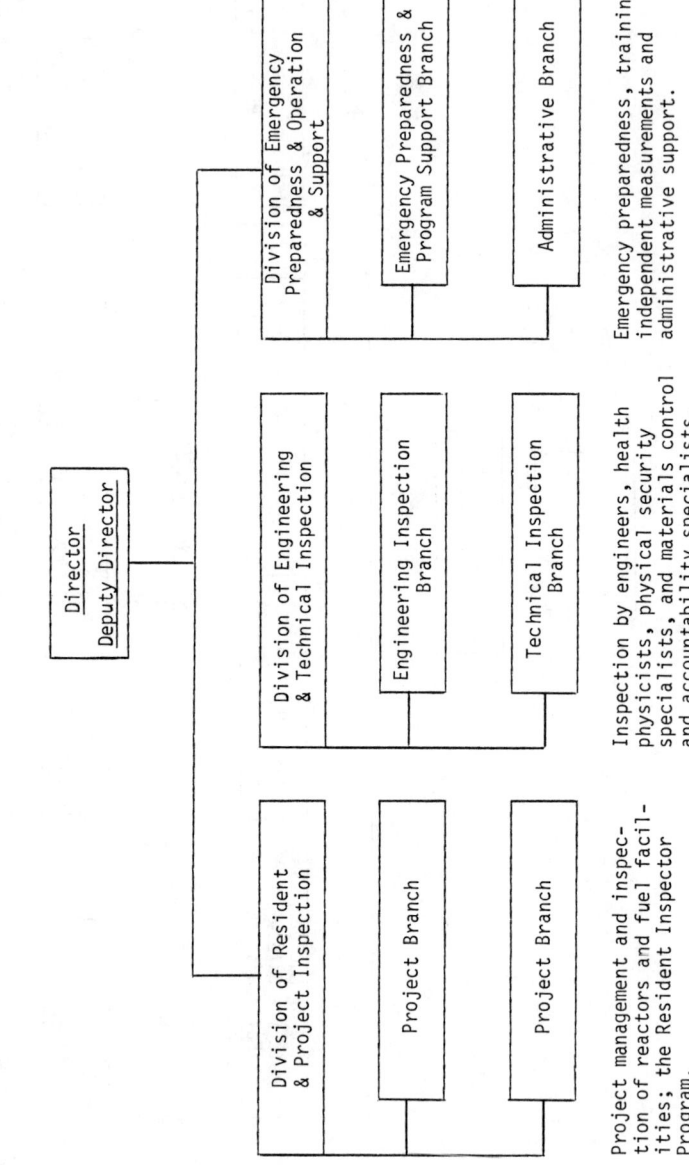

The office of Inspection and Enforcement (I&E) is composed of a HQ staff and five regional offices with inspection specialists supported by an administrative staff. The principal responsibility of the I&E headquarters staff includes policy-making by senior I&E management, program development and technical support as well as an interface function between the Inspection and Enforcement organization and the Nuclear Reactor Regulation organization. This interface involves the communication and resolutions of problems found in the field that deal with construction, design, quality assurance (QA), and other project related activities. Within the region there are three groups that have responsibility for the inspection at a construction reactor site. There is a project group that has the overall responsibility for coordinating the Inspection Program as well as conducting inspections in certain areas. There is an engineering and technical inspection group that has the responsibility for carrying out the technical inspections such as in civil engineering, welding, nondestructive examination, and electrical engineering. There is an investigative group that has responsibility for the investigation of any matters received from allegers that have the potential for compromising the quality of construction. Within the project group there is a further breakdown into region-based inspectors and site-based inspectors. Several years ago a Commission decision was made to put on certain sites, inspectors who would be based on the site and who could follow the day-to-day construction activities during the time the plant was being built. The main responsibility for this individual is to maintain inspection continuity; to become intimately familiar with the various organizations on site; to observe how they interface; to be sensitive to the complaints of workers who may have identified some problems relating to construction quality and to observe the overall effectiveness of the management, how it operates, directs and supports the construction activities at the site. The regional-based project inspectors on the other hand are more involved in the coordination and integration of the Inspection Program. They act as a liaison between the various groups within the region and headquarters on problems identified at the site. They review or initiate reviews of reports involving technical problems received from licensees. They are involved in the tracking of the completion of certain portions of the Inspection Program and generally assist in the overall implementation of the Inspection Program. There is, within the region, numerous other staff members that have peripheral support roles and I do not intend to get into their responsibilities at this time.

REACTOR CONSTRUCTION PROGRAM

As was implied earlier, the goal of the NRC Reactor Construction Program is to assure that reactors are built properly and operate safely. The methods used in obtaining that confidence involves such activities as:

a. Performing effective and efficient inspections

b. Conducting timely investigations

c. Establishing greater presence on site

d. Increasing the number of independent measurements and direct observations

e. Following up on licensee reported events

f. Providing regulatory feedback to other organizations within the NRC

g. Conducting periodic appraisals of licensee performance

The accomplishment of the stated goal is brought about by three types of construction inspection programs. The first is the <u>Routine Program</u>. This program is a defined program that involves the evaluation of licensee and its contractor's QA programs; it involves direct observation of work, and the examination of Quality records. The second program type is known as the <u>Reactive Program</u>. Under this program NRC inspectors pursue any allegations that are received from members of the public or the construction crafts to the point where there is or is not substantiation and any required corrective action. Under this program special events or special problems are followed. Examples of these involve being on site in a very timely manner in connection with the damage of safety-related equipment due to dropping of equipment; the failure of concrete during the process of post-tensioning; the damage of electrical equipment due to fires. In other words, the intent is to obtain a very timely assessment of the problem and its associated causes because of the very special significance of the matter. The third type of inspection program, which I do not intend to get into except to mention, is the <u>Vendor Inspection Program</u>. This is a program conducted by a group within the NRC Region IV (Dallas) office. The responsibilities of this group are to go to selected vendors and manufacturers in the country to inspect the manufacturers' or suppliers Quality Assurance Programs and in some cases, the quality of the product. Vendors inspected include those that manufacture large vessels, pumps, pipes and valves, electrical components, instruments, Architect Engineers, and Nuclear Steam System Supplies.

ROUTINE INSPECTION PROGRAM

It should be emphasized again that it is the licensee's responsibility to construct and operate the facility safely and in compliance with construction permit provision, license provisions and regulatory requirements. IE inspections are not designed to duplicate or substitute for a licensee's management controls established as a part of his quality verification system. The elements used in developing detailed inspection requirements include regulatory requirements, regulatory guides, industry standards and codes, and experience. The findings from both the nuclear industry and NRC quality verification and inspection programs and the technical judgment of engineers and scientists both in industry and in the NRC are also factored into the development of NRC inspection programs.

The routine inspection program is broken down into two phases - a preconstruction phase and a construction phase. During the preconstruction phase the NRC, both the NRR staff as well as the I&E staff, meet with the licensee. At that time, the licensee is informed of the inspection and licensing programs, the kinds of activities conducted by the NRC inspectors and NRR project managers. The Enforcement Program is also discussed along with the various communications that result from an inspection. One of the earliest determinations that is made, even prior to docketing of an application, is the evaluation of the design review and procurement related quality assurance controls that the licensee has developed. This is done by a team of inspectors who go to the licensees' engineering offices. Inspections of a nuclear steam system supplier or architect engineers are also made depending on the arrangement of the project contracts. Upon making a favorable finding, that the licensee's quality assurance program in the areas of design review and procurement appear adequate, this information is provided to NRR for the purposes of docketing the application. Following the docketing, the IE inspectors conduct another team inspection in which they look at the adequacy of the program related to the other QA criteria. Basically, they look at the organization, its ability to construct a plant; its ability to control the quality of construction; its ability to conduct independent audits; its ability to support both technically and managerially the construction of the project as well as other factors which will be involved during the subsequent years of construction. Upon a favorable finding that the licensee has in fact both the staff and program - this information is again provided to NRR as part of the basis for the issuance of a construction permit. Upon issuance of a construction permit, the licensee is allowed to construct their plant.

The NRC has developed an inspection program that covers the more important activities associated with the construction of the reactor facility. These include civil, mechanical, electrical and welding activities. The regional staff, as I previously mentioned, consists of both the project inspectors as well as the engineering specialists. These individuals follow the construction from the foundation work through and including construction testing. In carrying their inspection responsibilities inspectors use the Inspection Program to conduct their inspection activities.

The program consists of many procedures covering specific construction activities. These procedures identify the specific items which should be inspected as well as provide guidance as to the sample size, frequency of inspection and the acceptance criteria of the construction work which they are inspecting. As stated earlier, the fundamental areas inspected are:

a. The existence of proper QA/QC procedures governing the work

b. Direct observation of the work in progress

c. A review of the Quality records of work

The inspections include interviews with construction, QA/QC and management personnel. Following each inspection, the inspector meets with the licensee and the contractor representatives and discusses their findings. In the case where problems have been found, these are highlighted in the inspection report issued at the conclusion of the inspection. It is incumbent on the licensee to inform the NRC of the corrective action that is taken and the reactor inspector later follows up to see that in fact the actions have been taken. Where problems are significant and involve serious safety implications, meetings are held with licensee corporate representatives. In no case does an inspector leave the site without identifying any problem to somebody in a position of authority in the licensee's organization.

REACTIVE PROGRAM

As stated before, another type of program we have is the <u>Reactive Program</u>. As the term implies, I&E actions taken are in reaction to some special problem or some allegation. In this type of followup where some member of the public or a worker has alleged problems with construction quality an investigator based in the region is utilized along with the technical inspector. The allegers, whether they be craftsmen or members of the public, are interviewed by the investigator and sworn statements may be taken. It should be pointed out that a substantial part of our effort to date is conducted in conjunction with the followup of allegations. It should also be pointed out that some allegations may involve situations in which material false statements or other criminal actions may have occurred. In this case, the Department of Justice may become involved in the investigation and could lead to criminal sanctions against individuals or organizations. A large portion of our inspection effort is directed to the reactive program.

LESSONS LEARNED

It is through the inspection process that I&E discharges its organizational responsibility for the evaluation of the licensees' construction activities. The findings made are factored into the program through recommendations to change NRC regulations, regulatory guides or other licensee commitments. The use of this feedback process is important. When a potential safety issue is identified as a result of completing one or more inspection requirements, inspectors or other IE personnel must determine whether a matter affecting safety exists and whether other substantial, additional protection is required. The NRC has learned much through the years of involvement in the inspection of construction activities as well as operating activities with licensees. As a result of this learning the NRC has revised its inspection programs and manner in which it interfaces with licensees. For instance, we learned that it is important to conduct early reviews of Quality Assurance programs of licensees as well as to assess the resources of the utility to construct a nuclear reactor. We do this through a series of saturated inspections in which we look at a fairly large sample size of activity, organizational authorities responsibilities and controls to see whether the program is put together properly and managed by competent individuals. It should be kept in mind that many of these assessments are judgmental; however, we feel that the staff of the NRC is competent and possesses the proper experience levels to make such assessments.

Without our assessment of management controls we specifically make determinations as to the awareness of site problems by management in determining the licensees' involvement in both of those matters. It follows then that when the licensee does determine that he has a problem, we look at his actions with respect to stopping work, and with respect to the timeliness of the resolution of the problems. It goes without saying that the better performers are not only aware of problems but are on top of them and are not intimidated or bashful about taking strong corrective actions.

Our Inspection Program is a dynamic one. We try to factor in better ways of assessing the quality of construction work and we continually strive to work with licensees in preventing problems before they occur. Our management provides us full support and unfortunately, we do not have the resources we would like to have to conduct our program. Regardless of this, you can rest assured that the NRC is thoroughly convinced that a plant has been constructed in accordance with its application and its commitments before it is issued an operating license.

HAVE WE EFFECTIVELY ADJUSTED
TO NUCLEAR REGULATION ?

by

Robert J. Washabaugh[1]

ABSTRACT

Imposition of Federal Regulations on the Nuclear Industry required significant changes to the traditional management systems utilized in the Construction Industry. This paper discusses the Industry's response to these requirements and provides recommendations for improvements.

INTRODUCTION

The effectiveness of any organization's performance must be measured by its ability to meet specified goals. The goals of the Nuclear Power Industry are to engineer and construct a nuclear facility that conforms to the Regulatory requirements for minimum cost within the specified time frame.

The period from 1969 to 1980 has seen the industry pursue the goal of Regulatory conformance, but it has also seen the manual manhours necessary to construct a Nuclear Power Plant escalate by a factor approaching three, and the time frame from ground breaking to fuel loading increase from an average of 46 months to 130 months. This trend must be arrested and reversed if nuclear power is to be accepted as an economic energy source. This trend must be reversed without taking issue with those requirements that demand the highest achievable quality and safety of design.

The reversal of this trend requires that we appraise and adjust those actions we have taken in response to two of the major factors responsible for the increase in manhours and the resultant plant stretchout. These factors are:

1. The exponential increase in the complexity of plant design resulting from the search for maximum safety.

2. The significant changes to the management controls resulting from the imposition of Appendix B to 10CFR50.

---

[1]Manager Quality Assurance Department, Duquesne Light Company

Both these factors have experienced a continuous evolutionary growth in complexity and have challenged accepted traditional practices and existing management techniques.

COMPLEXITY OF DESIGN

The exponential increase in the complexity of plant design has been reflected in corresponding increases in the engineering details required to construct the plant. The dramatic increase in the Engineering/Design workload, that this additional detailing requires, has resulted in a massive intrusion of the efforts of Engineering/Design into the allotted construction time and has accentuated the traditional* incompatability that normally exists between the efforts of Engineering/Design and those of Construction in power plant development.

The encroachment of the Engineering/Design effort into the allotted construction time causes both groups to be submitted to intensive schedule pressures, disturbs the established construction assembly logic and permits the Engineering/Design effort to become the primary controlling factor in construction achievement.

The accentuation of the traditional incompatability of effort causes value judgements to be made by either Engineering or Construction regarding the adequacy of available information. These value judgements are often found to be in error when system designs are finalized, resulting in latent design changes requiring the expenditure of significant numbers of unscheduled manual manhours to correct.

For many years, large projects have utilized the Design-Construction Management concept. This concept formulated to compact a project's time frame from plant authorization to completion permitted the dovetailing of Engineering activities with those of Construction. The utilization of this management concept in Nuclear Power Plant development has provided results opposite to those for which it was formulated. Its utilization has made it difficult, if not impossible, for Construction to control the manual manhours expended or to establish accurate time frames for work performance.

---

*This incompatability finds its roots in the fact that the finalization of plant design depends on the finalization of individual system designs, whereas the completion of the construction effort is based on the completion of plant areas or buildings. In other words, maximum construction efficiency can only be achieved if the design of all systems in an area or building are complete prior to the start of construction.

If Construction is to regain its continuity of effort and reach acceptable levels of productivity, major Engineering activities must be extracted from the allotted construction time. The extraction of these activities will require those responsible for the management of Engineering to make significant adjustments in their present management systems and techniques. The thrust of these adjustments should be:

1. The establishment of the complete scope of the Engineering and Design effort. This task must give recognition to the depth of detail now required for efficient construction.
2. The establishment of a detailed, realistically manloaded Engineering/Design schedule developed to meet the needs of Construction -- A schedule which peaks the Engineering effort prior to ground breaking.
3. The establishment of Engineering and Design production controls to assure that scheduled milestones are achieved.

The extraction of the Engineering/Design effort from the construction allotted time will place these activities in series. This array will only provide success if action is taken to reduce both the Engineering and Construction time frames. The effort cannot be limited to those associated with Engineering logistical support. The Construction Management must also examine their systems and the actions they have taken in response to the invocation of Regulatory requirements.

CONSTRUCTION MANAGEMENT CONTROLS

The imposition of Appendix B to 10CFR50, in 1970, imposed the requirements for formal management control systems on the Nuclear Construction Industry. The full impact of this requirement was not initially recognized. The fact that the 18 criteria, contained in the document, were labeled "Quality Assurance Criteria," perhaps led to this confusion. In the early stages, many construction people felt that this new formal system was a separate add-on, and the responsibility for its execution could be delegated to the new quality organizations whose existance was mandated by the criteria. With experience it soon became evident that this was not a separate add-on system, but consisted of systematic methods that had to be integrated into the existing management system. To provide some perspective of why difficulties were encountered in this integration effort, I would like to provide you with a definition of the existing construction management system.

<u>The existing management system consisted of a confederation of good construction men, knowledgeable in their field, accumulated for the purpose of executing a specific project. The interrelationships of these men had been established by a long successful tradition.</u>

To further understand the system, we must define what constitutes a "good construction man." The definition applies to all participants, management, supervision and craftsmen.

<u>"A good construction man maintains project productivity by anticipating and solving logistical or construction sequence problems, has the technical or construction experience to provide solutions to erection problems, if encountered, and does all this while achieving the quality desired."</u>

These two definitions clearly indicate that the existing system was informal. In other words, we were attempting to integrate a formal management program, for the achievement of quality, into an informal program consisting of hundreds of autonomous construction men tied together by their well established traditions. This effort was further complicated because the new system, defined by Appendix B, placed limits on the actions of construction, and some cases either diluted or removed entirely their ability to use judgment or make decisions. Since the evolution into this new management system was a reaction to Regulatory requirements and was taking place while construction progressed, many of the methods and techniques developed for this adaptation were reactive rather than anticipative. I would like to discuss two areas where a change in the nature of these methods or techniques from reactive to anticipative would improve project productivity.

Management Controls and the Orientation of Personnel

The orientation of construction personnel to nuclear project controls has been largely dependent upon the oral transmission of project requirements by construction supervision. This traditional technique fails to give assurance that the presentation is timely, comprehensive or standard. The effectiveness of the system is heavily dependent on the audit or inspection findings of various quality verification groups. Detection of failure to implement the controls is followed by various actions to gain the adherence of the delinquent group or individual. This reactive technique increases the number of verifiers required, is sluggish and impacts project productivity. If action is to be taken to change the approach from reactive to anticipative, we must gain a perspective of some of the problems faced by construction management.

Source of Labor and Its Stability.--Since nuclear represents only a minority of active construction projects, the majority of workers come from construction projects utilizing traditional management controls. Those having nuclear experience will be coming from a variety of other nuclear projects, each using a unique management control program. This force, beginning with a few hundred, will grow to thousands. The turnover will be significant due both to the changing disciplinary cycles of the project and its duration.

Program Applicability.--Almost all nuclear projects differentiate between the controls exercised over safety related as opposed to non-safety related work. This distinction, in effect, establishes two management systems -- one formal and the other informal. The possibility exists that an individual, during his project tenure, will float between these systems. Adjoining groups will be operating under varying degrees of control.

Requirement for QA Programs.--The invocation of the requirement that all contractors active on safety related work have an approved Appendix B Quality Assurance Program causes the implementation, in some cases, of dozens of management control programs. The content of these programs and their effectiveness is directly related to the contractor's acumen and his familiarity in operating in the nuclear project environment.

Experienced Nuclear Personnel.--Each of the major management organizations, active in nuclear construction, has developed his own set of management controls. An individual, moving from one organization to another, must become familiar with how his new employer has seen fit to meet the requirements. Since projects operate under different NRC construction permit requirements, and some management firms permit a greater latitude to the Project Construction Manager regarding the details, variations may occur from one project to another within the same firm.

Summary of Construction Management Problems.--Nuclear projects are composed of a large dynamic labor force that must understand and operate within management controls. The number of management controls, utilized by a project, varies from a minimum of two to perhaps dozens. The proliferation of programs makes it difficult to provide the necessary comprehensive orientation to incoming personnel and fails to provide the analytical tools necessary to detect impediments to productivity.

Proposed Actions

The actions required to solve our problems are evident if two simple premises are accepted:

1. Knowledgeable people can perform operations more efficiently and effectively than less knowledgeable people.

2. No operation can reach maximum efficiency if more than one management control system is utilized.

The management controls being utilized consist of construction procedures, procedures written in response to Regulatory requirements, whose primary purpose is to describe the methods used to achieve quality. These procedures should be accumulated on a discipline functional basis. (For example: for concrete, raw material acquisition and care, mixing, transportation, testing, preplacement, placement and postplacement activities.) The procedures should then be analyzed using flow diagram logic techniques to establish if quality requirements are identified and to assure that the production program flows logically without awkward impediments. The procedures should then receive further analysis to assure that they are written in detail to describe the responsibilities of all participants from managers to craftsmen. The purpose of this task is to complete the formalization of the Management Control System through the establishment of an integrated <u>Production Control Program.</u>

When the task of establishing this <u>Production Control Program</u> is complete, a matrix should be developed. This matrix should plot the responsibilities defined against the individuals delegated the responsibility and should be the basis of a formal task orientation supplied to all construction personnel upon their arrival at the site. A task orientation tailored to fit the specific reponsibilities of management, supervision and craftsmen.

Summary

The Nuclear Construction Industry must consider the uniqueness of its position in the Construction Industry. The regulations imposed upon it have negated the traditional informality that still prevails in balance of the Construction Industry. A nuclear project cannot reach maximum efficiency half formal and half informal. The formalization, started by Appendix B, must now be completed. The procedures developed for this effort must be used both for the transmission of quality requirements and as an analytical tool to discover impediments to project productivity. Methods must be developed that permit rapid assimilation of traditional construction people into this unique industry.

Considering the complexity of this formalization effort, the industry should consider the advantages that can be accrued by cooperating in the development of <u>Standard Production Control Programs.</u> Programs, to be used as a base, from which specific project programs should be developed for all discipline functional areas.

Project management should reconsider the practice of using more than one management control program. Many contractors have as much difficulty adjusting to the nuclear environment as the average craftsman. The confusion and redundancy inherent in the practice increases manpower requirements and contributes to a reactive environment.

TRADITIONAL RELATIONSHIPS

When either corporate policy or intervention by Government Regulating groups require reorganization, the most difficult management decisions revolve around the impact that these required changes have on existing traditional organizational relationships. If management fails to logically redistribute impacted responsibilities, awkward production situations are created, redundancy of effort is encouraged and paperwork is increased. The invocation of Appendix B to 10CFR50 impacted a traditional relationshp between Engineering and Construction. To understand this impact, we must review the process through which engineering information evolves.

The source of engineering information is not limited to that generated by the organization responsible for basic design effort. Progressively, procurement contracts are awarded to manufacturers and fabricators, all of which are involved in the plant design. The efforts of all these design organizations are out of chronological phase with one another. An effort to cross check all the details that interrelate during design evolution would create insurmountable schedule problems. Traditionally the discovery and resolutions of the anomalies engendered, by this out-of-phase design effort, was the responsibility of the construction organization, since this was the first time all the information and the hardware were accumulated into one complete package.

Problems

The advent of Criteria III, "Design Control" of Appendix B to 10CFR50 impacted this traditional relationship of design to Construction. No longer could Construction provide prompt solutions to these anomalies. In order to answer the problems created by this regulation, a paper was established. This paper, called by some a "Design Change Request" or "Design Coordination Report" or other titles, is transmitted by Construction to Engineering. The purpose of this paper is to provide evidence that the problem and its resolution has seen the necessary review and evaluation of those responsible for plant design. Thousands of these papers are generated during the plant construction life. The utilization of this system has imposed a significant paperwork burden on a project. Residual problems are created by their existence, since these documents must ultimately be reviewed for their impact on the required "as-built" information.

As indicated, Construction had been traditionally responsible for both the discovery and resolution of these anomalies. The system developed, however, only affected the resolution. The responsibility for discovery of the anomalies still resides with Construction. In many cases, the crews are assembled and equipment is at the location when the anomaly is discovered. The processing of the paper can either delay or abort the operation.

Proposed Action

Since accurate engineering information is one of the keystones of both quality and productivity, Industry should reappraise the approach utilized to conform to Criteria III. Consideration should be given to the establishment of a <u>Production Engineering Group,</u> as an arm of the designer, at each nuclear site. The organization should consist of qualified engineers and technicians of various disciplines. The responsibility of the group would be to accumulate, review, modify and supply corrected information to Construction. This concept releases Construction from its responsibility to detect the anomalies, reduces the burden created by paperwork, simplifies the inspection and construction planning process and provides corrected information for future "as-built" use.

Summary

Since the invocation of Appendix B caused both the introduction of new groups, such as Site Quality Control, and has diluted the responsibilities of existing groups, the Construction Industry should reappraise the functional organizational structure of a nuclear site. The thrust of this effort should be to separate site functions logically and to ascertain if the responsibilities for execution have been delegated in accordance with this logic regardless of traditional practices.

CONCLUSION

The Nuclear Power Industry is being subjected to external attacks on the basis of both technological safety and fiscal viability. The problems related to safety have been subjected to continuous review since the industry's conception. These efforts, to assure a safe technology, are insufficient to guarantee the industry's survival, efforts must be undertaken to reduce the ever escalating capital costs. These efforts must include a close examination of the management systems being used with the purpose of improving the efficiency of plant development.

The purpose of this paper is to describe some of the types of problems faced by the industry, it should not be inferred that any company or project is a victim of all the problems described or that all the problems have been identified. However, problems of the type described in this paper are incumbent, to one degree or another, in the management systems of all companies involved in nuclear power development. The record indicates that no organization has developed the ideal management system "Role Model" to be emulated. A cooperative effort should be undertaken to develop such a "Role Model." The successful approach used by the various Codes and Standards Committees, whose activities have clarified <u>what</u> must be done, must now be emulated in an effort to define <u>how</u> these activities are to be achieved.

All companies involved must recognize that the invocation of strict regulations has evolved a unique industry and that operating as a unique entity within a unique industry incurs disadvantages that far offset any advantages accrued by the maintenance of a proprietary approach.

These efforts should not await the acquisition of the "Holy Grail" of a stabilized design base or standard plant. Time is important. Industry survival is at stake.

SESSION II - CONCRETE MATERIALS: REQUIREMENTS AND PERFORMANCE

SESSION OBJECTIVES/SESSION CHAIRMAN SUMMARY

by

John C. Archer[1], M., ASCE

Objective of Session

This session focused on the industry codes and standards and government regulations controlling materials for concrete for construction of power generating facilities. This specific topic area is central to the concern of the sponsors of this speciality conference since the ever increasing cost and time required to construct power plants demands that the engineering profession, in industry and government, examine the codes, standards and regulations which, collectively, it has imposed to determine what improvements can be made. The expectation of the sponsors was that the quality of construction essential to the protection of the safety of the public and essential to the reliability of these critical facilities could be achieved at less cost and in less time. It was known that significant increases in cost and time of construction have been the direct result of the proliferation of codes, standards and regulations. It was suspected that much of this proliferation had resulted in only increasing cost and time to construct, without commensurate assurance of, or improvement in, quality. Thus the objective of this session was to identify those codes, standards and regulations applicable to concrete materials, to examine - through the experiences of the panel members and the attendees - the effects of these requirements, and to determine how some of the requirements might be modified, consolidated, possibly even eliminated, to reduce cost and schedule without compromising quality.

Session Chairman Summary

The distinct material constituents of concrete; cement, aggregates, admixtures, and water, provided a convenient and logical subdivision of this session on Concrete Materials to permit each panelist to address in some depth the requirements applicable to a specific constituent and to concentrate his discussion on the effects, beneficial or other, of those requirements in the performance of power plant construction work. Special constituents, fly ash, heavy weight aggregates, and grout, were separately addressed, as were the requirements controlling how the constituents are to be combined in concrete design mixes.

The panel members have ably discussed the conclusions that their individual experiences had led them to, as presented in the papers reproduced in this volume. The discussion between the session panel and the attendees consistently reinforced those conclusions, frequently with descriptions of jobsite problems caused by present specification requirements.

---

[1] Chief Civil Engineer, Burns and Roe, Inc., Oradell, N.J.

In general there appeared to be two pervasive problems to which most of the conclusions and recommendations are addressed: 1) That there presently exist at least three distinct codes controlling concrete applicable to power plant construction: ACI 318, ACI 349, and ASME B&PV Code Section III Division 2. In the case of nuclear power plant construction, all three of the codes are applicable, each to different plant features. The problem arises because there are differences and conflicts between these codes and because there are requirements in each code that are not useful in ensuring the integrity of the structure or, in some cases, not even possible of being achieved. 2) The second pervasive problem can probably best be characterized as a carry-over from the past before the advent of formal Quality Assurance/Quality Control programs. Specifications for concrete materials had evolved as knowledge, techniques, test methods, became available to assist the engineer in monitoring and controlling the constituents and their use to increase his level of confidence that the required strength and durability of concrete would be achieved. Prior to formal QA/AC procedures, the engineer's judgement prevailed; requirements and test methods and acceptance criteria were used, or not used, or altered, to suit actual job conditions and, importantly, changing job conditions, as the work progressed.

Consensus Recommendations*

In general, as detailed in the specific conclusions of the session papers in this volume, the solution to these problems has to be the consolidation of the conflicting codes into one code by working through joint code committees to ensure that the new code emphatically does not become one more code but does instead supersede the present codes. That code must recognize that present day major construction projects, especially central station power plants, will not be built without QA/QC procedures. The panel members have described crippling delays when requirements unrelated to strength or durability but only to cosmetic effects were enforced under a formal QA/QC program. Therefore the new code must distinguish between mandatory requirements and other requirements which are not essential to strength or durability.

The other requirements which serve a useful purpose and therefore should be retained as contract requirements should be separately categorized as subject to being waived by the engineer without an extensive program of justification by test. All requirements should be reexamined to evaluate whether they serve any useful purpose at all, and eliminated if they do not. Regarding the presently enshrined concept that there are one (or two) kinds of concrete for nuclear safety related structures and another kind of concrete for other structures, the concession, if any, to be made would be that more frequent testing would statistically provide the desired higher level of confidence that strength and durability were actually achieved. No other difference appeared to be justified. And, regarding frequency of tests, not only should the present conflicts between codes be eliminated but the finally selected frequencies should be realistic i.e. both necessary and reasonable. Some tests, particularly for aggregates, may be found to be entirely unnesessary as a function of time or quantity of the material used but should only be performed when a change of source is proposed.

---

*Based upon the Panel of Speakers/Audience discussion period at the end of the session.

Beyond the pervasive problems and solutons described above, the presentations and discussion endorsed the practice of thorough laboratory testing of constituents and design mixes using materials proposed for the work but noted frequently that present specifications fail to recognize that no laboratory tests can exactly duplicate job conditions and therefore, specifications should formally allow for design mix adjustments based on production work test results, initially and as conditions change, without requiring new laboratory testing of adjusted mixes and the associated costly delay.

Concrete Aggregate Requirements and Performance

by

Ward R. Malisch[1] Member, ASCE

Abstract

Nuclear construction code requirements for aggregate qualification and in-process testing are reviewed. Grading, deleterious substances, potential alkali reactivity, and soundness are discussed; reasons for controlling these properties are given and limitations of the tests are covered. Field experiences with implementation of aggregate testing requirements in construction of nuclear power generating facilities are described.

INTRODUCTION

In his 1976 paper entitled "How Soon is Soon Enough?"[1], Bryant Mather stated that "...if the cement and aggregates are tested by applicable ASTM standard methods and are found to meet the applicable ASTM specifications; if they are stored properly; if those used in the work correspond to those used in selecting the mixture proportions; if the mixture proportions are selected as intended; and if the concrete is batched according to the selected mixture proportions, properly mixed and sampled, then there is no significant probability that the results of properly conducted strength tests will yield a result that will be considered other than satisfactory." Of course, adequate strength isn't the only performance requirement to be met; resistance to freezing and thawing distress and to volume changes sufficient to cause cracking must also be provided. However, the point being made is that if the materials, proportions, and production are right, the concrete will be right. How do we know that the materials are right? In this paper, applicable code requirements for aggregates will be reviewed, the significance of the required tests will be discussed, and some of the difficulties in implementing the code requirements will be pointed out.

---

[1]Director of Educational Services, Concrete Construction Publications, Inc. Addison, IL 60101

CONCRETE AGGREGATE

AGGREGATE REQUIREMENTS

The requirements for aggregates found in section III, division 2 of the ASME Boiler and Pressure Vessel Code [2] can be divided into two general categories:

1. Material acceptance requirements
2. Requirements relating to quality control and frequency of testing.

The backbone of the sections on material acceptance requirements is ASTM C33 "Standard Specifications for Concrete Aggregates." [3] We'll first examine the portions of ASTM C33 concerning selected aggregate properties and discuss the significance of these properties.

Aggregate Grading

Requirements on grading are given for both fine and coarse aggregates, in terms of fixed number limits for maximum and minimum percent passing. For fine aggregate, a maximum permissible amount retained between any two consecutive sieves is also specified and in addition, the aggregate is to be rejected or proportioning adjustments are to be made if the fineness modulus varies by more than 0.20 from the value assumed in selecting proportions for the concrete.

Why are grading limits specified? Primarily because:

- grading can affect the water demand for a given slump such that with a fixed cement content, changes in grading may result in changes in strength
- grading can affect the finishability and pumpability of a concrete mixture.

The effect upon water demand is related to aggregate surface area, with finer aggregates requiring more water to produce a given slump. The well known effect of nominal maximum size aggregate, larger sizes of well graded material requiring less water, is widely used in concrete proportioning methods [4]; the effect results from surface area differences in aggregates with varying nominal maximum sizes.

The finishability and pumpability effects are most usually associated with changes in the amounts of the finer sand particles (minus no. 50) or changes in the amounts of coarser sand particles (plus no. 8) or finer gravel or crushed stone particles (minus 3/8 inch). Inadequate amounts of minus no. 50 and minus no. 100 sand in low-cement-content, non-air-entrained concretes can result in concretes which bleed excessively and are difficult to pump and finish. For concrete where a smooth surface texture is required, the Portland Cement Association [5] recommends fine aggregate with at least 15 percent passing the no. 50 sieve and 3 percent or more passing the no. 100 sieve. In ACI 304.2R-17, Placing Concrete by Pumping Methods [6], the recommendation is 15 to 30 percent passing the no. 50 screen and 5 to 10 percent passing the no. 100 screen for concrete pumped through less than 6-inch diameter lines. For air-entrained and/or higher cement content mixes, the minus 50 and 100 fraction doesn't have to be as high.

With regard to the particles between about 1/8 to 3/8 inch, Tuthill [7] has drawn attention to the harmful effects on workability, pumpability, finishability and response to vibration when the amounts of these sizes approach even the lesser amounts permitted by specifications. The problem is compounded when crushed material is being used. Tuthill suggests that the 20 to 55 percent passing the 3/8 inch sieve required for ASTM size number 67 coarse aggregate should be changed to 0 to 30 percent. He also advocates changing the permissible percent passing the no. 8 sieve for fine aggregate from 80 to 90 percent, again to control excessive amounts of the "pea gravel" sizes.

Aggregate Soundness

Fine and coarse aggregate soundness requirements are specified in ASTM C33 based upon either the sodium or magnesium sulfate soundness test. Soundness is used by Bloem [8] as a general term to describe an aggregate as a whole. The sulfate soundness test [9] was intended to furnish "...information helpful in judging the soundness of aggregates subject to weathering action, particularly when adequate information is not available from service records of the material exposed to actual weathering conditions." However, several investigators have stated that there is little or no theoretical or experimental support for the assumption that the sulfate test simulates exposure to freezing and thawing in concrete or gives a reliable indication of field performance [10, 11, 12, 13]. In fact, after a critical review of literature on methods of identifying aggregates subject to destructive volume changes when frozen in concrete, Larson et al [14] concluded that the sulfate test is too sensitive to test variables and doesn't measure or reflect the susceptibility of aggregates to damage by freezing and thawing. They recommended that "...sodium or magnesium sulfate soundness tests...should be deleted from specifications as soon as suitable replacements are available." Dolar-Mantuani [15], on the other hand, states that "If the [sulfate test] specification limits are used carefully together with sound engineering judgment (emphasis mine), the test is certainly useful because of the relatively short time needed to perform it." Bloem [8] also advised against setting inflexible acceptance limits for the sulfate loss and disregarding judgment in interpreting the results, primarily because of the lack of correlation between the sulfate loss and the performance of aggregates in field concrete.

Deleterious Substances

Dolar-Mantuani [15] classifies harmful substances into four groups according to their composition or physical properties.

1. Materials finer than the no. 200 sieve, including both naturally occurring fines in gravel and sand deposits and fracture fines produced by crushing gravel and rocks. Large amounts of these fines in an aggregate increase the water requirement of concrete due to surface area effects and the surface activity of argillaceous fines. Excessive drying shrinkage and reduced compressive strength of the hardened concrete result when too many fines are present.

2. Lumps or clay in the aggregate which neither break up during processing nor disintegrate easily during the mixing and placing of concrete. If they occur at the surface of hardened concrete, unsightly surface pitting and popouts can be caused by freezing and thawing, wetting and drying, or wear.

3. Friable, soft and lightweight pieces, including porous chert. Friable and soft particles break down easily, contributing to fines in the concrete and may cause surface pitting or scaling. Large quantities of soft particles can cause strength reductions and smaller quantities lower the abrasion resistance of concrete surfaces.

4. Organic impurities such as coal, lignite, plant roots, twigs, and other vegetable and animal materials. Organic impurities consisting principally of tannins retard setting and reduce concrete strength particularly at early ages. Small particles of coal and lignite may cause localized pitting, popouts, and staining of concrete.

Five test methods, ASTM C117, C40, C87, C142 and C123, are referenced in ASTM C33 to control deleterious substances.

Potential Alkali Reactivity

For both fine and coarse aggregates, ASTM C33 states that "...for use in concrete that will be subject to wetting, extended exposure to humid atmosphere, or contact with moist ground [they] shall not contain any materials that are deleteriously reactive with the alkalies in the cement in an amount sufficient to cause excessive expansion of mortar or concrete, except that if such materials are present in injurious amounts, the...aggregate may be used with a cement containing less than 0.6 percent alkalies calculated as sodium oxide or with the addition of a material that has been shown to prevent harmful expansion due to the alkali-aggregate reaction." In appendix XI, several test methods for evaluating potential reactivity of an aggregate are described with referenced ASTM test methods. However, these are qualified by the statement that they don't provide quantitative information on the degree of reactivity to be expected or tolerated in service. Users are cautioned that "...evaluation of potential reactivity of an aggregate should be based upon judgment and on the interpretation of test data and examination of concrete structures containing a combination of fine and coarse aggregates and cements for use in the new work."

Diamond [16] states that "The practical consequences of alkali-silicate attack are that the affected structure or structural members: (1) are placed under expansive stresses, (2) expand, (3) crack, (4) lose strength, and (5) by virtue of cracks reaching the exterior, become exposed to further deterioration from external sources. The problems generated by alkali-aggregate attack range from the purely cosmetic to the complete failure of the structure and the necessity for its replacement." He further states, however, that the expansion and cracking that make up the harmful features of the attack don't always occur, even when the alkali-silicate reaction takes place. Under many circumstances reaction does not lead to distress perhaps because (1) the concrete is sufficiently porous to provide local accommodation for the expansive effect, (2) the concrete is under a sufficiently strong compressive loading so that the expansive stresses are overbalanced, (3) the available water is too limited in amount to produce the required swelling of the reaction product, (4) the character of the reation product produced is such that the gel converts rapidly to a sol at relatively low water content and hence ceases to exert expansive stresses, (5) the reaction product produced is soon converted to a gel of limited swelling ability, or (6) various other factors, not fully understood, come into play. Furthermore, it has been well established that even with considerable reaction taking place, expansion and cracking may not be severe if there is either too much or too little of the reactive material or if the alkali content of the pore fluid is much less or much greater than some amount that produces maximum deterioration.

Test Frequency Requirements

Frequencies for aggregate tests are given in Table CC-5200-1 of the code for Concrete Reactor Vessels and Containments (ACI-359-80) [2] and frequencies for those tests discussed in this paper are reproduced in Table 1.

In Code Requirements for Nuclear Safety Related Concrete Structures (ACI 349-80) [17], section 3.3.4 - Testing Requirements reads as follows:

"3.3.4.1-Tests for full conformance with the appropriate specifications, including tests for potential reactivity, shall be performed prior to usage in construction unless such tests are specifically exempted by the specifications as not being applicable."

"3.3.4.2-A daily inspection control program shall be carried out during concrete production to determine and control consistency in potentially variable characteristics such as water content, gradation, and material finer than No. 200 sieve."

"3.3.4.3-Tests for comformance with ASTM C131, ASTM C289, and ASTM C88 shall be repeated whenever there is reason to suspect a change in the basic geology or mineralogy of the aggregates."

Discussion of Aggregate Testing Requirements

Several key points should be made regarding aggregate acceptance requirements, in-process testing, and frequency of testing.

The first of these has to do with aggregate grading and specifically the fixed number limits for maximum and minimum percent passing. As was indicated in the discussion concerning why grading limits are specified, grading variations may affect strength and workability. However, if a single grading test falls outside the limits specified it is not a certainty that concrete quality will be affected. As Tuthill [18] has aptly stated, "Rarely are the encroachments on designated number limits, which cause so much confrontation, delay and expense, ever sufficient to measurably diminish safety, strength, and serviceability." If, for instance a grading test shows that for ASTM size number 67 coarse aggregate the percent passing the 3/8-inch sieve falls below the specified 20 percent value, workability may actually be improved with no adverse effect upon strength. However, quality control personnel frequently have no options

Table 1 - Minimum Testing Frequencies for Aggregates

Requirements	Test Method	Frequency Initial (1)	Frequency After Field Experience (2)
Gradation	ASTM C136	Each 1000 cu yd concrete	Each 2000 cu yd concrete
Matl. finer than #200 sieve	ASTM C117	Each 1000 cu yd concrete	Each 2000 cu yd concrete
Organic impurities (3)	ASTM C40	Each 1000 cu yd concrete	Each 2000 cu yd concrete
Friable particles (3)	ASTM C142	Monthly during production	Every 6 months
Lightweight particles (3)	ASTM C123	Monthly during production	Every 6 months
Potential reactivity	ASTM C289	Every 6 months	
Soundness	ASTM C88	Every 6 months	

NOTES

(1) Initially during the development of the first 30 tests of production concrete and whenever the criteria for the alternate frequency [Note (2)] cannot be met.

(2) Whenever the average strength, fcr of the latest 30 consecutive compressive strength test results exceeds the specified strength f'_c by an amount expressed as $fcr = f'_c + 1.419 \ (f'_c/8.69)$

(3) Not required for crushed stone quarried from rock strata unless petrographic analysis indicates acceptable but significant quantities of these contaminants.

in this case-the aggregate is outside specification limits and, depending upon procedures for a specific job, work may be held up until retesting and/or corrective action is carried out.

A second point addresses the need for judgment to be applied in interpreting the results of aggregate tests. The validity of the soundness test has been questioned for many years on the basis of its failure to predict field performance of aggregates. Even if we accept the test as a helpful aid in assessing durability of concrete aggregates, however, engineering judgment is a requisite to intelligent application of results to decisions regarding acceptability. Similarly, judgment must be applied in evaluating potential reactivity of aggregates. It's perhaps disquieting to know that in some cases there are no hard, fast answers to a question concerning aggregate quality. But it's equally disturbing to see costly delays that result when judgment must be subordinated to the rigid application of inflexible specification requirments.

A further discussion of potential reactivity tests is in order to make a third point regarding aggregate testing as it affects progress toward job completion. The requirement that ASTM C289, Potential Reactivity of Aggregates be conducted at a fixed interval, whether 3 months, 6 months or yearly, presents unique problems with respect to action to be taken when failing test results are obtained. In my opinion, C289 should be used only as a qualification test, and never as a control test in the sense that results plotting in the deleterious or potentially deleterious zones can be cause for work stoppage. The difficulty in using C289 as an in-process test stems from the test involving reaction conditions (temperature and pressure) different from those at which concrete undergoes alkali-aggregate in the field. According to Diamond [16] it "...is in no sense a predictor of the extent of expansion which a given 'deleterious' aggregate might produce in a given concrete." Properly used, results that indicate deleterious or potentially deleterious aggregates serve as a "red flag" in the qualification process and can be supplemented by petrographic examination, ASTM C295, the mortar-bar test, ASTM C227, and service records where available. When used as an in-process test, the time lag between determining failing results and obtaining supplemental information, from mortar-bar tests in particular, may result in very costly construction delays.

The final point to be discussed is the need to differentiate between deviations from specified values for aggregate properties that will measurably affect the strength or safety of the structure and those that cause primarily cosmetic problems such as popouts and staining. Excessive amounts of clay-type minus 200 material in the aggregate on a job where water requirements in the field are controlled by slump can significantly affect the water-cement ratio as well as the air content. Immediate action to correct the problem would be justified. On the other hand, failure of the aggregate to meet the maximum percentage requirements for coal or lignite is, in nearly all cases, not sufficiently important to justify stopping work because the effects are primarily cosmetic. To personnel not familiar with the significance of individual tests, all test have equal importance. In the absence of guidelines indicating the relative seriousness of deviations from various ASTM C33 requirements, the action taken as a result of some deviations may be too severe.

FIELD EXPERIENCES WITH AGGREGATE REQUIREMENTS

As Willenbrock et al [19] have documented in their analysis of concrete quality assurance practices, differences in approach to aggregate testing were not uncommon at different sites with Construction Permit dates between 1968 and 1978. The detail in which procedures are written, the interpretation of specifications, and the frequency of testing requirements imposed prior to ANSI N45.25 all vary. The field experiences related here occurred during the period from May 1976 to August 1978 and are used to illustrate some of the points made in the previous discussion.

One of the common problems with in-process grading tests was the out-of-specification test result, usually for coarse aggregate, that delayed a scheduled concrete placement. Project procedures required that when a failing test result occurred, two check tests be performed immediately. If either of these two failed, concrete batching stopped until corrective action was taken. This consisted of purging the batching bins, taking a new sample and verifying that the grading was within specifications before work commenced. The time delays could be several hours, and when the deviation from specifications was a minor one, it's certain that a work stoppage was not warranted on the basis of effects upon concrete quality. This problem was partially circumvented by using a procedure that gave an "early warning" when grading was nearing or exceeding specification limits. At the end of concrete production each day, samples were drawn from the coarse aggregate bins for grading tests necessary prior to the next days pour. After splitting the sample, one test portion was placed in the oven for drying and testing the following morning in accordance with ASTM C136. The other half resulting from the last splitting operation was sieved immediately, without drying. The results were not record tests, but within 20 minutes they indicated whether the grading was well within specification limits or if a failing test was likely. If the test on the wet sample was close to or outside specification limits, the bin was purged, new samples were obtained, and the process was repeated. This gave a reasonable degree of assurance that the following days pour wouldn't be held up by aggregate grading problems.

A better solution to the problem of delays caused by grading variations would be either a statistically based grading specification using random sampling and an acceptance criterion based upon variability in grading, or simply the requirement that acceptance be based upon a running average of five consecutive tests.

A second problem that occurred was related to tests for potential reactivity that were carried out at six month intervals in accordance with the frequency still included in the code. The sand used on the project was borderline with regard to ASTM C289; results of acceptance tests fell inside the innocuous region. However, one six month test crossed over into the deleterious region and two check tests confirmed the original results. Non-conformance reports were issued and use of the sand in structural concrete was prohibited. Additional testing was requested to aid in dispositioning the NCR's and to verify, with tests by an independent testing laboratory, that the sand was in a nonconforming condition. However, since ASTM C33 permits the use of reactive aggregates with low-alkali cements, and low-alkali cements were being used on the project, the decision was made to release the aggregate for use. A requirement that petrographic analysis, ASTM C295, be included as an adjunct to the C289 test was instituted so that the effect of changes in C289 results could be better evaluated. The total time lost was about 7 working days. If low-alkali cement had not been in use and qualification of a new sand source was necessary, the delay would have been extremely costly.

A suggested requirement, rather than running ASTM C289 at 6 month intervals, would be to conduct petrographic examinations more frequently and use C289 only when the examination indicates a change in the basic geology or mineralogy of the aggregate.

The nuclear code requirements for aggregates have changed considerably since the code was first introduced. Test frequency requirements are more realistic and efforts have been made to eliminate unnecessary testing. Two areas that still need attention are the development of statistically-based specifications and the treatment of reactivity testing in general.

REFERENCES

1. Mather, Bryant, "How Soon is Soon Enough?" American Concrete Institute Journal, March 1976.

2. ACI-ASME Committee 359, Code for Concrete Reactor Vessels and Containments (ACI 359-80)

3. Standard Specification for Concrete Aggregates, ASTM C-33, Annual Book of ASTM Standards, Part 14.

4. ACI Committee 211, Recommended Practice for Selecting Proportions for Normal and Heavyweight Concrete (ACI 211.1-77)

5. Design and Control of Concrete Mixtures, 12th edition, Portland Cement Association, 1979.

6. ACI Committee 304, Placing Concrete by Pumping Methods (ACI 304.2 R-71).

7. Tuthill, Lewis H. "Better Grading of Concrete Aggregates," Concrete International Design and Construction, Dec. 1980.

8. Bloem, D.L., "Soundness and Deleterious Substances" Significance of Tests and Properties of Concrete and Concrete-Making Materials, ASTM STP 169A, 1966.

9. Standard Test Method for Soundness of Aggregates by Use of Sodium Sulfate or Magnesium Sulfate, ASTM C88, Annual Book of ASTM Standards, Part 14.

10. Mather, Katharine, "Relation of Absorption and Sulfate Test Results of Concrete Sands," Bulletin No. 144, ASTM, 1947.

11. Woolf, D.O., "Relation Between Sodium Sulfate Soundness Tests and Absorption of Sedimentary Rock," Public Roads, December 1927.

12. Cantrill, C. and Campbell, L., "Selected Aggregates for Concrete Pavement Based on Service Records, Proceedings, ASTM, Vol. 39, 1939.

13. Lang, F.C., "Deleterious Substances in Concrete Aggregates," Bulletin, National Sand and Gravel Association, April 1931.

14. Larsen, T., Cady, P., Franzen, M., and Reed, J., "A Critical Review of Literature Treating Methods of Identifying Aggregates Subject to Destructive Volume Change when Frozen in Concrete and a Proposed Program of Research," Research Special Report No. 80, Highway Research Board, 1964.

15. Dolar-Mantuani, L. "Soundness and Deleterious Substances," Significance of Tests and Properties of Concrete and Concrete-Making Materials, ASTM STP 169B, 1978.

16. Diamond, Sidney, "Chemical Reactions Other than Carbonate Reactions," Significance of Tests and Properties of Concrete and Concrete-Making Materials, ASTM STP 169B, 1978.

17. ACI Committee 349, Code Requirements for Nuclear Safety Related Concrete Structures (ACI 349-80).

18. Tuthill, Lewis H., "Quality attainment and common sense in nuclear concrete construction," Concrete International Design and Construction, March, 1979.

19. Willenbrock, Jack H., Thomas, H. Randolph, and Burati, James J., "A Comparative Analysis of Structural Concrete Quality Assurance Practices on Nine Nuclear and Three Fossil Fuel Power Plant Construction Projects," Final Summary Report. Department of Civil Engineering, Construction Management Research Series, Report No. 11, The Pennsylvania State University, 1978.

PORTLAND CEMENTS: REQUIREMENTS VS. PERFORMANCE

By

William F. Perenchio[1]

ABSTRACT

Portland cements generally are manufactured with great emphasis on quality control. Their composition and physical properties are governed by consensus standards such as ASTM C150 for the five general categories of portland cement. These are made from clinker burned in a rotary kiln which is then ground with a small percentage of calcium sulfate in the form of gypsum or anhydrite. They may or may not be interground with an air-entraining agent. Blended hydraulic cements as described in ASTM C595 also are available. They consist of blends of portland cement and either blast-furnace slag or a pozzolan. Either basic type may contain an air-entraining agent. Either type also may be modified to provide moderate heat of hydration. The pozzolanic blend may be modified further to provide low heat of hydration.

Slump loss occurs in all concretes. The rate of slump loss may be greatly increased in some cases when a chemical admixture is used. A state-of-the-art summary on this subject is presented plus a report on a new investigation. Chemical and physical tests of portland cement pastes, mortars, and concretes are described. Slump loss is ascribed to the forms of calcium sulfate in the cement and the interaction of the sulfates with the aluminate phases. These interactions are shown to be affected by the introduction of various types of chemical admixtures to a portland cement-water system.

SPECIFICATIONS FOR CEMENTS

ASTM C150 includes detailed requirements for chemical composition and physical properties for portland cements which will produce the general characteristics shown in Table 1. Typical situations and benefits for use of Types II through V are shown also. If the special characteristics of these four types are not needed for a particular job, Type I cement is normally specified.

ASTM C595 includes similar requirements for three kinds of blended hydraulic cements. These are called Portland Blast-Furnace

---

[1] Consultant, Wiss, Janney, Elstner and Associates, Inc., Northbrook, Illinois

TABLE 1

STANDARD PORTLAND CEMENTS

ASTM C150

Type I: GENERAL PURPOSE

Type II: MODERATE SULFATE RESISTANT

150 - 1000 ppm sulfate (SO_4) for seawater exposure

Moderate heat of hydration

Type III: HIGH EARLY STRENGTH

Early form removal

Winter concreting

Type IV: LOW HEAT

Mass concrete

Type V: SULFATE RESISTANT

1000 - 2000 ppm sulfate (SO_4)

Structures exposed to soils and waters of high sulfate concentration

Slag Cement, Portland-Pozzolan Cement and Slag Cement. To the basic Type designations of IS, IP and S, respectively, several suffixes are added to denote various modifications in properties. The suffixes and the special properties they represent are as follows:

> A - air-entraining
> MS - moderate sulfate resistance
> MH - moderate heat of hydration

The latter two are not used with Type S cement. A Type P also is available, which develops strength more slowly than Type IP, primarily due to a higher pozzolan content. This cement may also be available with any of the above three suffixes and one more -LH, denoting low heat of hydration. The pozzolan used in Type P or Type IP can be any material which has pozzolanic properties, that is, which will react with calcium hydroxide at ordinary temperatures to form cementitious compounds. However, it typically consists of a high quality fly ash locally available to the cement manufacturing plant.

A great deal more could be said here about the various types of cements and the possible modifications of each; however, that is not the primary purpose of this paper. This paper is concerned mainly with the problem of rapid loss in concrete workability, or slump, which can occur due to the characteristics of a cement or the interaction between the cement and various types of chemical admixtures.

CAUSES OF ABNORMAL SLUMP LOSS

The discussion which follows is based largely on a paper by Meyer and Perenchio (1) which was presented at a symposium on slump loss sponsored by the American Concrete Institute, held in New Orleans during October, 1977.

All concretes which eventually develop compressive strength lose slump as a normal consequence of the hydration reactions that are continuous, albeit varying in rate, from the time the cement and water first come into contact. This fact is well known to anyone who has worked with the material and, barring unusual circumstances, no serious problems develop due to this phenomenon. However, during hot weather concreting or because of peculiarities in the cement or interactions between the cement and some chemical admixtures, the rate of slump loss can be accelerated sufficiently to cause difficulties ranging from poor consolidation of the in-place concrete through rejection of a load or loads to complete shut-down of a job. In such cases, mere inconvenience can rapidly develop into economic chaos. Reports of instances such as these prompted the study mentioned earlier.

<u>Concrete Tests</u>

For the laboratory portion of this study, four portland cements and four water-reducing admixtures were selected, all of which had been involved in reported problems of excessive slump loss. Two of the cements were low and two were high in alkali content. Each alkali-level category contained a Type I and a Type II cement.

The following admixtures were chosen for the study:

WR_1 - Modified lignin sulfonate (ASTM C494, Type A)

WR_2 - By-product industrial sugar (ASTM C494, Type D)

WR_3 - Salt of polyhydroxy-carboxylic acid (ASTM C494, Type D)

SWR - Melamine-formaldehyde condensate (high-range water-reducer or superplasticizer)

The modifying agent in admixture WR_1 was triethanolamine (TEA). The SWR was included in this study for purposes of comparison because of its acknowledged high rate of slump loss. All admixtures were added with the mixing water at the dosage rates recommended by the manufacturers.

The concrete mixtures were prepared in accordance with ASTM C192, using a 1/2 cu ft (0.014 m^3) Lancaster tub mixer. Air entrainment was not employed (concrete air content was 1 percent to 2 percent). Slump loss was determined by measuring the slump at 20 minute intervals until it had declined from the initial level of about 5 in. (127 mm) to less than 1 in. (25.4 mm). Concretes were remixed for 1 minute immediately prior to each slump test. The water-cement ratio was fixed at levels of 0.4 and 0.5, by weight. The resulting cement contents were in the range of 570 to 700 lb per cu yd (339 to 416 kg/m^3). After all control concretes had been prepared and tested, no significant effect of water-cement ratio was observed; therefore, all subsequent concretes were prepared at the 0.4 ratio.

Generally, all of the concretes which contained a chemical admixture lost slump at a higher rate than did the control concretes without admixtures. Slump loss was most severe with the SWR with all of the cements except one. This was the high-alkali Type II cement, which lost slump most rapidly when used in combination with the by-product industrial sugar admixture WR_2.

A typical set of data for the other three cements is shown in Fig. 1. Notice that all of the admixtures increased the rate of slump loss over that of the control. The SWR caused the most severe increase. The reason for these increases in slump loss appear to be related to the ability of one of the cement constituents, tricalcium aluminate (C_3A*), to remove organic materials, such as these admixtures, from solution. Removal of the SWR may be more rapid. One

---

*Standard nomenclature used by cement chemists: C = CaO, A = Al_2O_3, S = SiO_2, F = Fe_2O_3, etc.

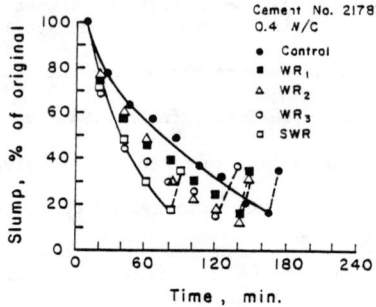

Figure 1. Slump versus time after start of mixing

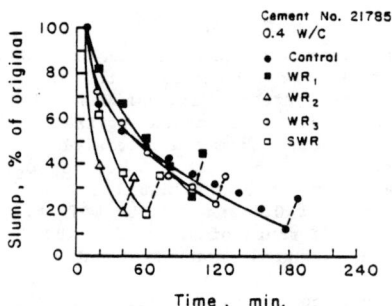

Figure 2. Slump versus time after start of mixing

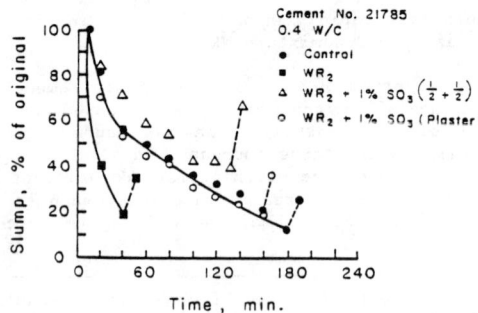

Figure 3. Slump versus time after start of mixing

study (2) showed that nearly 90 percent of a similar material could be removed in one minute, followed by partial release and a slower, long-time removal rate. Concretes containing a SWR contain much less total mix water than a concrete made without an admixture at equal slump; therefore, if most of the admixture is removed from solution, the effective dosage rate is reduced accordingly and the workability would be expected to decrease. The same mechanism may be acting in the cases of the other types of admixtures used in this study but, since they do not reduce the water requirement as much, loss of these materials from solution would not be expected to cause such severe workability reductions as that which occurs with the SWR. Another possible contributing factor is the hydration reactions which are occurring during this period, which chemically bind a portion of the mix water. This effect also would be expected to be more pronounced in a water-reduced concrete because, assuming equal reaction rates, a greater percentage of the available water would be tied up.

The same type of data for the high-alkali Type II cement is shown in Fig. 2. (The dotted line shown at the end of each curve represents the final slump following the addition of 2 gal per cu yd retempering water. No attempt was made to restore the mixtures to their original slumps.)

With this cement, admixture WR_2 caused more severe slump loss than the SWR, in spite of the fact that the other admixtures caused practically no greater slump loss than that shown by the control concrete. Obviously a different mechanism from those described above must be operating.

Fig. 3 shows the concrete slump-loss curves obtained with the same cement plus admixture WR_2 and the addition of 1 percent SO_3, either in the form of a 50/50 mixture of gypsum (dihydrate) and plaster of paris (hemihydrate), or entirely of plaster of paris. A substantial improvement in slump retention was realized in both cases. These data illustrate how a cement can vary in its response to different admixtures. This sensitivity is related to the form and amount of sulfate present.

Discussion of Results

As stated in the previous paper (1),

"All admixtures that affect the properties of the cement paste can be considered a means of manipulating the relative rates of hydration of the anhydrous phases - primarily tricalcium aluminate (C_3A) and tricalcium silicate (C_3S)."

This is a very important statement as regards abnormal slump loss. Cements which act normally have a proper balance between the sulfate available in solution and the amount of C_3A; insufficient balancing of the latter leads to flash setting, or very rapid hydration. This is not to be confused with false set. The difference between the two is described below.

False set is characterized by a rapid loss in workability, but if the concrete is mixed for an extended period or is remixed after the false set has occurred, substantially all of the original workability is restored. It is due to the presence of a small percentage of plaster ($CaSO_4 \cdot 1/2\ H_2O$) or soluble anhydrite in the cement. These usually result from sustained high temperatures in the cement grinding mill or storage of the cement at a somewhat lower temperature for a long period, in the cement silo. The plaster is formed from gypsum ($CaSO_4 \cdot 2\ H_2O$) which loses some of its water of crystallization when exposed to high temperatures. The cement constituents (primarily C_3A) rapidly combine with this free water, making it unavailable for reconverting the plaster to its original form, gypsum. The calcium sulfate remains in plaster form until water is added to the cement during the batching of concrete. After a few minutes the plaster is converted back to gypsum because cement is an efficient accelerator for the setting of plaster; however, this conversion produces gypsum in the form of needle-like crystals. These crystals are capable of causing a distinct stiffening of the cement paste, but the structure is very weak and easily destroyed by continued or delayed mixing.

Flash set is an entirely different phenomenon. It occurs when insufficient dissolved sulfate is present in the concrete mix water, allowing the C_3A to hydrate rapidly. The structure set up by the C_3A is much stronger and produces more solids, chemically binding much more water than does the plaster-to-gypsum reaction. This structure cannot be broken up by extended mixing, and is accompanied by the evolution of considerable heat. Although this rarely occurs with modern cements, it has happened. Contributing factors are fresh clinker, which has not been exposed to appreciable weathering, and the use of anhydrite ($CaSO_4$) as the source of SO_3. At normal temperatures, the solubility of anhydrite is lower than that of gypsum or plaster and because anhydrite is a much harder material than gypsum, it does not become as well distributed throughout the cement during grinding. Therefore, SO_3 is not supplied to the mix water rapidly enough to control the C_3A reaction properly.

The data shown in Fig. 3 indicate that flash set was responsible for the rapid slump loss induced in the high alkali Type II cement by admixture WR_2. The addition of 1 percent SO_3, either in the form of straight plaster or a 50/50 mixture of plaster and gypsum, brought the rate of slump loss of mixtures containing WR_2 back to the normal rate of the control concrete without any admixtures. Numerous papers have shown that such admixtures cause an acceleration in the rate of sulfate combined in the first few minutes.

Adding water-reducing or water-reducing, retarding admixtures late has a much different effect (3). If the cement and water has been in contact for as little as one minute before the admixture is introduced, the sulfate is able to complete most of its controlling function and less of the admixture is removed from solution. This can not only <u>reduce</u> the rate of slump loss compared to a control concrete, but can greatly extend the time at which the concrete reaches initial and final set compared to an immediate addition of the same amount of admixture.

Although the chemical compositions of the two cements represented in Figs. 1 and 2 were similar, their reactions to the addition of admixture WR_2 were markedly different. This is attributed to the higher alkali content of the cement of Fig. 2. Alkali is capable of accelerating the hydration of C_3A and the other major aluminate compound in portland cement, tetracalcium aluminoferrite (C_4AF), long considered to be a relatively dormant material. It is also possible that the high alkali cement contained anhydrite, which would supply SO_3 to solution at a slower rate than either plaster or gypsum.

This rapid hydration of C_3A after depletion of the available sulfate is shown in Fig. 4 (4). This figure shows that as long as gypsum (SO_3) is present, the C_3A has a slow reaction rate; however, when the gypsum is depleted, the remaining C_3A reacts quickly. Increasing the temperature increases the rate of these reactions, but some admixtures are capable of the same thing. Sugars such as that contained in admixture WR_2 are very effective in this regard.

As discussed previously, false set is caused by too much soluble sulfate rather than too little as in the case of flash set. Therefore, if sugar is capable of increasing the rate of combination of sulfate and C_3A, it should be capable of reducing or eliminating the effects of false set. Fig. 5 (based on data from Reference 3) shows that this is true. The top curve, representing the original cement, was completely cured of its false setting tendency by the addition of 0.05 percent of sucrose, ordinary table sugar. However, after this same cement was exposed to laboratory air at 50 percent R.H. for 24 hours, the false set became more severe and was only partially overcome by the addition of sugar. This illustrates that the surfaces of the C_3A had been rendered relatively inactive through exposure to atmospheric water vapor and carbon dioxide and would no longer react with sufficient sulfate to eliminate the false setting entirely.

A word of caution should be added here. Although the 0.05 percent dosage of sucrose appears small, it can have strong effects on the hydration rate of the cement compounds. This effect will vary with different cements; however, an unpublished study done by the author years ago showed that 0.20 to 0.40 percent sucrose was sufficient to essentially stop hydration of four cements with widely varying compositions. Concretes made with them had strengths of as little as 20 psi after 7 days. These concretes eventually developed almost normal strength, but most construction would be seriously delayed by such extreme retardation. Sugar is a good accelerator for the reaction between sulfate and C_3A, but once the sulfate is removed from solution, the remaining C_3A and C_3S are strongly retarded by the sugar.

This study also included the effect of calcium chloride and mixtures of sugar and calcium chloride on the strength development of concretes made with the four cements. The results showed that if calcium chloride was added at a dosage rate of at least ten times that of the sugar, essentially normal strength was developed. Fig. 6 shows the temperature rise of cement pastes prepared with no admixture, with 0.05 percent sucrose and with 2 percent calcium chloride.

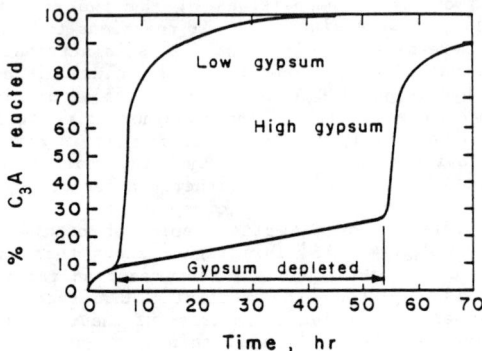

Figure 4. Effect of gypsum content on rate of C_3A hydration

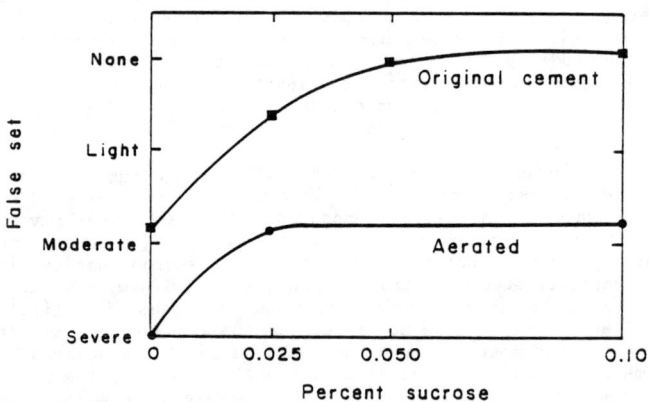

Figure 5. Effect of sucrose addition on false set

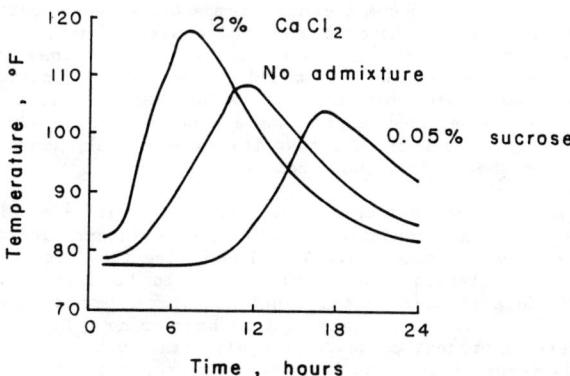

Figure 6. Effect of sucrose and calcium chloride on temperature curve during cement hydration

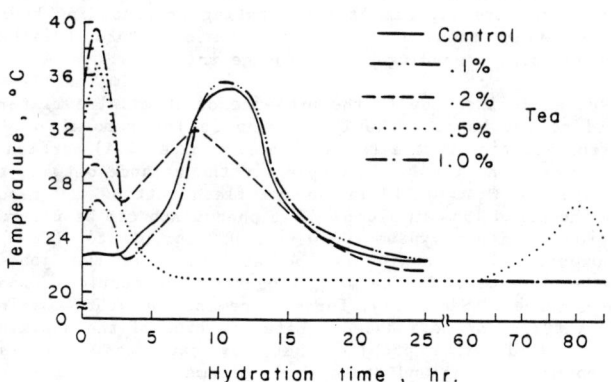

Figure 7. Effect of TEA concentration on temperature curve during cement hydration

The pastes were stored in a vacuum thermos. Temperatures were continuously recorded by means of thermocouples and a strip chart. Note that the sucrose reduces the maximum temperature attained and lengthens the time at which it is reached. The calcium chloride increases the maximum temperature and shortens the time at which it is reached. If we had had another paste sample containing both admixtures at these dosages, we would see that its curve closely approximates the curve for the paste without admixtures.

This approach, adding an accelerating agent to a water-reducing retarder, is used in the formulation of most Type A (non-retarding) water-reducing chemical admixtures. Triethanolamine (TEA) is sometimes used as an accelerator where chloride would be unsuitable. Fig. 7 (5) shows data similar to that in Fig. 6, for cement pastes made with and without various amounts of triethanolamine. The start of the normal setting process coincides roughly with the beginning of the second rapid temperature rise following the drop after the initial rise. A shift in the second temperature maximum, as shown in the dotted line for 0.5 percent TEA, illustrates that strength development can actually be retarded by the same material that acts as an accelerator at lower dosage rates. This peak represents heat given off during the hydration of C_3S, which is the major contributor to compressive strength development in concrete. This illustrates a further complication which can be introduced with the use of an admixture.

CONCLUSIONS

The effects of various chemical admixtures on the rate of heat evolution due to cement hydration and on the rate of slump loss are varied. Admixtures are capable of accelerating or retarding hydration and consequently strength development. One admixture, triethanolamine, can do both, depending on the dosage rate.

Increased slump loss due to the presence of chemical admixtures is attributed to two basic mechanisms. One is the removal of the admixtures from solution by the tricalcium aluminate (C_3A) portion of the cement. The second involves an upset in the balance between the soluble sulfate and C_3A. Mild or severe flash setting may result which causes the rapid loss in slump. This phenomenon can be reversed by the addition of either gypsum ($CaSO_4 \cdot 2 H_2O$) or plaster ($CaSO_4 \cdot 1/2 H_2O$); however, this should not be attempted at the jobsite because of the potential for causing long-term, deleterious expansions in the concrete. Similarly, for some cement-admixture combinations, false setting can be induced. Late addition of the admixture can sometimes avoid these problems but may cause a substantial increase in normal setting and strength development.

The material presented in this paper is not intended to cause alarm and an inordinate aversion to the use of admixtures. The intent is to explain some of the phenomena which have been observed. The most sensible approach to the use of admixtures is that recommended by ASTM C494, "Standard Specification for Chemical Admixtures for Concrete", which is, "whenever practicable, tests be made using the cement, pozzolan, aggregates, air-entraining admixture, and the mixture proportions and batching sequence proposed for the specific work ... because the specific effects produced by chemical admixtures may vary with the properties and proportions of the other ingredients of the concrete."

More detailed information on the early reactions of portland cement and water may be found in References 1, 4 and 5.

REFERENCES

1. Meyer, L. M., and Perenchio, W. F., "Theory of Concrete Slump Loss as Related to the Use of Chemical Admixtures," <u>Concrete International</u>, Vol. 1, No. 1, January, 1979, American Concrete Institute. Reprinted as Research and Development Bulletin RD069.01T, Portland Cement Association, 1980.

2. Personal communication from Dr. D. L. Kantro, formerly Principal Research Chemist, Portland Cement Association. Currently Director of Research, Master Builders Division, Martin Marietta Corporation.

3. Personal communication from N. R. Greening, formerly Director of Basic Research, Portland Cement Association. Currently Director of Research, Cement Division, Ideal Basic Industries.

4. Seligmann, P., and Greening, N., "Studies of Early Hydration Reaction of Portland Cement by X-Ray Diffraction," Highway Research Record No. 62, 1964, pp 80-105. Reprinted as Research Department Bulletin RX185, Portland Cement Association.

5. Lieber, W., and Richartz, W., "Effect of Triethanolamine, Sugar and Boric Acid on the Setting and Hardening of Cement," Zement Kalk-Gips, September, 1972, pp 403-409.

FLY ASH, ADMIXTURES, AND WATER QUALITY
by
Rupert E. Bullock, M. ASCE[1]

ABSTRACT

This paper discusses the sometimes conflicting requirements of standard specifications for fly ash, other admixtures, and mixing water for portland cement concrete. Consideration is given to the effects of those requirements on quality, quality control, and cost.

INTRODUCTION

This paper reviews standard specification requirements for fly ash, other admixtures, and mixing water for portland cement concrete and discusses the effects those requirements tend to produce on construction quality, costs, and schedules. Specific comparisons are made of the requirements of ANSI N45.2.5-1978; ASME Section III, Division 2 (ACI 359); and requirements used on Tennessee Valley Authority (TVA) projects.

FLY ASH

Qualification Requirements

Both ANSI N45.2.5 and ASME Section III, Division 2, require that fly ash meet the requirements of ASTM C618. Table 1 lists the requirements of ASTM C618 for admixture class F and comparable requirements used on TVA projects where they differ.

The ASTM specification was changed in 1978, generally a relaxation of some requirements, and TVA specifications were changed to conform except where it was judged that conformance would adversely affect quality control or excessively affect cost. The first item on which requirements differ is loss on ignition (LOI). This is presumed to be primarily carbon. Carbon absorbs air entraining agents (AEA). Some test results indicate high LOI results in extreme loss of entrained air with extended manipulation or mixing. Most concrete at major TVA projects is mixed at central mix plants with essentially constant mixing, but some is transit mixed. Adequate quality control can be achieved with 12 percent LOI (4), but we believe the 6-percent limit is advisable to minimize or reduce quality control

---

[1] Principal Civil Engineer, Civil Engineering and Design Branch, Tennessee Valley Authority, Knoxville, TN

problems. TVA collects fly ash for concrete admixture purposes only when a unit is running at constant load, and we have had no difficulty in meeting the 6-percent limit.

Table 1. Fly Ash Qualification Requirements

Chemical		ASTM C618F	TVA
$S_1O_2 + Al_2O + Fe_2O_3$	Min	70%	
SO_3	Max	5	
Moisture Content	Max	3	
LOI	Max	12	6
Optional Chemical			
MgO	Max	5	
Available Alkalies	Max	1.5	
Physical			
Fineness-Retained No. 325	Max	34	
Pozzolanic Activity Index			
Cement 28 days	Min	75	60
Lime 7 days	Min	800 psi	
Water	Max	105%	
Soundness-Autoclave	+	0.8	
Uniformity-Indv to 10 Tests			Adjust Mixes
Specific Gravity	±	5	For ±10%
Fineness	±	5	Same
(Percentage Points)			

The next item of disagreement is the pozzolanic activity index with portland cement. Note that the specification contains two measures of pozzolanic activity index and that these two measures do not necessarily correlate. Pozzolanic activity index is an indication of the strength producing properties of the fly ash. If concrete mixes were to be proportioned with the restriction that only equal volumes or weights of fly ash be substituted for some proportion of the weight of cement, then the minimim pozzolanic activity index stated by ASTM may be necessary. However, where fly ash is treated as an admixture and the proportioned amount increased while decreasing fine aggregate, then any pozzolanic activity index which provides a successful economic mix should be acceptable. Actually most TVA fly ash easily meets the 75 percent pozzolanic activity index requirement of ASTM C618. We do at times want to use a source which does not. A mix evaluating procedure should indicate the acceptability of the fly ash and its proportions. Such a method was presented by F. R. McMillan and T. C. Powers in 1934 in "A Method of Evaluating Admixtures" (5). Their method involved plotting constant strength contours in a field defined by cement content on one axis and fly ash content on the other. Figure 1 shows data from a recent TVA construction project. It can be observed that a large number of mixes would be required to make such a plot accurately. However, extreme accuracy is not required. The cost of cement plus fly ash per cubic yard is also plotted. The economy of this fly ash used with these materials is apparent. The most economic mixes occur where a tangent to a strength contour is parallel to the cost lines. Extreme accuracy is not required because there is little change in cost over a wide range in fly ash content. For a specified strength of 3000 psi at 28 days, one could use between 130 and 200 pounds per cubic yard of fly ash with between 320 and 300 pounds of cement with little change in cost. If 160 pounds of fly ash was used, then it could be said that 160 pounds of fly ash replaced 100 pounds of cement and some fine aggregate. McMillan and Powers made their tests before the advent of entrained air and concluded that strength was a criterion for durability regardless of the admixture used. Later research has shown that strength and air are criteria. I know of no inservice durability problems attributable to fly ash in approximately 25 years of TVA use.

The only other items on which requirements differ are the uniformity requirements. Even though this is listed as a qualification requirement, it can only be evaluated by inprocess tests. A 5-percent change in fly ash specific gravity in a mix containing a high proportion of fly ash would change the yield approximately 1/4 of 1 percent. Of course a 1-percent change in air content will change the yield 1 percent. We have some test data that indicate a 5 percentage point change in fineness will change a 28-day mortar strength by approximately the same change as would result from a 1-percent change in air content. If a TVA construction project receives fly ash from a single source, the specific gravity will not change more than 5 percent; but the fineness will occasionally change more than 5 percent. For such a change the ASTM C618 specification would result in a nonconformance report for a concrete change of little significance. No other concrete component has such uniformity requirements. ASTM C150 has no uniformity requirements on portland cement.

If a construction project is obtaining fly ash from a source which has a forced outage and the uniformity requirements prevent a change to another source, the impact on construction would be appreciable. The least cost option would be to discontinue the use of fly ash and use concrete costing approximately $2 per cubic yard more. For larger members, this would also increase hydration temperatures and generally increase cracking; and for some materials, higher specified strengths could be difficult to achieve. Consequently, our specifications have specific provisions for mix adjustments for changes in specific gravity and fineness.

In-Process Requirements

In-process test requirements for fly ash are listed in table 2.

Table 2. Fly Ash In-Process Tests

	ASTM C311	ASME	TVA
$SiO_2 + Al_2O_3 + Fe_2O_3$	2000 tons	1000 tons	1000 tons
SO_3	2000	1000	1000
Moisture	400	--	200
LOI	400	200	200
MgO			1000
Available Alkalies		1000	1000
Fineness	400	200	3 trucks 75
Pozz Index			
Cement	2000	--	1000
Lime	400	--	1000
Water Requirement	2000	--	1000
Soundness	400	1000	1000
Specific Gravity	400	1000	200
Reactivity-Cement Alkalies		1000	

ANSI N45.2.5 adopts the test frequencies in ASTM C311 by reference. Test frequencies for optional requirements of ASTM C618 have been left blank. My interpretation would be that if they were required the frequency would be at 2000-ton intervals. ASME Section III, Division 2, tabulates test frequencies. The lines indicate qualification requirements which are not subject to in-process tests. I understand that the working group judged these tests as inappropriate for quality control purposes. The test for reactivity with cement alkalies, which was optional for qualification, is a required in-process test at a 1000-ton frequency. This test should be of interest only if an alkali-aggregate reaction is a potential problem. We agree that LOI and fineness are important quality control tests. The moisture test adds little to the LOI test, and the specific gravity needs to be monitored. A footnote in ASTM C618 says that soundness test results are significant in block or shotcrete mixes. I haven't seen any research on application of the test.

OTHER ADMIXTURES

Other admixtures considered are air entraining admixtures and those termed chemical admixtures.

Qualification Requirements

Table 3 compares requirements.

Table 3. Other Admixtures

Qualification

ANSI N45.2.5: ASTM C260 or C494.
ASME Section III, Division 2: Same with not more than 1% Cl ion and not contribute more than 5 ppm to total concrete constituents.
TVA: Same with not more than 1% Cl ion and for C494 agents a water reduction of at least 3% in a specific mix containing fly ash.

In-Process

ANSI: Each shipment for chemical composition, pH, and specific gravity.
ASME: Each shipment or lot for infrared spectrophotometry, pH, and solids content.
TVA: Same as ASME plus Cl ion and specific gravity.

ANSI references the standard ASTM specifications without modification, ASME adds the restriction that the admixture not contain more than 1 percent chloride ion by weight and that the admixtures not contribute more than 5 ppm chloride ion to the total concrete constituents. The significance of the 5 ppm is not known. In a later ASME section the chloride content of the cement paste is limited to 400 ppm. I assume chloride is intended to mean chloride ion rather than compound. Research indicates that corrosion is affected by the soluble chloride ion content of the cement paste. A proposed revision of ACI 318 has a limit of 600 ppm by weight of cement in prestressed members. TVA also limits to 1 percent chloride ion in admixture but has applied the limit on total chloride ion only on grout or concrete in contact with prestressed steel. We also add a provision in our specifications requiring a 3 percent water reduction in a specific mix containing fly ash. Some commonly sold water reducers which comply with C494 for non-fly ash mixes fail this requirement.

Although not included in the table, we have recently made provision for use of high range water reducers or superplasticizers. A requirement is added to ASTM C494 that the freeze-thaw test requirements be met with specimens made with at least 7 inches of slump since we believe it may be more difficult to maintain the entrained air at this slump.

In-Process Requirements

Table 3 also lists in-process test requirements. I presume the ANSI requirement for chemical composition can be interpreted to mean determine the infrared spectrum. Note that ASME Section III, Division 2, does not require a chloride ion determination. The only in-process test difficulty that I am aware of has been with an air entraining agent which contained more solids than is permitted by the optional uniformity requirements of ASTM C260. When its use was initiated air content adjustments were necessary.

FLY ASH/ADMIXTURES/WATER QUALITY

WATER

Qualification Requirements

Almost any reasonably clean natural fresh water or drinkable water is suitable for concrete mix water, and thus no specific limits have been fixed for deleterious material. An acceptance standard of a maximum of 10-percent reduction in compressive strength has been used as a stand-alone requirement for years. Nuclear specifiers have strived for more. Table 4 lists some requirements.

Table 4. Water

	ANSI	ASME	TVA
Qualification			
Solids		max 2000 ppm	
Chloride ion		*	
Time of set	x		
Initial		±10 min	
Final		±1 hr	
Strength	x	-10%	-10%
Soundness	x		
In-Process			
Solids	monthly	6 months	
Chlorides	monthly	monthly	2 months
Time of set	monthly	6 months	
Strength	monthly	6 months	As required
Soundness			

*Chloride content of cement paste not to exceed 400 ppm.

ANSI N45.2.5 requires that the construction specification list the checked items. Soundness again refers to the autoclave expansion test for cement. I do not know of any research on how this applies to water. ASME Section III, Division 2, provides the generally accepted value for strength, limits on time of set, and a maximum dissolved solids value. This value has been found to be borderline when dissolved carbonates were present, and it is believed that it was originally advanced as a guide to the necessity to make other tests rather than as a specification limit. However, Steinour (7) has suggested that other tests should be made when dissolved carbonates exceed 1000 ppm. TVA uses CRD-C400, a Corps of Engineers' specification, as a water requirement. This specification suggests chemical tests but only requires the stated strength level. The application of the other requirements would only increase testing costs.

In-Process Requirements

ANSI requires monthly tests for dissolved solids and chlorides even though qualification tests were not required. It does not require in-process tests for soundness. ASME has a 6-month frequency on some tests, presumably indicating a low probability of specification violation. TVA has required chemical tests at 2-month intervals or anytime a change is suspected in the belief that changes in water quality would be preceded by an observable physical phenomena easily observed by quality control personnel. Strength tests are to be made only when chemical tests indicate that they are advisable. Generally, test personnel have chosen to make the strength tests at 2-month intervals. I do not know of any failures. Application of the ANSI requirements would have no effect on quality but would increase testing costs.

RECOMMENDATIONS

All the standard specifications require sufficient testing for these materials except that ANSI N45.2.5 may be considered insufficient where the materials are to be used with prestressing steels. Requirements for fly ash and water are more than sufficient. The result is primarily an increase in costs for testing. Only rarely will the requirements impact other construction costs.

Change can best be obtained through work with existing committees producing voluntary consensus standards. The existing conflicts between standards illustrates the difficulties that would be produced by additional standards.

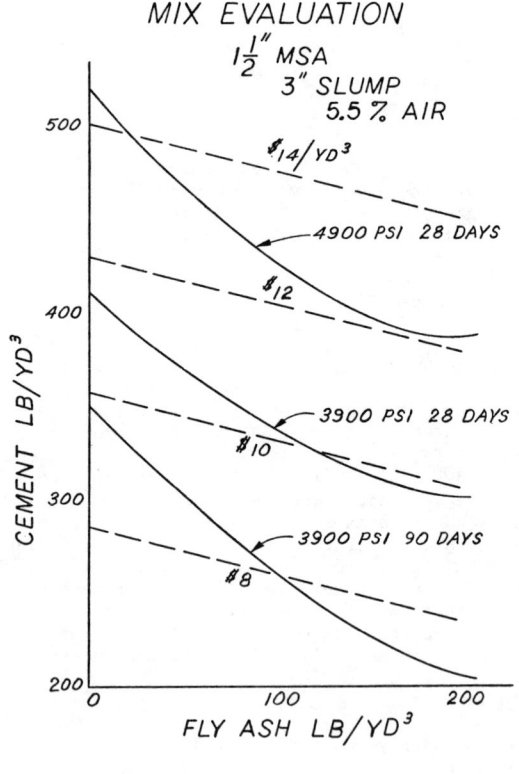

FIGURE 1

REFERENCES

1. ASME Boiler and Pressure Vessel Code Section III, Division 2, Code for Concrete Reactor Vessels and Containments, The American Society of Mechanical Engineers, 345 East 47th Street, New York, New York 10017.

2. ANSI N45.2.5 - 1978, Supplementary Quality Assurance Requirements for Installation, Inspection, and Testing of Structural Concrete, Structural Steel, Soils, and Foundations During the Construction Phase of Nuclear Power Plants, The American Society of Mechanical Engineers.

3. American Society for Testing and Materials, 1916 Race Street, Philadelphia, Pa. 19103; Standard Specification:
   ASTM C260 Air-Entraining Admixtures for Concrete
   ASTM C494 Chemical Admixtures for Concrete
   ASTM C618 Fly Ash and Raw or Calcined Natural Pozzolan for use as a Mineral Admixture in Portland Cement Concrete
   Standard Methods of:
   ASTM C311 Sampling and Testing Fly Ash or Natural Pozzolans for use as a Mineral Admixture in Portland Cement Concrete

4. T. G. Clendenning and N. D. Durie, "Properties and Use of Fly Ash from a Steam Plant Operating under Variable Load," ASTM Proceedings V62, 1962, pp 1019-1037, American Society for Testing and Materials.

5. F. R. McMillan and T. C. Power, "A Method of Evaluating Admixtures," Journal of the American Concrete Institute," March-April 1934, pp 325-344, American Concrete Institute, P. O. Box 19150, Detroit, Michigan 48219.

6. ACI Standard Building Code Requirements for Reinforced Concrete (ACI 318-77), American Concrete Institute.

7. H. H. Steinour, "Concrete Mix Water - How Impure Can It Be?" Journal of the Research and Development Laboratories, Portland Cement Association, 5420 Old Orchard Road, Skokie, Illinois.

MIX DESIGN REQUIREMENTS FOR
NUCLEAR POWER PLANTS

BY

GERALD R. MURPHY[1] and CHAMAN L. GROVER,[2] M. ASCE

ABSTRACT

Engineers responsible for concrete mix designs on nuclear projects must be cognizant of codes and federal regulations applicable to their project. The engineer must be aware of anomalies and conflicts between applicable documents regarding design requirements. He must choose wisely which requirements are most applicable to the project and document the choices clearly in the projects' Safety Analysis Report. Three existing ACI codes i.e. ACI, 318, 349 & 359 supplemented by Regulatory Guides and ANSI standards, establish criteria for the design and construction of nuclear power plants. The engineer should select one set of criteria for strength and durability and establish one set of material requirements and mix designs which will satisfy these codes. Field experience should be utilized to select mix proportions in lieu of laboratory trial batches whenever such experience exists. Initial proportions should not be based on 1200 psi overdesign but rather some reduced value which reflects the ability of the QA program to assure material uniformity and adherence to the batching, mixing, placing and testing procedures.

INTRODUCTION

The concrete mix design, production, and placement activity on a nuclear power plant project is monitored and controlled intensively starting with the selection of concrete materials and ending with the finished concrete. The designer, concrete material suppliers, testing agency, concrete manufacturer and the constructor are constantly inspected and audited to assure their adherence to commitments made to attain quality concrete while complying with the design and quality

---

[1] Manager, Concrete Technology, Power Group, Brown and Root, Inc., P. O. Box 3, Houston, Texas, 77001

[2] Engineer, Concrete Technology Group, Brown and Root, Inc., P. O. Box 3, Houston, Texas, 77001

MIX DESIGN REQUIREMENTS

assurance requirements imposed by the governing Codes, Standards and Government regulations. Undoubtedly the design of a suitable concrete mix puts immense onus on the Engineer as he is confronted with the varied requirements imposed by applicable Codes and Standards pertaining to different parts of a nuclear power plant.

The objective of this paper is to present and discuss only those portions of the Codes and Standards applicable to nuclear power plant construction which delineate the requirements for the design of concrete mixes. The differences, conflicts, anomalies and similarities existing between the concrete mix design requirements of these Codes and Standards will be discussed, and recommendations to resolve the associated problems will be made.

The discussion will be limited to the selection of proportions for normal weight concrete and will assume that all precautions have been addressed to assure within reason the continuity of material supply from the same source for the duration of construction which in some cases may extend to ten or more years.

The construction of current nuclear power generating facilities requires between 300,000 to 600,000 cubic yards of concrete the vast majority of which will be considered safety related and as such will be subjected to very stringent and comprehensive Quality Assurance and Quality Control Programs during the construction phase.

The Quality Assurance Programs as required by 10 CFR 50, Appendix B (5)[1] invoke controls during all phases of concreting activities associated with the construction of a nuclear power plant that have not previously been invoked or enforced on the concrete industry. Whether or not all of the controls resulting from the Quality Assurance Programs actually result in increased concrete quality could certainly be the subject of a very lengthy debate. Notwithstanding, these controls provide the Engineer responsible for designing or approving a concrete mix a much higher degree of confidence than his counterpart in the non-nuclear industry, that the approved mix will be continually duplicated in the actual construction.

As a result of the Quality Assurance Program controls are imposed on the following: plant design, qualification of material suppliers, testing laboratory, constituent materials, and the production of concrete.

---

[1] The italic numbers in parenthesis refer to the list of references appended to this paper.

The plant design based on the applicable Codes, Standards and Regulations establishes the strength and durability requirements which will be delineated in the project specifications along with other concrete properties such as heat rise, workability, density etc. Material suppliers, and testing laboratories must be qualified and have a Licensee approved Quality Assurance Program. Constituent materials must be qualified by the Material Supplier or the testing laboratory to meet the requirements of the project specifications.

The Engineer responsible for witnessing and evaluation of laboratory design and testing of concrete mixes, in addition to being cognizant of all applicable Codes and Regulations, is required to be a Certified Level III Concrete Inspector meeting the requirements of ACI359, Appendix VII.

APPLICABLE CODES & STANDARDS

Individual project Safety Analysis Reports delineate the specific codes and standards which are applicable to that project. Currently there are three Codes and one Standard Specification which are generally applicable to the design and construction of a nuclear power plant. They are as follows:

- Code for Concrete Reactor Vessels and Containments, (ACI 359) (1)

- Code Requirements for Nuclear Safety Related Concrete Structures (ACI 349) (2)

- Building Code Requirements for Reinforced Concrete (ACI 318) (3)

- Specifications for Structural Concrete for Buildings (ACI 301) (4)

In addition to the above, Regulatory Guides issued by the Nuclear Regulatory Commission modify code requirements or endorse various American National Standards (ANSI) which may supplement or modify the Quality Assurance requirements for concrete mix designs. The extent to which these Regulatory Guides and ANSI Standards are applicable are usually covered in the project's Safety Analysis Report.

The necessity of invoking ACI 301 on any future construction of nuclear power plants may not be justifiable. The three remaining documents, in addition to the project specification, will adequately provide the necessary requirements.

DISCUSSION

As the titles of the three Codes imply each pertains to specific portions of a nuclear power plant. ACI 359 is applicable for Concrete Reactor Vessels and Containments; ACI 349 for Safety Related Structures except those covered in ACI 359; ACI 318 for all buildings except those covered in ACI 359 and ACI 349. As is generally the case when three committees set out to establish requirements for essentially the same item they do not always agree. This is the case when the mix design requirements are compared. However the differences, anomalies and conflicts which exist between the mix design requirements contained in these Codes are certainly not insurmountable and can be minimized by taking the position that one mix design per aggregate size, and strength class will be applicable for use in safety and nonsafety related structures.

The concrete mix design requirements of the three previously mentioned Codes relative to strength and durability are presented in Tables 1 and 2.

Durability Considerations

The three Codes, ACI 318, ACI 349 and ACI 359 have identical requirements for freeze-thaw durability namely that the water-cement ratio shall not exceed 0.53 and that all concrete shall be air-entrained within given air content limits depending upon maximum aggregate size. The requirements for exposure to sea water, i.e. the water-cement ratio shall not exceed 0.45, and permeability (water tightness), i.e. the water cement ratio shall not exceed 0.50, are the same in ACI 318 and ACI 349 whereas the requirements of ACI 359 are more conservative and reflect the previous ACI 318-71 requirements of permissible maximum water-cement ratios for sea water exposure and permeability as 0.44 and 0.48 respectively. These differences could conceivably result in two different design mixes being developed for one project to satisfy commitments to the applicable Codes. One hopes reason would prevail to prevent such an occurrence.

The ACI 318 and ACI 349 requirements which address exposure to injurious amounts of sulfates (as $SO_4^=$) require that the water cement ratio not exceed 0.50 and that sulfate resisting cement be used. ACI 318 does not qualify the type of cement where as ACI 349 specifies a Type V or permits the use of a blend of Type F fly ash and cement meeting the expansion requirements of Table 2A of ASTM C 150 when the cement is tested in accordance with ASTM C 441. ACI 359 again appears more conservative in that a maximum water cement ratio of 0.44 is required for exposure to injurious concentrations of sulfate containing solutions. ACI 359 further requires, depending upon the sulfate (as $SO_4^=$) concentration, that either a Type II, IV, V, IP (MS) be used or that for severe exposure only

Type V may be used unless other means are used to prevent sulfate attack. However, ACI 359 does not establish any additional guidelines or recommend alternates as to acceptable "means" which could prevent any deleterious effect on concrete from sulfate attack.

In addition to ACI 359 being more restrictive regarding exposure to sea water, permeability and sulfates a restriction is placed on the chloride content of the concrete mix design by placing a maximum limit of 400 ppm of chloride by weight in the cement paste. ACI 318 or ACI 349 do not impose any limits on the chloride content of concrete mix designs.

Strength Considerations

As is the case with durability requirements discussed above there are differences, anomalies or conflicts between the three Codes regarding the requirements for the selection of mix proportions, based on strength considerations.

ACI 359 requires that the selection of mix proportions be based on laboratory trial batches only whereas ACI 349 and ACI 318 permit the selection of proportions based on either laboratory trial batches or field experience. ACI 318 will permit, with permission, the selection of proportions based on the Code published water-cement ratios, however, for reasons noted in the commentary to ACI 318 this option should not be considered in the selection of mix proportions for a project of the size or complexity as that of a nuclear power plant. The option of using field experience to establish the preliminary mix proportions is generally not available because the majority of nuclear projects are supplied from on-site concrete production facilities erected specifically for the project and as such prior history for the production facility and materials do not exist. However this option should be used for any subsequent mix designs which may need to be developed during the course of construction after sufficient information on the production facility has been established and there has not been any significant change in the source of materials.

It must be pointed out, however, that when the option of using "field experience" is exercised and the basis for the calculated standard deviation was an analysis of a relatively high strength mix (above 6000 psi) an overdesign of 1200 psi would be required any time the standard deviation was above 600 psi inspite of the fact that the coefficient of variation could be below ten percent which is generally considered an excellent control.

Based on the requirements of the Codes and the above observaions the laboratory trial batches should be the basis for selecting the original proportions for the concrete mixes to be used on a nuclear project. Numerous differences, conflicts or anomalies exist between the three Codes regarding the requirements for the selection of mix proportions based on laboratory trial batches.

MIX DESIGN REQUIREMENTS 175

ACI359 has recently approved (effective Dec. 11, 1980) a Code Case (N-297) which eliminates the requirement that the proportions be selected based on a 1200 psi over design and only requires that the proportions be selected to produce concrete that has an average strength which assures that the probable frequency of a test more than 500 psi below f'c will not exceed 1 in 100 and that the probable frequency of an average of three consecutive tests falling below f'c will not exceed 1 in 100. This change does not lessen the assurance that adequate strength will be available but permits the Engineer responsible for the selection of proportions to take advantage of the fact that the concreting activities will be subjected to a stringent Quality Assurance Program from the time of material selection through the mixing, placing and curing of the concrete. Both ACI 318 and ACI 349 require that the required average compressive strength shall be 1200 psi greater than f'c. All codes permit subsequent reduction in the required average strength based on an analysis in accordance with "Recommended Practice for Evaluation of Compression Test Results of Concrete (ACI 214-77)."

ACI 318 and ACI 349 require that the laboratory trial batches used for selection of proportions have air contents within $\pm$ 0.5 percent and the slump within $\pm$ 0.75 inches of the maximum permitted by the specifications whereas ACI 359 requires that the air content be within $\pm$ 0.5% of the mean of the total air content and the slump be within $\pm$ 0.75 inch of the mean of the range designated in the Construction Specification. It is pointed out here that Regulatory Guide 1.136 (11) which endorses portions of ACI 359 has taken exception to the wording and issued supplementary guidelines, in Regulatory Position C1, which basically endorses the wording of ACI 318.

ACI 359 requires that the proportions including water cement ratio be established on the basis of laboratory trial batches in accordance with ACI 211.1, "Recommended Practice for Selecting Proportions for Normal and Heavy Weight Concrete." Neither ACI 318 nor ACI 349 invoke this specific requirement.

ACI 359 requires that Construction Specification define the applicable concrete properties as listed in Table CC-2231-1, and that these properties be measured prior to construction. Neither of the other codes have such requirements. Some of the properties listed in Table CC-2231-1 are: slump, compressive strength, shrinkage coefficient, density, creep of concrete in compression, maximum temperature rise in concrete, etc. The requirement that the applicable properties be determined prior to construction could be challenged except in the case of the first two properties especially in light of the facts that the original proportions are based on a required overdesign of 1200 psi, (until the previously mentioned code case (N-297) relaxed this requirement), and that the vast majority of the concrete actually placed will have

different proportions resulting from reduced cement contents and
field adjustments for workability and not the proportions in the
laboratory trial batches. If creep data had to be measured "prior"
to construction the tests would have to be initiated approximately
the same time as the licensing activity for the plant. Design
engineers must have direct input into the specifications regarding
the necessity of measuring these properties. It is questionable
whether or not the results of tests made on one laboratory trial
batch should be used as a basis for design rather than values
developed from averages of many tests of similar materials and
proven thru prior use.

All Codes require that a curve be established which defines
the relationship between the water-cement ratio and strength of the
materials to be used. All codes require a minimum of three points
to establish the curve. The laboratory mixes used to establish the
required points should also be used to evaluate workability seeing
as there are the imposed requirements that the slump and air content of the mixes must be within certain limits. The requirement to
establish a water-cement strength curve is in a very large number
of instances redundant because the requirements for durability will
establish the water-cement ratio.

RECOMMENDATIONS

The following recommendations should eliminate some of the
potential problems which could result from the Code differences,
conflicts or anomalies discussed previously.

- Safety Analysis Report Commitments

It is recommended that the commitment for design of concrete
mixes in the Safety Analysis Report should be made only to ACI
359 as it adequately covers all the requirements.

- Limited Singular Classification of Concrete

A position should be established that similar requirements
apply to safety and non-safety related concrete in so far as selection and qualification of materials, selection of mix proportions
for strength, durability and workability, receipt and storage of
materials and the production of the concrete, is concerned.

- Water-Cement Ratio Curves

It is recommended that rather than address the minimum requirements, "that three points establish the curve for each mix", a
curve for each aggregate size and admixture, which encompasses all
strengths required on the project, be established based on 6 or 7
points and that the specimens be tested at 7, 14, 21, 28, 60 and 90
days.

It is strongly recommended that consideration should be given to specifying later design ages for strength requirements.

The cognizant Engineer should have the prerogative to select concrete mix proportions based on some overdesign value less than the 1200 psi overdesign required by ACI 318 and ACI 349.

- Field Adjustments of Mix Proportions

It is recommended that the above series of mixes be duplicated in the field at the on set of Construction at which time they should be adjusted to assure maximum workability. The same strength history should be developed as was done for the laboratory mixes. These verification mixes should serve as the basis for any subsequent mixes that may need to be developed during the course of construction and eliminate the need for any future laboratory trial batches.

Field adjustments must adequately address the frequently occurring "slump loss" problems which often are attributable to a cement admixture reaction.

- Use of High Range Water Reducers

Strong consideration should be given to the use of high range water reducers to facilitate the adjustments to the laboratory mixes in the field in order to mitigate the use of additional cement.

The use of high range water reducers to enhance the workability of established mixes should be permitted based upon acceptable strength test results.

- Mixes for Determination of Concrete Properties

The Concrete properties referred to in ACI 359, Table CC-2231-1 and required by the specification should be measured on the adjusted mixes and not on the "original" laboratory mixes.

- Economic Considerations

Economy although always commendable is pennywise and pound foolish as it applies to the selection of concrete mix proportions for a nuclear project. The reduction of cement content or non use of an admixture at the expense of workability should not be permitted. The necessity for repair of one placement because the concrete was unworkable could conceivably consume any economies realized by omission of admixtures or reduced cement contents. The inadvertent use of non-safety related concrete, made with cheaper materials from an unqualified source, into a safety related structure, could cost untold sums if it had to be removed. Thus the limited singular classification of concrete is recommended.

Regulatory Endorsement and Standard Updating

Regulatory Guide endorsements and positions should reflect more closely the current issues of the Codes and American National Standards they endorse. For example, Regulatory Guide 1.94 Rev. 1 (4/76) endorses ANSI N45.2.5 (74), the current ANSI standard is 1978, other examples are Regulatory Guide 1.142 (4/78) endorsement ACI 349 (76), the current issue is ACI 349 (80), R.G. 1.142 (4/78) references ANSI N 101.6 (72) and ANSI N 101.4 (72) which in turn are endorsed by R.G. 1.69 (12/73) and R. G. 1.54 (6/73). American National Standards Institute requires that action be taken to reaffirm, revise or withdraw the standards no later than five years from date of publication.

Table 1 - Concrete Mix Design Requirements - Durability Considerations

Environmental Condition	ACI 359	ACI 349	ACI 318
1. Watertight Concrete			
(a) For Exposure to Fresh Water	Max. $\frac{W}{C}$ = 0.48	Max. $\frac{W}{C}$ = 0.50	Max. $\frac{W}{C}$ = 0.50
(b) For Exposure to Sea Water	Max. $\frac{W}{C}$ = 0.44	Max. $\frac{W}{C}$ = 0.45	Max. $\frac{W}{C}$ = 0.45
2. Concrete Subjected to Freezing Temp.			
(a) Max. $\frac{W}{C}$	0.53	0.53	0.53
(b) Air Content Limits	Same for ACI 359, ACI 349 and ACI 318 (Air content range is dependent upon aggregate size)		
3. Concrete subjected to Sulfate Exposure	Concrete that will be exposed to injurious concentrations of sulfate - containing solutions shall conform to maximum water-cement ratio of 0.44 and be made with sulfate -	Concrete that will be exposed to more than 0.2 percent water soluble sulfate or ground water with a concentration exceeding 1000 ppm sulfate shall be made with Type V	Concrete that will be exposed to injurious concentrations of sulfate - containing solutions shall be made with sulfate - resisting cement, and in addition, the water-

Table 1 - Concrete Mix Design Requirements - Durability Considerations - Cont'd.

Environmental Condition	ACI 359	ACI 349	ACI 318
3. Cont'd. (Sulfate exposure)	resisting cement as stated below: i) For concrete in contact with soils containing 0.10 to 0.20% water soluble sulfate (as $SO_4^=$), Type II, IV, V or Type IP (MS) cements shall be used. ii) In areas where the structure may come into contact with soils having more than 0.20% water soluble sulfate, or ground water with a sulfate concentration exceeding 1000 ppm, Type V only shall be used for concrete in contact with the ground unless other suitable means are used to prevent sulfate attack and concrete deterioration.	cement or a blend of Type F fly ash and cement meeting the maximum expansion requirements of Table 2a in ASTM C 150, when tested in accordance with ASTM C 441, and in addition the water-cement ratio shall not exceed 0.50 by weight.	cement ratio shall not exceed 0.50 by weight for concrete made with normal weight aggregate.

Table 2. – Concrete Mix Design Requirements - Strength Considerations
(Extracts from the Codes)

ACI 359	ACI 349/ACI 318[1]
The proportions, including water-cement ratio, shall be established on the basis of laboratory trial batches in accordance with ACI 211.1 and strength tests shall be made in accordance with ASTM C39.	Concrete proportions, including water-cement ratio, shall be established on the basis of field experience or laboratory trial batches with materials to be employed.
The proportions shall be selected to produce an average strength at the designated test age exceeding f'c by at least 1200[2] psi when the air content is within $\pm$ 0.5% of the mean3 of the total air content allowed and the slump is within $\pm$ 0.75 inch of the mean3 of the range designated in the Construction Specification.	Where a concrete production facility has a record, based on at least 30 consecutive strength tests that represent similar materials and conditions to those expected, required average compressive strength used as the basis for selecting concrete proportions shall exceed required f'c at designated test age by at least:
A laboratory curve shall be established showing the relationship between water-cement ratio and compressive strength. The curve shall be based on at least three points representing batches which produce strengths above and below that required. Each point shall represent the average of at least three specimens tested at 28 days or other ages as specified in the Construction Specification.	400 psi if standard deviation is less than 300 psi 500 psi if standard deviation is 300 to 400 700 psi if standard deviation is 400 to 500 900 psi if standard deviation is 500 to 600 psi
The maximum permissible water-cement ratio for the concrete to be used in the structure shall be that shown by the laboratory curve to produce the average strength mentioned above or in the following paragraph	If standard deviation exceeds 600 psi, concrete proportions shall be selected to produce an average strength at least 1200 psi greater than required f'c.

Table 2. - Concrete Mix Design Requirements - Strength Considerations
(Extracts from the Codes) - Cont'd.

ACI 359	ACI 349/ACI 318[1]
unless a lower water-cement ratio is required by the durability requirements. Using ACI 214, the amount by which the average strength must exceed f'c may be reduced to an appropriate level below 1200 psi after sufficient test data becomes available from the job to indicate that at the lower average strength, the probable frequency of a test result more than 500 psi below f'c will not exceed 1 in 100 and that the probable frequency of an average of three consecutive tests below f'c will not exceed 1 in 100.	When laboratory trial batches are used as the basis for selecting concrete proportions, strength tests shall be made in accordance with ASTM C39. When laboratory trial batches are made, air content shall be within ± 0.5 percent and slump within ± 0.75 inch of maximums permitted by the specifications. A curve shall be established showing relationship between water-cement ratio (or cement content) and compressive strength. Curve shall be based on at least three points representing batches which produce strengths above and below required average compressive strength. If concrete construction facility does not have a record based on 30 consecutive strength tests representing similar materials and conditions to those expected, required average compressive strength shall be 1200 psi greater than f'c. Maximum permissible water-cement ratio (or min. cement content) for concrete to be used in the structure shall be that shown by the curve to

Table 2. — Concrete Mix Design Requirements — Strength Considerations
(Extracts from the Codes) — Cont'd.

NOTES	ACI 349/ACI 318[1]
Note 1 — ACI 318 has the same requirements as ACI 349 except that ACI 318 permits the additional option that concrete proportions may be based on the water-cement ratio limits of Table 4.5 of ACI 318-77.	produce the average strength mentioned above unless a lower water-cement ratio or higher strength is required by special exposure (durability) requirements.
Note 2 — A code case, N-297 effective Dec. 11, 1980 permits lesser values than 1200 psi to be used for overdesign providing the strength assurance is maintained such that not more than 1 in 100 tests will be more than 500 psi below f'c and that the frequency of an average of three consecutive tests below f'c shall not exceed 1 in 100.	Mix proportions by water-cement ratio shall not be permitted without field experience or laboratory trial batches.
	After sufficient test data becomes available from the job, ACI 214 may be used to reduce the amount by which the average strength must exceed f'c below that indicated above, provided:
Note 3 — As per R.G. 1.136, Rev. 2, June 81, Position C1, the word "mean" is recommended to be replaced by the word, "Maximum".	1) Probable frequency of strength tests more than 500 psi below f'c will not exceed 1 in 100.
	2) Probable frequency of an average of three consecutive strength tests below f'c will not exceed 1 in 100.

REFERENCES

1. ASME Boiler and Pressure Vessel Code, Section III, Division 2, (ACI 359), "Code for Concrete Reactor Vessels and Containments," 1980 including summer 1981 addenda.

2. Code Requirements for Nuclear Safety Related Concrete Structures (ACI 349), 1980.

3. Building Code Requirements for Reinforced Concrete, ACI 318, 1977 including 1980 supplement.

4. Specifications for Structural Concrete for Buildings, ACI 301, 1975.

5. Quality Assurance Criteria for Nuclear Power Plants and Fuel Reprocessing Plants, 10 CFR 50, Appendix B.

6. ACI 359 Code Case N-297, "Alternate Rules for Test and Strength Requirements of Laboratory Trial Batches of Concrete Mixes," effective Dec. 11, 1980.

7. ANSI N45.2.5, "Supplementary Quality Assurance Requirements for Installation, Inspection, and Testing of Structural Concrete, Structural Steel, Soils and Foundations During the Construction Phase of Nuclear Power Plants," 1978.

8. ANSI N101.6, "Concrete Radiation Shields, " 1972.

9. US NRC Regulatory Guide 1.69, "Concrete Radiation Shields," December 1973.

10. US NRC Regulatory Guide 1.94, "Quality Assurance Requirements for Installation, Inspection, and Testing of Structural Concrete and Structural Steel During the Construction Phase of Nuclear Power Plants," Rev. 1, April 1976.

11. US NRC Regulatory Guide 1.136, "Materials, Construction, and Testing of Concrete Containments (Articles CC-1000, 2000, 4000, 5000, 6000 of ACI 359)," Rev. 2, June 1981.

12. US NRC Regulatory Guide 1.142, "Safety-Related Concrete Structures for Nuclear Power Plants (other than Reactor Vessels and Containments)," April 1978.

CONCRETE MATERIALS: REQUIREMENTS VS. PERFORMANCE

by

Richard A. Bradshaw, Jr.[1]

ABSTRACT

This paper provides a review of the material requirements and the application of materials for high density concrete, cement grout for general use, cement grout in prestress applications and the particular considerations relative to the control of deleterious substances in cement grout as used as a corrosion inhibitor for prestressing tendons in prestressed concrete containment structures. The codes, standards and regulations applicable to nuclear power plant construction are reviewed to identify requirements unique to nuclear construction and to identify conflicts or similarities and actual or potential impact on application and implementation of these requirements during construction as well as potential or actual impact on such factors as construction quality, cost and schedule.

PART I

-HEAVY WEIGHT AGGREGATES-

Concrete, in addition to being an excellent structural material, also possesses the required characteristics for both neutron and gamma ray attenuation, which qualifies it as an excellent radiation shielding material. Its ease of construction, low initial cost and low maintenance cost are also favorable considerations when choosing a material for radiation shielding. The efficiency of concrete as used for radiation shielding depends mainly on its density and uniformity.

The density of the concrete is controlled within required limits by the selection and use of an appropriate heavy aggregate as a substitute for a part or all of the normal weight aggregate usually used in concrete. Maximizing the density and therefore the shielding capability of the concrete is especially important where space is a major consideration (Ref. 1, Pg. 424).

---

[1]Manager Civil Technology, Technical Services Division, Daniel International

Heavy weight aggregates are used to provide concrete with densities exceeding 300 pounds per cubic foot, however, the normal range of densities obtained is from 210 pcf to 240 pcf. This range will produce concrete with a density approximately 60% greater than concrete produced with normal weight aggregate (Ref. 1, Pg. 420).

The uniformity and effectiveness of the shielding properties require that the concrete form a homogeneous mass with uniform distribution of aggregate particles. Since excessive settling, segregation or formation of pockets during placement can be factors in reducing the shielding effectiveness of the concrete, placement methods and techniques must be optimized so as to minimize detrimental results.

The primary use of heavy weight aggregate in nuclear power plant construction is in concrete used for radiation shielding. The principal codes, standards and regulations applicable to the use of high density concrete in nuclear power plant construction are: U. S. Nuclear Regulatory Commission Regulatory Guide 1.94 (Ref. 2), ANSI N45.2.5 (Ref. 3), ASME Section III, Division 2 (ACI 359)(Ref. 4), ACI 349 (Ref. 5), and ACI 304.3R (Ref. 6). These documents consistently specify that heavy aggregate for use in concrete shall conform to ASTM C637, "Standard Specification for Radiation Shielding Concrete" (Ref. 7).

The composition and specific gravity of heavy weight aggregates, as specified in ASTM C637 are listed in Table 1.

Table 1

Predominant Constituent	Class of Material	Chemical Composition of Principal Constituents	Specific Gravity of Available Aggregates
Serpentine	crushed stone	$Mg_3Si_2O_5(OH)_4$	2.4 to 2.65
Limonite	crushed stone (hydrous iron ore)	$(HFeO_2)_x \cdot (H_2O)_y$	3.4 to 3.8
Goethite		$HFeO_2$	
Barite	gravel or crushed stone	$BaSO_4$	4.0 to 4.4
Ilmenite	crushed stone (iron ore)	$FeTiO_3$	4.2 to 4.8
Hematite		Fe_2O_3	
Magnetite		$FeFe_2O_4$	
Ferrophosphorus	synthetic	Fe_nP	5.8 to 6.3

(Ref. 7, Pg. 3)

ASTM classifies these aggregates as natural mineral aggregates consisting of serpentine, limonite, goethite, barite, ilmenite, hematite and magnetite and synthetic aggregates such as ferrophosphorus. Not included in ASTM C637 are steel punchings and iron shot, which are also used as heavy weight aggregate to produce high density concrete. Fine aggregate may consist of natural sand, manufactured sand or a combination of both. Coarse aggregate may consist of crushed ore, stone or synthetic products or combinations of mixtures of these materials. Magnetite and ilmenite are the heavy weight aggregates most commonly used to produce high density concrete utilized for radiation shielding in the United States.

The procedures for measuring, mixing, transporting and placing concrete containing heavy weight aggregates are similar to those used for conventional concrete construction, however, special considerations are required primarily due to the increase in weight per unit volume. Some of these considerations are:

1. <u>Supply</u> - The number of sources of heavy weight aggregates are limited, therefore these sources can be distant from the jobsite. Typical sources for heavy weight aggregates in North America are as listed in Table 2.

Table 2

HEAVY AGGREGATE	SOURCE
Ilmenite	Quebec
Limonite-Goethite	Utah, Mich.
Serpentine	Calif., Que.
Magnetite	Nev., Wyo., Mont.
Barite	Tenn., Nev.

(Ref. 6, Pg. 94)

In addition, hematite can be obtained from sources in Brazil, South America. The cost of heavy weight aggregate, excluding freight, can be expected to be approximately 20 times that of normal weight aggregates. Also, depending upon the location of the source relative to the jobsite, the transportation of heavy weight aggregates can be a significant cost factor.

2. <u>Handling</u> - Aggregates should be shipped, handled and stockpiled in a manner which will minimize loss of fines, contamination and degradation. Specifications often require that aggregates, which are to be stored for a period exceeding two weeks, are to be covered to prevent contamination. Experience in shipping 3/4" (ASTM Grading No. 67) ilmenite coarse aggregate a distance of approximately

1200 miles by rail indicated that degradation occurred in excess of that normally anticipated for normal weight granite or limestone aggregates. This level of degradation was apparently due to a combination of the high density and abrasiveness of the material. Adequate compensation for degradation must therefore be made based on the method and distance of transportation.

3. <u>Mixing</u> - Due to the increased weight of the concrete, it is necessary to reduce the quantities batched and mixed so as not to overload the mixer. Reduction in mixing capacity can approach 40% based on concrete produced with a heavy weight aggregate with a specific gravity of 4.5 as compared with concrete produced with a normal weight aggregate with a specific gravity of 2.70.

4. <u>Formwork</u> - Due to the higher pressures exerted by the increased density of concrete containing heavy weight aggregates, formwork for high density concrete must be stronger than comparable formwork for normal weight concrete. An increase in form capacity and cost of 100% to 150% can be experienced particularly when revibration is required.

5. <u>Placement</u> - Concrete containing heavy weight aggregate is more susceptible to segregation during placement. Segregation will cause variations in strength as well as variations in density, which in turn will adversely affect the shielding properties of the concrete. Placement must therefore be closely controlled to insure uniform density and to minimize segregation. Consolidation of high density concrete must take into consideration the smaller effective radius of the action of internal vibrators, thus requiring that vibrators be inserted at closer spaced intervals than would be necessary for normal weight concrete. To assure removal of entrapped air and water and to assure maximum aggregate to aggregate contact, specifications often require that high density concrete be vibrated completely a second time after initial vibration. Revibration often involves an interruption of placement to provide a time delay between the initial and second vibration. This interruption, which may be as much as two (2) hours between lifts, will result in a significant reduction in placement rate when compared with placement of normal weight concrete.

CONCLUSION

As an alternative to using high density concrete for the purpose of radiation shielding, the most economical approach, when space is not a critical factor, is to increase the thickness of the concrete to obtain the desired radiation protection. In addition to the factors as discussed above, which impact both cost and schedule, other factors such as the relative small quantity of high density concrete, the special preparations and the general disruption of the concrete operation must be considered when placing high density concrete. However, the use of high density concrete for shielding is by far the preferable approach, considering both cost and schedule, rather than providing shielding utilizing such methods as lead brick or lead or steel cladding.

CONCRETE MATERIALS

PART II

-CEMENT GROUT-

Cement grouts can be classified into two general categories: 1) General purpose grouts and 2) nonshrink or expansive grouts.

General purpose grouts are primarily used where shrinkage is not a major consideration, whereas nonshrink grouts are used where maximum contact or bearing and therefore minimum or no shrinkage is a primary consideration. A grout is regarded as being nonshrink if after hardening at an age of 28 days its volume is not less than its initial volume (Ref. 12, Pg. 1). In this case the initial volume is measured immediately after molding while the grout mixture is still of a plastic consistency.

Common uses for both general purpose and nonshrink grouts are:

1. Grouting of equipment bases and column base plates
2. Anchor bolt and rock bolting grouting
3. Preplaced aggregate concrete
4. Grouting of tendons
5. Foundation grouting

Both general purpose and nonshrink grouts can be categorized as pre-packaged or proprietary grouts and field mixed grouts and generally consist of a mixture of cement, fine aggregate, water and admixtures. Cement can be any of the types as specified in ASTM C150, however the same type and brand of cement used for construction is generally used for field mixed grouts. In some cases the amount of fine aggregate is limited where the grout must be sufficiently fluid so as to penetrate and fill all the voids in the medium being grouted. Admixtures may include any of the air entraining, water reducing, set retarding and pozzolanic admixtures commonly used in concrete, however special admixtures such as aluminum powder or grout fluidifiers are also used. Aluminum powder of the proper fineness and quality is added to the grout to produce expansion by the generation of minute bubbles of hydrogen gas. A grout fluidifier is used to inhibit early stiffening, to hold the solid constituents in suspension and to produce controlled expansion prior to initial set and will generally conform to Corps of Engineers Specifications CRD-C566 (Ref. 13, Pg. 1).

The principal specifications covering cement grout for nuclear power plant construction are ACI 359 (Ref. 4) and CRD-C621, formerly CRD-C588, (Ref. 12). ACI 359 includes requirements for both general purpose grouts and expansive grouts as well as detailed requirements for cement grouts for grouted tendon systems whereas CRD C621 primarily addresses nonshrink grouts. Detailed material requirements for cement grout for grouted tendon systems are discussed in Part III of this paper.

In addition to providing detailed material requirements for cement grout, ACI 359 addresses the proportioning of grouts as follows:

> "The proportions of materials for general purpose grout shall be based upon trial mixes using the same type and brand of cement, fine aggregate and admixtures as will be used for construction."

ACI 359 continues on the subject of compressive strength:

> "The compressive strength of the grout shall be established by 2" cubes molded, cured and tested in accordance with ASTM C109, except that if expansive grout is used the tests shall be performed in accordance with CRD-C588 (CRD-C621), Corps of Engineers Specifications for Nonshrink Grout" (Ref. 4, Pg. 145).

ACI 359 specifies the permissable materials and establishes general design requirements, but allows the designer to establish such parameters as compressive strength, expansion, shrinkage, fluidity and time of set based on individual project design requirements.

In contrast, CRD-C621 establishes minimum or maximum requirements for these parameters, as indicated in Table 1, with minimal consideration for material specifications or proportioning.

Table 1

Expansion, %, 3, 14 and 28 days	not greater than 0.4 at any of these ages
Expansion %, at 3 and 14 days	not greater than expansion at 28 days
Shrinkage, %, 28 days	none
Compressive strength, psi (MPa) 7 days, min. 28 days, min.	2500 (17.2) 5000 (34.5)
Time of final setting, hr. max.	8
Fluidity High fluidity (Fluid)	10-30 sec
Moderate fluidity (Flowable)	124-145 flow
Minimum fluidity (Plastic)	100-125 flow

(Ref. 12, Pg. 1)

CRD-C621 states that "any product of any composition that performs when tested so as to comply with the requirements of this specification may be acceptable for use" (Ref. 12, Pg. 1).

The intended use of cement grout will not only influence its compressive strength and expansive or shrinkage characteristics, but will also to a great extent dictate its method of placement. Grout is generally placed by one of three methods:

1. Poured or flowed into place

2. Packed in place as with "dry pack"

3. Pumped as with pressure grouting

The method of placement will affect factors such as flowability and time of set or "pot life", therefore these factors must be adequately addressed during design of the grout mixture and selection of grout materials.

CONCLUSION

Proprietary or prepackaged grouts have been extensively specified and successively used in the construction of nuclear power plants in the United States. These grouts, due to their proprietary nature, generally conform to CRD-C621, which is a performance oriented specification with minimum consideration for constituent material requirements. Due to the proprietary nature of prepackaged grouts, manufacturers are seldom specific relative to the material composition of the grout, in particular the grout additives. ACI 359 provides specific requirements for cement, aggregate and admixtures which in effect appears to eliminate proprietary grouts to be used for code construction.

PART III
-CEMENT GROUT IN PRESTRESS APPLICATIONS-

INTRODUCTION

Portland Cement grout can be used as a corrosion inhibitor for prestressing tendons in prestressed concrete containment structures. In addition to providing corrosion protection, a grouted tendon system also provides a degree of bond between the tendons and the surrounding concrete (Ref. 14, Pg. 1).

Cement grout for grouted tendon systems generally consists of a mixture of cement, water, and one or more admixtures. Sand may be used in cases where the gross inside area of the tendon tube or sheathing relative to the area of the tendon is sufficiently large to permit penetration of the grout into all spaces and not adversely affect grout expansion and bleeding.

The principal codes, standards and regulations which delineate the requirements for grout constituents for nuclear power plant construction are NRC Regulatory Guide 1.107 (Ref. 14), ASME Section III, Division 2/ACI 359 (Ref. 4) and ACI 349 (Ref. 5). Regulatory Guide 1.107 and Subsection CC of ACI 359 are primarily concerned with containment structures, whereas ACI 349 is primarily concerned with safety related concrete structures other than containment structures, which form a part of a nuclear power plant. Requirements as included in these documents are also based in part on requirements as included in the PCI Post-Tensioning Manual (Ref. 18). Since the requirements as included in the PCI Manual primarily address conventional construction rather than nuclear power plant construction, these requirements when included in this discussion are included only for comparison purposes. In addition to those requirements currently contained in the Winter 1980 Addenda to the 1980 Edition of ACI 359, additional proposed requirements are included in Committee Action JC5-59 (Ref. 15). The primary purpose of this committee action is to more thoroughly address the requirements relative to grouting of tendons including those requirements as contained in Regulatory Guide 1.107. A comparison of these codes, standards and regulations indicates certain similarities and differences in the grout material specifications. Material specifications for cement, aggregate, water and admixtures are compared in this discussion.

CEMENT

The basic specifications for cement for grout are ASTM C150, "Specification for Portland Cement" and ASTM C595, "Specification for Blended Hydraulic Cement," however each document differs slightly in the types of these cements which are permitted. Regulatory Guide 1.107 indicates that cement shall be ASTM C150, Type I or Type II or other types as dictated by climatic or environmental conditions. ACI 359 currently specifies ASTM C150, Types I, II, III, IV or V whereas the proposed change to ACI 359 specifies ASTM C150, Types I, II, IV or V or ASTM C595, Type IP, IP(MS) or IP(MH). ACI 349 indicates ASTM C150 or ASTM C595 cements excluding Type S and SA.

AGGREGATE

The use of sand in grout for grouted tendons is optional. Its use is permitted when the void space between the tendon and the sheathing is sufficiently large to permit the penetration of the grout into all spaces while obtaining required expansion and sedimentation or bleeding characteristics. Both Regulatory Guide 1.107 and the current ACI 359 code as well as the proposed change to ACI 359 indicate that aggregates shall meet the requirements of ASTM C33 with a provision that the gradation may be adjusted to comply with the size of the opening and the method of placement. ACI 349 indicates that when sand is used it shall conform to ASTM C144, "Specification for Aggregate for Masonry Mortar" and also allows for adjustment in gradation to obtain satisfactory workability. The primary difference between ASTM C33 as specified in ACI 359 and ASTM C144 as specified in ACI 349 is in the grading of the aggregate with masonry sand being the finer material.

WATER

The basic requirement for water as stated in all of the documents is that the water be clean and free of injurious amounts of materials which may be harmful or deleterious to either the prestressing steel or the grout. Additional requirements as stated in each of the documents are as follows:

Regulatory Guide 1.107 - Water contaminated with 1 ppm of hydrogen sulfide is prohibited. The water is to be qualified by making comparative tests in accordance with ASTM C191, Time of Setting of Hydraulic Cement by Vicat Needle and ASTM C109, Compressive Strength of Hydraulic Cement Mortars.

ACI 359 and the proposed change to ACI 359 states that "Water..... shall be clean with a total solids content of not more than 2000 ppm as measured in accordance with ASTM D1888, Tests for Particulate and Dissolved Matter in Water. The mixing water shall be tested for chlorides (Cl^-) as determined by ASTM D512, Chloride Ion in Industrial Water and Industrial Waste Water." (Ref. 4, Pg. 142). In addition, comparative tests in accordance with ASTM C191 and ASTM C109 are required. It should be noted that no limit is stated for chloride content in that this test result is not used for the purpose of qualification of the water supply but is used in determining the total chloride content of the grout mixture.

ACI 349 requires that "water used in mixing concrete shall be clean and free from injurious amounts of oils, acids, alkalies, salts, organic materials, or other substances that may be deleterious to concrete or steel. In addition, the mixing water for prestressed concrete or for concrete which will contain aluminum embedments, including that portion of the mixing water contributed in the form of free moisture on the aggregates, shall not contain deleterious amounts of chloride ion." (Ref. 5, Pg. 8)

ADMIXTURES

The use of admixtures in grout for grouted tendons is predicated on their demonstrated ability to improve the properties of the grout such as increasing workability, reducing or controlling bleeding, preventing water separation when pumped at high pressures, entraining air, expanding the grout or reducing shrinkage. In addition to these criteria, Regulatory Guide 1.107 indicates that harmful substances in the admixtures should be kept to a minimum and prohibits the use of calcium chloride as an admixture. The Regulatory Guide also cautions about the use of aluminum powder to produce expansion, due to the potential danger for hydrogen attack on the tendon steel.

The requirements for admixtures as contained in the current ACI 359 code indicate that air entraining admixtures shall conform to the requirements of ASTM C260, that fly ash and pozzolanic admixtures conform to ASTM C618 and that chemical admixtures such as water reducers, retarders and accelerators conform to ASTM C494. In addition,

ACI 359 requires that admixtures containing more than 1% by weight of chloride ions are prohibited on prestressed containments and admixtures contributing more than 5 ppm by weight of chloride ions to the total grout constituents are prohibited. ACI 359 allows the use of aluminum powder, conditional upon the performance of adequate tests to demonstrate the suitability of the mix for the purpose intended. The proposed change to ACI 359 in addition to including all requirements as previously indicated also includes the provision for special grouting admixtures, including those containing gels or gelling agents for the purpose of controlling loss of water from the grout.

The basic material specifications for air entraining admixtures, fly ash and pozzolanic admixtures and chemical admixtures as contained in ACI 349 are the same as those contained in ACI 359, however ACI 349 includes the additional stipulation that "admixtures.....known to have no injurious effects on the grout, the steel or the concrete may be used" (Ref. 5, Pg. 57). ACI 349 also specifically prohibits the use of calcium chloride. It should be noted that although ACI 359 does not specifically prohibit the use of calcium chloride, the fact that it is not listed as an acceptable material in the code is in effect a prohibition of its use. ACI 349 further states that "admixtures containing chloride ions shall not be used in prestressed concrete" (Ref. 5, Pg. 57). ACI 349 does not address the use of aluminum powder.

CONCLUSION

The requirements for grout constituents as included in the principal codes, standards and regulations applicable to nuclear power plant construction as reviewed in this discussion are similar in a number of instances, however dissimilarities do exist which could have an impact on construction. No significant differences are noted in the requirement for cement. Even though the Regulatory Guide lists only ASTM C150, Type I and Type II cements it also includes a provision for use of other types as may be dictated by climatic or environmental conditions. Sand however is specified as conforming to ASTM C33 in ACI 359 but to ASTM C144 in ACI 349. The finer gradation as specified by ASTM C144 appears to be preferable when compared to the coarser gradation as specified in ASTM C33 due to penetration, expansion and bleeding considerations. Both codes do, however, include provisions for adjustment of gradation for workability and placement considerations, therefore the differences in the standards relative to grading are considered to be insignificant from both a control and construction standpoint.

The requirement that water be tested for solids content and chloride content is considered to be standard practice. The determination and limitation of sulfides as required by Regulatory Guide 1.107, although not required in either ACI 359 or ACI 349, can be readily accomplished by existing standard test methods, if required. In the case of ACI 349 where no limits are recommended, the designer should consider only those factors which are critical to the integrity of the tendon system so as to establish a practical level of control for chemical requirements for water. The designer should also delineate the standard analytical test methods for the determination of the chemical constituents.

The specifications for air entraining, water reducing, retarding and pozzolanic admixtures are basically the same as for concrete. Although ACI 359 limits the chloride content of admixtures to 1% maximum, admixture manufacturers have indicated no problem in meeting this limit. It should be noted however that although the chloride content of admixtures is usually extremely low, it would be difficult if not impossible to demonstrate that the admixture contained <u>no chloride ions</u> as required by ACI 349. Some chloride content, even if extremely low would be anticipated. Analytical methods have not been sufficiently developed to allow accurate detection of extremely low levels of chloride content in admixtures. This requirement is not considered to be realistic in consideration of the QA/QC controls for construction of nuclear power plants.

PART IV

-CORROSION PREVENTION SYSTEMS-

The prestressing tendon system of a prestressed concrete containment structure is a principal strength element of the structure. Since the ability of the containment structure to withstand the events postulated to occur during the life of the structure depends on the functional reliability of the structure's principal strength elements, any significant deterioration of the prestressed elements due to corrosion may present potential risk to the public safety. It is therefore important that any system for inhibiting corrosion of the prestressing element possess a high degree of reliability in performing its intended function (Ref. 14, Pg. 1).

A comparison of the requirements as included in Regulatory Guide 1.107 (Ref. 14) and the current ACI 359 (Ref. 4) as well as the proposed change to ACI 359 (Ref. 15) indicates certain similarities and differences in identifying and controlling deleterious substances in grout constituents.

Regulatory Guide 1.107 states that "the grout (whether freshly mixed or hardened) should not cause chemical attack on the prestressing elements through its interaction with the material of the tendon steel, the material of the anchor hardware or the material of the duct" (Ref. 14, Pg. 2). The Regulatory Guide continues with the following discussion of deleterious substances and their potential effect on the tendon system:

> "Various deleterious substances have been reported as potential sources of corrosion of prestressing steel. Most of the reported failures of prestressing elements have been attributed (a) to the presence of chlorides in the atmosphere or in the constituents of grout or (b) to the presence of hydrogen sulfide in the atmosphere. Nitrates and sulfates generally found in mixing water have been theorized to be potential sources of stress corrosion of prestressing steel. However,

it has been reported that, in a concrete environment, oxygenated anions such as sulfates and nitrates do not exhibit intense corrosion properties. It has also been reported that most of the chlorides are neutralized during the hydration of portland cement. The threshold values below which these substances will not participate in initiating corrosion have not been established. Hence a safe and prudent approach would be to make sure that these substances are limited to the <u>lowest practical levels</u> (underline by author) in grout constituents. The use of water contaminated with hydrogen sulfide (1ppm) should be prohibited."

Regulatory Guide 1.107 continues,

"The protective mechanism of grout is primarily dependent on its ability to provide a continuous alkaline environment around the tensioned steel elements. The natural alkalinity of the primary product of cement hydration (i.e., calcium hydroxide) tends to be at a pH value of 12.5. The effectiveness of alkaline environment may be reduced by the leaching of alkaline substances with water, by reaction in an acidic or sulfide-containing environment, or by the presence of oxygen and chloride ions. The ability of chloride ions to develop corrosion increases with decreasing alkalinity of the calcium hydroxide solutions. Thus it is advisable to monitor the pH value of the in-place grout under actual field conditions and ensure that it remains above a value where the passivating effect of the grout is not reduced by the available chloride ions in the composition of the grout" (Ref. 14, Pg. 2).

In consideration of these concerns, Regulatory Guide 1.107 establishes the following limits for deleterious substances:

Chloride - 100 ppm (200 ppm if pH is maintained above 12)
Nitrates - 100 ppm
Sulfates - 250 ppm (excludes sulfates in the form of sulfur trioxide as a cement component)
Sulfides - 2 ppm
pH - maintained above 11.6 (12 if the allowable chloride content is 200 ppm)

The regulatory guide also recommends that the "Quantities of these substances in the grout constituents be determined <u>individually for each of the constituents</u> by the applicable ASTM methods and expressed in parts per million parts of water in the grout composition" (Ref. 14, Pg. 2). It should be noted that the current practice as stated in ACI 201 is to express chloride content as a percent by weight of cement.

The chemical requirements for cement grout for grouted tendon systems as stated in the current ACI 359 are as follows:

"The chloride and nitrate content of the grout constituents shall be controlled so that the following limits for freshly mixed grout are not exceeded:

a) Chlorides as Cl^- - 300 ppm maximum

b) Nitrates as NO_3^- - 100 ppm maximum"

The proposed change to ACI 359 states these requirements as follows:

"The total chloride and nitrate content of the grout combined constituents shall be controlled so that the following limits for cement paste (pozzolans, admixtures and water) are not exceeded:

a) Chlorides as Cl^- - 100 ppm

b) Nitrates as NO_3^- - 100 ppm"

The primary differences between the current code and the proposed change are the reduction of chloride content from 300 ppm to 100 ppm corresponding to limits as stated in Regulatory Guide 1.107, and the limiting of the determination of chloride and nitrate content to the pozzolan-admixture-water portion of the grout, thus eliminating the cement and aggregate from consideration.

In comparing the regulatory guide and the proposed change to ACI 359, four (4) primary differences are identified:

1. ACI 359 does not include limits for pH

2. ACI 359 does not include limits for sulfates

3. ACI 359 does not include limits for sulfides

4. ACI 359 does not require that all constituents be controlled

These differences are basically attributed to the identification of those substances which should be considered as being deleterious to the integrity of the tendon system and the establishment of the lowest practical levels at which these substances can be measured and controlled based on known standard test methods available in the industry.

The position of ACI 359 on each of these points is summarized as follows. Relative to the question of pH, it is not anticipated that the pH of the grout in the tendon duct will fall below 11.6. Irregardless, there is presently no known standard method for measuring the pH of either a freshly mixed or hardened grout. Any sulfate from sources other than the cement will be tied up chemically by the cement and therefore should be of minimal concern. With regard to sulfide, the concern would be for an exposure of the tendon to a sulfide environment rather than for sulfide present in the grout. At the present time, there are no known standard test methods available to accurately measure sulfide content of cement and admixtures within the limits proposed. The cement and aggregate have been deleted from the portion of the grout on which the chloride and nitrate content is determined recognizing that the chloride content of the portland cement and the total chloride content of the aggregates are indeterminates using existing standards for test methods.

CONCLUSION

Although the primary concern for prevention of corrosion of the tendon system should not be minimized, it is critical that those substances and the levels of those substances, which are deleterious to the system, be identified to the extent possible. It is also of primary concern, particularly from a construction standpoint, that standard test methods are available to uniformly and consistently measure the levels of deleterious substances being controlled. The establishment of practical levels of control and the availability of standard test methods is particularly critical in consideration of the system of QA/QC controls inherent in the construction or nuclear power plants. The establishment of unmeasurable levels of substances, even though these substances may be considered to be deleterious, could in effect exclude grouting as a method of corrosion protection for tendon systems.

REFERENCES

1. STP-169B, Significance of Tests and Properties of Concrete and Concrete-making Materials, Chapter 26, "Radiation Effects and Shielding"

2. U. S. Nuclear Regulatory Commission Regulatory Guide 1.94, Quality Assurance Requirements for Installation, Inspection and Testing of Structural Concrete and Structural Steel during Construction Phase of Nuclear Power Plants

3. American National Standards Institute N45.2.5, Supplementary Quality Assurance Requirements for Installation, Inspection and Testing of Structural Concrete, Structural Steel, Soils and Foundations during the Construction Phase of Nuclear Power Plants

4. ASME Section III, Division 2 - ACI 359, Code for Concrete Reactor Vessels and Containments

5. ACI 349, Code Requirements for Nuclear Safety Related Concrete Structures

6. ACI 304.3R-75, High Density Concrete: Measuring, Mixing, Transporting and Placing
7. ASTM C637, Standard Specification for Radiation-Shielding Concrete
8. ASTM C638, Standard Description Nomenclature of Constituents of Aggregates for Radiation Shielding Concrete
9. ACI 309, Recommended Practice for Consolidation of Concrete, Chapter 14, "Heavy Weight Concrete"
10. U. S. Nuclear Regulatory Commission Regulatory Guide 1.69, Concrete Radiation Shields for Nuclear Power Plants
11. American National Standards Institute M101.6, Concrete Radiation Shields
12. CRD-C621, Corps of Engineers Specification for Nonshrink Grout
13. CRD-C566, Corps of Engineers Specification for Grout Fluidifier
14. U. S. Nuclear Regulatory Commission Regulatory Guide 1.107, Qualifications for Cement Grouting for Prestressing Tendons in Containment Structures
15. ACI 359 Joint Committee Item No. JCS-59, Grouting of Prestressing Tendons
16. U. S. Nuclear Regulatory Commission Regulatory Guide 1.136, Materials for Concrete Containments
17. U. S. Nuclear Regulatory Commission Regulatory Guide 1.103, "Post-tensioned Prestressing Systems for Concrete Reactor Vessels and Containments"
18. Prestressed Concrete Institute, Post-tensioning Manual, Chapter 3, Section 2, Recommended Practice for Grouting of Post-tensioned Prestressed Concrete

SESSION III - QUALITY ASSURANCE/QUALITY CONTROL PRACTICES

SESSION OBJECTIVES/SESSION CHAIRMAN SUMMARY

by

William F Mercurio[1] M, ASCE and James H Olyniec[2] M, ASCE

Objective of Session

An increasing number of government regulations controlling the granting of construction and operating permits for nuclear projects, and the social responsibility undertaken in the building of such projects warrants a strong quality assurance/quality control program on the part of constructor, engineer, and owner. This session discusses the implementation of QA/QC programs with an emphasis on the regulatory requirements for QA/QC, organizational practices, training and qualification of inspectors, documentation practices, and information processing. By examining these areas, the session objectives are to (1) provide information on how various companies and organizations implement these regulations, and (2) identify methods by which implementation of these regulations can be made more effective and productive.

Session Chairman Summary

The first step in the implementation of Appendix B 10CFR50 is the organization for QA/QC functions. Whereas the electrical and mechanical disciplines had some familiarity with existing quality control programs imposed by Underwriters Laboratory, NEMA, ASME, etc, the civil design and construction industry was unaccustomed to such restraints. As such, the implementation of new regulatory requirements placed on nuclear design and construction was somewhat traumatic. An organization or company was offered two basic choices. The quality organization could be fully integrated into the company's basic organization and become a full-fledged partner with design, procurement, and construction. Companies that addressed this philosophy were basically successful. On the other hand, companies who resisted the intent of these new regulations and complied only with the letter of the law soon developed adversary relationships between production and quality. Although these are the extremes, most quality assurance/quality control organizations have a mixture of both philosophies.

---

[1] Supervising Engineer, Ebasco Services Inc., New York, NY

[2] Supervisor, O&CEU, Tennessee Valley Authority, Hollywood, AL

QUALITY ASSURANCE/CONTROL

In establishing a quality program, five basic rules are essential: (1) management support for the program (2) a defined scope for the quality program (3) clearly defined lines of authorization and communications (4) simplicity and (5) staffing with competent personnel.

Management support for the quality program is mandatory if the program is to have any chance of success. The key to this management support is to include quality as a full-fledged partner. This, in turn, requires that quality be included in the decision making process. Any program without a clearly defined scope will make it that much more difficult to clearly define the lines of authority and communication. Whereas the scope of the program must meet certain minimum regulatory requirements, economic considerations must be given to the work activities of the entire project. For example, it is generally more economical to buy all reinforcing steel for a nuclear project as safety related category I material. The added cost of the material that would be used in non-safety related applications is more than offset by the quality control costs that would be required to segregate two different classes of material onsite. The scope of the program must guard against unnecessary restrictions. Ideals or preferred practices should be defined as such and should not become rigid requirements. The lines of authority and communication should clearly distinguish and separate the reporting responsibilities for those who are responsible for <u>attaining</u> quality from those who are responsible for <u>verifying</u> quality. The authority for performing a function should be kept at the lowest possible level required to perform a task and to approve a task. No program can possibly anticipate every situation that develops during the course of construction. As such, the quality program should be established to facilitate improvements and revisions. No matter how well the program is planned, without adequate staffing of competent personnel it will not be successful. Proper training and indoctrination is essential.

The "basic quality organization" would fit a standard condition if such a condition existed. It would be responsible for a full quality program encompassing the 18 criteria of Appendix B 10CFR50 and charged with doing work on a multiple discipline basis. A quality assurance manager would supervise three basic groups; quality assurance, quality control, and quality records. Some facets of this basic organization may become too large or too complex for an individual supervisor. In this case, the three groups to the basic organization would be further split as necessary, for example, receiving QC supervisor, reorder supervisor, etc. A quality organization may also be organized by discipline rather than by function. Under each discipline there is a quality supervisor that is responsible for both quality assurance and quality control for his discipline. A limited function quality organization may be established to cover a particular situation. In this type of situation, the program would be tailored to fit the specific requirements identified in the scope.

Once the quality organization has been established, the implementation of criterion 2 of Appendix B 10CFR50 requires the indoctrination and training of personnel performing activities affecting quality. Regulatory Guide 1.58 describes the method acceptable to the NRC for meeting these requirements for qualification and training of inspectors. It does this by endorsing, with some exceptions, the industry standard ANSI N45.2.6. This ANSI standard provides the specific requirements that virtually all inspector qualification and training programs in the industry are written to address.

The qualification of personnel employed as nuclear plant construction inspectors is based on three factors; experience, education, and performance evaluation. Experience and education are normally brought to the job by the individual. Experience in the construction field and more importantly in the nuclear regulatory climate is especially advantageous. The formal education an individual receives is a valuable asset and can aid in the performance of the inspector by providing a basis for analyzing the inspection problems he encounters daily. The performance evaluation process includes initial evaluation to certify or recertify an inspector and the monitoring of on-the-job performance to verify continued satisfactory performance.

In developing a training program it is first necessary to determine the applicable code and standards to which the program will be designed. The importance of this identification is quickly recognized when one considers the numerous codes and standards and various revisions to them and the multitude of cross-references between standards and revisions. Once it has been determined which requirements apply to the training program, the next step is to develop procedures for the program's implementation. Several points should be considered in writing the training procedures; (1) applicability, (2) delineation of responsibilities, (3) references, (4) requirements for initial certification, (5) maintenance of proficiency, (6) recertification requirements, (7) factors affecting the status of certification, and (8) transfer of certifications from previous employer.

Initial precertification training and indoctrination covers general topics such as the organization, purpose, and objectives of the total quality control program, and specifics like the preparation and use of inspection plans. Jobsite training is provided by inspection supervisors and discipline engineers. Once the training program is implemented, periodic proficiency training in order to keep personnel updated on regulatory changes and procedural revisions is required. At least once a month a one hour minimum training session with a predetermined agenda is held to discuss inspection plans, documentation requirements, applicable codes and standards, etc.

Perhaps the most important portion of the training program is record management. Certification records are of primary importance in quality assurance and must be continually updated. Some points to consider in establishing record requirements include (1) keep the records as simple as possible, (2) require documented proof of previous certifications, training, and experience, and (3) establish length of storage and storage requirements. Records of this recertification training should be included in personnel certification files.

Actual civil engineering unit documentation practices on a nuclear construction project were detailed during the session. The implementation of 10CFR50 at the construction project is traced through the Project Safety Analysis Report, the Office of Engineering Design and Construction Quality Assurance Manual, the Division of Construction QA Program Manual, and finally to the 115 site-specific quality control procedures covering every construction activity associated with safety related buildings or equipment. Specific documentation practices addressed include rebar, embedments and concrete formwork, concrete placement, structural steel fabrication and erection, protective coatings, cadweld inspection, post-tensioning vendor surveillance, bolt anchors set in hardened concrete, backfill materials placement, work releases, control of temporary installations or omissions, anchor bolt freeze protection, and bending of partially embedded reinforcing stee.

The application of computers in civil documentation can vary from nothing to the use of computer cards for the documentation itself. Experiences at several nuclear sites has shown that although the completion of QC inspection records is required, the capability of knowing the status of all incomplete inspections or a verification that the documentation has been satisfactorily completed is essential for timely system operation and records turnover. Even with a QC program that results in timely inspections and documentation, the complexity of the work, the large number of items to be inspected, and the number of inspections on each item warrant more than a casual assurance that, in fact, all inspections have been completed. A good computer program uniquely lists every feature to be inspected and not only assists the tracking of the inspection status and documentation itself, but also offers a valuable scheduling and planning tool for the remaining work. Other computer uses, although not resulting directly in the inspection documentation, monitor the in-process activities associated with complex installations. This is accomplished by monitoring individual problems, work releases, field change requests, etc., associated with the ongoing installation.

The result of required quality assurance and quality control practices invariably results in a large volume of information and records. Information processing, methods, and applications have taken on added importance in the nuclear field. In the very broadest sense the terms "information" and "information processing" cover three phases of "information management": (1) creating information, (2) processing information, and (3) retaining information. In phase 1, information is created to meet day to day operational needs. In phase 2, information is processed using available equipment. During phase 3, the information is retained or disposed of through the records management systems of records retention, indexing, filing, microfilming, automated information retrieval and inactive records storage. Recent developments in information processing technology have demonstrated the potential for dramatic improvement in productivity. The work processing area has evolved from the manual typewriter. The reprographics area has evolved from the precopier period through the xerographic process onto the computer output microfilm. The telecommunications area has available computer driven interconnect systems, satellite communications, electrical mail and fiberoptics.

The data processing area has continually new developments appearing in computer technology and electronics. Much of this new information processing capability has arrived in today's business offices in bits and pieces. Most equipment bought in a hodgepodge fashion is designed to do specific office information chores. However, these machines could perform much more effectively if they could interact with one another. Therefore, proper planning and controls should be placed up front in the records program.

The objective of a good records program is to develop the program details, prepare the work plan, and implement the modified and improved system. The customary approach is to begin with a feasibility study which identifies problems. Equipment recommendations would parallel the development of the scope of the records program and the development of the work plan. The scope of the records program should address the following functions: (1) records flow and filing techniques, (2) record identification and indexing methods, (3) records turnover, (4) drawing and specific control, (5) existing equipment, (6) technical and operational support center, (7) personnel requirements and training programs. The final task is to implement the records management program with the following characteristics: (1) consideration of user needs, (2) easily administered and understood, stood, (3) cost beneficial both in terms of installation and operation, (4) takes advantage of new equipment and records management techniques.

Consensus Recommendations*

(1) Management support is required for an effective quality assurance/quality control program.

(2) Training programs must be established for the training of inspectors, clearly defining their duties and responsibilities. This training should include emphasis on a proper work attitude to insure that not only are all technical requirements met, but that problems are resolved in a timely manner with engineering personnel.

(3) To avoid duplication of training and to take full advantage of previous inspectors' experience, employers should provide, upon request, a detailed history of an employee's work experience and certifications.

(4) Criteria for turnover of documentation and records to the client should be established at the beginning of the project. This will allow the establishment of the inspection procedures to meet the requirements. More importantly, it will allow for proper planning and implementation of a good management informations system program.

---

*Based upon the Panel of Speakers/Audience discussion period at the end of the session.

ORGANIZING FOR QA/QC FUNCTIONS

by

Clyde L. Hawn

ABSTRACT

This paper provides an explanation of how a company or firm initiates the process of incorporating Quality Assurance and Quality Control functions into their method of operations. It will explain how to organize for those functions, how the new organization can best serve the company or firm, and will finally give specific examples of Quality Assurance and Quality Control organizations that may serve as a guide to use in setting up a new Quality Assurance group within a company of firm.

QUALITY IN THE CIVIL INDUSTRY

When Appendix B, 10CFR50, was imposed upon the nuclear industry in 1971, the first question that was raised by most companies was, "Why are we being subjected to this unnecessary control?" Of all the industries in the United States, perhaps the hardest hit was the Civil Design and Construction Industry. For more years than most of us can remember, these people had been designing and building structures that were and still are more than adequate for their intended purposes. Of course, there were a few failures, but certainly nothing that indicated any basic flaw in either the design or the construction of these structures. During the period of this construction (almost from the inception of the building industry) the civil industry had been free from restraints. Occasionally a really hard-nosed building inspector would be encountered, and he would insist on building strictly in

---

[1]Quality Assurance Manager, Ebasco Services, Inc.

accordance with the plans and specifications, but for the most part, a large degree of flexibility was allowed in civil construction. Everyone, or almost everyone, connected with the industry knew what they were building, how to build it, and that if the building failed, they were responsible for the failure. The electrical and menchanical/piping fields were not subjected to quite the trauma as the civil industry at the imposition of Appendix B. 10CFR50. These groups had for years been working under the quality programs imposed by Underwriters Laboratories, NEMA, or the ASME Boiler and Pressure Vessel Code, with the resultant outside inspection agencies. But to the Civil Construction Industry, it was the start of something new.

In spite of the excellent record that had been achieved by the Civil industry in general, suddenly the U.S. Government imposed a mandatory requirement that the industry have a Quality Program, and specified what that program must include by defining 18 criteria that would be used as a basis for evaluating the program. With the imposition of this program, industry took one of two basic courses.

The first approach was one of trying to understand what was required of the industry, and see how this could be incorporated into the scheme of operations. An attempt was made to comply with the intent of the requirement. The Quality organization was fully integrated into the company's basic organization, and became a full-fledged partner with design, procurement, and construction. The companies that adopted this general philosophy were basically successful. As time went by, they were able to improve the interface between Quality and the rest of the company, and compliance with the Quality Program became a corporate fact of life. There were problems, to be sure, but most of these problems were solved.

The second approach was one of resistance. This philosophy was manifest by the attitude of, "Why do I have to put up with this requirement? I have been successful for over 25 (or some such number) of years, and did fine without a bunch of inspectors looking over my shoulder telling me what I'm doing wrong and holding up my work." Usually, to comply with the letter of the law, these companies formed a group called Quality Control or Quality Assurance, put them off to one side, paid lip service to them, and refused to integrate them into the company activities. This very quickly developed into an adversary attitude by both Quality and the other groups within the company. Quality was playing the role of a policeman, not as a part of the building process. The overall program was not successful, and no one group has come out the winner.

Obviously, the foregoing is a simplistic treatment, and no one company fits into either bracket. The attitudes and methods of any company is somewhere between the two extremes cited above. However, where a concerted attempt was made to incorporate Quality as a working part of the company, the success and acceptance of the program has been much greater than where the inclusion of Quality has been resisted.

Of course, the Quality organizations themselves contributed to the problem of acceptance. All too often the Quality personnel were working under the attitude that "Quality is not concerned with cost or schedule" or "The work may meet the requirements specified by the designer, but I would like to see it better." This attitude on the part of the Quality personnel did little to encourage other groups within the company to accept them as a working partner.

Today, within the Civil industry, we find that generally all companies engaged in nuclear power plant construction have matured to the point where they accept Quality as a fact of life within the nuclear industry. Quality personnel have accepted the fact that all they can require is adherence to the designer's requirements, not perfection. Designers have gotten to the point where they do not fly into a rage when a Quality engineer asks for documented clarification of an engineering requirement, and Quality personnel are beginning to realize that, while their primary responsibility is to insure the plant built properly, they do have some responsibility for cost and schedule. Again, the real world conditions do not show the above as a black or white situation, but some shade of grey. The goal now is to improve the Quality functions so they contribute to the maximum extent possible in assuring that nuclear power plants are built correctly, on time, and within the budgets that have been set.

ESTABLISHING THE ORGANIZATION

Including a Quality program for the first time in any company should be approached in the same manner as you would if you were going to initiate any other new activity. Certain basic rules must be established at the inception if your inclusion of this new activity is going to be successful. These are:

1. Management support for the program
2. A defined scope for the Quality program
3. Clearly defined lines of authority and communications
4. Simplicity, with adequate provisions for revision
5. Staffing with adequate numbers of competent personnel.

Let's examine each of these in detail.

Management Support for the Program

Management support for the Quality Program is mandatory if the program is to have any chance of success. This does not mean that management must always side with the quality department. It means that all levels of management understand what the purpose of a quality department is, and works with the quality department in a manner that allows them to achieve their purpose. It means that upper management is committed to the Quality Program for the company as it is stated in the company Quality Manual. Where quality programs have failed, the greatest number of failures can be traced directly to upper management's lack of support for the program. All too often, upper management does not include quality in the decision-making process, and then reacts negatively when the quality department raises questions on the adequacy of decisions that have been reached and actions implemented based on those decisions. Many times, quality does not know what is going on until work has been initiated, and some error has been made. Then quality is put in the positions of identifying a deficiency that possibly could have been avoided had quality been involved in the decision-making process. One specific example can possbily illustrate this better than any further discussion. On one power plant, the supplier of concrete reinforcing steel had a tensile testing machine capable of 150,000lb. pull. His contract required him to pull full diameter specimens of rebar, and required him to furnish, among others, number 14 and 18 reinforcing steel. Obviously, he could not pull full diameters on these larger bars with a 150,000lb. machine. When asked if it would be acceptable to pull reduced diameter bars, both construction and enginerring agreed to the acceptability of the program. They had reviewed ASTM A-615, the governing standard, and it allowed to use of reduced diameter specimens for tensile testing. So the contractor proceeded, and ultimately produced and shipped over 18,000 tons of bar so tested. The quality organization, upon receiving the steel, was forced to place the material in a "Hold" status, as there exists a regulatory guide published by the USNRC that requires full diameter testing of concrete reinforcing steel used in nuclear power plant construction. Had quality been involved in the original decision, they could have alerted both construction and engineering that they were proceeding down the wrong path. However, at that time, there was no corporate requirement to include quality in such decision-making, and the corporate attitude was to keep quality out of everything.

So the first key to a successful program is to include quality as a full-fledged partner in the firm. It won't solve all the problems, but it will avoid a lot of costly and time-consuming errors. How can this be accomplished? Upper management must insist that middle and lower management work with the quality department. The general tendency of many construction personnel is to make a decision when a problem is presented. That is their job. Make decisions and keep the work moving. No quality person should attempt to prevent or overrule that responsibility. But construction or engineering, or procurement should be sure that their decision is going to meet the overall requirements of the project, and not create a bigger problem than the one they were trying to solve initially. Without management support for the quality program, too many decisions are made that violate one or more of the requirements imposed on the plant, and quality was not aware that the decisions were being made.

A Clearly Defined Scope for the Quality Program

Next, define the scope of your program. Usually, there is some form of contract, purchase order, or work authorization that will specify the minimum program you must have. Study it carefully and understand it. Then analyze your operations to see if you want to include more than the minimums in your program. While this sounds ridiculous on the surface, it is good basic business. You may be furnishing concrete reinforcing steel to a nuclear project, in which you are contractually bound only to control that material to be used in the nuclear safety-related buildings.

Believe me, it is cheaper and easier to order and control all the reinforcing steel as if it were all going into the safety-related structure than to implement the controls necessary to segregate and account for two classes of rebar. The same goes for aggregates, cement, weld rod, and a multitude of other materials. Even moving crews of workmen around a jobsite can cause problems if they have two sets of rules to follow. So analyze what you want your program to do, and publish a clearly defined scope of work.

The other thing to consider as you define your program is to keep it simple. Analyze what approach you have used in the past. Does this meet the current requirements you are trying to attain? If so, use it. Is there something that can be improved upon? Modify to show the improvement. Above all, do not put unnecessary restrictions on your activities. The number of companies that have become bitter about nuclear construction because of quality program requirements is almost matched by the number of companies that have themselves committed to a program that is in excess of requirements. Putting in phrases because they sound nice, or adding requirements on your activities above those specified in the hopes that it will expedite approval of your program is dangerous and foolhardy. If your program is subject to approval of a higher agency, usually that agency will only notify you of the areas where your program is deficient, not where it exceeds requirements. But be assured that once you commit to meeting a requirement, you will be expected to comply with that commitment, and will meet some resistance when you later find the commitment is unnecessary, and try to revise your program to eliminate it.

Most good civil companies and firms have had a good approach to how to design and build. What has been needed to meet the requirements of Appendix B, 10CFR50, is a written definition of how they are doing and intend to continue doing what has made them successful.

Clearly Defined Lines of Authority and Communication

Next, clearly identify who is responsible for doing what. Remember, Quality is a two sided coin. There are those who are responsible for <u>attaining</u> quality, and those who are responsible for <u>verifying</u> quality. Don't attempt to give an individual responsibility to attain quality without the authority to direct work activities. Those who verify quality should be, and in nuclear construction <u>must</u> be, different from the attainers. The verifiers also must have a clear authority to report their results accurately without fear of reprisals. Certain functions on nuclear plant construction must be retained in the QA organization. These are tests, examinations, inspection, vendor qualifications, audits, document reviews, corrective action, and control of nonconforming items. The remaining functions can be distributed to other organizations, if you so desire. Just be sure the responsibilities are clearly defined. As a general rule keep the authority for performing a function at the lowest possible level required to perform a task or approve a task. The higher the level for performing an activity the more time it will take. Also, keep the number of people involved in making a decision to a minimum. On one project, all nonconformances must be presented to a nonconformance review board for approval of a disposition, and that board has six members, four of them managers. The costs of administering that program are astronomical, due to delays in processing nonconformance reports. At the most, three people should be involved in resolving nonconformances, and in some cases, only one or two, and none need be at the managerial level.

Simplicity, with Adequate Provisions for Revision

Keep the program as simple as possible, consistent with meeting the requirements imposed on you by your contract or other work authorization. All too often, with the best of intentions, a firm or company will commit to meet a program that far exceeds the basic requirements imposed on them by their client. Examples of this are things like stating that all welding on a given project will be performed by welders that are qualified per ASME Section IX or AWS D1.1. While this may be desireable, a welder tacking together temporary reinforcing steel supports certainly does not need to be qualified. If you commit him to be qualified, then you will be expected to qualify him, and have the documents proving he is so qualified. One company committed, back some years ago, to have all his welding inspectors qualified to AWS QC-1, when all that was necessary was that they be certified per ANSI N45.2.6. At the time he made the commitment, there were only 47 AWS QC-1 certified welding inspectors in the entire United States. It was literally impossible for him to comply with his own program requirements at the time he made the commitment, and the commitment was far beyond the requirements of his contract.

Adequate Staffing with Competent Personnel

No program, whether Quality, Design, or Construction, will have any prospect of success unless it has the personel to make it work. What constitutes a good Quality type individual? First, he must understand Quality as a philosophy and science. His is primarily the job of preparing a day-to-day history, adequately documented, of the work activities. This includes not only accepting those activities that are in accordance with the program requirements, but identifying those activities that are discrepant. To do this properly, he must know specifically what the requirements are, both from a quality program standpoint, and from what the designer has imposed. He must know and understand the various standards and codes used to manufacture and build the product, be it reinforcing steel, concrete, or building construction. Then he must keep abreast of new regulatory requirements, if they are imposed on the work in progress.

More importantly, he must be a person who wants to get involved in helping "make it right the first time," and not a "gotcha" artist who delights in letting other people make mistakes and then identifying the mistakes to management via nonconformance report or corrective action request. This does not mean to imply that he will not identify errors. It means the good quality man is more concerned with helping prevent errors than in letting them happen so he can identify them. Of course, to be any assistance in helping eliminate errors, the Quality personnel must be a part of the initial planning to perform a task, as was mentioned previously. It also means the good quality individual must be intimately familiar with what is being produced, how it is produced, and why certain steps are taken in the production. In short, he must be as familiar with the manufacturing and/or construction techniques as the foreman, supervisor, or superintendent directing the work. Being able to quote a standard or code verbatim, but not knowing what it means is next to worthless.

Adequacy is a more difficult term to evaluate. How many Quality people does it take to do a job? The numbers of people you need to assign to quality will depend on what you are trying to make or build. A guide to numbers is that your production or construction work should not be held up unnecessarily while waiting for a quality man to perform his necessary duties as defined by your program. This assumes that the quality personnel are diligent, and apply themselves to staying up with the work forces. Remember, not only do the quality personnel have to perform some function, but they must document the results of their inspection, surveillances, tests, or measurements. So they have to be allowed time to perform the function and to generate the necessary documents. If your work is properly scheduled, and the schedule is distributed so all departments can plan their work around the schedule, and if your personnel are actively busy, but you find you are waiting repeatedly for quality services, you probably are understaffed. Conversely, if you find your quality personnel are not busy most of the time, either on quality activities, or the related documentation, you are probably overstaffed. Of the two, probably a slight overstaffing is preferred to understaffing. It is foolish economy to unnecessarily hold up a concrete placement, with twenty or thirty craftsmen standing around idle and drawing pay for want of one more inspector to perform a preplacement inspection and release.

One more comment on quality personnel. Good quality people are hard to find. If you have them, and want to keep them, support them when they are doing their job. Remember the original requirement, that the program must have management support to be successful. We in the quality business expect people to argue with us. That is one of the normal reactions we expect when we challenge the acceptability of an item. However, excessive personal attack or verbal abuse is not acceptable. Many people, good people, have left a a company because of verbal abuse they have been subjected to when they were only trying to do their job. There aren't enough good Quality people in the business right now to allow any company or firm to loose them due to some shop foreman or construction supervisor heaping a pile of garbage on them for doing their job.

TYPES OF ORGANIZATIONS

So far we have discussed how to go about organizing to perform Quality functions, and what it takes to have a good program. Now lets look at some specific types of organizations. Before we start looking at the organization charts, we need to define some the responsibilities of the positions you will find on these charts. Lets further realize that you may want to use other terminology, or other assignments of responsibility for specific tasks. With that in mind, here are the duties and responsibilities

1. Quality Assurance Manager - is responsible for the overall administration of the Quality Program. He directs the activities of his three supervisors, and generally establishes the quality attitudes of the project or job.

    He interfaces with the client's quality personnel, subcontractor's quality personnel, regulatory personnel, etc. He is the advisor of top management of any adverse trends in quality, and is usually the final approval authority on procedures and instructions. He is the final authority on reviewing conditions for reportability under 10CFR50.55 (e) and 10CFR21.

2. Quality Assurance Supervisor - is responsible for supervising the Quality Assurance Engineers in the performance of their duties. He makes the audit schedule, reviews audits and surveillances, and perform a trend analysis of nonconformances. He is usually the clearinghouse for client or regulatory audits, assigning responsibility for responses, and forearding them to the originating agency.

3. Quality Control Supervisor - is responsible for supervising the planning and performing of inspections and tests. He is responsible for the administration of training programs for inspectors, and usually is the person responsible for certifying the inspectors.

4. Quality Records Supervisor - is responsible for supervising the Quality Records vault. He is responsible for establishing the method of filing records so they are safe, and retrievable. He supervises the records specialists in the review and filing of the records. He prepares the records for turn over to the client at the completion of the project.

5. Quality Assurance Engineers – perform audits, surveillances, and procedural reviews. They review procurement documents, qualify suppliers, review specification changes, and generally assure overall compliance with regulatory and program requirements.

6. Quality Control Engineers – plan and supervise the activities of the inspectors. They are responsible to perform reviews of inspection results, and verify the acceptability of activities. They initiate nonconfomance reports, and assure nonconforming items are properly controlled. If there is a testing laboratory, it is under the supervision of a Quality Control Engineer.

7. Inspectors – are responsible for performing the first line inspection, and recording the results of the inspections. They send the results of the inspections to their descipling Quality Control Engineer for review and evaluation.

8. Quality Records Specialists – are responsible for the final review and filing of quality records.

With those definitions understood, let us now look at some basic Quality Organizations.

Basic Quality Organization

To describe a basic quality organization for a nuclear power plant, examine attachment "A". This is probably the type of organization that would fit a standard condition, if such a condition existed. This is the type program that would be put into effect if you were responsible for a full quality program, encompassing the entire 18 Criteria of Appendix B, 10CFR50, and were charged with doing work on a multiple discipline basis. As you would expect, there is a Quality Assurance Manager, and three groups, Quality Assurance, Quality Control, and Quality Records, all headed by a supervisor. Under Quality Control there are Quality Control Engineers for each discipline, with inspectors reporting to the Quality Control Engineers. Similarly, there are Quality Assurance Engineers for each discipline. The duties and responsibilities for these people would be as previously described. This seems to be the form of Quality Program most often arrived at, whether it is most appropriate or not. Now let's see how we can modify this to meet specific needs.

Limited Function Organization

Now look at Attachment "B".

For this organization, the following assumptions apply:
1. You are only responsible for the supply and delivery of concrete.
2. Your aggregate facility is located some distance from the batch plant.
3. You are required, by your contract, to forward all completed quality records to your client immediately after their review and acceptance.

In this type situation, your quality program can be reduced to only cover those activities necessary. There is no need to have a large staff of Quality Assurance Engineers, and there is no need to have a Quality Records vault, or large staff of records specialists. Duties and responsibilities for Quality Assurance and Quality Records have been assumed by your client. Very possibly, only one person is needed in each of these areas. The Quality

Assurance Engineer would primarily be concerned with performing internal audits of your overall program to insure it is continuing to meet your program requirements. In addition, he would review and approve supplier Quality programs, from such suppliers as cement manufactureres and admixture manufacturers. He would review and approve purchase orders, and possibly perform some sort of surveillance activities over your production at both the batch plant and aggregate plant.

The Quality Records Supervisor would review all records to see they are complete and correct, and promptly forward them to your client. He would, in this case, have no responsibility for establishing any filing system, or for assuring the records are retrievable for any customer or regulatory auditors.

Quality Control would probably need one QCE at the batch plant and one at the Aggregate Plant, due to the distance separating them. Each QCE would be charged with the responsibility for developing inspection planning for his facilities, for directing the inspection activities in accordance with the planning, and for reviewing the completed inspection and test reports to insure the necessary requirements were being met. Where discrepant conditions were detected, he would identify them, and insure the discrepant material was properly controlled until the disposition had been assigned by the designated agency.

The inspectors would perform the necessary inspections and tests in accordance with the planning provided by the QCE's, and document the results of the inspections.

QA/QC FUNCTIONS

Basic Organization with Split-out Functions

Now examine Attachment "C". Where, due to unique situations, some facet of the basic organization becomes too large or too complex for one group to handle, due to any number of reasons, you may have to split up one or more groups. Assume the following:

1. The receiving inspection areas (mulitple) are located at some distance from the main project or construction area.
2. There are mulitple suppliers that must be approved, and travel to their facilities is necessary to monitor on-going work.
3. Quality Assurance is required, by your contract, to review and approve all purchase orders, revisions to purchase orders, and design changes.

Then you may want to split up both Quality Control, and Quality Assurance, and assign one additional supervisor to each department. Quality Control would be split into Construction Quality Control and Receiving Quality Control. Because of the distances between the various receiving and storage areas, and the construction site, it may be impossible for one supervisor to excerise the necessary controls over both areas. Similarly, due to the abnormally large amount of work, Quality Assurance is split. One group, Quality Assurance Engineering, is responsible for reviewing and approving all purchase orders, changes to purchase orders, and design changes. They are also responsible to qualify suppliers, and review supplier activities in their shops to insure the suppliers are meeting their requirements. The Audit group is solely responsible for performing audits, of your own activities, and of your subcontractor and suppliers.

Organization by Discipline

Attachment "D" shows and organization by discipline, rather than by function. Under each of the disciplines, there is a Quality Supervisor that is responsible for both the Quality Assurance and Quality Control for his discipline. This can be a viable approach to a problem, provided you adequately identify where the interface boundary is between the disciplines. Traps to avoid on this type organization are many, and I don't really recommend it. However, it has been successfully used.

SUMMARY

In summary, I would like to emphasize the following points. First, review your requirements as stated in your work authorization, be it contract, specification, Safety Analysis Report, or whatever. As simply as possible, define how you intend to meet those requirements. Define your organization, including lines of authority and responsibility. Make your program flexible, with provisions for easy and rapid revision. Staff your department with adequate numbers of competent personnel. And finally, get upper management support for your program. Without it you have no chance for success.

QA/QC FUNCTIONS

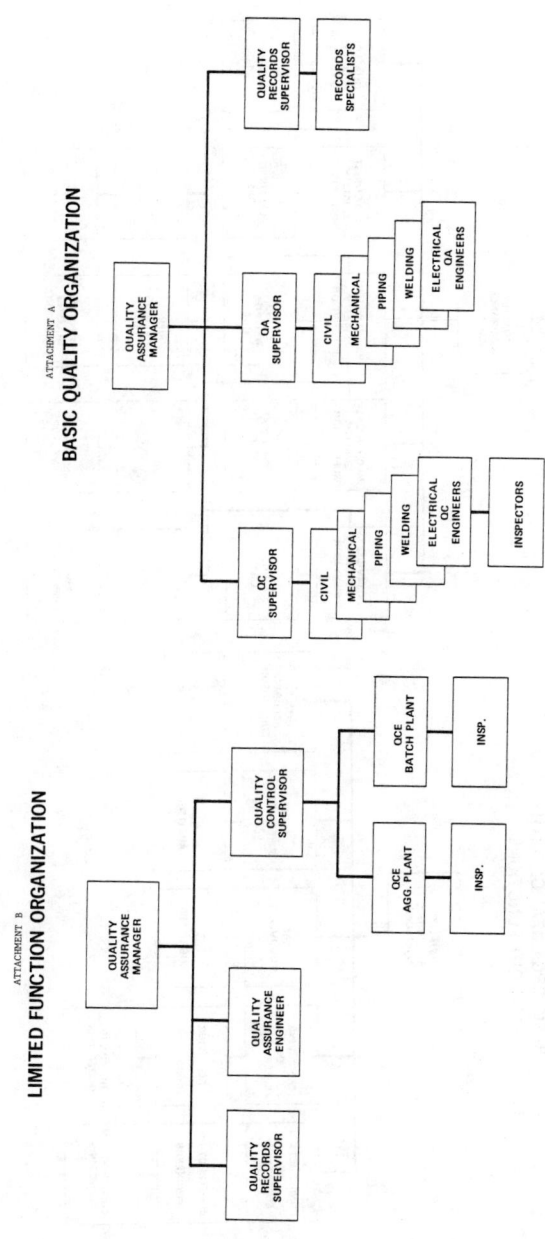

POWER GENERATION FACILITIES

BASIC ORGANIZATION WITH SPLIT-OUT FUNCTIONS

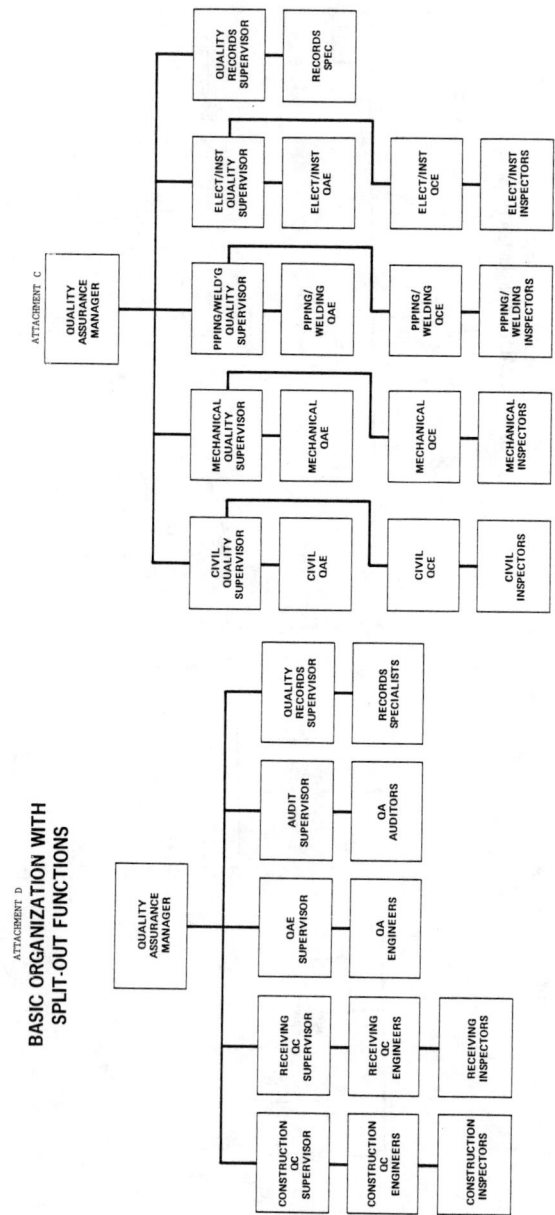

QUALIFICATION AND TRAINING OF INSPECTORS

by

Donald R. Johnson[1] and Dennis L. Vanderpol[2]

Abstract

The background requirements and actual practices used for the qualification and training of personnel responsible for quality verification inspection during the construction of nuclear power generation facilities are discussed. The factors used to evaluate the qualifications of inspection personnel are described. A typical method for training inspectors is also presented.

Requirements

The requirements for qualification and training of nuclear power plant construction inspectors can be traced to several sources, i.e., the federal government, industry, utility and constructor. The extent to which these requirements are imposed on a project is provided in the appropriate contract documents.

Government Requirements

Title 10 of the Code of Federal Regulations, Chapter 1, Part 50, Appendix B - Quality Assurance Criteria for Nuclear Power Plants and Fuel Reprocessing Plants (commonly called 10CFR50 App B or the 18 Criteria) is generally considered the top document in terms of requirements for all quality assurance functions, including qualification and training of inspectors. There are no specific qualification and training requirements in 10CFR50 App B, however, criterion II does state in part, "...(QA) program shall provide for indoctrination and training for personnel performing activities affecting quality..."

The Nuclear Regulatory Commission (NRC) has developed a series of documents called regulatory guides. Regulatory guides are issued to describe specific methods acceptable to the NRC staff for complying with the general requirements of the Commission's regulations. Regulatory Guide 1.58 describes a method acceptable to the NRC for meeting their requirements for qualification and training of inspectors. It does this by endorsing, with some exceptions, the industry standard ANSI/ASME N45.2.6.

---

[1] Manager of Quality, Bechtel Power Corporation, Richland, Washington

[2] Chief Construction Engineer, Bechtel National Inc., San Francisco, CA

Industry Requirements

The nuclear industry, acting through the consensus committee process for American National Standards which is composed of representatives from utilities, architect/engineers, constructors, suppliers and NRC, has issued an ANSI standard to specifically address inspector qualifications and training. That document, ANSI/ASME N45.2.6-1978, <u>Qualification of Inspection Examination and Testing Personnel for Nuclear Power Plants</u>, provides the specific requirements that virtually all inspector qualification and training programs in the industry are written to address.

Another industry standard which establishes requirements for inspector qualification and training is ANSI/ASME NQA-1-1979, <u>Quality Assurance Program Requirements for Nuclear Power Plants</u>. This recently issued standard (which is a consolidation of N45.2-1977 and several daughter standards, including N45.2.6-1978) uses a combination of Basic Requirements, Supplements and Appendices to cover inspector qualification and training. Since few organizations have had NAQ-1-1979 imposed on them, and because the requirements of NQA-1 (Basic Requirement 2, Supplement 2S-2, and Appendix 2A-1) are essentially the same as N45.2.6-1978, this paper is structured to describe a method of satisfying N45.2.6-1978.

Client Requirements

From our point of view, the client is normally a private or public utility serving as owner/operator of a nuclear power generation facility. Often times, additional requirements are imposed on the project constructor by the client as a result of his experience or special needs concerning inspector qualifications and training.

Constructor Requirements

Finally, the constructor may self impose special or unique requirements on the qualification and training of his inspectors for a particular project or series of projects. This may be predicated on past experience from other projects, or unique needs or skills which the constructor wants to develop among his inspection personnel.

Qualification

The qualification of personnel employed as nuclear power plant construction inspectors is based on three factors, i.e., experience, education and performance evaluation. The first two an individual "brings with him." The third is the process used by the employer to determine the inspector's level of knowledge/proficiency obtained from his experience and education, as well as the employer's pre-evaluation orientation and training program.

Our basic experience/education requirements for qualification come directly from the ANSI N45.2.6 recommendations. However, we recognize that variations from these requirements are not only common but necessary to address the diversity of backgrounds in experience and education we encounter in our "new hire" inspectors. This need to recognize and accept variations in the experience/education requirements is well established in the N45.2.6 standard which states in part,

> "These education and experience recommendations should be treated to recognize that other factors may provide reasonable assurance that a person can competently perform a particular task. Other factors which may demonstrate capability in a given job are previous performance or satisfactory completion of capability testing."

Experience

Experience is listed first because we believe it is more important than formal education in providing the basis for overall qualification. Our inspectors come from a variety of backgrounds, and they have gained their prior inspection experience on all kinds of heavy construction projects, including highway, dam, commercial, industrial, power plant, transit system and airport construction. We look for individuals who are first of all construction "work wise" and secondly have previously performed and documented quality verification inspections.

Education

The formal education an individual receives is a valuable asset, and can aid the performance of the inspector by providing a basis to analyze the inspection problems he encounters daily. College level work in an engineering, science or technology curriculum often provides the "problem solving" technique so helpful to the inspection process. What do I know (the design documents); what don't I know (the extent that the work meets the design documents); and how can I best get from the known to the unknown (perform inspections and document the results to verify conformance of the completed work to the governing specifications and drawings).

Performance Evaluation

Our performance evaluation process includes initial evaluation (to certify or re-certify an inspector) and the monitoring of on-the-job performance to verify continued satisfactory performance.

The "initial" evaluation consists of both an oral examination and a performance demonstration. The basis of this evaluation is our total quality control program and inspection procedures which the inspector will use in the field while performing actual verification inspections.

Appendices A and B show the Oral Examination Record and Performance Demonstration Record which we use for our evaluation. The checkpoints are entered by the examiner to identify the elements covered during the evaluation.

The completed Appendices A and B plus the inspection documents used during the performance demonstration provide the objective evidence upon which we certify the qualifications of our inspectors.

Training and Indoctrination

Our training and indoctrination program begins with each person newly assigned to the project before he is certified and permitted to perform inspections. It doesn't end until the project is completed. The initial pre-certification training and indoctrination covers both general topics like the organization, purpose and objectives of the total quality control program, and specifics like the preparation and use of inspection plans. All of the jobsite training is provided by inspection supervisors and discipline engineers.

Procedures and Methods

At least once each month, a one-hour minimum training session with a predetermined agenda is held. These training sessions are attended by the inspectors involved with the activities being discussed, which are consistent with work operations in progress. Typical subjects are the use of inspection plans, documentation requirements, applicable codes and standards, processing of nonconforming items, surveillance over subcontractors, calibration of measuring and test equipment, and other related topics.

Inspection Plans

Orientation and training in the application and use of each new inspection plan is provided to the inspectors prior to implementation of the plan in the field. The inspection plan serves double duty as a lesson plan for this training. The training is conducted in sufficient detail and repeated with sufficient frequency to assure the inspectors fully understand the inspection and documentation requirements as well as the acceptance criteria before beginning the designated inspections.

Training Records

A complete file is maintained at the jobsite to document the training and indoctrination received by each inspector. This file shows the number of sessions attended by each inspector, the subject covered in each session, the name of the instructor, the length of the session and the date the training was received.

Conclusions and Future Directions

- Requirements for the qualification and training of inspectors are derived from government, industry, client and constructor needs.

- Qualification is based on a combination of experience and education and a performance evaluation of individual skills and abilities.

- A continuous program of inspector training and indoctrination is not only an asset but a necessity.

The qualification and training of inspectors has undergone significant changes over the years to reach the formal level of detail and procedural requirements which we find today. As illustrated by this paper, the nuclear construction industry has established a complex and costly system of qualification and training for our inspectors. Whether or not it is too complex or too costly is a matter of subjective judgment.

In the past, when the system failed (a requirement missed or an inspection error made) we often times changed the system thereby making it more complex and costly to implement, instead of working with the individual to improve implementation. If we maintain this type of ratcheting approach, we will price our qualification and training program (and nuclear power) right out of the market.

INSTRUCTIONS FOR COMPLETING APPENDIX A

BLOCK NO.	ENTRY INFORMATION
1	Enter the candidate's last name, first name and middle initial.
2	Enter the candidate's Bechtel Employee Number
3	Enter the level of certification the candidate is to be evaluated for.
4	Enter the date the oral examination was completed.
5	Enter the Bechtel Project Number or home office where the evaluation was administered.
6	Enter the quality control activity covered in the quality control instruction or procedure shown on line 7, e.g., Receiving Inspection, Installation Inspection, Calibration Testing, Contract Surveillance, etc.
7	Enter the quality control instruction or procedure identification number and title used to conduct the evaluation.
8	The purpose of the oral examination is to ensure that candidate has the necessary knowledge to satisfactorily perform his assigned duties in accordance with the requirements of the construction quality control program. The examiner should ask questions concerning the candidate's responsibilities for performing quality verification or surveillance inspection activities in accordance with the specific requirements contained in the quality control instruction or procedure listed in block 7. List each requirement covered as a separate check point. If the candidate is being examined for Level II Certification, the examiner should also ask questions concerning the construction quality control program. List each section of the construction quality control manual covered as separate check points.
9	Enter "S" for satisfactory. Enter "U" for unsatisfactory.
10	If the candidate has demonstrated the ability to perform satisfactorily, enter the words: "Recommend Certification." If, during the course of the oral examination, some additional areas of study are considered advisable, the areas shall be identified, along with a completion date for accomplishing the additional study. If the candidate fails to complete the additional areas of study within the time allowed, the candidate's certification shall be suspended until the areas of study are satisfactorily completed. If the candidate fails to demonstrate the ability to perform satisfactorily, enter the words: "Do not recommend certification at this time." Explain why and recommend needed training.
11	Enter your signature.
12	Enter your level of certification.
13	Enter the expiration date of your certification.

ORAL EXAMINATION RECORD

NAME [1] _____ EMPLOYEE NO. [2] _____

EVALUATED FOR LEVEL [3] _____ DATE [4] _____ LOCATION [5] _____

QUALITY CONTROL ACTIVITY [6] _____

QUALITY CONTROL INSTRUCTION OR [7] _____
PROCEDURE

_____

CHECK POINTS [8]	RATING [9]
1.	
2.	
3.	
4.	
5.	
6.	
7.	
8.	
9.	
10.	

REMARKS [10] _____

_____

_____

_____

_____

SIGNATURE [11] _____

LEVEL OF EXAMINER [12] _____

EXAMINER'S CERTIFICATION
EXPIRATION DATE [13] _____

SEE REVERSE SIDE FOR INSTRUCTIONS

APPENDIX A

INSTRUCTIONS FOR COMPLETING APPENDIX B

BLOCK NO.	ENTRY INFORMATION
1	Enter the candidate's last name, first name and middle initial.
2	Enter the candidate's Bechtel Employee Number
3	Enter the level of certification the candidate is to be evaluated for.
4	Enter the date the performance demonstration was completed.
5	Enter the Bechtel office or project number, where the evaluation was administered.
6	Enter the quality control activity covered in the quality control instruction or procedure shown on line 7, e.g., Receiving Inspection, Installation Inspection, Calibration Testing, Contract Surveillance, etc.
7	Enter the quality control instruction or procedure identification number and title used to conduct the evaluation.
8	Enter the primary quality verification and/or surveillance inspection activities in the performance demonstration.
9	Enter "S" for satisfactory. Enter "U" for unsatisfactory.
10	If the candidate has demonstrated his ability to perform satisfactorily, enter the words: "Recommend certification." If, during the course of the performance demonstration, some additional areas of study are considered advisable, the areas shall be identified, along with a completion date for accomplishing the additional study. If the candidate fails to complete the additional areas of study within the time allowed, the candidate's certification shall be suspended until the areas of study are satisfactorily completed.
	If the candidate fails to demonstrate the ability to perform to acceptable standards, enter the words: "Do not recommend certification at this time." In addition, enter specific recommendations for training to be completed prior to reevaluation.
11	Enter your signature.
12	Enter your level of certification.
13	Expiration date of examiner certification.
GENERAL NOTE:	The purpose of the performance demonstration is to ensure that the candidate can satisfactorily perform his assigned duties and activities in accordance with the requirements contained in the quality control instruction or procedure listed in block 7. List each requirement covered as a separate check point. The examiner shall require the candidate to actually perform the activities or tasks required by the quality control instruction or procedure. See example below for typical check points.

CHECK POINTS: EXAMPLES FOR CONDUCTING RECEIVING INSPECTION PERFORMANCE DEMONSTRATION.

- REVIEW OF INSPECTION CRITERIA REQUIREMENTS
- DOCUMENTATION REVIEW & TRACEABILITY ASSURANCE
- VISUAL INSPECTION
- MARKING & TAGGING
- WITNESSING AND DOCUMENTING TEST RESULTS
- PREPARATION FOR STORAGE
- SUPPLEMENTARY RECORDS REVIEW
- REPORTING & PROCESSING EXCEPTIONS
- CONTROL & PROCESSING OPEN INSPECTION RECORDS

CHECK POINTS: EXAMPLES FOR CONDUCTING INSTALLATION INSPECTION PERFORMANCE DEMONSTRATION

- REVIEW OF INSPECTION CRITERIA REQUIREMENTS
- COMPLETING THE PREREQUISITES
- IN PROCESS SURVEILLANCE INSPECTION REQUIREMENTS
- FINAL INSPECTION REQUIREMENTS
- DOCUMENTING QUALITY VERIFICATION INSPECTIONS & TEST
- PREPARATION & REVIEW OF SUPPLEMENTARY RECORDS
- REPORTING & PROCESSING EXCEPTIONS COMPLYING WITH CALIBRATION REQUIREMENTS
- CONTROL & PROCESSING OPEN INSPECTION RECORDS
- FINAL INSPECTION REQUIREMENTS

PERFORMANCE DEMONSTRATION RECORD

NAME [1] _____ EMPLOYEE NO. [2] _____

EVALUATED FOR LEVEL [3] _____ DATE [4] _____ LOCATION [5] _____

QUALITY CONTROL ACTIVITY [6] _____

QUALITY CONTROL INSTRUCTION OR [7] _____
PROCEDURE

_____

CHECK POINTS	RATING
[8]	[9]
1.	
2.	
3.	
4.	
5.	
6.	
7.	
8.	
9.	
10.	

REMARKS [10] _____

_____

_____

_____

_____

SIGNATURE [11] _____

LEVEL OF EXAMINER [12] _____

[13] _____

SEE REVERSE SIDE FOR INSTRUCTIONS

APPENDIX B

QA/QC TRAINING, CERTIFICATION/RECERTIFICATION

by

Kevin T. Kimmel[1]

ABSTRACT

Establishing a program which meets Quality Assurance Training requirements poses many problems. Several approaches to solving these problems are presented in this paper.

QUALITY ASSURANCE TRAINING

Quality Assurance (QA) has become an increasingly important aspect of the nuclear industry in the last decade. The field has gained particular attention since the TMI Unit 2 accident. QA has experienced rapid growth in meeting the demands of the nuclear industry in the post-TMI regulatory environment. This growth, coupled with constantly changing regulations, has resulted in increased needs for trained QA personnel.

The proper development and implementation of a training program is necessary to meet organizational needs and regulatory requirements. More and more emphasis is being placed upon using "properly qualified personnel" to perform tasks affecting quality; the "proper qualification" of the personnel demands an adequate training program.

QA training contains several inherent problems. Ideally, a training program should be the result of planning and assessment. Realistically, it is more often a reaction rather than a planned action. Results are needed immediately in many cases, and the organization frequently will develop a program which just meets the regulatory requirements, particularly since training funds come from the organization's overhead dollars. Training may not always receive the attention necessary because it is an overhead function and quite frequently, if company

---

[1]Indoctrination and Training Coordinator, Quality Assurance Division, Gilbert/Commonwealth, Reading, PA

budgets dictate reducing overhead, the training organization is the first area of scrutiny. These problems must be considered when developing a training program which will continue to meet regulatory requirements.

Basis for QA Training Requirements

In developing a QA/QC training program which adequately addresses regulatory requirements, it is necessary to develop an understanding of the current codes and standards used by industry for training. A quick review of the more commonly used codes and standards will show the relationship between each regulation, and the applicability to QA.

Code of Federal Regulations. A QA training program is not only a result of organizational needs; it is required by federal regulations. The Code of Federal Regulations (CFR) forms the foundation for all QA training requirements. 10CFR50 gives the requirements for obtaining a permit to construct and a license to operate a facility which produces or uses special nuclear material. (This includes nuclear power plants.) In applying for a construction permit the licensee must submit a Preliminary Safety Analysis Report (PSAR) to the NRC for approval. The PSAR is in part a description of the type of facility the licensee intends to build and must contain a description of the licensee's plan for the "...organization, training of personnel, and conduct of operations..." (10CFR50 Para. 50.34 a 6) and a description of the QA program. In order to obtain an operating license for the facility, the licensee must submit a Final Safety Analysis Report (FSAR) to the NRC. The FSAR is a description of the facility as-built and must include the licensee's personnel qualification requirments.

Appendix B of the CFR specifies personnel training in the following criteria:

II. QA Program. "The program shall provide for indoctrination and training of personnel performing activities affecting quality, as necessary, to ensure that suitable proficiency is achieved and maintained.

IX. Control of Special Processes. "Measures shall be established to ensure that special processes, including welding, heat treating and nondestructive testing are controlled and accomplished by qualified personnel..."

XVII. Records. "...shall include closely-related data such as qualifications of personnel..."

XVIII. Audits. "...audits shall be performed...by appropriately trained personnel..."

In effect, 10CFR50 Appendix B requires: 1) program indoctrination of all personnel; 2) training and certification of inspection and test personnel; 3) training and certification of audit personnel; and 4) maintenance of personnel training and certification records.

Both 10CFR50 and Appendix B require that the licensee ensure that proper indoctrination and training of personnel is accomplished; however, neither 10CFR50 nor Appendix B specifies how this training is to be accomplished nor what the scope of the training is to be. This absence allows considerable program flexibility. The licensee has the freedom to develop a program (which the NRC has to approve, remember!) to suit his individual needs.

Not being specific can be as dangerous as overspecifying. Appendix B, in allowing for flexibility, has now allowed licensees to develop any type of approach to training they choose without guidelines. While these programs might all be in compliance with Appendix B, each would have to be reviewed separately in detail by the NRC prior to approval and issuance of a construction permit or an operating license. This process would be very time-consuming and would definitely not be cost-effective.

A much more effective approach would be to develop a standard format for use in meeting training requirements, resulting in:

1) Quicker training program reviews by the NRC -- once the original format was approved.

2) Assurance of adequate training personnel because all would receive training in accordance with a standard accepted program.

3) Increased flexibility in interfacing between AE's licensees, construction firms and vendors since QA training programs would be compatible.

4) Cost-effectiveness -- Organizations could give new employees credit for training received from previous employers avoiding unnecessary "retraining".

<u>Industry Standards</u>. Several industrial training standards have been developed which offer uniform methods for licensees to use in meeting the training requirements specified in 10CFR50 and Appendix B. However, several problems inherent in the development and application of these standards become apparent which must be taken into consideration when developing and implementing a QA training program:

1) A standard is developed only as the need presents itself, and the standard therefore usually addresses a specific program area (i.e., inspector training, NDE personnel training, or auditor training). This results in numerous standards being required for a complete training program, rather than one overall industry standard.

2) The development and review of a standard takes considerable time: a) to identify the specific need; b) to develop a standard which

adequately addresses the need; and c) for the NRC to review and approve the standard. Thus, it may be several years between the time the need is identified and the time the NRC approves a standard method of meeting the need.

3) As the standard is used, it may become apparent that certain areas require addressing which are presently not included in the standard, or (partly due to the time required for development/approval) certain requirements may no longer be necessary. Moreover, (again partly due to the long development/ approval time) new situations may require changes to the standard in order that they may be adequately addressed. This results in revisions to the standards.

4) Some standards may invoke other standards and may amplify the requirements of these other standards. In this case, compliance with a standard may vary with its applications.

The process becomes quite confusing since there are several standards with several revisions (which may all be approved by the NRC) with various modifications, depending upon the standard's application. It can easily be seen how establishing and managing a QA/QC training program can be a full-time job for an experienced staff of personnel.

Several standards are available which have been approved by the NRC as acceptable methods for training. These include several American National Standards Institute (ANSI) publications, American Society for Nondestructive Testing (ASNT) standards, and American Society for Mechanical Engineers (ASME) codes.

EXISTING TRAINING PROGRAMS

ANSI

In an attempt to develop a standard means for complying with 10CFR50 QA program requirements, ANSI published N45.2 which was reviewed and approved by the NRC as an acceptable method for implementing a nuclear QA program.

After developing N45.2, ANSI went on to develop a series of daughter standards to N45.2 which addressed specific program areas and offered standardized means for meeting specific program needs. Of particular interest to QA training are N45.2.6 and N45.2.23.

ANSI N45.2.6 describes a method for the qualification and certification of inspection, examination and testing personnel. The standard describes responsibilities associated with three levels of qualification and the educational and experience requirements for each level. This standard specifies that training shall include indoctrination of personnel to various codes and standards with which they will be working.

ANSI N45.2.23 describes a method for the qualification and certification of auditing personnel. This standard lists training, education and experience requirements for qualification and certification of personnel performing duties as lead auditors and gives requirements for the training and qualification of personnel performing duties as members of an audit team.

ASNT

ASNT Recommended Practice No. SNT-TC-1A provides a recommended program for the training, qualification and certification of NDE personnel. This standard allows the licensee to modify his NDE training program to meet his specific needs. ANSI N45.2.6 and ASME requires NDE personnel to be certified in accordance with this standard.

ASME

The ASME Boiler and Pressure Vessel Code is a code designed to standardize the design, fabrication and testing of pressure vessels and references other standards to be used for personnel training. 10CFR50 requires certain components of nuclear facilities to be designed, fabricated and tested in accordance with specific editions of this code. Section III Division 1 applies to initial design and fabrication, and Section XI applies during operational in-service testing of systems and components. Both sections require NDE personnel to be certified in accordance with SNT-TC-1A; however, the code modifies the requirements of SNT-TC-1A. These amplifying instructions vary with the code year and the specific code addenda.

Section III Division 2 applies to the construction of concrete reactor vessels and containments. Appendix VII to Division 2 gives requirements for the qualification and training for concrete inspection personnel. Appendix E (a nonmandatory appendix) gives guidelines for a program for the training, evaluation and certification of Level I and Level II concrete inspection personnel.

Regulatory Guides. Regulatory guides are the guidelines published by the NRC which describe methods acceptable to the NRC for implementing regulations. Several of these guides have been published which give the NRC staff positions on various codes and standards and may be used with the codes and standards they endorse to assure compliance with NRC regulations.

PROGRAM DEVELOPMENT

Commitment to Codes and Standards

In developing a training program it is first necessary to determine the applicable codes and standards to which the program will be designed. The importance of this identification is quickly recognized when one considers the numerous codes and standards and various revisions to them and the multitude of cross-references between standards and revisions. When developing program procedures, consider meeting the most restrictive requirements established by the codes and standards which the organization must meet. Different standards often address different areas of training and very frequently overlap (e.g., NDE training requirements, SNT-TC-1A and ASME Sections III and XI). It is often convenient in these areas of overlap to adopt the most restrictive requirements, thus ensuring that minimum requirements will be met.

It may also be advantageous to design the program to meet two revisions of the same standard (e.g., NDE training and certification in accordance with both 1975 and 1980 revision to SNT-TC-1A) in order to avoid revising the training program if the later revision is adopted in the future.

Procedures

Once it has been determined which requirements apply to the training program, the next step is to develop procedures for the program's implementation. The training procedures should be written by personnel experienced in applicable codes and standards. Since experienced personnel are not always available, as a minimum, the procedures should be reviewed by the persons who will be directly working with them (such as designated corporate examiners).

Several points to consider in writing the training procedures are:

1) Applicability -- to whom do these requirements apply? This should be clearly stated in the procedure to avoid confusion.

2) Responsibilities -- it is important to spell out who does what under the procedure so that duties are clearly understood.

3) References -- the regulations, codes and standards, or organizational procedures which are the basis for this procedure or provide additional information or clarification on this procedure should be listed. This list should include any procedures which are referenced or invoked by this procedure.

4) Requirements for initial certification under this procedure -- include necessary experience, training and education (and proof of these!).

5) Maintenance of proficiency -- what is necessary to maintain certification under this procedure once initial certification is attained. Include information about required periodic evaluations, testing, etc.

6) Recertification requirements -- what must be done if the initial certification lapses or expires. This may require complete recertification, proof of proficiency, or managerial recommendations.

7) Any factors affecting the status of certification -- this may include any required periodic physical examinations such as those specified by SNT-TC-1A, ANSI N45.2.6 and ASME.

8) Transfer of certifications from previous employer -- are any certifications from previous employers acceptable? What documentation is required of previous certifications?

Training Organization

It is important that the training program be given the proper management support. This point cannot be overemphasized. 10CFR50 Appendix B Criterion I requires that organization personnel "...performing quality assurance functions shall report to a management level such that the required authority and organizational freedom, including sufficient independence from cost and schedule, when opposed to safety considerations, are provided." Since the training organization's primary function is to ensure that the personnel performing quality related activities are properly qualified and adequately trained, the training organization is performing a quality assurance function and should have sufficient organizational freedom to carry out its function. If the training organization is answerable to lower levels of management, the training organization will be susceptible to guidance from many different sources, and conflicts of interest between cost, corporate needs and regulatory requirements will invariably develop. Having the training organization answerable to a sufficiently high level of management will also avoid (or at least control) instances of "crisis certification" (certified body needed to fulfill a commitment yesterday).

It is necessary that the training organization have a centralized location for program control, particularly if the company has several branch offices. This centralized location should have total responsibility for program policy, procedure generation and revision, and record control and retention.

In all organizational structures it is necessary to have someone overseeing operations -- someone "in charge" to give direction and guidance. If the program involves several different types of certifications (i.e., auditor, NDE, ASME, and ANSI N45.2.6) in which company representatives are designated as responsible for each type of certification, the representatives should be answerable to an overall program coordinator. Having more than one company representative responsible for one type of certification allows for several program interpretations at once and will result in inconsistencies in certification procedures and certification documentation.

Records

Perhaps the most important portion of the training program is record management. Certification records are of primary importance in Quality Assurance and must be continually updated. Remember, you only have as many certified people as you can prove on paper.

A few points to consider in establishing record requirements:

1) Keep records as simple as possible. This may be a bit of a trick because records are auditable. A balance between simplicity and adequacy must be reached. There may be a tendency to lean towards overdocumentation. This surplus can present a problem since as records become more complex the errors and discrepancies will increase in proportion to the complexity.

2) Require documented proof of previous certifications, training and experience. Again, if audited you can only prove what you have on paper.

3) How long should records be maintained, and what are the storage requirements? ANSI N45.2.9 gives guidelines for record retention and specifies retention times for various records. If the organization is involved with several projects (as in the case of AE firms), it may be advantageous to maintain records for a lifetime on microfilm rather than establish retention times for individual project records -- particularly if certified employees are involved in several projects.

In-Service Training

Once the training program is implemented, consideration should be given to periodic proficiency training in order to keep personnel updated on regulatory changes and procedural revisions which may affect them. This update may be accomplished through formal presentations or through "required reading." Records of this training should be included in personnel certification files.

SUMMARY

Inherent problems of a QA training program must be addressed in the initial program planning stages. In order to be effective and to continue meeting regulatory requirements, the training program must have adequate management support, must be staffed by properly qualified personnel and, above all, must have complete records of personnel certification. Meeting these conditions will result in an auditable training program which will satisfy QA Training requirements.

CIVIL ENGINEERING UNIT DOCUMENTATION PRACTICES

by: James H. Olyniec[1], M. ASCE

ABSTRACT

Quality control documentation is the life-line to plant licensing. Quality built into the work will insure that construction will be allowed to proceed. However, documentation provides the final link to an operating permit. Straightforward, easy to understand procedures that clearly define the inspection and acceptance criteria provide the groundwork for acceptable documentation. This paper presents the documentation practices for selected civil engineering inspection activities during the construction of the Bellefonte Nuclear Plant. Additionally, other selected documentation practices at this and other TVA nuclear construction sites are reviewed.

INTRODUCTION

The Tennessee Valley Authority (TVA) is an independent agency of the Federal Government established under the TVA Act of 1933. The Act empowers TVA to develop the natural resources of the Tennessee River Valley, an area encompassing parts of seven southeastern states. To this end, the construction and operation of an integrated power network is essential. Early power facilities in TVA's network generally consisted of hydro-electric stations, followed by coal-fired steam-electric stations, and more recently nuclear facilities. With four nuclear units in operation, and nine more under construction at six different sites, TVA is in the midst of a concentrated design and construction program. The Bellefonte Nuclear Plant, located near Scottsboro, Jackson County, Alabama, is a two-unit plant using Babcock & Wilcox pressurized water reactors, with a system generating capacity of 2664 MW. Design and construction activities are being handled primarily by TVA personnel, as is the case with the majority of TVA projects. All quality assurance and quality control activities are handled by TVA. Construction operations began in late 1974, and fuel loading of unit 1 is scheduled for early 1983.

GENERAL

10CFR50 was established by Congress to govern the licensing of nuclear power plants. Appendix B specifically addresses 18 quality assurance criteria that require implementation before a license is granted. TVA's Safety Analysis Report (SAR) for the Bellefonte Nuclear Plant committed all construction activities to comply with 10CFR50 Appendix B. Implementation of this committment is as follows.
The Office of Engineering Design and Construction (OEDC) established a document entitled OEDC Quality Assurance Program Requirements Manual for Design, Procurement, and Construction (PRM). This PRM delineates policy, responsibilities, requirements, and commitments for the Quality Assurance Program to be applied during the

---

[1] Supervisor, Office and Civil Engineering Unit, Bellefonte Nuclear Plant, Tennessee Valley Authority, Hollywood, Alabama.

design, procurement, and construction of TVA nulcear power plants. The major sections of the PRM are: (1) internal requirements, (2) external requirements, (3) commitments, (4) tables, and (5) cases. There are three matrix type tables. The first table is a listing by title and identifying numbers, of quality related codes, regulations, standards, regulatory guides and TVA documents to which a TVA QA program commitment has been established in licensing documents for each of the nuclear plants. The second table contains primarily the 18 criteria of 10 CFR 50 Appendix B. Each of the detailed requirements addressed in these 18 criteria is referenced to applicable external requirements, TVA commitments and internal requirements. The third table defines policy and organizational responsibility within OEDC and with interfacing TVA divisions for implementation of 10CFR50 Appendix B, and other applicable federal regulations.

Implementation of 10CFR50 Appendix B requirements and the OEDC PRM for construction activities is controlled by the Division of Construction QA Program Manual and Division of Engineering Design General Construction Specifications. The QA Program Manual consists of Quality Assurance Program Policies (QAPP) and Quality Assurance Procedures (QAP). The QAPP's outline general direction for implementation of each of the 18 criteria of Appendix B to 10CFR50, in addition to other activities (e.g. security, fire protection, housekeeping, etc.). The QAP's are selected implementing procedures for the QAPP's to provide consistency within TVA for certain activities (e.g. field change requests, nonconformance reporting, evaluation and selection of suppliers, etc.). The Engineering Design General Construction Specifications have been developed over the years and are applicable to certain TVA construction activities. Many specifications have been revised to better outline inspection and documentation requirements for nuclear plant construction. Some of the General Specifications covering civil activities include:

- Plain and reinforced concrete (G2)
- Formwork for concrete (G8)
- Rolled earthfill for dams and power plants (G9)
- Selecting, specifying, applying and inspecting paint and coatings (G14)
- Masonry (G21)
- Bolt anchors set in hardened concrete (G32)
- Repair of concrete (G34)
- Roller compacted concrete (G48)
- Grouting and dry pack of base plates and joints (G51)
- Cadweld splices in reinforcing bars (G52)
- Surface preparation, application, and inspection of special protective coatings for nuclear plants (G55)

At the project level, the Bellefonte Construction Engineer's Organization has developed detailed Quality Control Procedures (QCP's) which implement all the requirements of the higher level documents. There are presently 115 of these site-specific QCP's covering every construction activity associated with safety related buildings or equipment. Some of the QCP's are strictly governed by the General Specifications while others are governed by Quality Assurance Program Policies or Quality Assurance Procedures. A few QCP's were developed to control activities known to affect quality, even though not specifically addressed in a higher level document. The matrix shown in Table 1 lists major QCP's associated with the civil engineering activities, and the

TABLE I QCP MATRIX

QCP	Description	Inspection	Control of Work Activities	G-Spec	QAPP	QAP
2.1	Rebar, Embedments & Concrete Formwork	X		G2, G8		
2.2	Structural Steel	X			X	
2.4	Protective Coatings for Concrete & Carbon Steel Surfaces	X		G55		
2.6	Cadwelding Inspection	X		G52		
2.7	Prestressing System for Primary Containment	X				X
2.8	Bolt Anchors Set in Hardened Concrete	X		G32		
2.11	Masonry	X		G21		
5.1	Backfill Material Placement	X		G9		
5.2	Batch Plant Inspection	X		G2		
5.3	Concrete Placement	X		G2		
5.4	Concrete Curing & Repairing	X		G2		
5.5	Grouting and Drypack	X		G51		
5.10	Free Moisture & Gradation of Fine and Coarse Aggregate	X		G2		
5.12	Concrete Slump & Air Content Testing	X		G2		
5.13	Surveillance of Site Contractor Ready Mix	X				X
10.2	Drawing Control		X			X
10.4	Nonconforming Condition Reports		X			X
10.5	Field Fabrication Orders		X		X	
10.6	Work Release		X		X	
10.8	Control of Temporary Installations or Omissions		X		X	
10.11	Calibration of Measuring and Test Equipment		X		X	
10.14	Anchor Bolt Freeze Protection	X			X	
10.16	Bending of Partially Embedded Reinforcing Steel	X		G2		
10.23	Qualification of Protective Coatings Applicators		X	G55		
10.25	Qualification of Cadwelders		X	G52		
10.26	Quality Control Investigation Reports		X			X

UNIT DOCUMENTATION PRACTICES

controlling requirement for each QCP.

DOCUMENTATION PRACTICES

Rebar, Embedments, and Concrete Formwork - QCP 2.1
Concrete Placement - QCP 5.3

Whereas the inspection criteria for rebar, civil embedments, and concrete formwork are contained in QCP 2.1, the documentation of these inspections is contained on the Concrete Pour Card (Figure 1), an attachment to QCP 5.3, Concrete Placement. This QCP is one of the several that implement the extensive requirements of General Construction Specification G2, Plain and Reinforced Concrete. This General Specification first originated in 1939 for use primarily on dams, and is used today for all concreting operations.

The criteria established in QCP 2.1 originated from AISC, ACI, General Specification G2, Plain and Reinforced Concrete, and G8, Forms for Concrete Masonry. Detailed inspection criteria are contained in QCP 2.1 for reinforcing steel placement, structural steel embedments, anchor bolt placement, bending of headed concrete anchors, and concrete forming and alignment. In addition to the civil documentation, the concrete pour card provides inspection documentation for all engineering disciplines for all embedded items in the concrete and for the placement of the concrete itself.

The engineering design drawings locate concrete construction joints. From these drawings, the site prepares a concrete progress drawing. This is usually an isometric view of the area, floor, or building showing the concrete features and identifying each separate concrete pour with a unique number. No embedded features are shown on the progress drawings. When concrete is scheduled, a card is made out with the required information entered at the top. As the work is completed, the craft foreman signs off the card, signifying the work is ready for inspection. The engineering units then inspect their respective features using criteria contained in their specific QCP's (i.e., embedded piping, embedded conduit, reinforcing, formwork, etc.), and sign off after satisfactory inspection. The civil engineering shift engineer signs the bottom of the card after all engineering signatures are present; this releases the pour card to the Materials Testing Unit (MTU). The remaining inspection for concrete placment is done by MTU to the criteria set forth in QCP 5.3. These include preliminary inspection of the planned pour for such items as surface preparation, cleanliness, form tightness, etc., a review of the environmental conditions, both present and forecast, and an inspection of the placing equipment. Additionally, detail criteria are set forth in QCP 5.3 for inspections during concrete placement relating to such things as free fall, vibration, lifts, cold joints and consistency. The card is signed off by the MTU inspector at the successful completion of the pour. An MTU inspector remains with every concrete pour throughout the concrete placement activity.

Structural Steel - QCP 2.2
Field Fabrication orders - QCP 10.5

With the exception of major framing systems in several buildings, all structural steel fabrication is done by TVA forces in shops located on the jobsite. This fabrication includes all embedded plates, H&V duct supports, electrical cable tray supports, pipe supports and restraints, platforms, and small areas of floor and building framing. To control this fabrication, field fabrication orders (FF) are prepared in accordance with QCP 10.5 as each drawing is received on site that calls for field fabrication. The FF (Figure 2) contains a take-off of the items to be fabricated and the required materials, along with an attached weld assignment sheet. The materials section is used to control material issue and use. A materials control computer program tracks material requirements, receipts, issues and balance to insure adequate supplies. No fabrication of structural steel is allowed to begin until issue of the approved FF, which also contains the craft assignments for fabrication and installation.

QCP 2.2 addresses two areas: (1) the inspection of all non-ASME shop fabricated structural steel items and (2) the inspection of installed structural steel features for which the Civil Engineering Unit has responsibility. Both welding and civil inspections are performed on shop fabricated items to the criteria set forth in QCP 2.2 and documented on the FF Examination Record (Figure 3). Issues from the shop area are controlled to insure all items have been inspected prior to release. Fabrication of ASME Code items is controlled by a separate QCP and handled by the Mechanical Engineering Unit.

Structural Steel installations controlled and documented by the Civil Engineering Unit by QCP 2.2 are only a part of the total structural steel installation. Because of the close association between the supports and main features themselves, the electrical cable tray supports are inspected by the Electrical Engineering Unit; the H&V duct supports and integral pipe supports are inspected by the Mechanical Engineering Unit; and all pipe hangers are inspected by the Hanger Engineering Unit. Civil Engineering Unit responsibilities for structural steel installation inspection include all building framing, miscellaneous platforms, and steel and non-integral pipe supports. Inspection documentation is contained on the Structural Steel Installation Inspection Record (Figure 4).

Protective Coatings for Concrete - QCP 2.4

Documentation of Service Level I features is accomplished using the records shown on Figures 5 and 6. All work operations require inspection; hold points are established for prework, surface preparation, each coating or surfacing operation, and curing. As seen from the records, separate documentation is established for the surface perparation and coating application. Additionally, final acceptance of the coated areas is documented on a "checklist" record.

Cadweld Inspection - QCP 2.6

Inspections are performed on all cadwelding operations. Utilizing manufacturers instructions in this procedure, Figure 7 shows the documentation filled out for each cadweld. Representative samples are sent off for tensile testing, with separate test documentation coming from the lab. Prior to concrete encasement, all cadwelds must be acceptable and documented. For cadwelds not originally shown on the drawings, Field Change Requests are initiated to add them to insure compliance with the NRC Reg Guide for showing all cadweld splices on the drawings.

Prestressing System for Primary Containment - QCP 2.7

This QCP serves two functions: (1) outlines requirements and contains documentation for TVA construction activities (i.e., embedment of tendon sheathing and bearing plates in the concrete, etc.) and (2) outlines requirements for TVA personnel in the surveillance of the prestressing contractor's QC personnel for activities which the contractor performs (i.e., tendon placement, stressing, etc.). The contractor's onsite QC personnel perform and document their inspections in accordance with their approved QA Program. For TVA surveillance activities, documentation is accomplished by signing and dating the contractor's QC record itself, and noting what specific surveillance activity was accomplished (i.e., witness operation, record review, etc.). Although specific frequencies of inspections are established for the contractor's QC inspectors, TVA's level of surveillance activities are left to onsite determination.

Bolt Anchors Set in Hardened Concrete - QCP 2.8

General Construction Specification G32 establishes the extensive detailed requirements for the installation and testing of bolt anchors (expansion shell anchors, wedge bolt anchors, and grouted anchors). This specification was originally issued in 1972 and is currently in its sixth revision. QCP 2.8 implements the requirements of this specification.

Documentation of the testing requirements of this QCP is shown on Figure 8. This record provides documentation for tests performed on randomly selected anchors from a representative "lot". Since the civil engineering unit performs these tests on all features except pipe hangers, the completed reports are sent to the other engineering units whose features were installed (i.e., Mechanical Unit for H & V duct supports or mechanical equipment, etc.). The engineering units use this report as part of their checklist for final acceptance of their feature.

An Anchor Spacing Variance Report provides documentation for anchors that cannot be installed to meet the spacing requirements, and for which design approval has been obtained for a spacing variance. This facilitates resolution of construction problems, while providing documentation of the unusual condition.

Other requirements contained in G-32 and outlined as part of the installation procedure in QCP 2.8 are actually inspected and documented under other QCP's. A few examples: (1) drilling of holes for grouted anchors is controlled and documented on QCP 10.6 Work Release; (2) all grouting operations on grouted anchors are documented on QCP 5.5, Grout and Drypack; (3) bolt engagement and tightening of bolts/nuts and inspection of gaps between surface mounted plates and concrete surface are documented on the QCP associated specifically with the installed feature (e.g. QCP 2.2 for structural steel).

Backfill Materials Placement - QCP 5.1

Documentation for this QCP consists primarily of standard charts/tables/worksheets associated with the ASTM standards. Additionally, an inspector's report documenting the field operations and inspections within a specific area is completed for each shift.

Work Release - QCP 10.6

This procedure was initially developed to control the drilling or chipping of concrete and the cutting or welding of steel that was required during the course of the construction activities, but was not specifically addressed on the design drawings (e.g., welding of temporary lugs for rigging, drilling of concrete for wedge bolt anchors, etc.). Additionally, this procedure controlled rework done to completed installations which had completed QC documentation. As additional work processes were identified that required engineering controls, they were incorporated into this procedure, so that it has now evolved into a major document controlling several activities associated with civil, mechanical, and electrical features of the plant construction.

The engineering unit whose feature is being installed initiates the work release for any required special work. The release is routed to all engineering units for approval, special instructions, and noting prework inspection hold points, follow up inspection, and any affected QA records. Any required prework inspection must be completed prior to the work beginning. Required followup inspections and replacement concrete for chipped out areas must be completed prior to closing out the work release. Figure 9 shows the work release form.

A few sample uses of this form are as follows: (1) welded attachments for rigging purposes, including the removal and cleanup of the temporary weld; (2) all drilling of concrete, done primarily for installation of wedge bolts and grouted anchors, and for installation of added sleeves. Where specific criteria have been established for cutting rebar (i.e., for installation of bolt anchors), the work release provides documentation for cut rebar, which is forwarded to engineering design; (3) chipping of concrete, insofar as exposing rebar is concerned. This is done primarily to aid in installation of bolt anchors to relocate them as necessary to avoid cutting more rebar than allowed; (4) digging in the powerhouse vicinity with powered equipment to insure protection of buried facilities; and (5) sand blasting for protective coatings to insure protection of electrical and mechanical equipment in the area.

A central log is maintained on all work releases. To date, over 13,000 releases have been issued.

Control Of Temporary Installation Omissions (TIO) - QCP 10.8

This procedure is used by all engineering units and establishes a control over a non-permanent condition in a system or structure if it is necessary to bypass or omit components or equipment to facilitate construction or testing of installed equipment.

The majority of the Civil TIO's involve temporary omissions. There are generally two main areas. The first area is where work on a civil drawing is essentially finished, where further work is not scheduled in the near future, and a drawing revision is issued adding a new item or feature. In order to insure this added item is not forgotten, a TIO is inititated identifying it as a temporary omission. The second area is where almost all the work shown on the drawing is finished and inspection has been completed. However, since the remaining work will not be done soon, it is desirable to complete the inspection documentation. This is possible by completing the TIO for the incomplete work and noting the TIO number on the QC inspection record that is filled out for the balance of the work shown on the drawing.

A central log of TIO's is maintained, with a separate letter designation prefix for each engineering discipline. Every two months, the outstanding Civil TIO's are listed and forwarded to the construction personnel for appropriate scheduling of the identified work.

Anchor Bolt Freeze Protection - QCP 10.14

Although not specifically addressed by a higher level QA document, this inspection activity is needed to avoid possible damage to concrete due to water collecting in anchor bolt sleeves and freezing and expanding. Control and documentation of this is accomplished by Figure 10. This report is initiated whenever sleeved anchor bolts are encased in concrete. Freeze protection, consisting of styrofoam pellets put in the sleeve and the sleeve top covered to keep out water, is accomplished and documented. Thereafter, during the winter months, inspections are performed and documented every 30 days to insure the freeze protection is still in place and is adequate. Once the sleeves are grouted, or the location is not subject to freezing, the documentation is closed out.

Bending of Partially Embedded Reinforcing Steel - QCP 10.16

Onsite fabrication of reinforcing steel is accomplished using cold bending. However, for partially embedded bars requiring bending because of construction interferences, hot bending is required. The requirements are outlined in QCP 10.16, and the documentation of the required approvals and the actual bending is accomplished on Figure 11.

USE OF COMPUTER PROGRAMS

On the Bellefonte project, extensive use is made of computer programs for civil features assigned to other engineering units. These features include cable tray supports, H&V duct supports, and seismic pipe hangers. These three features are assigned to three different engineering units and while cable tray supports and H&V duct supports utilize the same computer program, the pipe hangers use a much more sophisticated program.

TVA's "Universal Program" is used for cable tray supports and H&V duct supports. Essentially, this program provides the following information: (1) a unique identifier for each hanger/support which is assigned in the development of weld maps and which the Welding Engineering Unit also uses for their weld identification; (2) the applicable design drawing; (3) field fabrication number, code class, system, or other specific item input by the engineering unit; (4) description of item; (5) a listing of the different required inspections that must be performed; and (6) a status of the required inspections. Computer cards are generated for each different inspection (or associated groups of inspection activities covered by the same QCP) on each unique feature. As the inspections are performed, the preprinted computer cards themselves are signed off and become the final documentation. The computer printout shows an * by each completed inspection. If rework is required after final inspections, a "B" card is input, thus signifying a new inspection is required. The versatility of the program allows for numerous field sorts. For example, to assist in as-constructing a drawing, a sort can be made by design drawing, listing all unique features on that drawing and showing the inspection status of each feature. A different sort by system or building can be accomplished for use in determining inspection status on a particular system prior to operational release.

Whereas the above programs are run approximately once per week on a scheduled basis, the pipe hanger computer program utilizes online computer time with instantaneous input and output display on CRT. This online feature is crucial to maintaining an accurate and up-to-date status. There are about 50,000 hangers, each with 37 separate data points, resulting in about 2 million entries. The 15 relatively fixed data points include the hanger number, location, type, class, etc. The 22 changing data points which are input daily include inspection status, work hold numbers, several types of problems identification numbers, various status codes for material, installations, drawings, etc. Documentation filled out by field personnel provides this basis for CRT input. The true benefit of this program is the various types of output available, which include: hangers available for fabrication or installation, by work package, activity number, area, or type. CRT displays are used by craft and engineering personnel to assist them in scheduling their work.

Although the Civil Engineering Units located at the other nuclear construction sites have inspection documentation similar to what has been decribed in this paper, the use of computer programs varies somewhat. Although each site's documentation is in full compliance with all the requirements, the differences in use of the computer are a reflection of the degree of importance that each site places on "documentation verification." It is not surprising then that the Sequoyah Nuclear Plant which recently received its full power permit has utilized the computer the most extensively. Sequoyah uses the "Universal Program" for the majority of it's civil inspection activities. The next plant to load fuel is presently establishing a "verification" program for most of its civil inspections utilizing a word processor. Although the output features are similar to those described for the "Universal Program," the input consists of a manual entry of a code letter next to the listed feature after the completed documentation has been physically located in the QA records vault. As new features are completed and documented, the manual entry into the word processor is then made at that time. Other than the use of a special computer program for concrete materials and compressive strength test cylinders, Bellefonte and the remaining plant sites have made little use of the computer for civil documentation. However, efforts are now underway to implement the "Universal Program" for Bellefonte civil documentation. It will be used not only for QC inspection verification but also for as-constructed drawing status, field fabrication order completion status, and for monitoring status of remaining concrete pours.

CONCLUSION

The Bellefonte documentation practices described have been revised and refined during the six years of construction. These revisions and refinements have resulted in additional controls and instructions necessary to insure quality construction.

Even with a QC program that results in timely inspections and documentation, the complexity of the work, the large number of items to be inspected, and number of inspections on each item warrant more than a casual assurance that, in fact, all inspections have been completed. This increased level of assurance, or "verification", is being developed within TVA by the increased use of computers and word processors. The versatility of a good computer program not only assists tracking the inspection status and documentation itself, but also offers a valuable scheduling and planning tool for the remaining work. Increased productivity is the final result of a well conceived computer application for civil engineering documentation.

UNIT DOCUMENTATION PRACTICES

BNP-QCP-5.3 R2
Attachment A
Addendum No. 3

CONCRETE POUR CARD

POUR NO. _____ TO EL _____ LOCATION _____

CRAFTS	FOREMAN	ENGR. UNIT REP.	DATE	ACC. PER BNP-QCP
Carpenter				2.1 R
Boilermaker				
Mechanical				6.1 R
Electrical				3.1 R
Piping				6.1 R
Reinforcing				2.1 R
Struct. Embd.				2.1 R
Sheet Metal				
Lines & Grades, Embedments, Forms				2.1 R
Cleaning				5.3
Welding				

Remarks _____

Shift Engineer _____ Date _____ Time _____

FRONT

INSPECTOR'S REPORT

BNP-QCP-5.3 R2
Attachment A
Addendum No. 3

INSPECTION	ACCEPTABLE		
Surface Preparation and Cleanliness	( )	Mix Number	_____
Form tightness	( )	Type Finish	_____
Environmental conditions	( )	Foreman	_____
Condition of placing equipment	( )	Date Poured	_____
Conveying and placing operations	( )	Cu. yds. Placed	_____
Deposit of concrete	( )	Time: Started ____ Comp. ____	
Layer thickness	( )	Remarks	_____
Concrete compaction	( )		
Illumination	( )		

( ) The above concrete placement and inspection was done in accordance with BNP-QCP-5.3, and is acceptable.

( ) The above repair poured with replacement concrete was done in accordance with BNP-QCP-5.3, and BNP-QCP-5.4,R___ and is acceptable. Attach. D of BNP-QCP-5.4 has been completed.

( ) The above grout or drypack operation was done in accordance with BNP-QCP-5.5,R___ and is acceptable. Attach. A of BNP-QCP-5.5 has been completed.

MTU Representative_____ Date _____

BACK

Figure 1 - Concrete Pour Card

BELLEFONTE NUCLEAR PLANT
FIELD FABRICATION ORDER

BNP-QCP-10.5 R4
Attachment A

☐ Permanent Fabrication
☐ Construction Fabrication

TO:

FF _____
Sheet _____ of _____
Date _____
By _____

SUBJECT:

☐ Safety Related
☐ Non-Safety Related

STRUCTURE/SYSTEM _____ UNIT _____

Drawing No.	Item No.	Description	Quantity Required	Quality Level

MATERIAL REQUIRED

Description	Quantity	Wt. (lbs)	Source

_____   _____
Const. Supt.'s Office   Const. Engr.'s Office

All items on this FF are complete _____
                                           CEU Representative

Figure 2 – Field Fabrication Order

UNIT DOCUMENTATION PRACTICES

Figure 3 – FF Examination Record

FRONT

BNP-QCP-2.2 R11
Attachment A
Sheet 1 of 2

STRUCTURAL STEEL INSTALLATION INSPECTION RECORD

Drawing No(s). _____

Structure _____ Unit _____

Description _____

Inspections:	Acceptable	N/A
Correct item no., mark no., etc.	☐	☐
Lines and grades	☐	☐
Verify FF fabrication acceptance	☐	☐
Welds present	☐	☐
Bolt torquing or tensioning (excluding expansion anchors and grouted anchors) - If applicable, see Attachment A1	☐	☐
Expansion Anchors & Grouted Anchors		
Correct size or equivalent size	☐	☐
Spacing (Section 6.4.1)	☐	☐
Anchor Spacing Variance # (if applicable) _____		
WB or SSD test lot # _____ Report # _____ (Attachment B, QCP 2.8)	☐	☐
Bolt engagement and tightness for SSD (Section 6.4.4)	☐	☐
WB projection and maximum attachment thickness	☐	☐
GA pull test lot # _____ Report # _____ (Attachment B, QCP 2.8)	☐	☐
GA nut tightness (Section 6.4.7.1)	☐	☐
Gap between attachment and concrete	☐	☐
Grout required? Yes ☐ No ☐		
Shims required? Yes ☐ No ☐		

BACK

BNP-QCP-2.2 R11
Attachment A
Sheet 2 of 2

Notes or Corrective Action
_____
_____
_____
_____

I certify the above structural steel installation has been inspected in accordance with BNP-QCP-2.2 and is acceptable.

Unit _____ Inspector _____

Date _____

cc: WRU

Figure 4 - Structural Steel Installation Inspection Record

UNIT DOCUMENTATION PRACTICES

BNP-QCP-2.4 R 5
Attachment B

BELLEFONTE NUCLEAR PLANT
COATING APPLICATION RECORD

No. _____

System/Structure _____ Unit _____
Reference Drawing _____
Location _____

1. Ambient Conditions:
   Temp., Ambient _____ °F   Relative Humidity _____ %
   Temp., Surface _____ °F   Therm. # (Dry Bulb) _____
   Temp., Dew Point _____ °F   Therm. # (Wet Bulb) _____
                                  Therm. # (Surface) _____
   Time Since Previous Coat _____
   Mixing In Accordance With Paragraph 7.3.5 _____

2. Substrate:
   Concrete _____ Masonry _____ Steel _____ Other _____
   Attachment A Completed and Area Released for Coating
   Inspector _____ Date _____ Time _____

3. Coating Application Information:

Coating Primer/ Intermediate/ Top	Product Batch No.	DFT (Mils) Min./Max.	WFT (Mils) Measured	% Solids	Actual	Measuring & I.D. No. Instrument

Method of Application _____ Applicator _____
4. Curing Complete ☐ psi, ID # _____
5. Adhesion _____ Mode of failure _____
Remarks _____

I certify that the above features were inspected for compliance with procedure and criteria in BNP-QCP-2.4 and are acceptable.

Inspector _____ Date _____

Figure 6 - Coating Application Record

BNP-QCP-2.4 R 5
Attachment A

BELLEFONTE NUCLEAR PLANT
CONCRETE/STEEL SURFACE PREPARATION RECORD

No. _____ Date _____
System/Structure _____ Unit _____
Reference Drawing _____
Location _____

1. Substrate Type:
   Concrete _____ (Age ≥ 28 Days)
   Masonry _____ Steel _____ Other _____

2. Base Conditions: _____ (Sandblast / Hydroblast Release No. _____)
   Equipment Protected _____
   Surface smooth, dry, clean, and free of contamination _____
   Surface Previously Coated _____

3. Method of Surface Preparation:
   Blasted Cleaned _____ Other (Detail) _____
   Type & Size Abrasive _____
   SSPC Spec. (Steel Only) _____
   Anchor Pattern Obtained _____
   Dusted or Vacuumed _____
   Blasting Air Free of Contaminents _____

4. Remarks: _____

I certify that the above features were inspected for compliance with criteria in BNP-QCP-2.4 and are acceptable.

Inspector _____ Date _____

Distribution:
1 - Applicator
1 - QCGRU (Original)
1 - CEU

Figure 5 - Surface Preparation Record

Attachment B
BNP-QCP-2.6 R4

CADWELD SPLICE INSPECTION DATA SHEET

Unit _____
Building _____
or System _____

Dwg. No. _____

Location of Splices

Splice Number	Type (P or S)	Bar Size	Position (V, H or D)	Location	Splicer	Inspection					Tensile test **	Splice Acceptable?		Inspector	Date
						Splicing Crew	Cleanliness and Dryness	Bar Deformation and Gap	Sleeve Position	Filler Metal *		No (Disposition - Failed Visual Inspection, tensile test, replaced, etc.)	Yes		

* Inspection results acceptable per procedure if checked (✓)
** Forwarded to SMEL for testing if checked (✓)

Figure 7 - Cadweld Splice Inspection Data Sheet

UNIT DOCUMENTATION PRACTICES

Figure 8 - Bolt Anchors Test Report

POWER GENERATION FACILITIES

Figure 9 – Work Release

UNIT DOCUMENTATION PRACTICES

BNP-QCP-10.16 R2
Attachment A
Addendum No. 2

BELLEFONTE NUCLEAR PLANT

REINFORCING STEEL BENDING PERMIT
TENNESSEE VALLEY AUTHORITY

Location of Bars: Pour No: _____ Permit No. _____
    Grid Location: _____ EL: _____

Location of Bend: EL: _____
    Grid Location: _____

Minimum diameter of bend allowed and minimum distance from concrete surface to the beginning of the bend:

Bar Size:	4	5	6	8	9	11	14S
Minimum diameter (inches):	3.0	3.75	4.5	6.0	9.0	11	17.5

Other by R&D, if specified: _____

Quantity: _____ and type of bar _____ bent

Approved: _____ _____
    EN DES ENGINEER   CONST CONTACT ENGINEER

Date: _____

BEND OR STRAIGHTENING DATA:

Preheat at bend (1100° to 1200° F.) ☐ Yes ☐ No ☐ NA
Temperature at concrete surface (<500° F.) ☐ Yes ☐ No
Examination for Cracks ☐ Accepted

I certify the above bending permit is complete in accordance with BNP-QCP-10.16 and is acceptable.

_____
O&CEU Inspector

_____
Date

Figure 11 – Rebar Bending Permit

BNP-QCP-10.14 R2
Attachment B
ADDENDUM NO. 1

ANCHOR BOLT FREEZE PROTECTION REPORT

Report No. _____

Project: _____
Structure/System _____ Unit _____
Drawing Number: _____
Feature: _____
Location (or Pour No.): _____
Elevation: _____

Freeze protection for the above anchor bolts was provided on

_____
Date

_____
CEU Representative

Periodic Inspection Checklist

Date	Inspector's Signature	Freeze Protect. Yes	Freeze Protect. No	Remarks

All anchor bolts on this report ( ) have been grouted ( ) are not subject to exposure to freezing temperatures and this report is complete.

_____
Date

_____
CEU Representative

Figure 10 – Anchor Bolt Freeze Protection Report

INFORMATION PROCESSING METHODS AND APPLICATIONS

by

Dennis D. Millican[1]

ABSTRACT

The acquisition of power generation facilities brings with it the necessity to recognize the existence of, and to cope with, the many and various records generated during each phase of the life of the power plant -- from initial planning through design, procurement, manufacturing, installation and construction, start-up and preoperation, operation and maintenance. (See Figure 1.) The number of records generated over the life of a power plant is approximately ten to twelve million for nuclear plants and fossil plants generate much less. (Table 1 represents a typical quantity of nuclear record pages.) This paper addresses combining the functions of planning, control and technology to solve information handling problems.

INTRODUCTION

The electric power industry is becoming more aware that these records must be controlled to ensure safekeeping, retrievability and cost benefits to the Owner. These records are kept, in some cases, for

---

[1] Formerly: Manager, Information Services, Gilbert/Commonwealth
Currently: President, Dennis Milican & Associates, Information Management Consultants, Jackson, Michigan

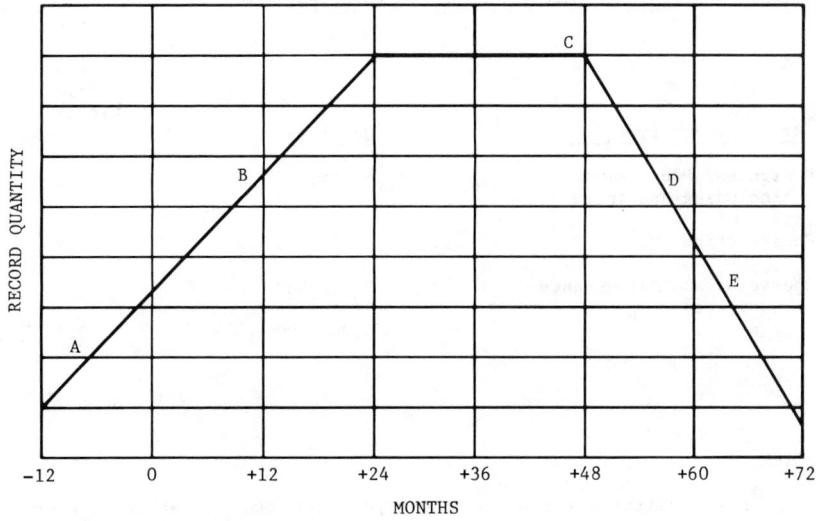

LEGEND:

A = Resident Construction Manager On-Site

B = Full Construction Forces On-Site

C = Cold Hydro

D = Fuel Load

E = Commercial Operation

Figure 1. Record Flow at Site

Table 1. **Typical Quantity of Record Pages per Power Plant Unit**

Record Type	Quantity	4 Drawer File Cabinets Required
Design and Procurement (including drawings)	4,265,000	395
Construction	2,500,000	232
Operation and Maintenance (40 years)	3,235,000	300
	10,000,000	927

NOTES:

[1] Some utilities are projecting twice this amount, yet the conservative number above still shows the magnitude of information dispersal.

[2] A microfilm storage system will use five percent or less of the space taken by file cabinets.

the length of the plant life and in other cases, on a permanent basis. Reference to these records is very active, especially during the construction and operating phases. During outages for maintenance and refueling, it is critical to have records available within the shortest possible time to ensure cost-beneficial operations and rate stability.

The various regulations on record maintenance also helped the utility business to become aware of information control, especially in nuclear power plants. The 18 criteria of Appendix B of 10CFR50 "Quality Assurance Criteria for Nuclear Power Plants and Fuel Reprocessing Plants" establish the requirements for Quality Assurance, including record requirements in the nuclear power industry. For the past several years, ANSI/ASME N45.2, "Quality Assurance Program Requirements for Nuclear Facilities," and its daughter standards have served as an expanded definition of the Appendix B requirements. Now a new document has been published which sets forth requirements and nonmandatory guidance titled ANSI/ASME NQA-1-1979, "Quality Assurance Program Requirements for Nuclear Power Plants." This document contains an introduction, basic requirements, supplements and nonmandatory guidance correlated to the 18 criteria of Appendix B. The document is based upon the contents of ANSI/ASME N45.2.-1977 and the daughter standards, N45.2.6, N45.2.9, N45.2.10, N45.2.11, N45.2.12, N45.2.13 and N45.2.23.

Another document, USNRC NUREG-0996, "Functional Criteria for Emergency Response Facilities," (published February 1981) describes the facilities and systems to be used by nuclear power plant licensees to improve responses to emergency situations. This document was issued to establish criteria for the Nuclear Regulatory Commission staff to evaluate whether an applicant/licensee meets the requirements of 10CFR50, Appendix E, Article IV.E.8 and Appendix A, GDC19. It is not a substitute for the regulations and compliance is not mandatory. However, this document outlines, among other items, the need to provide better information and the need to have a complete and up-to-date repository of plant records and procedures at the disposal of Technical Support Center[1] and Operation Support Center[2] personnel to aid in their technical analysis and evaluation of emergency conditions. The TSC and OSC personnel must have ready access to up-to-date records, operational specifications and procedures. All of these records must be available in current form and updated as necessary to ensure currency and completeness.

TECHNOLOGY TODAY

Let us look at the combination of planning, control and technology to comply with information management requirements in the electric power industry today. In the very broadest sense, the terms "information" and "information processing" cover three phases of Information Management.

---

[1] The Technical Support Center (TSC) is an on-site facility located close to the control room that shall provide plant management and technical support to the reactor operating personnel located in the control room during emergency conditions.

[2] The Operation Support Center (OSC) is an on-site area separate from the control room and the TSC where licensee operations support personnel will assemble in an emergency.

Phase 1. Creating information,

Phase 2. Processing information, and

Phase 3. Retaining information.

In Phase 1, information is _created_ to meet day-to-day operational needs. Management information is generated to allow managers to keep abreast of day-to-day operations through financial reports and other related data. Technical information is generated to eventually lead to purchase of equipment and materials and the actual installation and construction of the physical power plant. Information is also generated to meet federal, state and local regulatory requirements.

In Phase 2, the information is _processed_. In this stage, word processing, reprographics, electronic telecommunications, and data processing operations are used.

During Phase 3, the information is _retained_ or disposed of through the records management systems of records retention, indexing, uniform filing, microfilming, automated information retrieval and inactive records storage.

Moreover, in Phase 2, recent developments in information processing technology have demonstrated the potential for dramatic improvement in productivity and in the capability to produce the tools necessary to meet the regulatory technical, business and records requirements in today's nuclear power plants. Such developments include examples from each of the information processing operations.

The <u>word processing</u> area has evolved from manual to electric typewriters through the single element, Selectric typewriters, Mag Card machines, the Floppy Disk systems - all the way to shared logic systems, computer host systems and laser technology. Each evolution <u>(if it has been properly planned)</u> has reduced cost, provided greater availability of plant data and provided faster access time.

The <u>reprographics</u> area has evolved from the precopier period through the xerographic process, console copiers and automatic offsets to the integrated systems that are available today, such as the Xerox 9700; IBM 6640, IBM 6670; and Computer Output Microfilm (COM), which allows information to be printed out and distributed at a very fast rate straight from a "mini" or mainframe computer. Computer Output Microfilm (COM) prints information at a fast rate and offers microfiche that contains up to 270 page-images of information. The advantages of COM are: 1) printing at computer speed and 2) reduced material costs. Among other savings are those in postage and shipping that allows the mailing of a 105 x 148mm fiche (with 270 page-images) at the first-class rate of an ounce. Multiplied throughout the year, these can have a powerful impact on a firm's profits.

The <u>telecommunications</u> area has available computer-driven interconnect systems, satellite communication, electronic mail, facsimile sub-minute transmission and fiberoptics. All of these, again with proper planning, reduce costs.

The <u>data processing</u> area also has many new developments in computer technology and electronics. As the cost of human labor rises, the cost of computer power decreases. Office machines, powered by microcomputers the size of one's fingernail, now offer capability that nearly a room full of machinery provided a few years ago. This greater capability means that computer power can be distributed throughout the business office. Instead of being concentrated in data centers, computer power can now be at the command of those who use it -- the managers and the office workers who support them.

However, much of this new electronic capability has arrived in today's business offices in bits and pieces. Indeed, most offices have what amounts to a hodgepodge of equipment -- typewriters, text editors, intelligent terminals, printers, memory devices, telex equipment, calculators, facsimile machines and copiers -- from several different manufacturers. This equipment is designed to do specific office information chores and do them efficiently. But these machines could be much more effective if they could interact with one another, if they could communicate. Many companies are developing hardware/software systems for all machines to communicate. This will be on the market in the near future.

The integration of micrographics with computer technology is opening the way for an increasing number of applications in the nuclear power plant environment. For example, computer-assisted retrieval (CAR) allows an operator to retrieve a desired image document in 30 seconds or less.[3]

The new technologies available today can reduce costs and cope with the multitude of records, if <u>proper planning and controls</u> are placed up front in the records program.

PLANNING AND CONTROLLING YOUR RECORDS PROGRAM

Many aspects of a good records program are probably already in place or under development in your organization. The objective is to develop program details, prepare the work plan and implement the modified and improved system. The customary approach has been to begin with a "feasibility" study. This study identifies problems and provides mainly equipment and microfilming recommendations. In most cases, the report actually reinforces what is already known, but it states the facts in a more organized way.

---

[3]The phase "computer-assisted Records Management" refers to the use of a main-frame computer and/or mini-computer to store indexing information in a data base that is interactively accessed by the use of an on-line CRT.

While one must define problems and make recommendations, the report or plan one prepares should include a list of major tasks that must be completed in order to achieve a viable information management system. The report should provide work breakdown support, complete with the necessary tasks for each fragment. The development and implementation plan must show the order in which tasks should be completed and the relationship of the tasks during the development and implementation stages. This information is laid out to ensure orderly progression in the building of one's system. This is necessary to avoid some of the serious problems faced by organizations which have purchased equipment prematurely or initiated tasks before the basics were in place, only to have their cost double or triple due to major retrofits - to say nothing of the cost of lost user credibility.

Here are a few examples of lack of planning and control to demonstrate the point: First, the <u>executive</u> syndrome. "Give me all the facts," and "more information means better management." This syndrome leads to proliferation of information at all levels of management. Management receives more information than it can use - and at a large cost. Secondly, there is the <u>roundhouse</u> syndrome. "This equipment will do anything you want it to do." So it leads to more equipment, then to more labor, which equals more information, which equals more equipment, etc., and the roundhouse effect becomes apparent. This syndrome is present in many organizations. Finally, is the <u>time capsule</u> syndrome. When in doubt about what to do with a sheet of paper, file it, thereby leading to more record storage, equaling more space, equaling more labor. Again, the solution to these problems lies in planning and control.

RECORD SYSTEM REVIEW

The first function in a records program, as has been seen, should be to carry out an assessment of the present situation, including an objective appraisal of the records program with respect to systems and procedures and objectives. Particular emphasis should be placed on establishing the basis for interfaces among the vendors, the contractors and the Owner.

Formal work sheets should be prepared during the review process and should address the following functions:

1. Records Flow and Filing Techniques:

    What procedures are in place to regulate the volume of records processed, including uniform systems for preparation and distribution?

2. Records Identification and Indexing Methods:

   What record type lists are in place and how will these records be indexed to provide retrieval?

3. Records Turnover:

   How are records to be turned over to the Owner from vendors and contractors? Record requirements should be clearly identified and the process for determining whether the requirements are satisfied should be established. The following major items should be addressed:

   a. Development of a Records Listing for record types described in specifications, in contractor and vendor procedures, and in sample document packages.

   b. Review of individual record forms for adequate traceability to the work breakdown structure (which includes both the plant hardware and activities performed).

   c. Design of record indexes to assure positive identification and cross-referenced retrieval.

   d. Procedures for the collection of the various record types, and written transfer plans for records transfer to the retention center.

4. Drawing and Specification Control:

   What systems are established for the control of drawings and specifications, including receipt and distribution, change control, indexing, storage, and retrieval?

5. Existing Equipment (Data Processing and New Equipment - Micrographics):

   What is presently available and what will be needed to interface smoothly with user requirements and existing procedures in the areas of word processing, computer-assisted retrieval and micrographics?

6. Technical and Operational Support Center:

   What plans are prepared for support facilities, including existing and proposed equipment, record storage equipment, micrographic equipment, record volumes, personnel requirements and facility layout plans?

7. Personnel Requirements and Training Programs:

   What personnel are required to support the Records Management/Document Control program and what educational programs and on-the-job training programs are available?

After the review has been completed, the organization will have the necessary information to prepare an implementation plan based on the data collected during the review. The completion of the plan will facilitate a successful Records Management/Document Control program. Recommendations should be provided and these should be supported by clearly defined tasks and steps required to implement the recommendations.

All recommendations should consider:

1. Retrieving needed information promptly,
2. Reducing professional search time,
3. Having convenient access as required,
4. Locating records by descriptive terms,
5. Locating records by means of a clear filing system,
6. Determining current version or status,
7. Tracing chain of related events,
8. Ensuring records received as committed,
9. Ensuring adequate and accurate indexing,
10. Reducing manual filing and searching,
11. Providing summary lists as required, and
12. Providing flexibility to meet future requirements.

At the completion of the records review, a presentation to management should be made describing the plan, system design and the benefits that will be derived upon its implementation.

The final task(s) are to implement a Records Management Program with the following characteristics:

1. Consider user needs,
2. Can be readily administered and easily understood,
3. Is cost-beneficial, both in terms of installation and operating costs, and
4. Takes advantage of new equipment and records management techniques yet is simple and direct.

CONCLUSION

Much of the technology is already at hand to realize white-collar productivity gains and to meet our record requirements. Recent developments in information processing technology have demonstrated the potential for a dramatic improvement in productivity and capability of providing services to ourselves. The rapidity and magnitude of these developments could result in far-reaching changes in traditional procedures and practices. It is important, however, that a company examine present programs, establish controls, and adopt these technologies as part of a <u>unified</u> plan.

Remember, the human tendency is to prematurely purchase equipment and have costs double and triple. This usually decreases credibility because the up-front system planning does not fall easily into place.

In any case, I advise that none deter the progress of information processing by clinging to old habits. Do not be reluctant to try the new techniques and skills. If proper planning and controls are in place, the new technologies will meet a company's information management needs in a most cost-beneficial manner.

SESSION IV - CONCRETE TEST REQUIREMENTS AND PROCESS CONTROL PRACTICES

SESSION OBJECTIVES/SESSION CHAIRMAN SUMMARY

by

Joseph F. Artuso[1] M, ASCE

Objective of Session

The objective of this session is to present code requirements regarding testing and process control practices and to review experiences during implementation on various Nuclear Construction Plant Projects. The ASME Section III Division 2, ANSI N45-2.6, ACI 318, ACI 301, ASTM C-94 and ASTM C-33 requirements are discussed. Suggestions are made by each author regarding changes to the codes which will improve both quality control and construction performance while at the same time providing proper structural integrity.

Session Chairman Summary

The papers presented in this session pertained to the testing and quality control of concrete during the construction of nuclear power plants. The testing requirements included qualification of concrete materials and the in process testing to determine conformance to the code or standard governing the construction. The current requirements were presented and experience with the specific standard were included in the author's presentation. All of the papers contained suggestions for improvements to expedite construction without causing any detrimental consequences on the quality of construction.

Two of the papers elaborated on the development of two primary standards used in the construction of nuclear power plants. ASME Boiler and Pressure Vessel Code, Section III Division 2 (ACI-359) "Code for Concrete Reactor Vessels and Containments," pertained specifically to the code requirements for Construction of Containments and Reactor Vessels. The Quality Assurance Standard ANSI/ASME N45.2.5 "Supplementary Quality Assurance Requirements for Installation, Inspection and Testing of Structural Concrete, Structural Steel, Soils and Foundations" during the construction phase of nuclear power plants pertained to the quality assurance requirements for the balance of Plant's safety related structures. Both papers included information on the development of the standard and evolution of changes. The application of the ASME Section III Division 2 Standard has definite requirements which cannot be easily changed during construction. Governmental jurisdictional authorities enforce all requirements

---

[1] President, Construction Engineering Consultants, Inc., Laughlintown, PA

and use Third Party authorized nuclear inspectors to verify conformance. The code requirements cannot be changed except by addenda (Revisions) or Code Cases. This is a time consuming procedure which may impact construction.

The ANSI N-42-2.6 standard provides some modification of testing requirements if specifically covered by the project construction specifications and approved by the authorities.

Efforts are being made to use the same testing and qualification of material requirements in both standards for uniformity and avoidance of confusion during construction.

Two of the papers pertained to specific requirements for specialized tests and properties of concrete addressed in the standards. These were "Experiences with Special Potential Volume Change Test Requirements of ASTM C-33 and ASTM C-342" and "Creep and Shrinkage Studies of Concrete Mixes for Prestressed Concrete Nuclear Containment Structures."

The paper on "Creep and Shrinkage of Concrete Mixes for Prestressed Concrete Nuclear Containment Structures" included actual test procedures and results used to evaluate the specific properties. This data was used by the designer to assure that properties of the specific materials used on a specific project were compatible with the design criteria.

The paper on "Potential Volume Change Test Requirements of ASTM C33 and ASTM C-342" provided a discussion of the tests used to assure that excessive volume changes of the concrete used on the specific project would not occur. The conclusion of the testing program indicated that the specific test method was not applicable for the materials used in the project. However, the tests provided valuable information which enabled the elimination of unsatisfactory material.

Another paper dealt with securing samples for in-process testing of concrete materials and concrete. Emphasis was placed on the importance and methods of securing representative samples for reliable determination of properties of the materials. A case was made for correlation testing to provide expediency during construction. The author proposed greater use of correlation testing and advocated that the construction specification govern the engineering criteria for sampling.

The final paper contained criteria and methods used to achieve proper quality with inspection and testing. Suggestions were made on how to set the tone of quality control to assure that quality is achieved without detrimental impact on construction progress. A specific project was cited where a high order of quality construction was achieved without causing construction delays and coincidentally complying with requirements of the standards and regulatory agencies.

Consensus Recommendations*

The discussion which followed included comments on the test methods concerning

---

*Based upon the Panel of Speakers/Audience discussion period at the end of the session.

applicability, excessive requirements and inflexibility of the requirements.

The primary consensus was the inflexible characteristic of specific code requirements. It was felt that a means should be provided to allow an evaluation of any non-conformance by the Engineers and an engineering resolution with corrective action. The ASME code requirements, historically cover metal parts, and therefore corrections of deviations can be more easily accomplished by removal (cutting out) and replacement. This type of corrective action can be more easily adapted to welded joints and pipe sections.

The concrete industry utilizes various methods of evaluation such as load tests or core tests of a placement that may have had deviation to determine whether the structure will function as designed. It is suggested that requirements such as curing and temperature be placed in construction specifications to enable legal corrective actions when constructing under specific codes that have legal ramifications.

CONTROL OF CONCRETE MIXING AND TESTING
J. R. Wells
Member, ASCE

ABSTRACT

A general review of the testing requirements of concrete being used in nuclear power plants is presented. Concrete usage at nuclear power plants are many and varied. Most of the regulatory requirements for testing of concrete involve what is termed "safety-related" concrete. This would include concrete for the pressure vessels, containment building, auxiliary building, equipment supports, and shielding walls. Of course, some characteristics of the concrete are more important in a given use than others. For example, concrete used in a prestressed containment must have different characteristics than concrete used for a foundation for the reactor vessel. It is important that the testing requirement reflect these important characteristics.

INTRODUCTION

There are many regulatory bodies who have to be satisfied during the design and construction of a nuclear power plant. The primary body having jurisdiction is, of course, the Nuclear Regulatory Commission. The NRC has recognized several codes and standards to be followed. The most important ones are as follows:

1. American Society of Mechanical Engineers Boiler and Pressure Vessel Code, Section III, Division 2.

2. ANSI/ASME N45.2.5, Supplementary Quality Assurance Requirements for Installation, Inspection, and Testing of Structural Concrete Structural Steel, Soils, and Foundations During the Construction Phase of Nuclear Power Plants.

3. American Society for Testing and Materials, C-94, Standard Specification for Ready-Mixed Concrete.

4. U. S. Nuclear Regulatory Commission, Regulatory Guide 1.94, Quality Assurance Requirements for Installation, Inspection and Testing of Structural Concrete and Structural Steel During the Construction Phase of Nuclear Power Plants.

5. American Concrete Institute, C-349, Code Requirements for Nuclear Safety-Related Concrete Structures.

COMPARISON OF VARIOUS CODES AND STANDARDS

As you can probably imagine, it is very difficult to build a nuclear power plant with so many separate codes and standards covering the same subject.

---

[1] Corporate Quality Assurance Manager, Duke Power Company, Seneca, S.C.

Part of the problem was due to different end uses of the concrete and different perspectives of the several codes and standards writing bodies. For example, it appears to me that ASTM C-94 was developed primarily to be used by a purchaser of ready-mixed concrete. It seems that its major use was that of a procurement specification. In fact, in the scope it says "This specification does not cover the placement, consolidation, curing, or protection of the concrete after delivery to the purchaser. Yet, there have been instances where this specification has been forced on a constructor of a nuclear power plant even though the constructor did not purchase concrete but mixed it himself. The ASME Code, Section III, Division 2, was developed for concrete vessels and containment building; but there have been times when the regulators tried to impose the testing requirement specified in this document to all concrete, whatever its end use was. There are but two examples of what I believe to be the misuse of a code or standard. There are many more than could be cited. We must all remember that each code or standard was developed for a particular purpose; and as long as we use it for that purpose, good results will usually be obtained. But I will have more of this in my recommendations.

ANSI/ASME N45.2.5

For the past few years, I have been chairman of the Work Group responsible for writing N45.2.5. Most of this paper will deal with this standard. The topic of the paper is testing requirements, so that will be primarily what will be covered. A general background, however, is necessary in order to lay the foundation for the testing aspects.

In April of 1970, the American National Standards Committee N45 on Reactor Plants and their Maintenance established a Subcommittee N45-3 to guide the preparation of nuclear quality assurance standards. The subcommittee was responsible for establishing guidelines and policy to govern the scope and content of the various standards; monitoring the status of standards in process; recommending preparation of additional standards; and final approval of standards prior to their submittal to the N45 Committee for balloting. In November 1970, the N45-3 Subcommittee established an Ad Hoc Committee N45-3.5 on Quality Assurance Requirements for Civil and Structural Work. In September 1971, this Ad Hoc Committee was changed to a working group. In October 1972, the N45-3 Subcommittee was renumbered N45-2 and the Work Group was renumbered N45.2.5. In 1975, the N45-2 Subcommittee was reorganized into the ASME Committee of Nuclear Quality Assurance. The scope of N45.2.5 is to set forth the supplementary quality assurance requirements for installation, inspection, and testing of nuclear safety-related structural concrete, structural steel, soils, and foundations for nuclear power plant construction.

It was the goal of this Work Group to bring together in this one document all of the quality assurance requirements for civil works on nuclear projects. This was done by either putting the requirements actually in the standard or by reference to a recognized national standard. I believe that this goal has been very nearly met. There has been a very concerted effort made to not conflict with any standard but simply to bring together in one place the requirements and amplify on them where the need was perceived. Of course, the testing aspect is but one, albeit a very important one, part of the standard.

TESTING REQUIREMENTS

There are two types of tests to be done on concrete. The first is the qualification tests to determine the acceptability of the ingredients, and the second is the in-process tests made during the concreting operation. There seems to be general agreement that the purpose of qualification tests is to determine whether or not the material is suitable for the intended purpose. There seems to be some disagreement within the industry on the purpose of the in-process tests. But more about this later.

The committee responsible for N45.2.5 developed a table which outlined the required qualification test. This table is included as Appendix A to this paper. As you will note, this table has items other than concrete materials. The standard covers other civil-type construction; and, therefore, the table includes some of these things. This table of required qualification tests is really just a compilation of the ones generally required, and reference is made to several industry standards and specifications for test methods or test requirements. Additional tests may be required to qualify materials for special application. In the several years that this standard has been in use, very few comments or questions have arisen concerning the meaning of this qualification test table. This indicates to me that it is fairly well understood and agreed upon.

The in-process test requirements are not so easy. There have been many questions, comments, suggestions for change, and disagreements on meaning relating to Table B. This table is included as Appendix B of this paper. As I said, there seems to be some disagreement between knowledgeable people on the purpose of in-process tests for concrete. I have heard one school of thought that says the tests are for uniformity purposes only. The argument goes that the various controls placed on the mixing, transporting, and placing really control the quality of the concrete as placed and that strength tests can only assure you that you have some uniformity. On the other hand, I have experienced a situation where an engineer was justifying a low 28-day break by waiting to see if the 90-day cylinders broke above the specified values. This seems to me to be saying that in this engineer's logic the cylinders do indeed represent the in-place concrete and that if the cylinders break at 3200 psi, the in-place concrete probably is near the value.

This philosophical discussion may not seem important to you, but it became very important to some people during the writing of N45.2.5. There was considerable discussion on the place of sampling of the concrete. There are three key locations to be considered. These are the mixing point, the delivery point, and the placement point. The reason for testing becomes very important when you are deciding where the sample is to be taken. During the development of the standard, there were some who argued that the mixing point was the proper place to take strength samples because the reason for the tests was to insure uniformity of the mixed concrete. Others said that the samples should be taken at the point of placement. As is true in so many standards, a compromise was reached. Two key definitions were placed in the standard. I believe it would be good to put them in here exactly as they are in the standard.

1. Delivery Point - The point of discharge in the case of a truck agitator unit, or non-agitating unit when another conveying device is to be used to transport the plastic concrete to the placement point. Where a truck agitator unit is used in the transit of concrete, the delivery point and the mixing point are considered coincident when: (a) the delivery point is not more than a distance of two miles and an average of one-half hour in transit from the mixing point; and (b) the delivered concrete commences to be placed within an average of one-half hour from the time the transporting vehicle arrives at the delivery point.

2. Placement Point - The point of discharge of plastic concrete into the forms. Except for pumped concrete, the placement point and the delivery point are considered coincident when five minutes or less is used in transit of the concrete from the delivery point to the placement point.

By the use of the above definitions, the standard is really saying that strength test samples can be taken at the mixer plant if the delivery point and mixing point are coincident, using the definition given. This seemed to have eased the pain of some users who desired to take samples at the mixer plant. These users also have the option of doing correlation tests if the two points are not coincident. The use of the definition of placement point also allows the air content, slump, and temperature to be taken at the delivery point under the conditions specified. As I said, these two points caused the most controversy; and I believe that the committee worked out a good compromise while still obtaining adequate test results.

Now, let's take a look at Table B from ANSI N45.2.5. As you saw in Table A, there are a number of tests in Table B relating to things other than concrete. If you will just ignore these, we can concentrate on the concrete items. The Work Group attempted to list all of the essential tests and at the same time recognize that special tests may be needed. In this case, the engineer would need to specify these. Standard test methods have been listed such as ASTM or CRD. I will not go down the list of tests, but you can see what they are. The test frequency is listed and is considered to be a minimum. In some cases, the engineer may desire to increase the frequency. Of special note, you see an asterisk beside some of the test frequencies. There was much discussion on this point. As you note, the frequencies which have an asterisk beside them shall be considered minimum frequencies unless current documentary test data are available to establish complete confidence in conformance to specification requirements. For example, the Work Group saw no reason to require unit weight tests on a daily basis if the aggregate was consistent with the required unit weight and the quarry was the same. In these cases, we left it to the judgment of the engineer to document this fact and not do daily tests. Of course, some frequency of test is needed as a check. Many other tests are in this same situation.

FUTURE DEVELOPMENTS

As was stated early in the paper, there are several standards and codes with test requirements. The Work Group for N45.2.5 was asked to take a look at getting together with some of the others to eliminate any possible conflict on test requirements. Some of the Work Group members met with a number of the other committees; and after much discussion and debate, it was felt that Table B of N45.2.5 could be eliminated and a reference made to Table CB-5200-1 of Section III, Division 2, of the ASME Code. This table is shown as Appendix C. As you can see, there is very little difference in the two tables and it will make future changes easier. This change has been voted upon and approved and will appear in the next revision. With this change, this standard will probably serve industry for some time. There are no major changes anticipated.

RECOMMENDATIONS

I don't know of any substantial recommendation that can be made. My company has been using the standard on four separate construction sites with very good results. We have found the test requirements both helpful and reasonable to meet. Some judgment needs to be applied; but with that, the standard has served us well. I have had some discussions with people who would like to have variable test frequency requirements. For example, there may be a frequency of 50 cubic yards for temperature tests initially; but if these tests prove the temperature to be well within allowable, the frequency may be relaxed. Others have proposed a sliding scale of test frequency. For example, strength cylinders would be taken every 100 cubic yards for the first 200 yards, then every 300 for the next 600, and less frequent after that on large placements. This may be of some value, but how many placements go over 600 cubic yards and is this kind of accounting worth the effort? I doubt it. From my viewpoint, the standard is workable and I don't believe the extra refinements are of very much value. We all need to just work at making the concrete test program meaningful by sound judgments.

Table A Qualification Tests

Material	Test For	Test Method
Concrete Aggregates (Regular)	Compliance with ASTM C-33	As referenced in ASTM C-33, C-637, and C-330 respectively.
Concrete Aggregates (Heavy)	Compliance with ASTM C-637	
Concrete Aggregates (Light)	Compliance with ASTM C-330	
Cement	Compliance with ASTM C-150	As referenced in ASTM C-150
Admixtures	Compliance with ASTM C-260 or C-494 whichever is applicable.	Manufacturer's certification
Fly Ash & Pozzolans	Compliance with ASTM C-618	As referenced in ASTM C-618
Water & Ice	Compliance with specifications for effect on:	
	Compressive Strength	ASTM C-109
	Setting Time	ASTM C-191
	Soundness	ASTM C-151
Liquid Membrane Forming Curing Compounds	Compliance with ASTM C-309	As referenced in ASTM C-309
Sheet Materials for Concrete Curing	Compliance with ASTM C-171	As referenced in ASTM C-171
Concrete Mixes	Compliance with ACI 211	As referenced in ACI 211 (Note 1)
Preplaced Aggregate Concrete	Compressive Strength	CRD-C-84
Reinforcement	Physical properties of full section test specimen per ASTM A-615	One full section test in accordance with ASTM A-370 for each bar size.
Structural Steel & Prestressing Steel	Compliance with appropriate specifications, such as ASTM A-36, A-440, etc.	Manufacturer's certification
High Strength Bolts	Compliance with ASTM A-325 or A-490	Manufacturer's certification
Grout for Preplaced Aggregate Concrete	Flow	CRD-C-79
	Expansive Characteristics	CRD-C-81
	Bleeding Characteristics	CRD-C-82
	Water Retention and Unit Weight	CRD-C-80
Wood Piles	Compliance with appropriate specifications, such as ASTM D-25 or AWPA C3	Manufacturer's certification
Steel Pipe Piles	Compliance with ASTM, A-252	Manufacturer's certification
Steel Piles of Structural Shapes	Compliance with appropriate specifications, such as ASTM A-6 and A-36.	Manufacturer's certification
Concrete Piles—Precast, Cast in Place and Prestressed	Compliance with the specifications	As specified
Preservative Treatment of Wood Piles	Compliance with appropriate specifications, such as ASTM D-1760.	

Note 1: When splitting tensile strength tests are required for lightweight concrete mix, the method given in ASTM C-330 shall be used.

These in-process tests are only required where applicable to the item being tested.

Table B Required In-Process Tests

Material	Requirement	Test Method	Test Frequency
Concrete	Mixer uniformity	ASTM C-94	Initially and every 6 months thereafter
	Sampling method	ASTM C-172	
	Compression cylinders	ASTM C-31	
	Compression cylinders—preplaced aggregate concrete	CRD-C-84	
	Compression strength	ASTM C-39	2 cylinders for 28-day test from each 100 cu yd or a minimum of 1 set/day for each class of concrete
	Slump	ASTM C-143	First batch produced each day and every 50 cu yd placed
	Air content	ASTM C-173 or C-231	With each set of compression cylinders
	Temperature		First batch produced each day and every 50 cu yd placed
	Unit weight/yield	ASTM C-138	Daily during production*
	Unit weight for structural lightweight concrete	ASTM C-567	Daily during production
Grout	Compressive strength	ASTM C-109 (for expansive grout use CRD-C 589)	Daily during production
Grout for Preplaced Aggregate Concrete	Time of set	CRD-C 82	Daily during production
	Flow	CRD-C 79	Daily during production
	Expansion & Bleeding	CRD-C 81	Daily during production
Aggregate	Compliance with requirements for:		
	Gradation	ASTM C-136	Daily during production
	Moisture content	ASTM C-566	Twice daily during production
	Material finer than No. 200 sieve	ASTM C-117	Daily during production
	Unit weight of aggregate.	ASTM C-29	Daily during production*
	Fixed water and iron content of aggregates only for radiation—shielding concrete	ASTM C-637	Daily during production*
	Organic impurities	ASTM C-40	Daily during production*
	Flat and elongated particles	CRD-C-119	Monthly during production*
	Lightweight particles	ASTM C-123	Monthly during production*
	Soft fragments	ASTM C-235	Monthly during production*
	Specific gravity & Absorption	ASTM C-127 or ASTM C-128	Monthly during production*

Table B Required In-Process Tests (Continued)

Material	Requirement	Test Method	Test Frequency
Aggregate (Cont'd)	Los Angeles Abrasion	ASTM C-131 or ASTM C-535	Every 6 months*
	Potential reactivity	ASTM C-289	Every 6 months*
	Soundness	ASTM C-88	Every 6 months*
Water & Ice	Compliance with project specifications for effect on:		
	Compressive strength	ASTM C-109	Monthly
	Setting Time	ASTM C-191	Monthly
	Chlorides	ASTM D-512	Monthly
	Total solids	ASTM D-1888	Monthly
Admixtures	Chemical composition, Ph, and specific gravity	ASTM C-494	Composite of each shipment
Fly Ash & Pozzolans	Chemical & physical properties per ASTM C-618	ASTM C-311	As specified in ASTM C-311
Cement	Standard physical and chemical properties	ASTM C-150	As specified in ASTM C-183
Reinforcing Steel	**Physical properties of full section test specimen per ASTM A-615	ASTM A-370	One full section test for each bar size for each 50 tons or fraction thereof from each heat
Cadweld Reinforcing Bar Splices	Section 6.12	Secton 6.12.3	Section 6.12.4
Soil	Compaction Test	ASTM 698 or 1557, Method A, B, C, or D, as specified.	One for each 10,000 cu yd with at least one for each soil type, and when soil type is questionable
	Grain Size	ASTM D-422 hydrometer or sieve as appropriate	One for each compaction test
	Plasticity Index	ASTM D-424	One for each compaction test and when volume change characteristics are questionable
	Borrow Moisture	ASTM D-1556, 2167, 3017, or 2937, as specified	One for each soil type, one before each work shift, and when moisture content changes or is questionable
	Field Density Test	ASTM D-1556, 2167, 2922 or 2937 as specified	Test as specified in owner's specs and when compaction of soil type is questionable. Minimum every 10,000 sq ft
	Fines Content	ASTM D-1140	Every 100,000 sq ft

Note 1. See definition of In Process tests.

*These test frequencies shall be considered minimum unless current documentary test data are available to establish complete confidence in conformance to specification requirements
**Reduced section test specimen may be used for determination of the percentage of elongation.

TABLE CB-5200-1
TESTING FREQUENCIES FOR CONCRETE MATERIAL AND CONCRETE

Material	Requirements	Test Method	Frequency
Cement	Standard physical and chemical properties	ASTM C 150	Each 1200 tons
Fly Ash and Pozzolans	Chemical and physical properties in accordance with ASTM C 618	ASTM C 311	Each 200 tons
Aggregate	Gradation	ASTM C 136	Once daily during production[1]
	Moisture content	ASTM C 566	Twice daily during production
	Material finer than #200 sieve	ASTM C 117	Daily during production
	Organic impurities	ASTM C 40	Daily during production
	Flat and elongated particles	CRD-C 119	Monthly during production
	Friable particles	ASTM C 142	Monthly during production
	Lightweight particles	ASTM C 123	Monthly during production
	Specific gravity and absorption	ASTM C 127 or ASTM C 128	Monthly during production
	Chloride content	ASTM D 1411	Monthly during production
	Los Angeles abrasion	ASTM C 131 or ASTM C 535	Every 6 months
	Potential reactivity	ASTM C 289	Every 6 months
	Soundness	ASTM C 88	Every 6 months
Water and Ice	Compliance with CB-2223		
	Effect on compressive strength	ASTM C 109	Every 6 months
	Effect on setting time	ASTM C 191	Every 6 months
	Total solids	APHA-208	Every 6 months
	Chlorides	ASTM D 512	Monthly
Admixtures	Chemical composition	Infrared spectrophotometry, pH, and solids content in accordance with ASTM C 494	Composite of each shipment
Concrete	Mixer uniformity	ASTM C 94	Initially and every 6 months
	Sampling method	ASTM C 172	
	Compression cylinders	ASTM C 31	
	Compressive strength	ASTM C 39	One set of 2 cylinders from each 100 cu yd or a minimum of 1 set per day for each class of concrete
	Slump	ASTM C 143	First batch placed each day and every 50 cu yd placed
	Air content	ASTM C 173 or ASTM C 231	First batch placed each day and every 50 cu yd placed
	Temperature		First batch placed each day and every 50 cu yd placed
	Unit weight/yield	ASTM C 138	Daily during production

NOTE:
(1) Twice daily during production if more than 200 cu yd are placed.

SAMPLING OF CONCRETE AND CONCRETE CONSTITUENTS

By Douglas J. Haavik,[1] M. ASCE

ABSTRACT

Obtaining a truly random sample in accordance with an appropriate sampling plan is the first step to proper control of concrete and its constituents. Present standards and field practices encourage regular, frequent, probably biased sampling and testing and in some cases are both expensive and impractical to implement. With some modifications of standards and field practices, testing programs could provide improved effectiveness and economy.

INTRODUCTION

In order to have a valid quality control program, concrete materials must be tested in order to demonstrate conformance with specification requirements. In order to have an economical material production operation, for either concrete or its constituents, testing must be performed to allow in-process control based on the properties and variability of the materials being used. In today's nuclear power plant construction, comprehensive testing is performed to document conformance with specification requirements, but very little effort is applied to achieve economic efficiency. In any case process control is impaired unless the test is started using a sample that properly represents a greater quantity and was obtained by a practical procedure.

The need for proper sampling cannot be overemphasized. It is recognized by application of ASTM C 172, Standard Method of Sampling Fresh Concrete (1), as the required procedure for taking concrete samples under both ASME Section III, Division 2 (2) and ANSI/ASME N45.2.5-1978 (3). However, neither of these documents references a specific method for sampling aggregates, and variations exist in the specified requirements for cement sampling as well as other concrete constituents.

Correlation of material properties from test data is another area characterized by variable requirements between the standard documents. Evaluation of concrete compressive strength test results and concrete mixer uniformity is consistently covered, but correlation of properties from concrete sampled at different points is not adequately defined.

This paper will (1) examine the case for representative sampling, (2) describe methods of sampling including examples of methods currently in field use and (3) review the scope and usefulness of

---

[1]Engineering Specialist, Bechtel Power Corporation, Los Angeles, California

the correlation data generated from the sampled materials. The paper assumes a general familiarity with concrete production and construction. While the examples cited are not all-inclusive, they do illustrate the principles and current field practices.

REPRESENTATIVE SAMPLING

Purpose of Sample

Test data obtained from the sample are used to identify the characteristics of the larger amount of material from which the sample was obtained. Materials supplied and produced in concrete production are rarely, if ever, uniform. Thus, a proper concern is to use a sampling method that will reflect the uniformity, or lack of it, as accurately as possible, even though perfection in this regard is unlikely and proper evaluation of uniformity should be made using a planned series of samples. This analysis is a means for improvement of the production, processing and handling of materials which can generate more uniform materials, resulting in improved quality and economy.

However, if shortcut methods are employed that generate a biased or distorted sample selection any subsequent evaluation of the sample's test result is not only invalid, but may lead to a mistaken conclusion. Use of this test result with other cumulative data will then lead to further mistaken analyses and conclusions.

Sampling Plan

ASTM is moving towards the utilization of sampling plans which govern overall sample collection. Compared to current practice, the distinguishing feature is the attempt to insure collection and testing of a completely random sample, an element conspicuously lacking in today's practice. This is illustrated in Construction Practice Example 1:

> Construction Practice Example 1.--A quality control engineer writes a Non-Conformance Report stating that the specified sampling frequency of every 50 cu yd of concrete placed was violated because more than 50 cu yd of concrete was placed in the structure between samples, although samples were taken during the 0 to 50 cu yd and 51 to 100 cu yd ranges. Present specification requirements do not encourage a true randomness in sampling fresh concrete, for such a "violation" should be acceptable in a truly random sampling plan.

Specifications and standards that require sampling "every 50 cu yd" typically have the problem illustrated, since it is unclear if the requirement is to have one sample (A) taken at random within each 50 cu yd increment placed or (B) for every increment of 50 cu yd between samples. Method (A) is subject to criticism by those preferring the Method (B) interpretation. If Methods (A) and (B) are combined it results in unnecessary additional work. The usual result is to sample regularly in increments of 50 cu yd, offering the opportunity for bias in the sample when it is known that a "test" load is in production. This situation notoriously produces biased test results, given the human incentive to produce a favorable test record.

Appropriate remedies for this situation would be to require sampling plans that emphasize random selection of samples and specifically permit the Method (A) approach to be used.

It is also important to generate a practical plan that can be accomplished without excessive physical effort. If the plan does not incorporate these elements, its likelihood of being properly completed is small. To this end, there must be a willingness to construct physical facilities as needed.

METHODS OF SAMPLING

Concrete

Sampling is required to be in accordance with ASTM C 172, Standard Method of Sampling Fresh Concrete. The method covers sampling concrete from typical mixing, agitating, and non-agitating equipment. The basic procedure is to divert a complete stream of concrete at two or more intervals during discharge of the middle portion of the batch. The standard also gives limits for the amount of time between sampling and the start of fresh concrete testing and a minimum sample size. Methods of sampling from stationary mixers as well as truck mixers or agitators are given.

ANSI N45.2.5 specifies ASTM C 172 and additionally requires that concrete be sampled at the "placement point." This simply means the sample must be taken from the concrete truck or conveying device discharge unless a concrete pump is used or a correlation program is performed to relate the concrete properties between the specified and actual points of measurement.

The following examples illustrate current field practices:

> Construction Practice Example 2.--Concrete is sampled by the use of a flop gate in the discharge hopper of the central mixer. Fresh concrete property tests and concrete cylinders are made on the spot. Cylinders are stored in an air-conditioned "holding house" next to the test area or in a laboratory until they are stripped. They are then transported to the fog room. A correlation program is performed in the field to relate expected changes in fresh concrete properties to the batch plant.

Construction Practice Example 3.--Concrete is sampled
from mid-batch of the truck discharge at the delivery
point. A portion of the sample is retained to perform
fresh concrete property tests at the sampling point,
while the remainder is transported by truck to the
on-site laboratory, where concrete cylinders are made
and stored.

Construction Practice Example 4.--Between the batch
plant and construction site the concrete truck stops at
the test laboratory, where a concrete sample is taken
and fresh concrete property tests are run and cylinders
are made. This procedure is unacceptable because the
concrete sample is not taken from "the middle portion
of the batch" as specified by ASTM C 172.

Construction Practice Example 5.--Concrete is sampled
from the mid-batch of the truck discharge at the delivery
point. Fresh concrete property tests and concrete
cylinders are made on the spot. The cylinders are transported to the laboratory the following day, where they
are stripped and placed in the fog room.

Although these sampling requirements are simple and straightforward in concept, the Construction Practice Examples 2-5 show some
problems:

1. Productivity interruption -- Sampling devices or methods
that allow a sample to be taken without interrupting the flow of
concrete to the forms are rare. Central-mix plant manufacturers
typically do not provide a sampling device nor is provision made for a
device unless it is called for at the time of purchase; even then it
usually requires manual discharge at a reduced rate from the mixer to
obtain a sample conforming to ASTM C 172. Backfitting a sampling device
onto an existing central-mix plant is usually awkward and typically
results in a clumsy arrangement. Productivity interruption is usually
even more of a problem at the delivery or placement points, since at best
interruption of truck discharge to fill a wheelbarrow is necessary, at
worst it is necessary to devote substantial logistical effort to retrieving
a sample taken at the end of a pump line.

Even worse than productivity interruption is the potential disruption
to a smooth running concrete placement operation. Those experienced in
placing operations know that the highest quality results are obtained
when concrete is conveyed, deposited, and vibrated in a smooth, methodical, regular manner with a minimum of interruption. Requirements
for sampling should recognize this fact.

2. Initial curing -- If concrete cylinders are not fabricated in
close proximity of a temperature-controlled laboratory, it is necessary
to provide a temperature-regulated "holding house" to meet the ASTM C 31,
Standard Method of Making and Curing Concrete Test Specimens in the
field (1), requirement that "the temperature immediately adjacent
to the specimens (shall be maintained) in the range of 60 to 80 F
(16 to 27 C)" until the specimens are stripped. This requirement has

been met by providing a small air-conditioned trailer-sized structure that can be moved around the site as needed. Using curing boxes, damp sand, or wet burlap as mentioned in C 31 will not result in consistently meeting the temperature requirements; attempts to do so have consistently met with non-conforming conditions due to temperature. Also, the cylinders are often damaged, sometimes lost or vandalized. Vibrations from heavy equipment, particularly earthwork, have been known to influence cylinder properties. Field samples can be properly cared for only at excessive cost and effort, still with a high probability of being rejected as non-conforming.

3. Sampling from pumps -- The versatility of the pump line generally dictates that fresh concrete property tests cannot be made nor strength cylinders fabricated near the pump discharge. Thus it may be necessary to have a crane standing by to remove samples or to physically carry a sample out of the internal parts of a structure. On longer pump lines it may be difficult to identify the midload of a sample and if two trucks are discharging into a pump hopper simultaneously it is impossible. Concrete cannot be sampled from the end of a pump line on a regular and economical basis.

The preferred solution to all of these problems is to avoid sampling at any location other than the batch plant, assuming the minimal haul distance which is typical of major power plant construction. Such a sampling point provides the best overall chance to produce test results and specimens meeting specification requirements.

Aggregate

ASTM C 33, Standard Specification for Concrete Aggregates (1), requires that ASTM D 75, Standard Methods of Sampling Aggregates (1), be used to obtain test samples for verification of its requirements. However, if the specification requirement is given elsewhere (as in the job specifications or ASME Section III, Division 2) only about half of the methods cited in C 33 require that D 75 be employed to sample aggregates. Thus it would seem appropriate to require that D 75 be used, where appropriate, in the ASME and ANSI standards.

The D 75 Standard is a document which clearly defines procedures for sampling from conveyor belts and a flowing aggregate stream (bin or belt discharge). Sampling from stockpiles is discouraged and handled by requiring that a sampling plan be agreed upon for the given situation. The details are illustrated in Construction Practice Examples 6-8. In all these examples, three increments from different parts of the unit being sampled are needed to form a field sample.

Construction Practice Example 6.--Normal production aggregates are loaded onto a conveyor belt which is stopped while the sample is taken. A template device, shaped like the belt profile, is used to divide the sample from the remaining material on the belt. The sample is scooped off the belt and the fines are recovered by the use of a brush and pan.

Construction Practice Example 7.--A sample trolley has been fabricated consisting of a pan on wheels which can be rolled on a track under an aggregate discharge bin. With the pan in position, the hopper discharge gate is opened and closed in a time span which gives a full-stream-discharge sample without overflowing the pan. Another acceptable method is to restrict the gate opening (minimum width consistent with the aggregate size being sampled) to provide a flowrate that would allow the pan to cut through the stream.

Construction Practice Example 8.--For preliminary acceptance from the aggregate pit, samples are taken from a stockpile to check gradation in accordance with a sampling plan that requires removing a sample by cutting into the pile with a shovel. The plan gives good results in this case, and the main gradation control is based on samples withdrawn from the batch plant bins. It is recognized by all concerned that sampling from aggregate stockpiles should be avoided wherever possible.

The D 75 Standard does not prohibit the biased sample that can result from a bin that has been left standing overnight after being charged with moist aggregates. Moisture draining and evaporation can cause significant movement of fine materials. It is always a better practice to sample when production is in progress, than from locations where the aggregate has been idle for a period of time, especially if drying is involved.

Cement

For many years, the standard test frequency for portland cement was to take a sample and perform user tests every 1200 tons. This is a current requirement of ASME Section III, Division 2 and was a requirement of the 1974 edition of ANSI N45.2.5. As previously seen, this test frequency requirement does not provide a random sample and could lead to anticipation of the load that probably will be sampled.

The 1978 edition of ANSI N45.2.5 requires a test frequency "as specified in ASTM C 183," Standard Methods of Sampling Hydraulic Cement (1).

ASTM C 183 gives requirements for kind and size of samples, methods of sampling, amount of testing, acceptance, noncompliance and retest. Upon first reading, it appears cumbersome and complicated but should be workable when implemented. It includes the element of randomness necessary in an acceptable sampling plan. The C 183 standard is not yet in general usage because current projects are typically committed to the 1974 ANSI standard rather than 1978.

Cement samples are usually taken using a tube sampler. Such samplers may be slotted or not, depending on the length of the sampler and the size of the container being sampled. No problems should be expected in obtaining representative cement samples. The general methods of cement sampling also apply to fly ash and other pozzolans.

Water and Admixtures

No methods of sampling water and admixtures are specified in the standard documents. The common field practice is to draw a sample of the fluid from a representative point in the flow path. For admixtures, the samples are generally taken during delivery of the material.

CORRELATION OF SAMPLES

After a series of repetitive tests have been made, the data should be examined to determine what relation exists between the test results. Such information may be useful in making modifications in the production and usage of concrete and its constituents.

Uniformity of Concrete Strength Test Results

The ASME and ANSI documents include the use of ACI 214, Recommended Practice for Evaluation of Strength Test Results of Concrete (4). ACI 214 gives statistical procedures useful in interpreting the strength variation in concrete and includes a table on standards of concrete control, allowing a standardized evaluation to be made of the overall testing process. Information is also given which allows an evaluation to be made of the overall testing quality, to determine if poor results may be properly attributed to test or production procedures.

ACI 214 must be used to evaluate compressive strength test results in order to adjust concrete mixes. The uniformity of these results governs the amount the required average compressive strength must exceed the design strength. This system of controlling mix designs is now in general use and the experience with it is favorable. Significant cost reductions can be made through savings in cement and implementation of uniformity and control procedures which produce concretes of uniform, reliable strength.

SAMPLING OF CONCRETE

Uniformity of Concrete Mixers and Agitators

A correlation of samples is made when the test series is performed as specified in Annex A1, Concrete Uniformity Requirements of ASTM C 94, Standard Specification for Ready Mixed Concrete (1). The uniformity requirements of ASTM C 94 are cited in ASME Section III, Division 2 and ANSI N45.2.5.

A check of uniformity requirements requires the collection of two samples of concrete (after approximately 15% and 85% is discharged) from the mixer or agitator being evaluated. Each of the samples is tested for unit weight, air content, slump, coarse aggregate content, unit weight of air-free mortar and compressive strength at seven days. Requirements expressed as maximum permissible differences between test results are cited in the standard; five of the six results must meet the standard to achieve acceptable uniformity.

The uniformity test has long been in use and is accepted as being an adequate means of evaluating uniformity of concrete due to mixing time and condition of agitators and mixers.

Uniformity of Concrete During Delivery and Placement

ANSI N45.2.5-1978 requires samples for in-process tests of concrete shall be taken at the placement point (point of discharge of plastic concrete into the forms). The placement point and delivery point (usually the truck discharge) may be considered coincident if the transit time between these two points is less than five minutes (except for pumped concrete). For pumped concrete, samples must be taken at the placement point (pump discharge). For any other sampling schemes, a correlation testing program involving some duplicate testing of samples taken from another point in the process at regular intervals is required. The standard calls out further details of what correlation tests are required as well as the required frequencies, based on where the tests are to be taken and whether or not a concrete pump is used. By contrast, the ANSI 1974 standard requires that samples be taken from the pump line discharge if a pump is used. If other conveying equipment is used, samples may be taken at the truck discharge. Although the ANSI 1974 standard is endorsed as acceptable (but not required) by the NRC (Regulatory Guide 1.94)(5), the previous field practice examples on concrete sampling (Examples 2-5) indicate the general industry practice does not follow the standard. ASME Section III, Division 2 does not require correlation testing for pumped concrete.

The ANSI 1978 standard is not yet endorsed by the NRC and it is difficult to envision the industry using the standard as written in preference to any other reasonable program of testing. The need and purpose of frequent, regular correlation testing between various potential sampling points is undocumented. The random correlation work between sampling points performed in the field typically indicates that fresh concrete tends to moderately lose both slump and air as a function of time from leaving the mixer, as long as traditional, high quality concrete materials are used and no additions of material are made after the concrete is initially mixed. The basic principles of concrete technology lead to the conclusion that the concrete, as mixed, is going to maintain the same potential strength or the strength will increase due to loss of entrained air or loss of slump due to evaporation of water. This generally results in a better concrete in place than was originally produced in the mixer.

The ANSI 1978 standard also gives several potential programs of correlation testing between samples, but provides no guidelines for the evaluation of the test results. A major problem in trying to do so is the variability in field conditions, including haul distance, temperatures (both ambient and concrete), humidity, type of conveying equipment, delays in unloading trucks, etc. Not having a means of evaluation leaves the need for correlation testing on a regular basis questionable. The integrity of the production, transportation and conveying system can be better protected by the regular visual inspection of a knowledgeable inspector, who has the authority to require correction of deficiencies.

Good advice is given in ACI 304, Recommended Practice for Measuring, Mixing, Transporting, and Placing Concrete (4), which states:

> "---It frequently is desirable to sample at both the point of delivery to the pump and the point of discharge from the line and perform correlation testing to determine if any significant changes in slumps, air content, and other mix characteristics are occurring. If significant changes are found to be occurring, appropriate allowance should be made for them. Sampling for control testing should be at locations as specified in ASTM C 94 and C 172."

This simple statement gives adequate and appropriate guidance by suggesting that (1) correlation checks be performed at unspecified intervals, (2) appropriate allowances should be made for the expected significant changes in the concrete characteristics, and (3) control test samples should be taken from the batch plant or truck delivery discharge.

The concrete pump has proven itself an economical and versatile concrete placing tool and has consistently proven its ability to handle concrete without adversely affecting its properties, provided contact with aluminum is prohibited. Its record of reliable field use does not justify a complicated and expensive program of correlation testing, especially when variable field conditions preclude the possibility of a uniform standard of correlation evaluation.

CONCLUSION AND RECOMMENDATIONS

Present sampling practice in the nuclear construction industry is adequate to verify the quality of the concrete and concrete constitutents used in the work. The standards that direct this work, however, could benefit from revision to increase the overall effectiveness of the testing program by providing more meaningful results while decreasing the overall cost by specifying more appropriate test frequencies and requirements. Specific recommendations to accomplish these objectives are as follows:

1. Provide that samples be taken at random, in accordance with a sampling plan that reflects any previous experience with the material. Specify that samples must be taken to represent specific increments of material rather than maximum intervals of sampling (see Construction Practice Example 1).

2. Permit the primary control of fresh concrete to be based on sampling at central mixer or truck discharge (depending on delivery time). Permit any correlation testing for slump or strength (except C94 Uniformity Tests) to be based on a plan (including locations and frequencies) approved by the designer.

Improvements can also be made in field practices to meet the program objectives by implementing these recommendations:

3. Provide sampling equipment as required which will minimize interruption of concrete production and placement.

4. Avoid sampling fresh concrete under conditions that prohibit making fresh concrete tests and strength cylinders at the point of sampling. The most advantageous location is usually at the central mix plant.

The experience and recommendations reported in this paper reflect current nuclear power plant construction methods. They are generally applicable to other power plant construction provided frequency of testing is specified after consideration of the amount of testing needed along with the uniformity of the material being tested.

ACKNOWLEDGMENTS

The author deeply appreciates the efforts made by J. A. Cazares, D. E. Graham, S. B. Helms, R. N. Key, and L. H. Tuthill in providing constructive criticism after reviewing the paper. The support of R. J. Kosiba and Bechtel Power Corporation for the production and presentation of this paper is gratefully acknowledged.

REFERENCES

1. American Society for Testing and Materials. *1980 Annual Book of ASTM Standards*, Philadelphia, 1980.

2. American Society of Mechanical Engineers. *ASME Boiler and Pressure Vessel Code, Section III, Rules for Construction of Nuclear Power Plant Components, Division 2, Code for Concrete Reactor Vessels and Containments*, 1980 Edition including Summer 1981 Addenda, New York, 1980.

3. American Society of Mechanical Engineers. *Supplementary Quality Assurance Requirements for Installation, Inspection, and Testing of Structural Concrete, Structural Steel, Soils, and Foundations During the Construction Phase of Nuclear Power Plants*, ANSI/ASME N45.2.5-1978, New York, 1978.

4. American Concrete Institute. *ACI Manual of Concrete Practice -- 1980*, Detroit, 1980.

5. U.S. Nuclear Regulatory Commission. *Regulatory Guide 1.94--Quality Assurance Requirements for Installation, Inspection, and Testing of Structural Concrete and Structural Steel During the Construction Phase of Nuclear Power Plants*, Revision 1, April 1976, Washington, D.C.

SUPPLEMENTARY REFERENCES

1. Molnar, J. T. *Controlling Aggregate Properties for Compliance with Statistical Specifications*, Quality Assurance in Pavement Construction, ASTM STP 709, American Society for Testing and Materials, Philadelphia, 1980, pp. 19-27.

2. Proudley, C. E. *Sampling of Mineral Aggregates*, Symposium on Mineral Aggregates, ASTM STP 83, American Society for Testing and Materials, Philadelphia, 1948, pp. 74-87.

3. Shergold, F. A. *A Study of the Variability of Roadstones in Relation to Sampling Procedures*, Quarry Managers' Journal, London, January, 1963.

4. Warden, W. A. *Stockpiling of Aggregate for Gradation Uniformity* Circular 99, National Sand and Gravel Association, Silver Spring, MD, May, 1966.

STRENGTH AND SPECIAL PROPERTIES OF CONCRETE

James P. Allen, III[1]
Fellow, ASCE

ABSTRACT

There are a number of ACI documents that define the requirements for compressive strength testing of concrete. These standards have developed from an initial requirement to test each 250 cy of concrete to a present day requirement of each 100 cy. There appears to be a mistaken notion that concrete quality can be controlled by more frequent testing. However, concrete quality depends not only upon the ability to consistently process and handle the constituents during batching and mixing, but also on the success of transporting, placing, consolidating and curing the concrete. This paper suggests that the frequency of concrete testing be related to the ability to control strength as shown by statistical analysis. Further, it challenges code writing authorities to evaluate the methods available for early strength determination and nondestructive testing with the intent of recognizing these techniques as a means to comply with code requirements.

INTRODUCTION

Previous speakers at this conference have covered various topics on the control of constituent materials that make up concrete. It must be pointed out that any concrete is only as good as the constituents from which it is made, and the most essential aspect of producing good concrete is that the constituents, regardless of their individual qualities, are handled, processed, and mixed consistently to produce a uniform concrete. Additionally, there are a number of other factors which control the quality of a concrete structure that are relatively independent of the inherent qualities of the concrete itself. These include the final transportation, placement, consolidation, and curing of the concrete in place. An excellent concrete product can be produced that demonstrates all the proper attributes in the lab or in the field and still fails to produce a quality structure due to the independent variables mentioned above. Therefore, it is important to keep these factors in mind in this discussion of the properties of hardened concrete and what is being measured when these properties are examined. Basically, the evaluations that can be made are appropriate if the above four variables have been considered adequately. A conclusion can be drawn that says a properly designed concrete mix invariably will lessen the impact of problems involved in final transportation, placement, consolidation, and curing as compared to those that might be encountered with a less desirable mix.

---

[1]Chief Structural Engineer, Stone & Webster Engineering Corporation, Boston, MA

APPLICABLE CODES AND STANDARDS

The following codes of the American Concrete Institute (ACI) generally are used in nuclear power plant construction:

ACI 318 - Building Code Requirements for Reinforced Concrete

ACI 349 - Code Requirements for Nuclear Safety-Related Concrete Structures

ACI 359 - Code for Concrete Reactor Vessels and Containments, American Society of Mechanical Engineers, Boiler and Pressure Vessel Code, Section III, Division 2

To satisfy the quality assurance requirements on many plants, we also invoke the American National Standards Institute/American Society of Mechanical Engineers (ANSI/ASME) N45.2.5, "Supplementary Quality Assurance Requirements, for Installation, Inspection, and Testing of Structural Concrete, Structural Steel, Soils, and Foundations During the Construction Phase of Nuclear Power Plants." The Nuclear Regulatory Commission (NRC), in its Regulatory Guide 1.55, offers additional guidance and direction in the selection of other ACI documents that can be followed, such as ACI 301, "Specification for Structural Concrete for Buildings."

Currently in the area of concrete testing, there are numerous conflicts and variations between the requirements in these documents. It has been general practice to select the most stringent of the requirements contained in the various codes, and selectively use them for a nuclear project thus producing an unnecessarily expensive testing program. Identification and recommendations for industry resolution of several of these conflicts in a sound, technical manner is the aim of this presentation.

SPECIAL PROPERTIES OF HARDENED CONCRETE

The several documents previously mentioned also list various tests that can be made on hardened concrete to determine special properties. However, none of the codes have special requirements for performing such tests and the need for such testing programs is left entirely to the judgment of the engineer/designer. Perhaps the most comprehensive listing of special concrete properties that might be measured is contained in the ACI 359 code. This listing is given in Table 1.

The concrete property tests listed in Table 1 generally are used as material or concrete mix qualification tests rather than as production control tests. Certain of the tests, such as the coefficient of thermal expansion, creep of concrete in compression, and shrinkage coefficient, are relatively long-term tests and generally are performed only during the period of materials qualification or source selection. It is safe to say that these property tests are not normally of concern during concrete construction once materials and mix designs have been finalized. However, under high levels of compressive stress such as in large prestressed/post-tensioned structures or architectural structures having relatively small concrete sections, these properties may be of considerable significance when considering the long-term state of stress, loss of prestress, or in the case of more elaborate architectural structures, detrimental deflections due to creep. Suffice to say, these tests are not of general interest to the average concrete construction project or designer. However, they are considered (under the conditions above) during the design process and are particularly vital in the design of nuclear concrete reactor vessels and containments using post-tensioned design. In any case, compressive strength test results are useful as a measure of these properties of hardened concrete. Even though a specific relationship may not be exact, an approximation is helpful to the designer to assure that the concrete used conforms to the properties assumed in the design.

Two other properties not listed above, but which are of interest when the concrete becomes exposed to the elements, or especially to environmental conditions peculiar to nuclear facilities, are durability and impermeability of the concrete. In general, improvements in the compressive strength of the concrete will be beneficial to producing impermeable and durable concrete. Impermeability and durability practically go hand-in-hand in that the factors which contribute to a greater degree of impermeability of concrete also contribute to its greater resistance to the effects of freeze-thaw cycles. Both of these properties are enhanced by use of air-entrained, low water/cement ratio mixes made with sound aggregates of low porosity combined with proper placing techniques, thorough consolidation, and a conducive curing environment. Here again, these properties generally are evaluated in the materials selection and mix proportioning stage and usually are not monitored by control tests during construction. However, if the designer exercises proper control over proportioning of materials, batching, mixing and placing activities, he needs not be concerned with a periodic verification that the concrete being placed in the structure has these qualities.

CYLINDER COMPRESSIVE STRENGTH TESTS

The concrete cylinder compressive strength tests are, without doubt, the grandfather of concrete test procedures and are still the basic yardstick for acceptability of a concrete mix or a concrete structure. Basically, by considering that acceptable cylinder strength has been produced, we have concluded mistakenly that the structure will be adequate. What is known is only that acceptable materials have been used in the structure.

History of Cylinder Strength Testing

To review history somewhat, the yardstick for an acceptable level of testing the quality of concrete placed in major structures once was considered to be one set of tests for every 250 cu yd of concrete placed. This was the requirement up to and including the 1951 Edition of ACI 318. The 1956 Edition had the same requirement, but added that a minimum of one test per day be made. In 1963,

ACI 318 increased the frequency to 150 cu yd with a minimum of once per day, and it remains at that point today.

During the development of ACI 301, which was issued first in 1966, the committee increased the frequency of testing to one test for each 100 cu yd of concrete placed or fraction thereof. This was an apparent attempt to account for smaller sized concrete placements under minimum quality control that were intended for coverage by that code.

In the development of the ANSI/ASME N45.2.5 standard for concrete testing, a similar 100 cu yd criterion was adopted. With the advent of the ACI 349 and 359 codes, ACI 349 continued with the requirement of one test for 150 cu yd. While the ACI 359 committee initially started with that requirement, it subsequently increased the frequency to one set of tests for each 100 cu yd, to comply with the ANSI/ASME N45.2.5 requirements which were being applied to all nuclear work through endorsement by NRC Regulatory Guide 1.94. It is obvious that major uncertainty exists as to which set of test frequencies should be required.

Relationship of Cylinder Strength Tests and Structures

As far as the quality of a structure is concerned, more attention should be paid to the workmanship on the building rather than to the concrete itself. Also, the quality of the concrete produced, as stated previously, basically will be determined by the quality of the constituents and the control and uniformity thereof. Therefore, the current emphasis placed on increased control of the constituents and of the handling, batching, and mixing processes is the vital ingredient in producing consistent, quality concrete. The industry has increased control over the quality of the constituents and over the batching and mixing controls which generally were considered acceptable 20, 30 or 40 years ago, even though there was a history of long-lasting structures. However, at the same time, there seems to have been an additional, misguided, industry-prompted thrust in the direction of increased compressive strength testing in the hope that this would improve upon the quality of the concrete.

Just what does happen at a test frequency of each 100 cu yd or fraction thereof, but not less than one test per day? When the test data is reviewed from any project using such a test frequency, it can be shown that a compressive strength test represents not each 100 cu yd, but rather each 60 to 70 cu yd. This is due primarily to the fact that a test is required for any fraction over 100 cu yd and at least once per day regardless of the size of the incremental fraction or total concrete quantity in the placement.

Another aspect to examine is the recurring frequency of testing in regard to the size of the production batches of concrete and the time taken for production. With current automated batching, mixing, and delivery equipment, concrete is being placed at rates of 100, 150, or 200 cu yd an hour, and delivered in trucks that have capacities of 12 to 20 cu yd. This translates into a test on the order of once for every six to seven batches, which can occur in as short a period of time as 30 minutes. Such frequent testing is unreasonable, especially when conditions affecting concrete quality may not have had sufficient time to change.

Proposed Code Activities

Ideally, if the ingredients and the batching and mixing operations can be controlled, testing would be unnecessary except for periodic verification. Along this line, the ACI 359 committee has attempted, in conjunction with the ACI 349 and ANSI/ASME N45.2.5 committees, to develop a uniform approach to cylinder testing that will give us confidence in our concrete production operation, and not produce an abundance of data that have little bearing on our overall performance. By joint agreement the ACI 359 committee has attempted to reduce cylinder strength testing, by developing a two-stage test frequency. This change to the current ACI 359 code was announced in the May 1981 issue of CONCRETE INTERNATIONAL. Assuming no adverse comments, it will be included in the Winter 1981 Addenda to the ACI 359 code.

This code change requires a test frequency of one test per 100 cu yd with a minimum of one test per day for each class of concrete. When 30 consecutive test results show that the criteria for concrete acceptability are being met, the test frequency can be reduced to one test per 200 cu yd of concrete placed. We believe that this approach provides ample assurance of concrete quality on nuclear power plant projects where large amounts of concrete are being produced. It is the hope of the ACI 359 committee that ACI 349 and ANSI/ASME N45.2.5 will adopt this approach or refer to ACI 359 for their testing frequency requirements. These test requirements should be included in only one document; the other documents should refer to that one for the required testing. A further recommendation should be that ANSI/ASME N45.2.5 concentrate on the quality assurance aspects of inspection and testing and leave the test methods and frequency of test requirements to the applicable technical code. Mr. J. R. Wells, chairman of the ANSI/ ASME N45.2.5 committee, concurs with this concept. The ANSI/ASME committee became involved in this area of testing because there was no single compilation of required testing at that time that could be used to document quality control requirements.

ACCELERATED STRENGTH TESTS

This concept has been around for a while, but no official ACI code position yet has been taken. The compressive strength routinely determined at 28 days is nothing more than a prediction of the potential strength that a particular batch of concrete is likely to achieve, assuming that it is properly placed, consolidated, protected and cured. In this sense, it lends some confidence that the concrete in the structure will be adequate for its intended purpose. But, why has it been necessary to wait 28 days to get this information? Why couldn't we learn about potential strength sooner, especially with today's construction techniques capable of placing many cubic yards of concrete in multiple-story buildings in a short period of time?

There are three methods for which an ASTM standard is available to determine potential concrete strength in far less time than 28 days. These methods as delineated in ASTM C684 use either warm or boiling water, or an autogenous curing method which uses the heat of hydration within the specimen itself to accelerate strength development. Results from these methods generally are available in 24 to 48 hours. Another method which subjects a sample of fresh concrete to pressure and elevated temperature yields results in as little as five hours. Still another method determines the water and cement content of the concrete which, when evaluated along with conventional air content test results, can be related to 28-day strengths. The duration of this test is about 15 minutes and gives the advantage of being able to do something with the concrete within the forms while it is still in a plastic condition. A detailed presentation of accelerated test methods is contained in ACI SP-56, "Accelerated Strength Testing."

Present codes do not make allowances for using accelerated strength tests, and any attempt to use them is left to the discretion of the individual designer or engineer who would have to substantiate his position with the regulating authorities. These tests could prove useful in developing laboratory design mixes, wherein variations in design mixes could be established more quickly to reach a final design in less than the time required to go through a succession of 28-day curing periods. Of course this will require that a correlation be established for each project to enable an accurate 28-day strength prediction as influenced by such job-site variables as materials, testing personnel, test equipment, etc. This is an approach that could be incorporated into the provisions of the code, and thus allow designers and engineers another means of assuring that their structure is being constructed in a safe manner.

NONDESTRUCTIVE TESTING

In general, nondestructive tests imply no resulting damage to the concrete. Not all of today's so-called nondestructive tests truly meet that criterion since some do leave the surface damaged. However, repair of the damaged areas is considered aesthetic in nature rather than necessary from a load-carrying standpoint.

In general there are basically three types of nondestructive techniques. ASTM STP 169B, "Significance of Tests and Properties of Concrete and Concrete-Making Materials," provides a detailed description of these techniques. Sonic testing involves generating a sustained vibration and measuring the number of vibrations that occur over a unit of time. This method enables one to predict compressive strength in relation to a determination of modulus of elasticity. For a number of reasons, this method applies to small-sized specimens and is of little value in determining the characteristics of in-place concrete.

Pulse velocity tests involve the measurement of the time it takes for the impact of some kind of blow to the concrete to travel between two vibration pickups. In general, high velocities indicate quality concrete as opposed to very low velocities which indicate inadequate concrete. Further, pulse velocity techniques have been used in the past to detect the presence, and in some cases the magnitude, of cracks or voids. This technique is as useful in examination of in-place concrete as it is to laboratory specimens.

In the third category of nondestructive examination are in-situ tests. This category is further broken down into four methods: surface hardness, rebound, penetration resistance and pullout tests. The surface hardness testing method involves impacting the concrete surface with a specific mass and energy and subsequently measuring the indentation. The rebound method consists of a spring-controlled hammer which impacts the concrete. The rebound of the hammer is recorded and used to estimate the concrete strength. The results of neither of these two methods relate directly to compressive strength, but empirical correlations have been established between strength and either surface hardness or rebound.

The third method of in-situ test is the penetration technique which consists of a device which fires a probe into the concrete. The length of exposed probe is correlated to compressive strength.

The fourth method of in-situ test is the pullout test. This test requires the embedding of special steel rods prior to concrete placement. The force required to pull out the rod is believed to be a measure of the direct shear strength. This test does provide a direct measure of the in-situ shear strength from which the compressive strength can be estimated. In addition to having to preplan the location of these devices, the test does have a disadvantage in that the concrete surface must be repaired if tested to failure. However, loading the device to a predetermined level to indicate that some minimum strength of concrete had been achieved would eliminate this concern.

The industry consensus regarding any of the foregoing techniques is that they should not be considered indicators of absolute strength, but rather measures of relative strength within different parts of the same structure, or between different structures.

In-situ determination of concrete strength, either from cores taken from the actual structure or by nondestructive test methods, should be considered in our codes. The current ACI 318 code recognizes the use of cores taken from the structure to assess strengths that may have been in question as a result of low strengths from the cylinder test program. More work needs to be done in this area to allow the use of nondestructive testing to evaluate relative concrete strength in similar situations. There are certain instances in which nondestructive testing is being done quite successfully, rather than by testing a few cored specimens, with far less destruction to the structure and with more assurance that the structure is sound. Cores present problems because they are highly destructive to the structure, and are very difficult to obtain without causing damage to reinforcing and other embedded items. Further, because of these problems, cores generally are limited in number and thus provide data for evaluation from only a few isolated locations. Therefore, the nondestructive methods are much more effective, and the structure can be evaluated much more extensively than with a few core samples.

The ACI 359 committee has been given the challenge to develop a set of code rules for evaluating structures where the concrete cylinder strengths may not have met the requirements. Such a set of rules would enable the designer to establish the acceptability of the structure based on a combination of coring and/or nondestructive testing. This is well within the range of our current technology. It is hoped that the ACI 359 committee will be able to develop a set of rules that will be adapted readily by other codes and standards in the industry.

SUMMARY

In conclusion, the challenge to the code committees remains to establish truly meaningful test requirements which will guarantee the building of an acceptable structure and the production of uniform, quality concrete. A reassessment of the proliferation of testing methods must be undertaken, since these tests are based mistakenly on the ideas that: 1) more testing assures better quality, and 2) tests should be performed because they are simple and inexpensive. These ideas will needlessly increase the cost of a construction project and the time needed for its completion.

The time has come to recognize the sophistication of today's methods of materials control and batching and mixing control, and to reflect a level of confidence in these controls by establishing a combination of reasonable test frequencies for conventional, accelerated and in-situ concrete testing.

PROPERTIES OF CONCRETE

REFERENCES

American Concrete Institute

ACI 318, "Building Code Requirements for Reinforced Concrete," 1951, 1956, 1963, and 1977.***

ACI 301, "Specifications for Structural Concrete for Buildings," 1966 and 1972 (Revised 1975).***

ACI 349, "Code Requirements for Nuclear Safety-Related Concrete Structures," 1976.***

ACI 359, "Code for Concrete Reactor Vessels and Containments," July 1, 1980 (American Society of Mechanical Engineers, Boiler and Pressure Vessel Code, Section III, Division 2).***

"Concrete International," May 1981.

"Accelerated Strength Testing," Publication SP-56, 1978.

American Society for Testing and Materials

ASTM C39-72 (1979), "Test for Compressive Strength of Cylindrical Concrete Specimens."

ASTM C78-75, "Test for Flexural Strength of Concrete (Using Simple Beam with Third-Point Loading)."

ASTM C157-75, "Test for Length Change of Hardened Cement Mortar and Concrete."

ASTM C469-65 (1975), "Test for Static Modulus of Elasticity and Poisson's Ratio of Concrete in Compression."

ASTM C496-71, "Test for Splitting Tensile Strength of Cylindrical Concrete Specimens."

ASTM C512-76, "Test for Creep of Concrete in Compression."

ASTM C684-74, "Standard Method of Making, Accelerated Curing, and Testing of Concrete Compression Test Specimens."

ASTM STP 169B, "Significance of Tests and Properties of Concrete and Concrete-Making Materials," 1978.

American Society of Mechanical Engineers

ANSI/ASME N45.2.5 - 1978, "Supplementary Quality Assurance Requirements for Installation, Inspection, and Testing of Structural Concrete, Structural Steel, Soils, and Foundations During the Construction Phase of Nuclear Power Plants."***

U.S. Army, Corps of Engineers, Handbook for Cement and Concrete

CRD - C39-55, "Method of Test for Coefficient of Linear Thermal Expansion of Concrete."

CRD - C44-63, "Method of Calculation of Thermal Conductivity of Concrete."

U.S. Nuclear Regulatory Commission

Regulatory Guide 1.55, "Concrete Placement in Category I Structures," June 1973.

Regulatory Guide 1.94, "Quality Assurance Requirements for Installation, Inspection, and Testing of Structural Concrete and Structural Steel During the Construction Phase of Nuclear Power Plants," April 1976.

*** Document adopted as standard of the American National Standards Institute

TABLE 1

PROPERTIES OF HARDENED CONCRETE

Property	Test Method
Compressive strength	ASTM C39*
Flexual strength	ASTM C78
Splitting tensile strength	ASTM/C496
Static modulus of elasticity	ASTM C469
Poisson's ratio	ASTM C469
Coefficient of thermal conductivity	CRD-C44**
Coefficient of thermal expansion	CRD-C39
Creep of concrete in compression	ASTM C512
Shrinkage coefficient	ASTM C157

*American Society for Testing and Materials
**Corps of Engineers, U. S. Army, Handbook for Cement and Concrete

RELEVANCE OF ASTM C-342 VOLUME CHANGE TEST

Ashok J. Desai[1], M. ASCE

and

Allen J. Hulshizer[2], F. ASCE

ABSTRACT

This paper summarizes the results of an extensive program carried out to test concrete components used in the structures at the Seabrook Station, a nuclear power plant located on the New Hampshire Coast, approximately 40 miles north of Boston, Massachusetts. A gamut of qualification tests were performed on three cements and various aggregates to establish full compliance with all the requirements of ASTM C-150, Standard Specifications for Portland Cement, and ASTM C-33, Standard Specification for Concrete Aggregate. Certain combinations of cement and fine aggregates tested by the ASTM C-33 referenced method ASTM C-342, Standard Test Method for Potential Volume Change of Cement-Aggregate Combinations, exhibited expansions that were in a range that required further investigation to determine the suitability of these materials for project utilization.

Although ASTM C-342 is categorized as an aggregate test, this test produced information relative to an undesirable property of cement which went undetected in testing by all of the other standard ASTM cement and aggregate tests. This information lead to the restrictive use of cement which indicated highly questionable durability properties.

While there have been strong movements to eliminate ASTM C-342 as a standard test because of "problems" surrounding its initial development and evaluating its results, a recommended engineering evaluation has been developed to meaningfully interpret the test results so that the Standard can be used to examine cement and/or aggregate for deleterious products where durability properties of exposed concrete are of strong concern.

GENERAL INFORMATION

The Seabrook Station consists of a number of concrete structures partly buried and partly exposed to the atmosphere. Most of the underground portion of the structures have waterproofing membrane designed to prevent leakage of saline ground water. The inside temperature in the structures is generally less than 90°F, but in specific areas could occasionally be in the range of up to 120°F during the normal operation condition of the plant. The humidity inside

---
1. Structural Engineer, United Engineers & Constructors Inc., Phila., Pa.
2. Supervising Structural Engineer, United Engineers & Constructors, Phila., Pa.

the structures will range from 10% to 100%. The outside temperatures of the walls and mats buried below the grade level will be that of the surrounding material and rock, which is normally in the range of 50°F to 60°F. The outside temperature of the exposed walls will be in the range of -30°F to 100°F.

The estimated quantity of concrete for the Seabrook Station is in the order of 750,000 cubic yards. Testing was done on three different cements, four different fine aggregates and five different course aggregates.

TESTING OF CEMENT

Three viable cement suppliers were considered for the Seabrook Station. Since many structures are composed of large members, moderate heat of hydration Type II cement was specified. Additionally, Type II cement has moderate sulfate resistance which was a desired property for the Seabrook Station structures. Cements were tested for physical and chemical tests to assure that they meet ASTM C-150, Standard Specifications for Portland Cement.

Additional Project Requirements for Cement - In addition to the standard ASTM C-150 specification requirements for Portland Cement, limitations on free lime (CaO), total alkali ($Na_2O + 0.658 K_2O$) and magnesium oxide (MgO) were imposed to further assure concrete durability. The free lime content was limited to maximum of 0.7%.

It was also desired to limit the total alkali content to a lower value because reaction between alkal in cement and certain siliceous aggregates can result in a detrimental chemical reaction and corresponding concrete expansion. The general practice in the concrete industry is to impose this limit only when there is substantial evidence that the fine or coarse aggregates are reactive. Although fine and coarse aggregates for the Seabrook Project were selected from commercial sources having no known history of reactivity, the maximum upper limit of 0.6% was specified to avoid the remote possibility of adverse chemical reaction.

In addition, all three manufacturers could provide cement with MgO less than 4% which would provide less "FREE" portion of MgO (Periclase), and the potential to produce deleterious, expansive magnesium hydroxide ($Mg(OH)_2$).

All three manufacturers' cement met ASTM C-150 requirements and the additional project requirements for free lime and total alkali contents. It was concluded that the use of any of these cements would be satisfactory for the Seabrook Project provided it was confirmed that the deleterious alkali-aggregates chemical reaction would not take place when combinations of a particular cement and aggregate were tested for potential reactivity as outlined in the appendix to ASTM C-33.

REVIEW OF BASIC ASTM C-33 TEST RESULTS FOR SAND AND COARSE AGGREGATES

Initially four sources of sand and five sources of coarse aggregates were considered. The aggregate sources were evaluated in terms of pricing, delivery and quantity requirements, and at the same time, the materials were tested per ASTM C-33. The results of basic tests (excluding potential reactivity tests) for all aggregates satisfactorily met the ASTM C-33 requirements.

Long term potential reactivity tests were initiated for a selected sand and coarse aggregate in combination with three cements. Testing on other sources of sand and coarse aggregates was partially started so as to have alternate suppliers of materials available should the selected aggregates fail the potential reactivity tests.

EVALUATION FOR POTENTIAL REACTIVITY OF CEMENT AND AGGREGATE COMBINATIONS

General - In addition to testing the physical properties of cement and aggregates, it was necessary to confirm that the aggregates were chemically inert to avoid "alkali-aggregate" reactions which are generally brought about by the alkali characteristics of cement and/or water. The physical consequences of alkali-aggregate reactions are build-up of excessive stresses, expansion, loss of strength, and exterior cracking. By virtue of the cracks extending to the exterior, the structure also becomes exposed to further deterioration from external sources. The problems generated by classic alkali-aggregate chemical action range from the purely cosmetic to the complete failure of the structure.

Avoidance of the deleterious consequences of alkali-aggregate chemical reactions in concrete can be achieved by one or both of the following approaches:

a. By using cement with a low alkali content.

b. By using aggregates known to be non-reactive.

The ASTM Specification for Portland Cement (C-150) suggests that a limit of 0.6% on the total alkali content be specified when a cement is to be used in concrete with aggregates that may be reactive. The value of 0.6% was arrived at empirically by correlation of actual conditions of concrete distress. Cement tested for qualification and subsequent use on the Seabrook Project was limited to total alkali of 0.6%. Various fine and course aggregates were also selected from commercial sources with no history of reactivity. The most common tests to evaluate potential reactivity of cement and aggregate combinations are suggested in Appendix to ASTM C-33 which recommends the following approach:

> "A number of methods for detecting potential reactivity have been proposed. However, they do not provide quantitative information on the degree of reactivity to be expected or tolerated in service. Therefore,

evaluation of potential reactivity of an aggregate should be based upon judgment on the interpretation of test data and examination of concrete structures containing a combination of fine and coarse aggregates and cements for use in the new work..."

Accordingly, as recommended in ASTM C-33 Appendix, results of the following tests were used to assist in making the evaluation:

a) ASTM C-295, "Standard Recommended Practice for Petrographic Examination of Aggregates for Concretes"

b) ASTM C-289, "Standard Test Method for Potential Reactivity of Aggregates (Chemical Method)"

c) ASTM C-227, "Standard Test Method for Potential Alkali Reactivity of Cement Aggregate Combinations (Mortar Bar Method)"

d) ASTM C-342, "Standard Test Method for Potential Change of Cement-Aggregate Combination"

<u>Evaluation Based on Petrographic Examination of Sand and Coarse Aggregate (ASTM C-295).</u> Petrographic examinations of various sands and coarse aggregates under consideration were performed for two broad purposes:

a) Geologic identification and classification, determination of composition, and physical and chemical characteristics.

b) Identify the presence of constituents known to be capable of reacting with the alkali in cement.

The petrographic examinations of samples from four sources of sand and five sources of aggregates did not reveal any properties or constituents in any significant quantity which were known to have specific unfavorable effects in concrete.

Since interpretation is involved in using the results of petrographic examinations, it was considered essential to test cement/aggregates combinations with the other available test methods to assure that the material selected for the concrete is of desirable quality.

<u>Evaluation Based on Chemical Determination of the Potential Reactivity (ASTM C-289).</u> This test covers the chemical method for determining the potential alkali reactivity of aggregates, and their potentiality for producing abnormal expansion in concrete when used with high-alkali cement.

It is an empirical test developed by chemically testing a large number of aggregates and using the same aggregates in combination with a high-alkali cement in the mortar bar test (ASTM C-227).[1] The dissolved silica content was plotted with respect to the reduction in alkalinity, and a line drawn through the diagram to separate the innocuous aggregates, from the reactive aggregates, as indicated in Figure 1. If test results for aggregates fall on the left of demarcation line, the aggregate is considered innocuous and if it falls on the right of the demarcation line, the aggregate is considered potentially reactive.

VOLUME CHANGE TEST

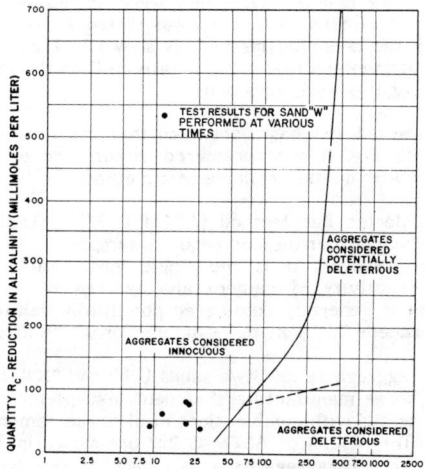

FIGURE 1 - GRAPHICAL ILLUSTRATION OF DIVISION BETWEEN INNOCUOUS AND DELETERIOUS AGGREGATES AND TEST RESULTS FOR SAND "W" AS PER ASTM C289.

TABLE 1
SUMMARY OF RESULTS FOR POTENTIAL ALKALI REACIVITY OF CEMENT AGGREGATE COMBINATIONS (MORTAR BAR METHOD)-ASTM C227

CEMENT SOURCE OF AGGREGATES	% EXPANSION OF MORTAR-BAR AFTER 6 MONTHS			
	TYPE-II, LOW ALKALI CEMENT OF			REMARKS
	MANUFACTURER-A	MANUFACTURER-B	MANUFACTURER-C	
A. SAND				
1. SOURCE-W	0.04	0.02	0.02	
2. SOURCE-X	0.04	NOT TESTED	0.02	
B. COARSE AGGREGATES				
1. SOURCE-M				
1 1/2"	0.04	0.03	0.02	
3/4"	0.03	0.03	0.02	
1/2"	0.04	0.03	0.03	
#8	0.03	0.03	0.02	
2. SOURCE-N				
#4	0.03	NOT TESTED	0.02	
#67	0.03	NOT TESTED	0.02	
#8	0.03	NOT TESTED	0.02	
3. SOURCE-O				
1 1/2"	0.05	NOT TESTED	0.02	
1"	0.04	NOT TESTED	0.02	
#8	0.04	NOT TESTED	0.02	
4. SOURCE-P				
1 1/2"	0.03	0.02	0.02	
3/4"	0.03	0.03	0.01	
#8	0.04	0.04	0.02	
5. SOURCE-R				
3/4"	NOT TESTED	0.02	0.01	
1/2"	NOT TESTED	0.02	0.01	

NOTE: 1) THE CEMENT/AGGREGATE COMBINATION MAY BE CONSIDERED TO BE POTENTIALLY REACTIVE IF THE EXPANSION AFTER 6 MONTHS IS IN EXCESS OF 0.1%.
2) FOR COARSE AGGREGATES, THE TESTING WAS DONE FOR VARIOUS GRADATION OF AGGREGATE E.g. 1 1/2" MEANS MAXIMUM SIZE OF AGGREGATE WILL BE 1 1/2".

Four sands and five coarse aggregates were tested by the chemical method. A typical result of Sand "W", which was tested at six different times from 1977 to 1980 by the chemical method, is shown in Figure 1. The test results of all samples of sand and coarse aggregates fell on the left of the line indicating that all aggregates were innocuous.

Since the concrete industry has recognized that the quick chemical test is not altogether conclusive, it was considered essential to test cement and aggregate combination with another reliable test method.

Evaluation Based on Mortar Bar Method (ASTM C-227). Unlike the quick chemical method, the mortar bar method requires samples to be examined for three to six months; however, it is the single most reliable method for evaluating potential reactivity of cement and aggregate. A cement and aggregate combination is generally considered potentially deleterious by this test if its expansion exceeds 0.05% at three months, or 0.10% at six months.

The five coarse aggregates and two sands ("W" and "X") were tested in combination with the three manufacturers' cement, except for certain aggregates which were not tested with combinations of all three cements since their use was in question. The results of ASTM C-227 are shown in Table 1. The indicated expansion for all aggregates at six months fell well below the prescribed limit and therefore, their innocuous character, as concluded by ASTM C-289 test, was further reinforced by the results of ASTM C-227 test.

Evaluation of Cement and Coarse Aggregate Combinations Based on ASTM C-342. Inasmuch as the concrete is to be utilized in structures associated with nuclear portions of the power plant, it was considered prudent to perform all of the methods recommended in the ASTM C-33 Appendix for evaluating potential reactivity of an aggregate, which included the less commonly utilized ASTM C-342. A single exception to the list was the non-applicable use of ASTM C-586 for testing Potential Reactivity of Carbonate Aggregates. ASTM C-342 method covers the determination of the potential expansion of a specific cement aggregate combination by measuring the linear expansion of a mortar bar which has been subjected to a variance of temperature and saturation under the prescribed conditions. Criteria for evaluating potential reactivity is suggested in Appendix to ASTM C-33 and is as follows:

"Cement aggregate combinations tested by this procedure whose expansion equals or exceeds 0.200% at an age of one year, may be considered* unsatisfactory for using concrete exposed to wide variations of temperature and degree of saturation with water."

(* EMPHASIS SUPPLIED)

The results of ASTM C-342 test between various sources of coarse aggregates and cements are presented in Table 2. The coarse aggregate from Source M was selected because it was economically competitive and, at the same time, combinations with each type of cement produced expansions much less than 0.2% suggested limit, further confirming the innocuous characteristics of the cement and coarse aggregate combinations.

EVALUATION OF CEMENT AND SAND COMBINATIONS BASED ON ASTM C-342

The results of phase 1 evaluation, utilizing combinations of two sources of sand and three manufacturers' cement, are shown in Table 2. It was concluded that Cement "C" in combination of Sand "W" and Sand "X", exhibited innocuous characteristics, consistent with the results of ASTM C-295, C-289, C-227. Although tests were conducted to qualify all the materials, it was determined that Sand "W" would be an economical and viable source for the Seabrook Project, provided that sand in combination of various manufacturer's cement was established to be innocuous after evaluating the ASTM 342 results. However, the results of the ASTM C-342 tests indicated that the combinations of sand from Source "W" and cement by either manufacturer "A" or "B" needed to be further investigated since these tests produced expansions which raised the question of potential reactivity. Since the Seabrook Project required a large quantity of cement, it was decided that cement from at least two suppliers should be qualified. It then remained to be confirmed that combinations of Sand "W" and either Cement "A" or Cement "B", when used in concrete, would not produce deleterious chemical reactions. For this reason, it was necessary to again thoroughly investigate Sand "W" in combination with Cement "A" or Cement "B" by the ASTM C-342 test procedure.

Investigation of ASTM C-342 Results of Sand "W" in Combination with Cement "A".

Since Sand "W" in combination with Cement "A" produced an "excessive" expansion, the test was repeated two additional times to make sure that there were no anomalies in the testing work. The results of the subsequent testing also indicated "excessive" expansion, which could be considered deleterious. Leading petrographers in the industry were involved to determine the cause of expansion. It was established that none of the samples contained aggregate particles which could be classified as potentially reactive with cement alkali and no evidence of cement alkali-aggregate reaction was observed in any of the three test mortar bars examined. However, the inspection of "failed" mortar bars did reveal the presence of Periclase (free crystalline MgO) in unhydrated remnants of cement grains in each bar. It is recognized that periclase, per se, is not a source of distress, rather, its hydration to form Brucite ($Mg(OH)_2$) which causes expansion that doubles the volume of the initial unreacted Periclase.

At this stage even though the exact mechanism which created expansions in excess of the suggested ASTM C-342 limits was not determined, Cement "A" was eliminated from further testing and subsequent use on Seabrook Project due to the uncertainties involved with Periclase content.

INVESTIGATION TO DETERMINE POTENTIAL REACTIVITY BETWEEN SAND "W" IN COMBINATION WITH CEMENT "B"

ASTM C-342 test for the Cement "B" and Sand "W" combination was initially conducted by an independent Testing Laboratory. The initial samples exhibited an expansion of 0.42% after the samples were kept continuously immersed in water for 12 weeks and thus exceeded the suggested ASTM C-33 Limit of 0.2% at one year. Further measurements of expansion were discontinued after 12 weeks.

In order to validate the results of the first ASTM C-342 test, a series of other tests were simultaneously conducted at four independent laboratories. The results of the four laboratories' findings indicated expansion in the range of 0.39% to 0.64% in less than a year, which confirmed the previous finding.

When the Cement "B" and Sand "W" combination produced expansion in excess of what is considered normal for all the tests, the following questions and concerns were raised:

a) Is excessive expansion the result of classical alkali-silica chemical reaction?

b) Does variation in the sand and cement proportion affect the amount of expansion?

c) Does variation in the alternate wetting and drying cycle at high temperatures affect the amount of expansion since the Seabrook structures are not subjected to these conditions?

d) What does ASTM C-342 detect which the other tests for potential reactivity did not detect?

e) Is Cement "B" and Sand "W" combination reactive even though ASTM C-295, ASTM C-289, ASTM C-227 tests did not indicate reactivity?

In order to answer the above concerns, a detailed evaluation was undertaken, including literature searches and discussions with various authorities in concrete, which are summarized as follows.

Petrographer's Findings. The samples of initial ASTM C-342 test of Cement "B" and Sand "W" showing "abnormal" expansion were forwarded to a leading petrographer in the industry to look for the presence of gel filled pores and fissures, evidence of chemical reaction, properties of the reaction products, and to examine for cracking.

The petrographic examination did not reveal any abnormalities, that is

a) no evidence of deleterious reaction between Cement "B" and Sand "W";

b) no evidence of any internal micro-cracking or fracturing of the hydrated cement matrix within the samples;

c) no presence of periclase or brucite, which is known to be a deleterious by-product of the chemical reaction;

d) no evidence of any other chemical reactions.

Also, an x-ray diffraction of two different mortar bars, one which expanded in ASTM C-342 tests and one that did not expand in ASTM C-227 test, was carried out and no significant differences, product of deleterious reactions or unusual products of hydration was found.

It was concluded that the expansion of test samples was not the result of any deleterious chemical reaction.

Portland Cement Association's (PCA) Findings. ASTM C-342 has been identified over the years as an aggregate test, falling under the requirements of ASTM C-33 for Concrete Aggregates. In the course of performing the ASTM C-342 test and evaluating the "abnormal" expansion, there appeared to be a more direct relationship to the performance of the cement at which time Cement Company "B", manufacturer of the specific cement, was approached to see if they could provide an explanation for the test results. As part of their investigative work on the problem, Cement Company "B" enlisted the services of PCA, who in turn looked at various aspects of the test (ASTM C-342) itself to evaluate its applicability and suitability to testing of cement and aggregate combinations for the Seabrook Project. The following two basic areas were addressed as part of PCA's evaluation to determine the relationship of this test to actual structures and exposures:

1) Variations in alkali content of cement

2) Variations in sand-cement proportions and temperature-humidity cycles

The specific results of Portland Cement Association's 12 months of testing are given in Table 3 and are summarized as follows:

a) ASTM C-342 was performed utilizing the combinations of Sand "W" and two other types of cements which PCA has historically standardized in their long term research testing. The other cements had alkali contents of 0.16% and 0.92% compared to the Cement "B" sample content of 0.44%. The results indicated that the expansion of the samples utilizing Sand "W" and the PCA's "standard" cement are less than 1/10 of that of the sample utilizing Cement "B" and Sand "W" combination. The comparative testing indicated that there is no apparent relationship between alkali content and the performance of Sand "W". The alkali content in the cement often acts as a catalyst to trigger expansion; however, Sand "W" in this case would have to be considered non-reactive since the higher alkali cement did not produce "abnormal" expansion.

b) Testing done over a 12 month period with seven variations in sand-cement proportions and temperature - humidity cycles resulted in maximum expansions in the order of 25% of the ASTM C-33 suggested limit of 0.2% in 12 months. These test results were in line with the work done by Mr. A. D. Conrow, which was the basis of ASTM C-342. The results of this examination clearly indicated

TABLE 2
PHASE I RESULTS OF ASTM C342 TESTS

CEMENT SOURCE OF AGGREGATES	% EXPANSION OF MORTAR-BAR AFTER 1 YEAR (UNLESS NOTED)			REMARKS
	TYPE-II, LOW ALKALI CEMENT OF			
	MANUFACTURER-A	MANUFACTURER-B	MANUFACTURER-C	
A. SAND				
1. SOURCE W	0.248 (12 WKS.)	0.42 (12 WKS.)	0.046	
	0.571	-	-	
	0.854	-	-	
2. SOURCE X	0.524	0.45 (16 WKS)	0.118	
3. OTTAWA SAND (INERT)	0.583 (24 WKS)	0.029	0.013	
B. COARSE AGGREGATES				
1. SOURCE-M				
1 1/2"	0.037	0.031	0.019	
3/4"	0.035	0.025	0.013	
1/2"	0.038	0.027	0.013	
#8	0.045	0.039	0.010	
2. SOURCE-N				
#4	0.039	NOT TESTED	0.019	
#57	0.048	NOT TESTED	0.016	
#8	0.028	NOT TESTED	0.016	
3. SOURCE-O				
1 1/2"	0.345	NOT TESTED	0.027	
1"	0.116	NOT TESTED	0.029	
#8	0.268	NOT TESTED	0.029	
3/8"	0.260 (12 WKS)	NOT TESTED	NOT TESTED	
4. SOURCE-P				
1 1/2"	0.126	0.051	0.015	
3/4"	0.032	0.032	0.012	
3/8"	0.033	0.060	0.014	
5. SOURCE-R				
3/4"	NOT TESTED	NOT TESTED	0.49 (32 WKS)	
1/2"	NOT TESTED	NOT TESTED	0.45 (36 WKS)	

NOTE: 1) THE CEMENT/AGGREGATE COMBINATION MAY BE CONSIDERED TO BE POTENTIALLY REACTIVE IF THE EXPANSION, AFTER 1 YEAR, IS IN EXCESS OF 0.2%.
2) FOR COARSE AGGREGATES, THE TESTING WAS DONE FOR VARIOUS GRADATION OF AGGREGATE Eg. 1/2" MEANS MAXIMUM SIZE OF AGGREGATE WILL BE 1/2"

TABLE 3 - C 342 TEST RESULTS BY PCA

MIX NO.	MATERIAL VARIABLE	TEST CYCLE	% EXPANSION OF SAMPLE AFTER 1 YEAR
1	PER C-342, USING NATURAL SAND W & CEMENT, BOTH FROM SAME SUPPLY AS USED FOR MORTAR BARS SHOWING EXCESSIVE EXPANSION	SAME AS ASTM C-342	.440
2	PER C-342, USING NATURAL SAND W & CEMENT WITH 0.16% ALKALI (PCA CEMENT)	SAME AS ASTM C-342	.013
3	PER C-342, USING NATURAL SAND W & CEMENT WITH 0.92% ALKALI (PCA CEMENT)	SAME AS ASTM C-342	.029
4	USE CEMENT SAND RATIO OF 1:4.50 INSTEAD OF SPECIFIED 1:2.25 RATIO. USE SAME WATER-CEMENT RATIO AS MIX No.1, AND CEMENT B WITH SAND W	SAME AS ASTM C-342	.038
5	PER C-342, EXCEPT REPLACE 15% OF CEMENT WITH -200 SIEVE SIZE QUARTZ SAND. USE CEMENT B AND SAND W	SAME AS ASTM C-342	.032
6	PER C-342, EXCEPT REPLACE 4+8 SIEVE SIZE SAND WITH -200 SIEVE SIZE QUARTZ SAND. USE CEMENT B WITH SAND W	SAME AS ASTM C-342	.051
7	SAME AS MIX No.1	28 DAYS-SOLUTION AT 73°F, 14 DAYS-SOLUTION AT 131°F, CONT'D IMMERSION IN SOL. AT 73°F TO AGE OF ONE YEAR	.029
8	SAME AS MIX No.1	28 DAYS SOL AT 73°F, 14 DAYS DRY AT 131°F, CONT'D IMMERSION IN SOL AT 73°F TO AGE OF ONE YEAR	.051
9	SAME AS MIX No.1	28 DAYS-SOL AT 73°F, 14 DAYS AT 73°F, CONT'D IMMERSION IN SOL AT 73°F TO AGE OF ONE YEAR	.018
10	SAME AS MIX No.1	7 DYS-SOL 73°F, 3 DYS. DRY AT 131°F, 4 DYS-SOL AT 73°F, 3 DYS DRY AT 131°F, 4 DYS. SOL AT 74°F, 3 DYS DRY AT 131°F, CONT'D IMMERSION IN SOL AT 73°F TO AGE OF ONE YEAR	.014

ALL DATA ARE THE AVERAGES FOR THREE COMPANION MORTAR BARS.
ALL MORTAR BARS MADE WITH SAND W.

that only a unique set of conditions relating to the following would produce the "abnormal" expansions with the Cement "B" and Sand "W" combination.

- o   A specific proportion of sand and cement,
- o   A specific wetting and drying temperature of 131ºF,
- o   A specific alternate wetting drying interval,
- o   A specific humidity during storage of sample before final testing.

Review of ASTM C-342 Test Results using Cement "B" and Sand "W" with Various Industry Consultants

a) *General.* A discussion was held with various experts and industry consultants to review the test results of Cement "B" and Sand "W" combinations, in particular the "failure" observance in the ASTM C-342 tests. It was recognized that the ASTM C-342 test was based on Mr. A. D. Conrow's studies of abnormal expansion of Portland Cement concrete, which was observed in Oklahoma, Kansas, Nebraska and Iowa (Great Plains). It was indicated that in the 1950's, about 5 to 7 years after being built, that concrete in many Great Plains highway structures developed serious cracking which in many cases required extensive repairs on bridges.[2] As a result of these concrete problems, investigative work commenced which lead up to the ASTM C-342 procedure which was to avert future deterioration. ASTM C-342 is an accelerated test developed to detect abnormal expansion assumed to be the result of a chemical reaction experienced in the limited geographical area of the Great Plains which was considered to be different in kind from the classical alkali-silica reaction.

It was indicated that the work of concrete researchers has revealed that the reaction suggested by Mr. Conrow is not unique to that part of the world but appears to be associated with a difference of environmental exposures peculiar to those structures. It was postulated that the hot sun and dry weather on the Great Plains drew moisture up through cracks in the slab, which evaporated at the top, leaving a concentrated solution containing the alkali from the whole thickness of the pavement in the top ½". This concentration made the structures built with low-alkali cement behave as if they were made with high alkali cement, which resulted in cracking of the pavement and bridge.

This particular phenomena was not recognized at that time and Mr. Conrow attributed the behavior to some kind of chemical reaction other than the classical alkali-silica reaction.

b) <u>Suggested Reason Why Cement "B" and Sand "W" Combination Exceeded ASTM C-342 Limit</u> - Since no chemical reaction was detected in the petrographic examination, it was suggested that the growth exhibited in the Cement "B" and Sand "W" combination may be associated with thermal incompatabilities and cement-aggregate interfaced bond failures which is related to a physical phenomena and not a chemical behavior.[3] A working hypothesis was established which explained that the behavior can be attributed to differences in thermal properties of aggregates and the mortar or cement paste. When considerable differences exist between thermal coefficients of aggregates and the mortar or cement paste, stresses may exist that are sufficient to overcome the interface bonding on isolated aggregate particles, creating minute gaps which become filled by the capillary attraction of moisture.[4] This was felt to be the contributor to the expansion by increasing the net volume of the mass by the trapped moisture.

Since the petrographic examinations of the test samples of Cement "B" and Sand "W" combination did not show the presence of brucite, or indications of other materials which may be the result of deleterious chemical reaction, it was concluded that the expansion produced by the ASTM C-342 test can be safely ignored. It was indicated that an added factor in favor of the use of Cement "B" lie in the fact that concrete industry experts are not aware of any concrete performing unsatisfactorily when it has passed the autoclave test, (ASTM C-151) as has Cement "B". Although there was no reason to expect poor performance from concrete made with Cement "B" and Sand "W", it was felt prudent to confirm that concrete made from those materials have satisfactory in-service records when exposed to conditions similar to that of Seabrook. On that basis no lesser performance should be expected from the Seabrook Structures.

RESULTS OF EXAMINATIONS OF THE STRUCTURES UTILIZING CEMENT "B"/SAND "W"

To confirm the conclusions, an in-service inspection was conducted of concrete structures utilizing those components and exposed to the conditions anticipated at Seabrook. Accordingly, the following structures were examined by the authors for overall deterioration or distortion or evidence of adverse cracking and spalling (see photographs):

1. Parking garage by the American and U.S. Air Terminal, Boston

2. Charles River Dam

3. Piers for I-93 elevated expressway

4. Parking structure for Commercial Building

5. Fossil Power Plant in the Boston area

6. South Cove tunnel and station.

PARKING GARAGE ROOF EXPOSED TO SNOW AND PONDING OF WATER.

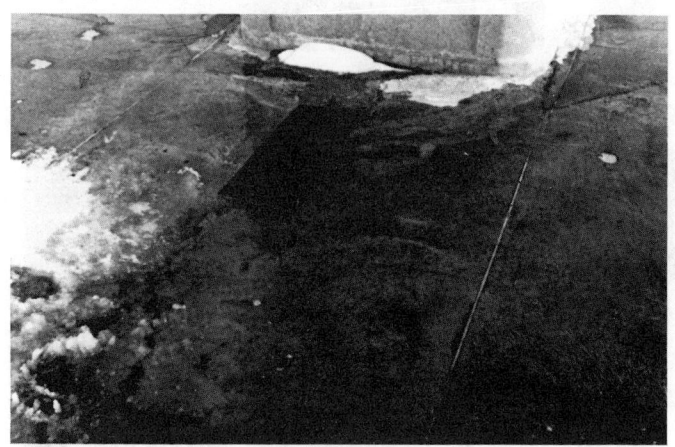

CLOSE VIEW OF PARKING GARAGE ROOF AREA SUBJECTED TO ALTERNATE WETTING & DRYING FOR A NUMBER OF YEARS BUT DEVOID OF ANY DISINTEGRATION

IN-SERVICE INSPECTION OF PARKING GARAGE STRUCTURE

FLOOR AREA OF POWER PLANT EXPOSED TO SEVERE EXPOSURE OF STEAM FROM DRAIN LINE.

CLOSER VIEW OF FLOOR INDICATED NO DISINTEGRATION OR DISTORTION.

IN-SERVICE INSPECTION OF POWER PLANT STRUCTURE

The examinations of the structures revealed no indication of any overall deterioration, including disintegration and distortion, or any evidence of extensive cracking or spalling. There was no apparent expansion or distortion of the type evidenced by ASTM C-342 sample bars. What was observed were normal imperfections such as localized hairline cracks, minor spalling, and dusting. None of the structures inspected showed any potential signs of instability or deterioration. The in-service inspection of the structures confirmed the earlier conclusion that the cement and the sand materials under consideration are suitable for the project use.

CONCLUSIONS

Numerous physical and chemical tests were performed on cements, and various sands and coarse aggregates were tested, including exhaustive individual potential reactivity tests. The methods of detecting potential reactivity did not provide quantitative information on the degree of reactivity to be expected or tolerated in service. To insure structural concrete integrity and performance, proper interpretation of test results and evaluation of their significance are of utmost importance.

During the lengthy course of evaluating the "abnormal" expansions exhibited with certain combinations of cement and sand, criticism has been raised by some as to the appropriateness of ASTM C-342 per se, and some would opt for its limited use or omission due to its idiosyncrasies.

Based on the authors' involvement in this "problem", it is their conclusion that the difficulties associated with the use of ASTM C-342 are not with the quality of the test but with the utilization of the results. ASTM C-342, through test produced expansions, has been shown to detect probable deleterious non-alkali-silica reactions, in cases where none of the other tests (ASTM C-227, C-289 and C-295) indicated abnormalities and therefore this test does possess certain merits.

It is the authors' conclusion that where none of the other tests indicate abnormalities and where conditions or the nature of a structure warrant investigative work by ASTM C-342, that the following be utilized in evaluating the test results.

1. If expansion of samples does not exceed 0.2% in one year, the materials should be considered satisfactory.

2. If expansion of samples exceed 0.2% in one year, examine the test samples petrographically for products of reactivity.

   A. If products of deleterious chemical reaction are present, either reject the material producting the reaction or, if appropriate, do further investigative work accordingly.

   B. If *no* products of deleterious chemical reaction are present, evaluate the intended service of the structure with respect to its compatibility with the temperature-humidity cyclic conditions simulated by ASTM C-342.

If the service conditions are <u>similar</u> to that of the test, the material(s) should be considered unsatisfactory.

If the service conditions are <u>different</u> to that of the test, the materials should be considered satisfactory if historical or in-service inspections records of the combined cement and sand indicate acceptable performance.

Elimination of the ASTM C-342 as a test procedure does not answer the problems it surfaces.

ACKNOWLEDGEMENTS

Seabrook Station is jointly owned by a number of utilities. Public Service Company of New Hampshire is the major shareholder and agent for the Owners. Yankee Atomic Electric Company is the Engineering Supervisor for the Owners. United Engineers and Constructors Inc. are the Architect-Engineers and Construction Managers for the plant. The major part of concrete component testing was done by Pittsburgh Testing Laboratory. Also, services of Construction Engineering Consultants Inc., Mr. E. M. Krokosky of Carnegie-Mellon, Mr. B. Mather of Corps of Engineers, Mr. D. Stark of Portland Cement Association, and Erlin Hime Associates were utilized.

REFERENCES

1. <u>Correlation Between Chemical and Mortar Bar Tests for Potential Alkali Reactivity of Concrete Aggregates</u> Reported by Bernard Chaikew and Woodrow J. Halstead, Division of Physical Research and Public Roads, Vol. 3, No. 8.

2. <u>Studies of Abnormal Suspension of Portland Cement Concrete</u> by A. D. Conrow. ASTM Proceedings, Vol. 52, 1952.

3. <u>Prediction of Concrete Durability from Thermal Tests of Aggregates</u> by E. C. Higginson and D. G. Kretsinger, ASTM Proceedings, Vol. 53, 1953.

4. <u>A Wetting and Drying Test for Predicting Cement-Aggregate Reaction</u> by C. Scholer. ASTM Proceedings, Vol. 49, 1949.

5. <u>Observations of the Performance of Concrete in Service</u> by Highway Research Board. Publication 309-01790-4.

UTILIZING TEST DATA FOR
GREATER QUALITY CONTROL

by

Reginald Coupland[1]

ABSTRACT

An approach is described for obtaining Inspection and Test data of sufficient accuracy to be utilized by the Engineer. A suggested role of the Engineer in obtaining credible data is also discussed, with particular emphasis on his interface with the Quality Control organization.

DEFINITIONS

The following definitions are used:

Engineering — The group responsible for the engineering, design and acceptability of the final plant.

Construction — The group responsible for producing, installing and fabricating material and components in accordance with the Engineer's requirements.

Quality Control — The group responsible for assuring that the results of tests, examinations and inspection meet the acceptance criteria of the Engineer.

Inspector — An individual who performs and records the results of tests, examinations and inspections.

INTRODUCTION

There are numerous prerequisites to be fulfilled in order to obtain accurate test/inspection results in a timely manner for long-term trend analysis or the short-term decision making process. These prerequisites can initially be divided into four main groups given as follows:

(1) Specification - defining the type of tests required and acceptance criteria.

---

[1] Energy Consultants, Inc. - Duquesne Light Company - Director, Site Quality Control

(2) Equipment - of sufficient accuracy to obtain and/or record results within the specified tolerance band.

(3) Procedures - to assure consistency of application and reporting.

(4) Personnel - trained to implement the procedures described.

If the test/inspection activity was being performed in relative isolation, a limited expansion of the above prerequisites would probably suffice in the obtaining of the needed results. However, in the environment of a power generation facility construction site, two other factors are considered. These are given as:

(5) The interface between Quality Control and Construction.

(6) The interface between Quality Control and the Engineer.

SPECIFICATION

The Engineering specification should be subject to a review with the aim of identifying any potential problem areas, conflicts, and any contradictions with licensee commitments that may have been made in the safety analysis report. A potential problem area is where the Engineer has specified his own requirements in the specification and has also specified that a certain Code will be applied. It is possible that with this type of approach that a Code statement may contradict a specification statement, and it is essential to establish which requirement has to be fulfilled. Other potential problem areas are in the wording of the specification that could be subject to different interpretations by Construction on one hand and the Quality Control on the other.

The practice that has occurred in some instances of waiting for a disagreement to occur between Quality Control and Construction before the Engineer intervenes to resolve the problem, should be avoided. When it is not avoided, disruption occurs in production, installation and in the important interface between Quality Control and Construction.

It is realized that under certain conditions, the engineering specification is available prior to awarding the contract to produce or install. In this event, every effort should be made by the Engineer to resolve or clarify points of interpretation and any possible conflicting statements with Quality Control. This would enable the Engineer to make suitable corrections to his specification, if he considered the points raised as being valid, prior to any discussions with Construction.

Experience has shown that there is a tendency for the specification to be written to suit the understanding of the writer rather than that of the user. Therefore, the emphasis has been placed on the need of the maximum "pre-commencement of Construction/Quality Control activities discussion" relative to the key document - the Engineering Specification.

EQUIPMENT

Within the scope of this paper the definition of equipment is given as measuring instruments, those tools and devices utilized to obtain and prepare samples for tests and comparators of any type.

When the Engineering Specification and its associated Codes have been fully analyzed and understood, then the required equipment should be obtained. The selection of the equipment by Quality Control and its limits and tolerances should be fully described with and known by the Engineer prior to purchase or delivery, if transferred from another site. The initial calibration results supplied by the manufacturer, supplier, or at receipt of the equipment should also be made known to the Engineer. This action assures that from the very beginning of any Quality Control program, the Engineer will be aware of the limitations of the equipment being used to supply the test/inspection results that he may wish to analyze.

It is strongly recommended that facilities and personnel be made available to assume the maximum amount of calibration capability on the site as possible. The utilization of outside sources such as the original manufacturer or an external testing laboratory should be kept to the minimum and only applied when it is wholly impractical to perform the required calibration on the site.

It is not the intent of this paper to describe in detail a calibration control program, but it is emphasized that the training of personnel involved in calibration activities is of the utmost importance.

The training program should be set up on the basis of specific training and certification for specific equipment calibration activities. Procedures for the calibration of equipment are considered essential, reflecting only the details of the activity to which they are applied. Generalization within the training and the procedures should be avoided.

The industry has available the independent inspection services of the Cement and Concrete Reference Laboratory (C.C.R.L.). These services assist in obtaining some level of credibility for the equipment and should be applied as soon as practically possible. The Engineer should take an active interest in the scope and particularly the results of this type of inspection and follow up any adverse condition reported until satisfactorily resolved.

It should be noted that pre-planning for this activity is required in order to obtain the inspection services within the C.C.R.L. Inspection Tour schedule.

PROCEDURES

This discussion applies to procedures utilized in the actual test or inspection, rather than those procedures previously referenced in the equipment section. There has been an unfortunate tendency to regard procedures as documents that are necessary to fulfill the requirements of Criteria V, Appendix B, to 10CFR50, rather than being a necessity to assure consistency of the performance of a given activity. The constant and continuous use of procedures to perform specific tasks is necessary because knowledge and repetition is no guarantee that the activity will be performed in a given sequence or manner.

Procedures are essential vehicles for the collection of information that is obtained from multiple sources. Typical sources of information are:

Engineering Specification
Codes - referenced in Specification
National Standards - referenced in Specification/Code
Engineering Standards - referenced in Specification
Drawings - including notes
Engineering/Drawing Change Documents
Production/Installation Procedures - for planning and sequence
Manufacturer's Instructions

In order to establish a standard Quality Control approach, the analysis of the various sources of information should be assigned to a specific group of qualified personnel. To allow an individual Inspector to analyze such information in an unknown and random manner to form a basis for his personal inspection plan is an uncontrollable and unacceptable condition.

The formation and content of the "information analysis" group is one of the keystones of a successful Quality Control program. The group should consist of engineers of the appropriate discipline who have a thorough knowledge of test/inspection functions. They are also required to have the ability to write procedures/instructions compatible with the education and ability of the Inspectors available. In addition, they should have the "engineering stature" to discuss engineering problems or queries with the Engineer on an equal level of expertise and understanding.

An important, but sometimes overlooked, aspect of a Quality Control procedure production cycle is the input of typical Inspector(s) who will use it in either the site laboratory or the field. This input can serve two purposes: one being a practical application or experience critique, the other allowing the user to have a sense of contribution to its authorship.

All Quality Control procedures should contain the form for reporting the results obtained. Clear instructions to the Inspector regarding the actions to be taken for both an acceptable or unacceptable test result should also be included. The Engineer must become involved in this aspect of procedure production prior to implementation. It is necessary that his future requirements of results analysis, review and retrieval are known in order to produce a report that highlights and reflects his areas of interest and method of retrieval. In addition, the transmittal path(s) should be established and the distribution organized.

It is recommended that multi-colored forms be used, each copy identifying the receiver by title and its end use. This approach can reduce the amount of copies produced. Avoid reproduction copies, this approach indicates a lack of knowledge of the end use of the reports and has a tendency to multiply the number of copies produced.

Having established a Quality Control procedure, a document distribution system is required. The system should have two basic goals:

(a) To have the procedure in the possession of the user when he performs his test/inspection.

(b) To assure that the procedure held by the user is the revision required for the test/inspection to be performed.

This activity is not as simple as it appears, generally procedures are issued from a single point to be contained within a collective, identified "volume". This distribution does not include every Inspector, for obvious reasons. The solution suggested is to have a primary distribution center as described, plus a controlled number of secondary distribution centers. The secondary centers operate to a program that allows controlled copies of individual procedures to be issued to Inspectors who are qualified and trained to perform the activity.

PERSONNEL - TRAINING

A training concept related directly to the test/inspection procedures should be implemented. This approach does not qualify structural, civil, welding, mechanical or electrical technicians/ inspectors. It does qualify personnel for an activity in accordance with a specific procedure. For example: Structural Cadwelding, Concrete Pre-Placement inspection, Hilti Bolt Installation inspection. The program should be based on a building block principle as the technician/inspector develops, so more building blocks are added to his qualifications. In some instances, certain building blocks are prerequisites for obtaining other qualifications.

With such a detailed training program, it is important to have a constant feedback of qualifications of personnel to the appropriate Supervisors who assign the personnel to their test/inspections. It is also required to have a system to identify changes in Quality Control procedures that require changes or additions to the existing training course.

It is recommended that the same group who have the responsibility to analyze the various engineering requirements and produce the test/inspection procedure have the responsibility for producing the training course for that procedure. By using this group, the delay in obtaining a training course for qualifying Inspectors following the issue of a procedure is kept to a minimum. This group and the Engineer had a strong interface during the Quality Control procedure development, and as the training course is based wholly on the procedure, a link has been made between the Engineer, the Quality Control procedure and the training of Inspectors.

An essential element in this program is the training administrator who is responsible for the assembly of the various training courses into a comprehensive training program. He also maintains the certifications of the personnel and transmits the essential feedback of qualifications to the previously referenced Supervisor.

Every training course requires an examination with a passing grade. The examination should consist of written and practical tests with the emphasis on practical application. For example: samples being made available for examination, acceptance or rejection, actual drawings to be interpreted, slump tests to be performed. When a passing grade is obtained, the questions answered incorrectly should be thoroughly discussed, corrected and understood by the Inspector prior to qualification.

This training concept is based on a centralized and standardized approach. This complements the standardized inspection approach. The program is dynamic, subject to change or additions when the engineering requirements are revised, improved techniques are devised and applied and as the construction phase progresses.

INTERFACE, QUALITY CONTROL - CONSTRUCTION

So far this paper has described four major factors in obtaining accurate test/inspection results, however, the effectiveness of obtaining those results in a manner that would minimize disruption to Construction progress and schedule has not yet been taken into consideration.

Quality Control activities, when related to Construction activities, fall into main categories:

(A) The activities that directly affect Construction progress. These activities occur as part of the process in a series fashion.

(B) The activities that indirectly affect Construction progress. These activities are those that are performed in parallel with Construction processes, with the possible exception of taking a sample for testing.

The results of either of these categories can, of course, affect the progress of Construction in the event that the results indicate some form of error or breakdown in the Construction program.

The objective of the Inspector is to be in the right place at the right time with the right equipment. In order to achieve this objective, Quality Control must be completely aware of the Construction work plan and schedule and to integrate their own work plan and personnel assignments with that of Construction. Construction procedures and programs should always indicate where and when a Quality Control activity is to occur.

It is not recommended that the full details of the Quality Control procedure be included in the Construction procedures or program. If the two programs are completely integrated, a delay in a change approval time cycle can occur. For example, the Construction program or procedure may need a revision that does not affect the Quality Control activities in any way, and yet Quality Control would be in the review circuit, causing an unnecessary delaying action. However, Quality Control procedures and requirements should be well known to Construction, at all appropriate levels. Local detail planning meetings are good arenas for the explanation and understanding of both parties' intentions.

An example of the need for integration of the two activities is in a cooling tower erection activity. This activity requires a close relationship between form system movement and the strength of concrete. Construction requires to commence work at a given time and needs to know the strength of the concrete. Therefore, the Quality Control program should consist of taking as many test cylinders as are required, testing them during the off hours to have the results available to Construction as needed. Plotting of the results against time, temperature, and other environmental conditions can minimize the amount of cylinders cast as the cooling tower erection system progresses and more knowledge is gained.

In conclusion, if maximum productivity is to be obtained, the compatability of the Quality Control and Construction work plans is essential.

INTERFACE, ENGINEER AND QUALITY CONTROL

As the Engineer will be relying on Quality Control to provide him with test/inspection results for his decision making process, a close relationship between the two groups seems logical. Unfortunately, in some instances, this relationship becomes very fragile, and on occasions, the two groups are in an almost adversary condition. The approach being suggested is offered to prevent that condition arising with the additional benefit of an advantageous effect upon Construction achievement of quality goals.

In the event that Quality Control reject material, it is possible that the Engineer will institute a detailed investigation into how the sample was taken, how the test or inspection was performed, and the qualification of the inspector who performed it. There is no objection to this type of investigation, however, the Engineer is urged to establish the credibility of Quality Control from the very beginning and be as concerned with their methods and personnel when the results of examinations and tests are acceptable, as he may be when they are not.

A typical scenario that can occur on a power generation facility site is at the beginning of a project. The specifications have been produced, Quality Control procedures have been established along with the associated equipment and personnel training, and Construction then applies itself to the placing of the first concrete. On a typical site as described, the early activities are usually concerned with areas that may not require in themselves a high standard of quality, such as base floors for warehouses, slabs, storage area preparation, sidewalks and similar, non-generating related structures. These activities, however, usually are required to be achieved under the same specification conditions that will be applied to the whole project. Quality Control apply themselves in accordance with the specification for the delivery and placement of concrete, utilizing their procedures. At this stage, Construction is just commencing their work, and equipment, personnel and control systems are unproven. Under these circumstances, it is likely that this concrete will not meet the specification requirements. Following this activity, the first interface of Quality Control with the Engineer is him accepting the concrete placed out of specification on the basis that it will fulfill its intended purpose.

Let us weigh the issue at stake relative to the apparently simple decision by an Engineer to accept concrete out of specification at the beginning of a six-year project. The two groups of people involved are Construction and Quality Control. From this initial decision, Construction could consider that it is not the details of the specification that are essential, but the engineering decision relative to the quality achieved versus the end use of the product that is important. Quality Control have had the requirements of the specification impressed upon them and could now consider that their efforts are treated lightly and the care applied to the tests and inspections was a waste of time.

Already the tone of the project has been set and unless the "tone" is changed, it will pervade every discipline to the detriment of all.

It is not suggested that the Engineer refuse to accept the concrete under the conditions described. It is being suggested that the Engineer makes a deliberate and determined effort to have the cause of the problem in the production cycle established and concur with specific corrective action to be effected before he will allow the production and placement of concrete to proceed. This action should be taken without regard to the final use of the product.

It is further suggested that during this initial period and at regular periods throughout the project life, the Engineer should take the opportunity to make himself known to Quality Control personnel. He should discuss their procedures and methods and any activity that may cause a future problem. He should emphasize to the Inspectors his reliance upon their work and their importance relative to the project's success.

The industry also has available the nationwide sampling program instituted by the Cement and Concrete Reference Laboratory (C.C.R.L.). The Engineer should encourage the management to have the Quality Control activities applicable to CCRL involved in this program. The performance of many laboratories are compared against given samples and high ratings give some credibility to the results obtained in the general work. Low or unusual results pinpoint areas that need attention. This program has the secondary effect of giving some form of competitiveness to the Inspectors involved. The association of the site laboratories with this sampling program is highly recommended.

CONCLUSION

The approach previously described emphasizes the need for the maximum interface between the Engineer and Quality Control prior to and immediately following the commencement of Construction and Quality Control activities. A strong link between the Engineer and the Quality Control group must be forged if the Engineer is to have any confidence in the test and inspection results that can demonstrate that the final product meets his specification requirements and indicate the achievements of the construction program, relative to its quality goals.

It appears that the trend of the 80's will be an increased use of computers with multiple terminals for the review and recording of test results.(1) If these trends occur, the human link between the personnel performing the test and inspections and the Engineer analyzing the same becomes even more important.

When the Engineer imposes his determination to have the specification requirements met at the beginning of the project, the later activities that do require the complete adherence to specification can be achieved with the highest productivity.

(1) "Test Data Reports by Computer Improve Quality Control of Ready-Mixed Concrete", by Donald R. Shelangoskie and Jerry R. Newmark of the Austin Company, Cleveland, Ohio; Published, Concrete International/July 1981.

CREEP/SHRINKAGE STUDIES:

CONTAINMENT STRUCTURES

By

Mauro J. Scali[1] and Donald W. Pfeifer[2]

ABSTRACT

The tests reported herein were performed as part of several extensive and detailed programs designed to evaluate the long-term behavior of concrete mixes, which were used in the construction of containment vessels at three separate nuclear power generating stations.

The studies were conducted over a two-year period and included an evaluation of the long-term compressive strength, modulus of elasticity, creep, drying shrinkage and thermal properties of both plain portland cement concretes and portland cement concretes containing fly ash. The specimens used in each study consisted of 6 x 12 in. and 6 x 20 in. concrete cylinders which were instrumented with three 10 in. gage lines, at 120 degree intervals, using Whittemore strain gage points. The specimens were then monitored periodically with a Whittemore strain gage to determine their creep and shrinkage characteristics under conditions of constant load and simulated environmental temperatures.

The results of such studies can be used to predict more accurately the long-term prestress losses associated with a particular concrete, due to the effects of creep and shrinkage.

INTRODUCTION

During the initial design phases for prestressed concrete containment structures, the necessary information concerning the physical properties of the hardened concrete such as creep, drying shrinkage, modulus of elasticity, Poisson's Ratio, and coefficient of thermal expansion are obtained from design assumptions, accepted standards or formula calculations. Although these values are based upon

---

[1] Petrographer, Wiss, Janney, Elstner and Associates, Inc. Northbrook, Illinois 60062

[2] Vice President, Wiss, Janney, Elstner and Associates, Inc. Northbrook, Illinois 60062

sound engineering logic or standard recommended practices of design, they may not totally reflect the actual long-term behavior of the concrete. As a result, these assumptions may be of limited use in predicting the actual long-term behavior of the concrete, under conditions of actual design loads or during periods of change in temperature, load or pressure. For these reasons, the designers and engineers involved in the construction and licensing of nuclear power plants may require that a long-term study be conducted to determine the actual physical properties of the concrete used during the construction of containment vessels. Such studies can be conducted under simulated laboratory conditions of anticipated design loading and temperature to provide a realistic basis for comparing the original design assumptions with the actual measured long-term physical properties of the concrete.

Since 1977, the firm of Wiss, Janney, Elstner and Associates, Inc. (WJE) has conducted a series of extensive two-year laboratory studies to determine the physical properties of concrete mixes which were used in the construction of containment vessels at three separate nuclear power generating stations. Each of these studies consisted of a comprehensive laboratory investigation of the effects of age, curing history, environmental temperature and constant creep loading upon the physical properties of different concrete mixes.

The purpose of this paper is to describe the test procedures and methods of evaluation which were used during each of the long-term studies. In addition, this paper will also present some examples of typical data which were obtained during one particular long-term study.

SCOPE OF INVESTIGATIONS

The purpose of the studies was to evaluate the physical properties of plain portland cement concretes (P.C.) and portland cement concretes containing fly ash (F.A.) over a two year period. Each study consisted of six general tasks, as presented in Table 1 which, in turn, required 26 separate test conditions (Set Numbers) to complete. An example of a typical testing program involving the 26 separate test conditions is presented in Tables 2, 3, and 4.

A major portion of each study was concerned with a determination of the long-term compressive strength, modulus of elasticity, Poisson's Ratio, creep and shrinkage properties of the concrete mixes. The only differences in the long-term studies were the aggregate sources, proportions of the concrete mix designs, the types of cement used (Type I vs. Type II) and the maximum temperature at which each study was conducted (105° F vs. 130° F). A brief description of each of the physical properties which were evaluated during the studies, together with their respective test ages, is given below:

1. Compressive Strength (ASTM C39) - The compressive strength of 6 x 12 in. cylinders were determined at 7, 28, 91, 182, 364, and 728 days, as outlined in Table 2. Additional compressive strength data were obtained as a

TABLE 1 – BASIC OUTLINE FOR EACH TESTING PROGRAM

Task No.	Test Procedure	Measured physical property
1	ASTM C469 ASTM C39 ASTM C642 ASTM C186	Static modulus of elasticity and Poisson's ratio Compressive strength Total evaporable water in concrete test specimens remains Heat of hydration pH
2	CRD C124 CRD C44 CRD C39	Specific heat Thermal conductivity Linear thermal expansion
3	ASTM C157	Shrinkage (length changes)
4	ASTM C512 ASTM C157	Elastic and creep strains
5	ASTM C512 ASTM C157 ASTM C469 ASTM C39	I. Unloading II. Elastic and creep strains recovery III. Reloading IV. Elastic and creep strains after reloading Static modulus of elasticity and Poisson's ratio Compressive strength
6	ASTM C469 ASTM C39 ASTM C642	Static modulus of elasticity and Poisson's ratio Compressive strength Total evaporable water in concrete test specimens remains

TABLE 2 — SCHEDULE FOR COMPRESSIVE STRENGTH, MODULUS OF ELASTICITY, POISSON'S RATIO AND EVAPORABLE WATER CONTENT

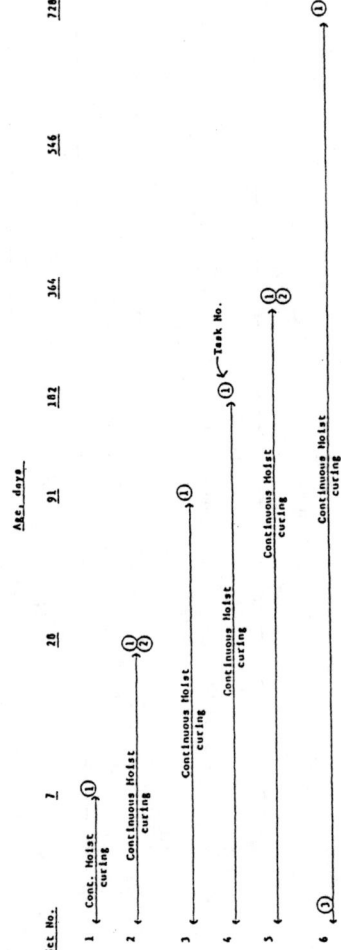

CREEP/SHRINKAGE STUDIES

TABLE 3 - SCHEDULE FOR SHRINKAGE TESTS AND SUPPLEMENT
TASK NOS. 1, 3 and 6 TESTS

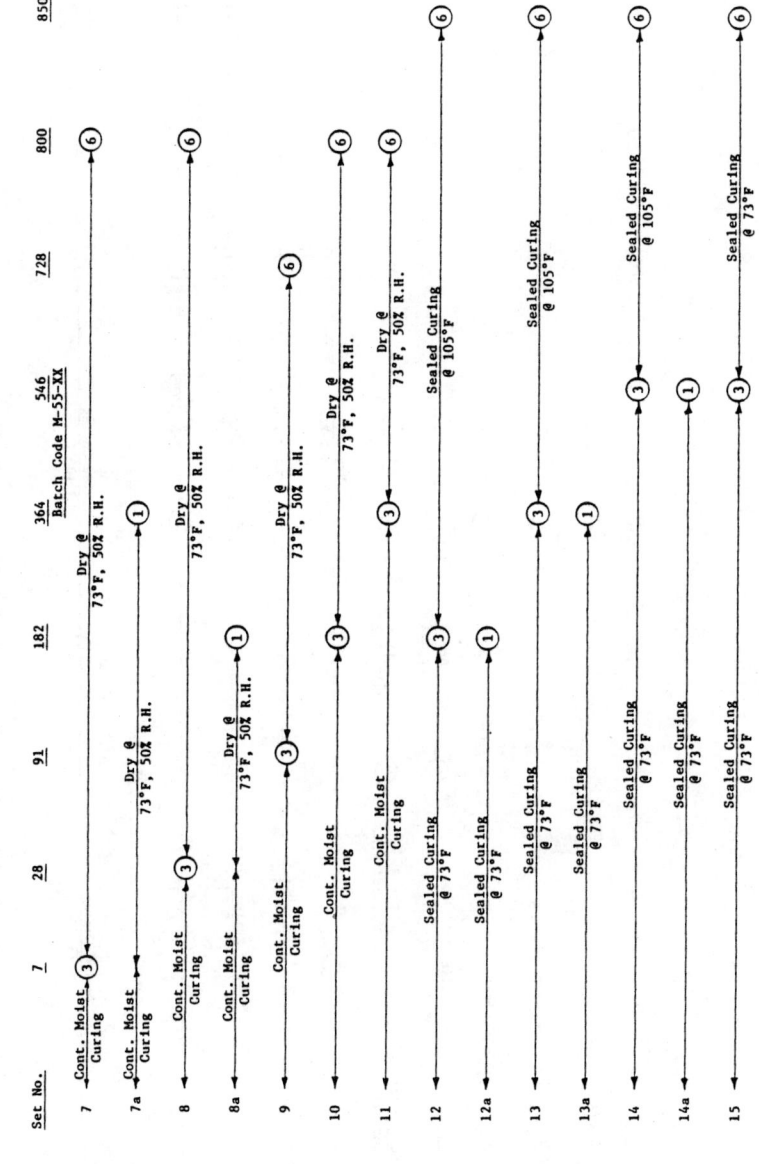

TABLE 4 - SCHEDULE FOR CREEP TESTS

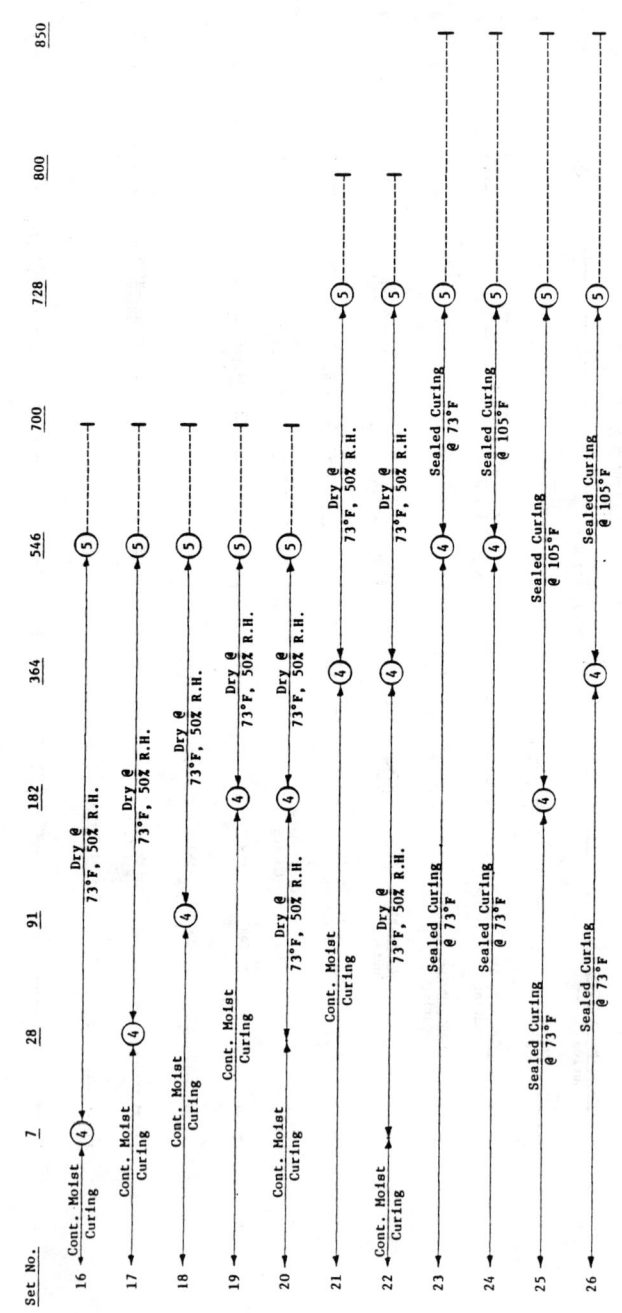

result of testing companion creep and shrinkage test specimens at the conclusion of their testing, as outlined in Tables 3 and 4 (Tasks 1, 5 and 6).

2. <u>Modulus of Elasticity and Poisson's Ratio (ASTM C469)</u> - In accordance with the Table 2 outline, the specimens were measured for modulus of elasticity and Poisson's Ratio at 7, 28, 91, 182, 364, and 728 days. Additional values were obtained as a result of testing companion creep and shrinkage specimens as outlined in Tables 3 and 4 (Tasks 1, 5 and 6).

3. <u>Drying Shrinkage and Autogenous Shrinkage (ASTM C157)</u> - The shrinkage characteristics were determined through a periodic monitoring of companion control cylinders for each of the creep specimens. The schedule followed during these shrinkage tests is outlined in Table 3 (Task No. 3).

4. <u>Creep of Concrete (ASTM C512-Modified)</u> - The creep characteristics were obtained through a periodic monitoring of test specimens, which were loaded at 7, 28, 91, 182, 364, and 546 days. The initiation of the creep tests was in accordance with the outline in Table 4 (Task No. 4). The applied stress was equal to a value of 40 percent of the compressive strength at the age of loading.

5. <u>Evaporable Water Content (ASTM C642)</u> - The evaporable water content of the hardened concrete was determined by testing one large remnant of a cylinder from each of the compressive strength test sets, as outlined above in Item No. 1.

6. <u>Heat of Hydration (ASTM C186)</u> - The Type I and II portland cements were tested to determine the heat of hydration properties of the different cements at ages 7 and 28 days.

7. <u>pH Determination</u> - The pH of the hardened concrete was determined through the testing of a remnant piece of concrete from the compressive strength test sets outlined above in Item No. 1.

8. <u>Thermal Properties</u> - The thermal properties of the hardened concrete were determined as outlined in Table 2 (Task No. 2). The thermal properties of hardened concrete which were determined at ages 28 days and 1 year included the following:

   a. Thermal Diffusivity - CRD-C36
   b. Specific Heat - CRD-C124 (adapted for concrete)
   c. Thermal Conductivity - CRD-C44
   d. Linear Thermal Expansion - CRD-C39

The above tests were conducted on specimens which had been stored under varying conditions of temperature and relative humidity. These environmental storage conditions were as follows:

* $73 \pm 3°F$ and R.H. of 100 percent (continuous moist curing)

* $73 \pm 3°F$ and R.H. of 50 percent

* 105 or $130 \pm 5°F$ and R.H. of 50 percent

The combination of various storage environments, and the required testing of the specimens at different ages resulted in a total of 26 separate test conditions. Each test condition or Set Number required from 3 to 5 individual specimens. The specimens contained in Sets 1-11 and 16-22 consisted of 6 x 12 in. or 6 x 20 in. unwrapped concrete cylinders. As a result, these specimens experienced drying shrinkage upon exposure to air in a 73, 105, or 130°F environment, at 50 percent relative humidity. However, those specimens in Sets 12-15 and 23-26 consisted of concrete cylinders which had been permanently wrapped in copper foil and sealed with solder. As a result, these specimens experienced only autogenous shrinkage, regardless of the temperature or relative humidity of the environment in which they were stored.

DESCRIPTION OF MATERIALS, EQUIPMENT, AND TEST PROCEDURES

Materials

All of the materials used during each study were obtained from on-site stockpiles. After sampling, the aggregates, water, portland cements and fly ash were stored in sealed containers to prevent contamination and assure a constant moisture condition.

The aggregates used for each of the studies consisted of crushed 3/4 in. limestones and natural sands, which were air-dried to a low and constant moisture condition, prior to mixing. The Type I and Type II portland cements were analyzed in accordance with ASTM C150. The results of the analysis indicated that each cement was in compliance with the standard physical and chemical requirements of the specification.

Mix Proportions

A total of ten batches of concrete were made to satisfy the specimen requirements for each program. Each batch was made in accordance with ASTM C192 and mixed in a 3 cu ft capacity Eirich Mixer. The average mix proportions, as calculated from the actual batch proportions for each of the mixes, are presented in Table 5, together with the average values of slump, plastic unit weight, and air content.

TABLE 5 - MIX PROPORTIONS FOR EACH LONG TERM STUDY*

Item	Long-term study			
	No. 1A	No. 1B	No. 2	No. 3
Cement (lb/cu yd)	611	510	582	668
Fly Ash (lb/cu yd)	-	96	119	-
Water (lb/cu yd)	295	288	251	261
Sand (lb/cu yd)	1400	1422	1222	1335
Stone (lb/cu yd)	1700	1686	1829	1663
Air Entraining Agent (oz/cu yd)	1.3	3.8	4.7	5.6
Water Reducer (oz/cu yd)	-	-	-	27.8
Cement Type	Type I	Type I	Type II	Type II
Net W/C (SSD)	0.483	-	-	0.391
Net $\frac{W}{C + FA}$ (SSD)	-	0.475	0.358	-
Slump (in.)	2.9	3.1	3.8	3.3
Air Content (%)	3.5	3.6	4.6	4.8
Unit Weight (lb/cu ft)	148.4	148.1	148.2	145.4

* SSD Weights

Specimen Preparation and Instrumentation

Generally, a total of twenty-two 6 x 20 in. cylinders (creep specimens) and fifty-one 6 x 12 in. cylinders (compressive strength and companion shrinkage-control specimens) were required for each study. The 6 x 12 in. specimens were cast in rigid and restrained plastic cylinder molds, while the 6 x 20 in. specimens were cast in specially constructed steel molds. Each steel mold consisted of two standard 6 x 12 in. steel molds which were coupled together and contained a 4 in. concrete plug. The use of 6 x 20 in. creep specimens rather than 6 x 12 in. specimens eliminated the need for additional non-instrumented concrete cylinder inserts between the ends of the specimens and the bottom and top bearing plates of the creep frame.

All specimens were stripped at age 24 hours. Following their removal, those cylinders which were designated as creep and shrinkage specimens were instrumented with three 10 in. Whittemore gage lines. The gage lines were located at 120 degree perimeter intervals and consisted of two gage points which were epoxied to the concrete. The creep and shrinkage specimens which were wrapped in copper foil and sealed with solder were wrapped and sealed prior to being instrumented. All specimens were stored under wet burlap, until the above described preparation and instrumentation was completed in order to prevent any moisture loss. Following their preparation and instrumentation, all specimens were placed in their respective storage environments.

Environmental Storage Conditions

Continuous moist curing was provided by total immersion of the unsealed specimens in thermostatically controlled water tanks. The temperature of the water was maintained at 73 $\pm$ 3°F.

Those unsealed and copper-sealed specimens which were to be stored in air at 73°F and 50 percent R.H. were placed in a climate controlled room designed for this purpose. The storage of unsealed and copper-sealed specimens in air at an elevated temperature of 105°F or 130°F and 50 percent R.H. was accomplished through the use of environmental heat chambers which were specifically designed and constructed for the long-term studies. The temperatures and relative humidities of these rooms were monitored periodically in order to insure that a constant temperature and humidity was maintained.

Creep and Shrinkage Tests

At the predetermined test ages (see Tables 3 and 4), the creep and shrinkage specimens were transferred from their curing environments to a designated testing environment. The creep specimens were then placed into specially constructed steel loading frames and a 50 ton hydraulic ram was installed above the top bearing plate. A predetermined stress of 0.40 f_c was then applied by the hydraulic ram which also compressed the load-maintaining springs. At this point, a set of strain readings was taken on each creep specimen to determine the elastic strain which had occurred as a result of the applied load.

The load was readjusted and the specimens were again read at the end of the first day of testing. Each time the creep specimens were read, the load was checked and, if necessary, a load adjustment was made. In addition, the companion shrinkage-control specimens were also read.

Regardless of the age at which each set of creep and shrinkage specimens were introduced into the test sequence, the above initial loading schedule remained the same.

TEST RESULTS

Typical results obtained during one long-term study involving an evaluation of a plain portland cement concrete (P.C.) and a fly ash concrete (F.A.) indicated the following:

Compressive Strength

Each of the concrete mixes had a 91-day design compressive strength requirement of 4500 psi. The results of the compressive strength tests for each mix are presented in Tables 6 and 7. An analysis and comparison of these results indicated that each concrete mix exceeded the 91-day design strength requirement by approximately 50 percent and that the compressive strength of the F.A. concretes was consistently higher than those of the P.C. concretes. This observation was true for both continuously moist-cured, and copper-sealed specimens. At age one year, the F.A. concrete mix had an average compressive strength of 8330 psi and 8530 psi for copper-sealed and continuously moist-cured specimens, respectively. In contrast to this, the P.C. concrete mix had an average one-year compressive strength of 7880 psi and 7690 psi for similarly copper-sealed and continuously moist-cured specimens.

The advantage of continuous moist curing for P.C. concretes for periods of one year or longer as compared to shorter moist-curing periods of 7 and 28 days was significant. The 7 and 28 day moist-curing periods resulted in one year strengths for P.C. concretes which were 24 and 15 percent lower, respectively, than those of concrete specimens which were continuously moist-cured for one year. However, the advantage of continuous moist-curing beyond 28 days as compared to an initial period of moist curing followed by an extended period of air drying was insignificant.

A plot of the compressive strength vs. age data for a P.C. concrete mix is presented in Fig. 1. A similar comparison of the influence of curing environment vs. age, upon the compressive strength properties of the F.A. concrete was not possible due to a lack of sufficient data.

Those P.C. concrete specimens which had experienced long-term creep loading achieved either equivalent or higher compressive strengths than those unloaded specimens which had not seen any applied load. A comparison of the compressive strength data, for loaded vs. unloaded specimens, indicated a 1 to 11 percent higher

TABLE 6 – AVERAGE COMPRESSIVE STRENGTH OF UNLOADED SPECIMENS FOR PLAIN PORTLAND CEMENT (P.C.) CONCRETE MIXES

Curing Condition	Compressive Strength, psi — Age, Days							
	7	28	91	182	364	607	728	789
Unsealed				(73°F)				
7 Day Moist, then 50% R.H.	4440	–	–	–	–	–	–	5670
28 Day Moist, then 50% R.H.	–	5290	–	–	5870	–	6400	–
91 Day Moist, then 50% R.H.	–	–	6760	–	6540	7440	–	–
182 Day Moist, then 50% R.H.	–	–	–	6900	7550	–	8090	–
364 Day Moist, then 50% R.H.	–	–	–	–	8140	–	–	8590
364 Day Moist, at 100% R.H.	–	–	–	–	7690	–	–	–
728 Day Moist, at 100% R.H.	–	–	–	–	–	–	8680	–
Copper Sealed								
364 Day at 50% R.H.	–	–	–	–	7880	–	–	–
728 Day at 50% R.H.	–	–	–	–	–	–	–	8190
Copper Sealed				(73°F and 130°F)				
182 Days at 73°F, 50% R.H. then 130°F, 50% R.H.*	–	–	–	–	7120	–	–	7720
364 Days at 73°F, 50% R.H. then 130°F, 50% R.H.*	–	–	–	–	–	–	–	8530
564 Days at 73°F, 50% R.H. then 130°F, 50% R.H.*	–	–	–	–	–	–	8350	–
Copper Sealed/Unsealed								
364 Days Sealed at 73°F, 50% R.H.* then Unsealed at 130°F, 50% R.H.	–	–	–	–	–	–	7020	–

*RH of Concrete Inside of Copper is 100%

TABLE 7 - AVERAGE COMPRESSIVE STRENGTH OF UNLOADED SPECIMENS FOR FLY ASH (F.A.) CONCRETE MIXES

Curing Condition	Compressive Strength, psi Age, Days							
	7	28	91	182	364	607	728	789
Unsealed				(73°F)				
364 Days Moist at 100% R.H.	-	-	-	-	8530	-	-	-
364 Days Moist, then 50% R.H.	-	-	-	-	-	-	-	10,100
Copper Sealed								
364 Days at 50% R.H.*	-	-	-	-	8330	-	-	-
789 Days at 50% R.H.*	-	-	-	-	-	-	-	8800
Copper Sealed/Unsealed				(73°F and 130°F)				
364 Days Sealed at 73°F, 50% R.H.*, then Unsealed at 130°F, 50% R.H.	-	-	-	-	-	-	7420	-

*R.H. of Concrete Inside of Copper is 100%

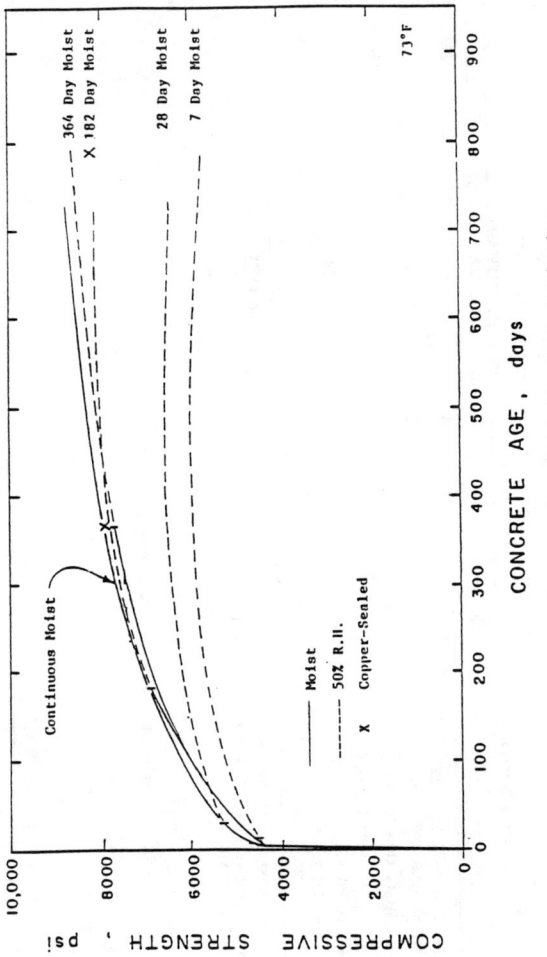

Fig. 1 - Compressive strength characteristics under different curing conditions

strength for specimens which had experienced some degree of creep loading, prior to being tested for compressive strength.

Modulus of Elasticity

The measured modulus of elasticity together with the various storage and curing environments for each concrete are presented in Tables 8 and 9. A brief summary of the test results is given below:

P.C. concrete mix - The measured one year data show an average modulus of 5.5 million psi, when 91 or 182 days of continuous moist curing was provided. When the period of continuous moist curing was extended to one year and two years, the modulus increased to 5.9 million psi. Similar values were measured for copper-sealed specimens which had remained sealed for one to two years at 73°F and 50 percent R.H. However, the modulus was slightly lower at age 789 days for those copper-sealed specimens which had been exposed to a 130°F environment following their initial curing at 73°F. The degree of reduction was from 4 to 8 percent and depended on the length of curing at 73°F which each specimen received prior to being exposed to 130°F.

The most significant decrease in modulus was observed for those copper-sealed specimens which after one year of curing at 73°F and 50 percent R.H. were unsealed and exposed to air at 130°F and 50 percent R.H. At age two years, the modulus of elasticity of these specimens was measured and compared to that of copper-sealed specimens which had been stored continuously at 73°F for the entire two year period. The result of this comparison was an observed reduction in the modulus of elasticity of approximately 10 percent.

F.A. concrete mix - At age one year, the modulus of elasticity for continuously moist-cured and copper-sealed specimens averaged about 6.3 million psi.

A reduction in modulus of approximately 21 percent was observed for those specimens which had been initially copper-sealed and cured for one year at 73°F and then unsealed and allowed to air dry at 130 ± 5°F and 50 percent R.H. This reduction was substantially greater than that observed for similarly cured P.C. concrete mixes. It should be noted, however, that in general, the fly ash concretes had a higher modulus of elasticity than those measured for the P.C. concretes at similar ages and under similar curing and testing conditions.

The measured values of modulus of elasticity in Tables 8 and 9 were compared with theoretical values calculated from the compressive strength and unit weight data for the P.C. and F.A. concrete mixes. The calculation involved the ACI equation in Section 8.5 of the "Building Code Requirements for for Reinforced Concrete" (ACI 318-77). The results of this comparison, showed that the measured values were from 5 to 10 percent greater than the ACI calculated values.

TABLE 8 - AVERAGE MODULUS OF ELASTICITY FOR PLAIN PORTLAND CEMENT (P.C.) CONCRETE MIXES (UNLOADED SPECIMENS)

Curing Condition	Modulus of Elasticity, 10^6 psi							
	Age, Days							
	7	28	91	182	364	607	728	789
Unsealed					(73°F)			
7 Day Moist, Then 50% R.H.	4.32	-	-	-	-	-	-	4.60
28 Day Moist, Then 50% R.H.	-	4.76	-	-	5.05	-	4.99	-
91 Day Moist, Then 50% R.H.	-	-	5.35	-	5.48	5.06	-	-
182 Day Moist, Then 50% R.H.	-	-	-	5.69	5.55	-	5.55	-
364 Day Moist, Then 50% R.H.	-	-	-	-	5.88	-	-	5.50
364 Day Moist, at 100% R.H.	-	-	-	-	-	-	-	-
728 Day Moist, at 100% R.H.	-	-	-	-	-	-	5.90	-
Copper Sealed								
364 Days at 50% R.H.	-	-	-	-	6.02	-	-	-
728 Days at 50% R.H.	-	-	-	-	-	-	-	5.94
Copper Sealed				(73°F and 130°F)				
182 Days at 73°F, 50% R.H. Then 130°F, 50% R.H.*	-	-	-	-	5.49	-	-	5.70
364 Days at 73°F, 50% R.H. Then 130°F, 50% R.H.*	-	-	-	-	-	-	-	5.50
546 Days at 73°F, 50% R.H. Then 130°F, 50% R.H.*	-	-	-	-	-	-	5.72	-
Copper Sealed/Unsealed								
364 Days Sealed at 73°F, 50% R.H.*, Then Unsealed at 130°F, 50% R.H.	-	-	-	-	-	-	5.34	-

*R.H. of Concrete Inside of Copper is 100%

TABLE 9 — AVERAGE MODULUS OF ELASTICITY FOR FLY ASH (F.A.) CONCRETE MIXES (UNLOADED SPECIMENS)

Curing Condition	Modulus of Elasticity, 10^6 psi Age, Days							
	7	28	91	182	364	607	728	789
Unsealed					(73°F)			
364 Days Moist at 100% R.H.	–	–	–	–	6.37	–	–	–
364 Days Moist, Then 50% R.H.	–	–	–	–	–	–	–	6.19
Copper Sealed								
364 Days at 50% R.H.*	–	–	–	–	6.33	–	–	–
789 Days at 50% R.H.*	–	–	–	–	–	–	–	6.40
Copper Sealed/Unsealed					(73°F and 130°F)			
364 Days Sealed at 73°F, 50% R.H.*, Then Unsealed at 130°F, 50% R.H.	–	–	–	–	–	–	5.06	–

*R.H. of Concrete Inside of Copper is 100%

Creep and Shrinkage

Regardless of which concrete mix was being evaluated, all of the creep tests were conducted at an applied stress of approximately 40 percent of the concrete strength at the time of loading. As a result, the stress/strength ratio for each of the creep tests was held constant. An analysis of the creep data, for the P.C. concrete mixes, showed that for similar periods of loading and environmental exposure (i.e., Sets 16, 17, 18, 19 and 21), the measured creep strain did not remain constant, even though the stress/strength ratio was held constant. This observation suggests that the creep characteristics are not only influenced by the stress/strength ratio but also by the compressive strength (maturity) at the age of loading. The applied stress and creep strain data from Sets 16, 17, 18, 19 and 21, have also been analyzed and plotted in Fig. 2 to show that the values of unit creep strain at a constant stress/strength ratio are significantly influenced by the compressive strength of the concrete at the age of loading.

The typical drying shrinkage characteristics of the plain portland cement concrete mixes which were allowed to air dry after an initial period of moist curing are shown in Fig. 3. This data indicates that the total drying shrinkage varied from approximately 300 to 700×10^{-6} in./in., as a function of the length of the initial moist-curing period. It was observed that a swelling or expansive strain of approximately 320×10^{-6} in./in. had occurred within those P.C. concrete specimens, which had been stored continuously in water for a period of two years (i.e., Set No. 6).

Results of pH Tests on Hardened Concrete

The pH of the P.C. and F.A. hardened concretes was determined for each of the various environmental storage conditions and testing ages. The test results showed that the pH of the P.C. and F.A. concrete mixes varied very little as a function of testing age or environmental storage condition. The average values of pH, for each of the concretes, were 12.56 and 12.59, respectively.

Evaporable Water Content of Hardened Concrete

The total evaporable water content of the P.C. and F.A. concrete mixes was determined for each of the environmental storage conditions and testing ages. The results showed that the evaporable water content for the P.C. concretes varied from 1.1 percent for copper-sealed specimens to 7.3 percent for continuously moist-cured specimens. The average evaporable water content of the P.C. concrete specimens was 4.5 percent.

In contrast to the above values, the total evaporable water content of the F.A. concretes varied from 2.1 to 6.1 percent, with an average value of 3.6 percent.

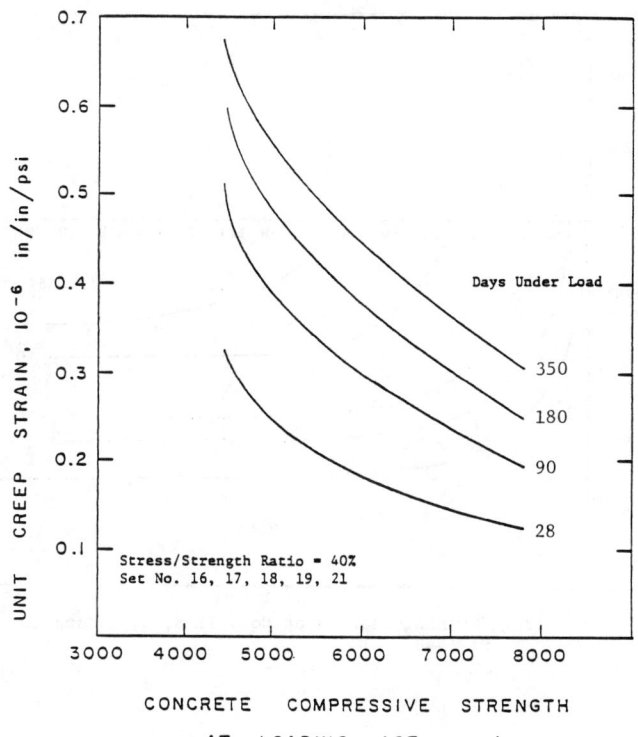

Fig. 2 - Unit creep strain versus concrete strength at loading under constant stress/strength ratio

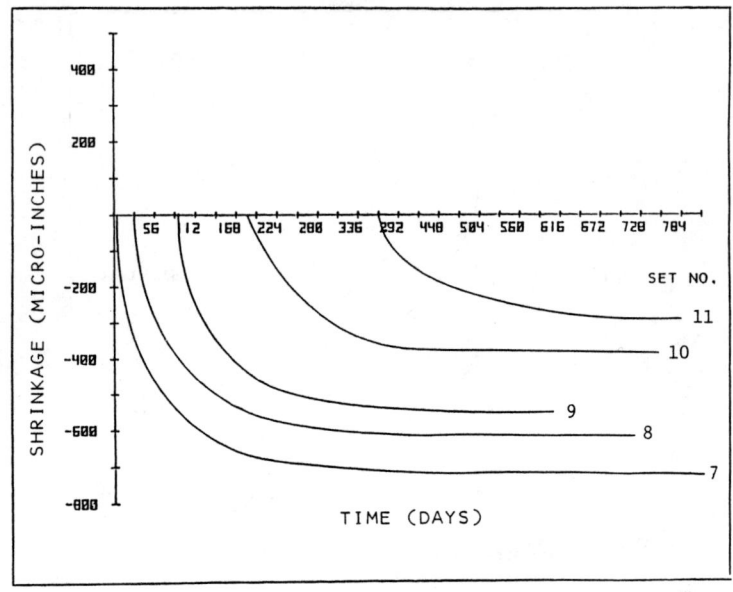

Fig. 3 - Shrinkage data—Set Nos. 7, 8, 9, 10 and 11
Mix 446

Thermal Properties of the Hardened Concrete

The thermal diffusivity, specific heat, coefficient of linear thermal expansion and thermal conductivity properties of the hardened concretes were determined. These tests were conducted at ages 28 days and one year for the P.C. concrete mixes and only at age one year for the F.A. concrete mixes. The results are presented below.

Item	P.C. Concrete 28 days	P.C. Concrete 1 year	F.A. Concrete 1 year
Thermal Diffusivity (ft^2/hr)	0.0292	0.0284	0.0299
Specific Heat (BTU/lb-°F)	0.243	0.250	0.239
Coefficient of Linear Thermal Expansion (micro in./°F)	5.19	5.73	5.35
Calculation of Thermal Conductivity (BTU/ft^2-hr-°F/in.)	12.7	12.8	13.0

An analysis of the strain data from the drying shrinkage control specimens exposed to an environment of 130 $\pm$ 5°F and 50 percent relative humidity indicated that as a result of this exposure, the shrinkage control specimens experienced an average expansive strain of approximately 320×10^{-6} in./in. Based upon this value and the net change in temperature, the calculated coefficient of thermal expansion for the P.C. concrete mix was 5.66×10^{-6} in./in./°F. This calculated value was very similar to the measured value for the P.C. concrete, at age one year (5.73×10^{-6} in./in./°F).

SUMMARY

This paper describes the testing programs and testing procedures which Wiss, Janney, Elstner and Associates, Inc. has used in the last decade to study the physical properties of concrete mixtures used in nuclear containments. These comprehensive laboratory investigations concerned the effects of age, curing history, temperature and humidity of the environment and creep loading on the physical properties of the concrete over a two year period.

The data from these long-term studies resulted in the following typical conclusions:

1. The 91 day design strength of the concrete was generally exceeded by 50 percent with or without the use of fly ash.

2. The two year compressive strengths were typically 7,500 psi to 10,000 psi for copper-sealed specimens or

when 28 days or more of continuous moist curing was provided for unsealed specimens

3. Copper-sealed specimens which were provided no additional water during their curing attained similar two-year compressive strengths when compared to continuously moist-cured specimens.

4. The use of the A.C.I. modulus of elasticity equation resulted in calculated modulus values which were typically 5 to 10 percent less than those measured.

5. The concretes exhibited relatively low creep characteristics when loaded at realistic ages of 90 days or more. This is in part due to the high compressive strength of the concrete, at the age of loading.

6. The concretes exhibited drying shrinkage values of 300 to 700 x 10^{-6} in./in. depending upon the length of the initial moist-curing period.

7. The relatively low creep and shrinkage characteristics of the concretes, at ages one and two years, indicated that the amount of prestress loss due to these phenomena would be far less than that normally anticipated for prestressed concretes, at earlier ages.

SESSION V - STEEL FABRICATION PRACTICES, CONCRETE
ANCHORAGE SYSTEMS, PLANT MODIFICATION AND RETROFIT EXPERIENCE

SESSION OBJECTIVES/SESSION CHAIRMAN SUMMARY

by

Stephen R. Toth[1] M, ASCE

Objective of Session

The objective of the session was to present actual work experiences on the various subjects and to promote the idea of modifying overly restrictive codes and specifications. Only through a coordinated effort of a field construction staff merging with the assigned engineering disciplines can the total cost of any project be reduced. What each paper succeeded in presenting was a combined engineering/construction effort that has reduced costs yet has maintained the highest quality standard which is needed in the nuclear construction industry.

Chairman Summary

Concrete anchorage systems, in light of NRC Bulletin 79-02, has caused the nuclear industry to reanalyze its position on the use of "drilled-in" anchorage systems and to explore alternative methods of providing future capabilities for backfit projects. Various commercial products are available on the market to assist the engineer in his design of any specific anchorage requirement.

In conjunction with the design of any base plate, the construction representative has certain responsibilities. The responsibilities center on providing the design engineer with information concerning the "as-built" conditions of existing facilities, potential interference problems, access limitations, alternate location proposals, and suggested ways of using anchorage systems not related to "drilled-in" anchors. Usually, due to time restraints, the interface needed between design and construction is somewhat limited which inhibits the communication effort between the two parties.

The design engineer must ensure, through his efforts, that inspection parameters involving the QA/QC organizations are concisely and completely delineated through the specification. The engineer, because of possible cost saving realizations, should not be overly restrictive in choosing parameters;

---

[1] System Superintendent, Generation Construction, Northeast Utilities Service Company, P.O. Box 128, Hartford, CT 06101

however, the inspection attributes must include embedment depth, torque values hole size/drill size, which are field attributes, as well as those attributes consigned to the steel fabricator. The design engineer, being aware of field conditions, should employ the necessary expertise to assure that the design considerations are totally satisfied. The anchor bolt spacing criteria provides the adequate distance between bolts that permit a sufficient capacity to be developed. Without this capacity, a cone-shape failure results when the bolts are pulled from the concrete before the ultimate tensile capacity of the stud has been reached. The anchor bolt distance to the free edge of the plate is also extremely important, for it also relates to cone-shape failure vs. ultimate capacity of the stud. Coupled with this criteria is the ever-present problem of an unstiffened base plate secured with drilled-in anchor bolts that over-stressed those bolts closer to the point of load as opposed to bolts located on the extremities of the plate.

Upgrading of a MARK 1 torus to return it to the originally specified factor of safety is also a new challenge to the nuclear industry. New hydrodynamic loadings have been realized coupled with the "loss of coolant accident" (LOCA) understandings have required that additional supports, saddles, and restraints be added to the torus. In essence, the same type of problems are encountered as in the anchor bolt problem.

The steel fabricator can assist the design engineer, if properly and timely notified, of suggested ways to reduce costs on any project. The steel fabricator has the necessary expertise to suggest various types of templating, for example, which will enhance the mobility and installation ease of base plates, structural shapes, saddles, etc. The range of suggestions also run the gamut from information concerning the availability/non-availability of certain structural shapes, delivery and schedule problems, codes and standards that the fabircator has been certified to be in compliance with, recent audits performed by various organizations to ensure compliance, alternate yet satisfactory means of reducing elaborate designs, in-house coating capabilities, to the other end of the spectrum. That end includes the fact that the fabricator may not want a particular project but can recommend an alternate supplier/fabricator.

The extra-ordinary amount of auditing being performed on steel fabricators requires the QA organizations to be staffed to a much higher level than needed for the day-to-day quality related work. Besieged by countless numbers of architect/engineering firms, utilities and other users, the fabricators' organizations must be staffed in essence to provide two services. These are to provide the responsive questions to auditors and to provide the day-to-day QA/QC coverage on the shop fabrication cycle. This obviously has a direct cost on the finished product.

Audits may serve the purpose of assuring that the fabricator has the ability to perform work activities and even though the criteria in steel fabrication has and will become more stringent, the pride of workmanship and the Nondestructive Examinations performed on the work is really the summation of Quality.

The field civil engineer has an important, demanding role usually not recognized in the nuclear industry because of NRC mandates resulting in strict interpretations of existing codes and standards by "quality" related organizations. The role becomes far more important when the actual costs attributed to any project, new or backfit related, is considered. The role centers on construction innovations that have to be developed if nuclear construction will continue to hold an important position in the construction world. Field engineering has the responsibility to initiate, perform, and provide for documenting new tests to reduce schedule and cost impacts. These tests include, but are obviously not limited to, rebar locaters, use of various drill bits, concrete pumps, "drilled-in" anchor "pull-out" tests, future embed locations, etc. The field engineer, using his knowledge and experience, succeeded in changing specifications that would have placed impossible inspection and corrective action requirements on "nicked" rebar when using carbide steel bits for the drilling of anchor bolt holes in concrete. By using a simple, yet effective, field test, it was concluded that insignificant strength reductions occurred in the reinforcing bars. They were 0% to 2% for N8, N11, N14, and N18 rebar and from 3% to 6% for N5 rebar. What is more significant is that all the test samples exceeded the required ultimate tensile strength. The significance of the test cannot be overemphasized. Nicking or gouging of bar by carbide steel bits is not an area of concern. The same analogy cannot be used for other types of bits. Obviously, other types would require tests to be performed using them before they could be considered in the "hold harmless" sense. The field engineer also provides information to the design engineer in respect to loading criteria and fabrication areas for construction concepts that may include Steam Generator replacements of containment dome lifts.

Previously, discussions have centered on relatively smaller backfit projects. Replacement of Steam Generators poses a whole new set of problems to construction. The enormous size and weight coupled with radiation/contamination worries transforms the project into an extremely large construction undertaking. Not only are the normal radiation restrictions focused on, but the magnitude of the restrictions increases exponentially when considering the large work areas needed both inside and outside the Containment. Protection of the field forces doing the work is the foremost thought in everyone's mind. A totally new application of existing construction cutting, welding, hauling, and rigging tools and apparatus was suggested and proved very successful. During the replacement activities which consisted of installation of new, lower shells and refurbishment of upper shells. All the information relative to each operation was finalized into work packages which provided the management tool for an integrated and successful construction project. Each package contained the administrative controls, engineering review, references and drawings, description of work, materials list, schedule, construction and testing procedures, and the acceptance criteria. Without this management tool, the replacement of the Steam Generators would not have been completed within the cost and schedule restraints. It should be noted that the radiation exposure of personnel during the construction phase was more than offset by the calculated radiation exposure savings that will be realized during the operations cycle. The important working relationship among the field, home office, and QA/QC staffs also extends to any existing nuclear generating facility. When working on these operating plants, Health Physics, ALARA, Station Operating Orders and Procedures, etc. place an additional demand on the field engineer. He is saddled with the added responsibility of understanding and complying with circumstances and regulations not normally under the construction umbrella.

The construction concept of a containment dome lift is not new in the industry. The first dome lift was made in 1970. What is new is the wide range of acceptance of this concept as more and more utilities and other users continuously search for and approve fabrication efforts that reduce costs and improve schedules. The particular dome lift described during the session for Millstone 3 resulted in approximately $1 million in subcontract savings.

Benefits are also realized when working on or near the ground, such as conduit, cable, piping, sandblasting, and painting activities. What is more important is the time delays not experienced when any craft person forgets a tool or has to climb to the work station. The cost savings cannot be calculated into a dollar value because there isn't anything with which to form a comparison.

Certain construction and engineering precautions were followed before and during the lift. Calculations for total weight and weight distribution ensured the "picks" would not rotate because of a weight imbalance. Lifting lugs were magnetic particle examined before and after the three inch (3") test pick as a testing precaution which proved the welding adequate. Weather forecasts were constantly monitored because calm and stable conditions help dramatically during this type of operation.

Even with all the preplanning, the existing cylinder does not mate exactly with a dome section which are both 140' in diameter. There always remains a final amount of field cutting and fitting needed to ensure a proper make-up for welding parameters.

Consensus Recommendations*

1. Each new construction assignment brings with it a new list of challenges. The construction representative should be instructed to develop test methods to ensure that only cost effective practices are used. Because of the continuing expanding knowledge of construction techniques, the method used on past projects may not be the most economical method on future projects.

2. The construction industry, in light of NRC Bulletins, particularly 79-02, has demanded that suppliers redefine the parameters of their "drilled-in" anchor bolt systems. Construction representatives must insist that the engineering effort include enough parameters to ensure a proper, yet constructable installation. The QA/QC effort associated with such installations should provide the assurance yet not hamper the construction effort.

3. A concerted effort is needed to accept the results of formal audits conducted by various code committees to reduce the amount of independent audits conducted by others in the industry. The amount of time spent by suppliers and the auditors results in an increase in the cost of any purchase order yet does not assure a better product.

4. New rebar finders are not in the market place. Ease of anchor bolt installations would advocate that users explore each of the new techniques to see which one is best suited to their needs.

5. Welder qualifications, as expressed in AWS, ASME, Monitoring Specifications etc., need a reciprocity agreement among the various code committees. This would reduce the costs spent in welder qualification and enable the fabricator to use the time in a productive fashion.

6. Communication lines among engineering, construction, QA/QC, need to be strengthened during the preplanning of a construction undertaking to reduce costs and shorten schedules.

---
*Based upon the Panel of Speakers/Audience discussion period at the end of the session.

Field Installation of Concrete Anchorage Systems

by

James A. Flaherty,[1] and Louis J. DiLuna,[2] Member, ASCE

Abstract

Concrete anchorage systems are frequently used in nuclear power plant applications to restrain piping, pressure vessels, pumps, and other mechanical components. The behavior of mechanical component supports is often critical to the operation of a nuclear power plant. This paper will focus on the field installation of concrete anchorages. Typical field problems and practices and their impact on quality, cost, and scheduling will be discussed. Emphasis will be placed on expansion-type anchors (wedge, shell, sleeve), which have received considerable attention in the nuclear industry in recent months.

Introduction

The support and restraint of mechanical components in a nuclear power plant is ordinarily accomplished by anchoring to concrete. Piping, pressure vessels, and pumps are typically attached to reinforced concrete walls and floor slabs. The loads induced on the support may be due to deadweight, thermal expansion, seismic inertia, fluid-dynamic phenomena (water and steam hammer), and pipe-break. These supports may be vital to the safe operation of the plant in that they insure the operability of the component under abnormal conditions.

Over the past two years, the nuclear industry has become increasingly aware of the need to improve design and installation procedures for concrete anchorages. Testing programs, such as one described by Ciatto and Boentgen (Ref. 1), have provided new data concerning anchor bolt behavior, which is being incorporated into the design and installation procedures of the industry.

This paper will focus upon the use of concrete expansion anchors in nuclear power plants. A discussion of how field installation practices can impact design assumptions will be presented. The design and installation of the Mark 1 BWR Suppression Chamber Tie-down will be used as an example.

---

[1]Manager, Engineering Design and Testing, Teledyne Engineering Services, Waltham, Massachusetts 02254

[2]Senior Engineer, Teledyne Engineering Services, Waltham, Massachusetts 02254

Concrete Anchorages - Application in a Nuclear Power Plant

Two general types of concrete anchorages are used in a nuclear power plant: Expansion anchors, which are installed by drilling holes in concrete; and inserts, which are installed prior to the concrete pour and become an integral part of the slab. Expansion anchors typically secure a steel baseplate to a slab, upon which structural members, forming the support, are welded. Inserts, or cast-in-place type anchorages, typically are headed studs welded to a steel baseplate, set into the sides of form work prior to concrete placement. Again, structural members are welded to the baseplate. The major disadvantage of this type of anchorage is that it is nearly impossible to know, at the time of concrete placement, the location and design of mechanical component supports. Thus, these anchorages are often located at some regular intervals for later use by the support designers.

Supports using expansion anchors can generally be located close to the component so as to simplify the support design. Typical piping and vessel supports using concrete expansion anchors are shown in Figure 1. Three types of anchors are generally used: wedge, shell, and sleeve. These are shown in Figure 2. The wedge anchor is typically a stud, threaded at one end, and tapered out at the other end. A clip, located near the tapered end, is forced over the taper when the nut is tightened, and bears against the sides of the hole, creating the anchoring force. A sleeve anchor behaves similarly in that a sleeve is forced over the taper as the nut is tightened. The shell anchor uses a threaded sleeve which is inserted into the concrete, with a small wedge at one end. A threaded stud or bolt is inserted into the sleeve or shell, forcing the wedge down, and causing the sides of the shell to expand and bear against the concrete. Some shell anchor designs incorporate "teeth" on the end of the shell, and a drill chuck on the other end, such that the shell becomes the drill bit.

The typical nuclear power plant has several thousand mechanical component supports using expansion anchors. Anchor bolts generally vary from 1/2" (12.7 mm) diameter, 4" (101.6 mm) long, to 2" (50.8 mm) diameter, 3' (914.4 mm) long. The three basic types described above come in a great variety of styles, brands, and variations. The bolts are generally made of high-strength steel, and material traceability is <u>sometimes</u> available.

Design Assumptions Versus Field Installation

The component support designer must have an understanding of the component analysis and assumptions, a thorough knowledge of the behavior of the concrete anchorage including design parameters, and an appreciation of the limitations of the field installer. It is important to understand all the assumptions made during the analysis. If the piping analyst calls for a "rigid" support at a particular location, the designer must be certain he understands what is meant. The support designer must interact with the component designer and the group responsible for the field implementation of the support or modification. This failure to interact can create cost overruns, delays in schedules and ultimately poor quality workmanship.

The important design parameters for concrete anchorages using expansion anchors can be summarized as follows:

CONCRETE ANCHORAGE SYSTEMS

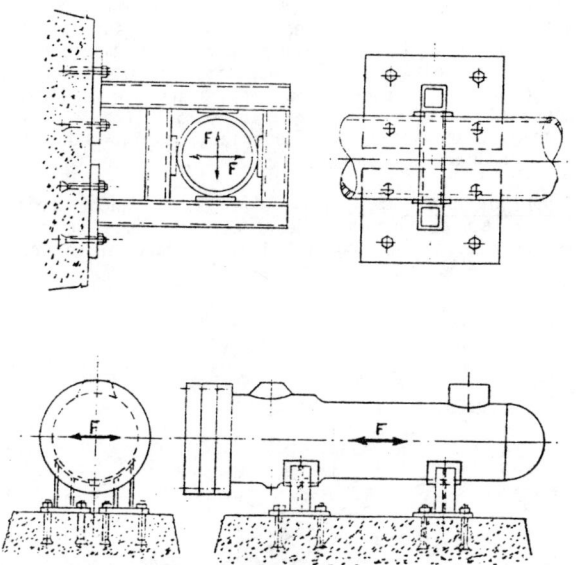

Figure 1. Typical Piping and Vessel Supports Using Concrete Expansion Anchors

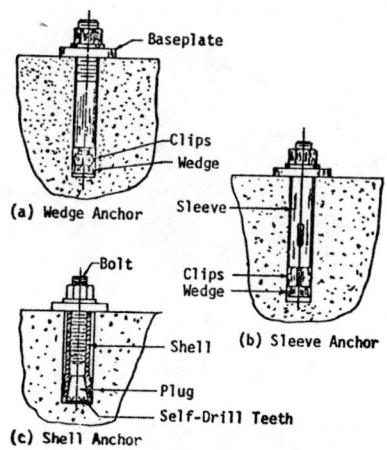

Figure 2. Wedge, Shell, and Sleeve Expansion Anchors (Ref. 1)

1. <u>Anchor Bolt Spacing</u> - Adequate distance between anchor bolts must be provided in order to develop sufficient capacity. Assuming a bolt is installed properly, two modes of failure can occur: the bolt can fail in tension (ultimate tensile capacity of the stud) or a cone-shape failure surface will develop in the concrete. These are shown in Figure 3a. If bolts are spaced too closely, these "cones" will overlap, causing reduced capacity of the anchorage system.

2. <u>Anchor Bolt Distance to Free Edge</u> - If the anchor bolt is located close to a free edge in the concrete, a reduction in capacity can also result, as in 1 above. This is shown in Figure 3b.

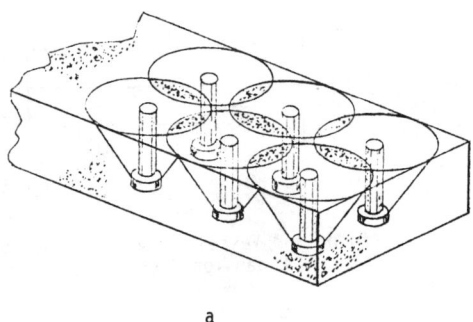

a

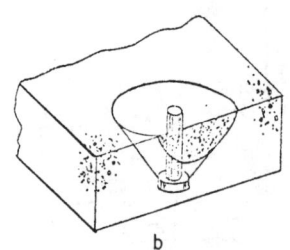

b

Figure 3.   Concrete Cone Failure Surfaces
          a.   Adjacent Bolt Cone Overlap
          b.   Free-edge Cone Overlap

3. **Anchor Bolt Embedment Depth** - The mode of anchor bolt failure, ductile (stud tensile capacity) or non-ductile (concrete shear cone) is often determined by embedment depth. If the bolt is embedded deep enough to preclude the development of a failure surface in the concrete, then the bolt stud will fail in tension. Insufficient embedment depth can result in drastic reduction in bolt capacity.

4. **Load Distribution and Shear-Tension Interaction** - The designer must consider how the loads are being transmitted from the steel baseplate into the anchor bolts. Bolts located further from the point of loading, in an unstiffened baseplate, may be receiving substantially less load than closer bolts. This effect is discussed in Ref. 2. For shear loads (loads parallel to the plate), the baseplate hole size-tolerances may determine which bolts actually resist the shear loads. That is, a bolt located in an oversize hole may see no load, and a bolt located in a very tight hole may see more load than the designer considered.

Obviously, each of the above design parameters are affected by field installation practices. Moreover there are a host of installation variables which can affect the capacity of concrete expansion anchors. Some of these are summarized as follows:

1. **Hole Size Tolerance and Drill Bit** - For each of the three types of expansion anchors - wedge, sleeve, and shell - hole diameter is very important in assuring rated capacity. An oversize hole may limit the amount of bearing that occurs between the wedge and the hole side. Although very little investigation has been done, it is generally felt that the surface condition of the hole may affect the capacity of the anchor bolt; that is, an extremely smooth hole may cause slippage of the anchor bolt. Thus, the type of drill bit used can be important. To this end, many manufacturers now require that drills and drill bits which are especially suited for their particular anchor bolts be used.

2. **Anchor Bolt Tightening** - The "wedge-action" generally employed by each of the three types of expansion anchor bolts is induced by tightening down the protruding stud or bolt. This tightening is important in "setting" the bolt, as well as providing some bolt preload, which some feel to be effective especially in low-level vibration environments. For large diameter bolts, tensioning is often recommended to insure that the bolt is set, and often to provide a known amount of preload.

3. **Measurement of Embedment Depth** - Actual anchor bolt embedment depth is a very important parameter, as previously stated. After the anchor bolt has been "set", embedment depth should be measured and recorded. Knowing the actual bolt length, the protrusion length above the concrete can be measured, subtracted from the total, to obtain the embedment depth. During nut tightening, or tensioning, substantial slippage can occur. Thus, measurements should be taken _after_ installation.

4. Concrete Reinforcing Bar - Generally, this is the biggest problem facing the expansion anchor installer. Most often, there are severe limitations placed on the cutting of rebar. From the anchor bolt behavior point-of-view, it is not clear how an anchor bolt will behave when the load-bearing wedge lies against a cut rebar. Most anchor bolt tests performed to date were done in unreinforced or lightly-reinforced slabs. The installer should attempt to clearly define how much, if any, rebar may be cut, and define an approval procedure for relocating bolts to avoid rebar.

5. Installation Verification - In order to verify the proper installation of anchor bolts, quality control hold points are often required. Care should be exercised to insure small details, such as removal of packing tape over the bolt wedge clips, are followed. It may be worth the time and cost to qualify an anchor bolt installation procedure much as one would qualify a welding procedure. That is, install anchor bolts in a test slab, and pull-out the bolts, measuring the achieved capacity and comparing it to the required capacity. This may eliminate questions concerning installation details <u>after</u> the job is completed.

6. Specifications, Procedures and Drawings - When preparing a specification, all aspects of construction must be covered. This includes material, design, fabrication, examination, testing and inspection.

   If at all possible, one should try to standardize the design or at least limit the type and size of bolts used. This would help to resolve quality control issues such as embedment lengths.

   Material procurement should encompass type of material, marking, and control of material through all aspects of construction. One should encourage the manufacturers to mark the bolts with a standardized marking system that enables the user to readily ascertain the length, diameter and type of bolt used.

   The choice of materials should include consideration of fracture-toughness, minimum and maximum ultimate tensile strength and yield strength.

   Design must include the effect of concrete strength, anchor bolt spacing, type and size of reinforcing steel. The design drawings must stipulate hole size tolerance, both for the drilled hole and baseplate. Plumbness requirements must also be given. Torquing requirements, as well as method of preload, should be stipulated.

   Anchor bolt installation procedures should be foolproof, yet simple. The type of equipment and tooling should be determined. Manufacturer's installation procedures and instructions should be reviewed and followed. This insures that the as-installed bolts meet manufacturer's tolerance(s) and therefore the basis for allowable design loads is not violated.

   Careful review should be given to manufacturer's recommendations.

   Wherever possible, on-site or in-situ tests should be performed to qualify the actual installation procedure. Based on the qualification test, adequate witness and hold points should be required.

7. <u>Codes and Standards</u> - There exists very little formal guidance concerning methods of installation of anchor bolts. The U.S. Nuclear Regulatory Commission Inspection and Enforcement Bulletin 79-02 dictates that verification of installation variables is required. Embedment depth, size and type of bolt, and the amount of preload, if any, are some of the variables which must be documented in quality control records in order to validate the installation.

This partial list of installation variables is typical of nuclear power plant applications. A knowledge of these areas by the designer and installer can make the installation process much smoother.

Example: MARK 1 BWR Suppression Chamber Tie-Down

The MARK BWR Suppression Chamber (Torus) Tie-Down is an example of the use of concrete expansion anchors for component restraint. A cutaway view of the MARK 1 BWR Reactor Building is shown in Figure 4. The

Figure 4. MARK 1 BWR Reactor Building (Ref.3)

suppression chamber is a toroidal-shaped steel vessel, about 30 feet (9.14 m) in diameter, with a major diameter of about 100 feet (30.48 m). This vessel is connected to the "light-bulb" shaped primary containment by approximately eight 7-foot (2.13 m) diameter pipes. Internally, these vent pipes are connected to a header system. The torus is typically about one-half filled with water, and downcomers protruding off the vent system are submerged in the pool. The primary function of the torus is to condense steam during a loss-of-coolant accident (LOCA). If a steam pipe were to break within the drywell, steam would escape down the vent pipes, into the torus header system, then into the pool through the downcomers.

The typical MARK 1 torus, as originally designed, consists of approximately 16 mitered cylindrical sections, made up of rolled plate, about 5/8" (16 mm) thick. These mitered sections, or bays, are welded together at the miters, and stiffened at this point by a ring girder. At each miter, a pair of columns, either pipe or wide-flange sections, support the torus.

Over the past several years, new hydrodynamic loadings have been uncovered and as a result the torus has been undergoing major modifications to return it to the originally specified factor of safety. The LOCA event is now understood to impose a severe pressure transient within the torus, causing large downward and upward forces. Improved support and restraint has been provided by adding a saddle under each ring girder, with anchor bolts tying the vessel into the floor slab. This tie-down is shown conceptually in Figure 5.

The problems faced by the designer of the tie-down portion of the saddle were as follows:

1. Anchor Bolt Selection - Based on the known loadings, it was quickly determined that a very substantial anchor bolt would be required. Also, anchor bolt material specification was a consideration, since this is a nuclear power application. In order to provide some degree of certainty as to the anchor bolt capacity, an installation procedure was developed by the designer, working with the bolt manufacturer, and tests were run to assure the rated capacity. The designer also was able to convince the manufacturer to provide alternate materials for some bolt parts, and to provide material certification to meet all applicable codes and standards.

2. Anchor Bolt Spacing - It was known by the designer in advance that installation would be a problem due to rebar cutting restrictions and headroom beneath the torus. Some flexibility in anchor bolt location had to be provided, but within tolerances such that cone overlap did not reduce the anchor bolt capacity. Thus, a "stand-off" design was used as shown in Figure 6. This design, being separate from the saddle baseplate, allowed thermal growth of the structure. Also, up-load was transferred to the anchor bolts without inducing bending moments in the bolts.

The actual field installation of this tie-down was accomplished at several MARK 1 BWR nuclear power plants with relative success. In each case, interfaces between the designer, installation contractor, and owner, to develop installation procedures and define deviation tolerances, made the installation process proceed very smoothly. Some interesting points concerning installation are:

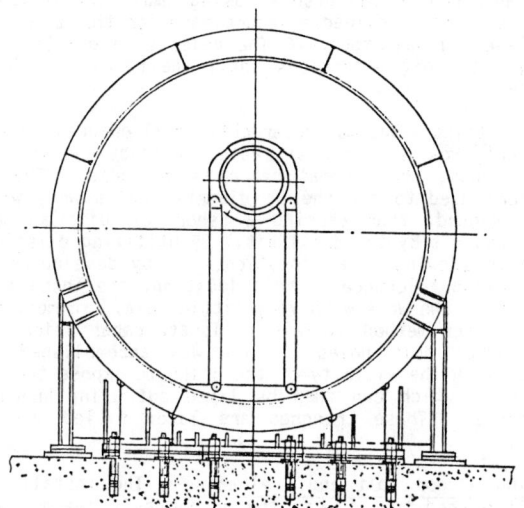

Figure 5. Saddle Tie-Down

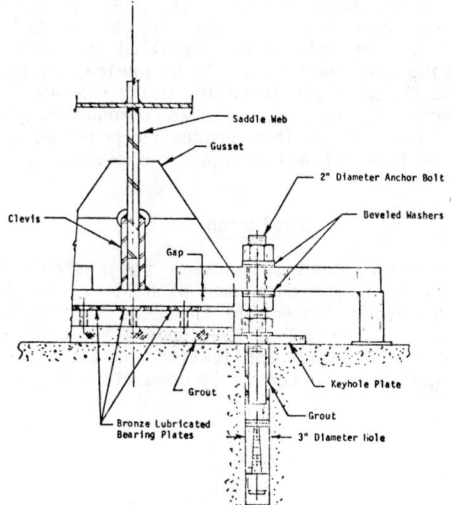

Figure 6. Tie-Down Anchor Bolt "Standoff"

1. Hole Drilling - A special drilling arrangement had to be devised in order to insure the hole was plumb within specified tolerances. The core-drill was secured to a template which was in turn anchored to the basemat using small shell-type expansion anchors. This provided a secure base for the drill and allowed for leveling adjustments. The holes were drilled in sections using drill bit extensions until the required hole depth was reached.

2. Rebar Cutting - Basemat rebar cutting allowances vary from plant to plant; however, in most cases, restrictions are severe. At many plants, the basemat rebar is grounded. Thus, the core drillers used to cut the 3" diameter holes were wired to the rebar ground, thus causing a "short-circuit" as soon as the rebar was hit by the drill bit. By utilizing existing concrete drawings showing rebar location, and by developing a location map showing tolerance on bolt location, the installer was able to perform the work with very little delay, in most cases.

   Another method used is to locate rebar prior to drilling the anchor bolt holes. This was accomplished by cutting trenches in the vicinity of the holes to expose the first layer of rebar, which can then be layed out using known points of reference. These trenches are later filled with nonshrink grout.

3. Setting Anchor Bolt and Preloading - The installer was faced with the need to: 1) set the anchor bolt "wedge" against the side of the hole, 2) apply substantial preload to the anchor bolt, 3) maintain a gap between the standoff and the saddle, and 4) grout the final assembly. In order to accomplish these tasks, an assembly, shown in Figure 6, was devised. The two-nut design allowed the bolt to be installed and torqued to provide setting using the lower nut. Then, preloading could be accomplished and the standoff installed using the beveled washers and shims. Grouting was accomplished through "keyholes" in the anchorage baseplate. This procedure provided a high-quality installation that met all design requirements.

Conclusion

The use of concrete expansion anchors for restraining mechanical components is widespread in nuclear power plants. A recent awareness in the industry has developed concerning the need to improve design and installation procedures. The MARK 1 BWR suppression chamber tie-down is an example of an expansion anchor application where designer, installer, and owner worked together to successfully complete the project.

References

1. Ciatto, R. D. and Boentgen, R. R., "Strength of Concrete Expansion Anchors for Pipe Supports", ASME Publication PVP-40, August, 1980.

2. DiLuna, L. J. and Flaherty, J. A., "An Assessment of the Effect of Plate Flexibility on the Design of Moment-Resistant Baseplates", ASME Paper 79-PVP-50, June, 1979.

3. General Electric Company Atomic Power Equipment Department. General Description of a Boiling Water Reactor, San Jose, California.

REPLACEMENT OF STEAM GENERATORS:
PWR NUCLEAR PLANT

Wallace G. Sanborn[1]
Member, ASCE

ABSTRACT

The steam generators were replaced in 1978 at the Virginia Electric Power Company's (VEPCO) Surry Power Station Unit 2, located south of the James River near Williamsburg, Virginia. Surry Unit 2 is a Westinghouse 3 loop 822 MWe unit that began commercial operation in May 1972. In 1975, the unit began experiencing steam generator tube leaks. The tube deterioration was brought about by corrosion products in the tube-to-tube support plate area resulting in tube denting and subsequent tube leaks. The tube leaks so affected steam generator performance and reliability that it became necessary for a major overhaul to restore performance and prevent future derating of the unit. VEPCO chose to overhaul the unit by replacing the lower shell section of the steam generators. This method consisted of cutting the steam generator shell just above the tube bundle, removing the upper and lower shell assemblies, including the tube bundle, installing a new lower shell and tube bundle, and replacing the original upper assembly.

This paper addresses the involvement of construction personnel in the preoutage planning and the construction techniques utilized in the rigging and handling of the steam generators; the cutting, machining and welding of the steam generators; and ALARA considerations.

PREOUTAGE PLANNING

The approach to performing this work was the use of "Technical Work Packages." A fairly detailed network type construction schedule was developed early in the conceptual phase. The network defined the sequence and specific Technical Work Packages necessary to perform the overhaul. This approach consisted of subdividing a large project into smaller tasks that could be examined, engineered, and planned to a level of detail that would assure minimal problems and delays during the outage. The Surry steam generator replacement resulted in the development of more than 100 discrete work packages.

---

[1]Senior Construction Manager, Stone & Webster Engineering Corporation Boston, Massachusetts 02107

The concept of the Technical Work Package was to combine into one document descriptions, reference drawings, sketches, specifications, procedures, materials, and equipment to perform the task. Radiation protection, quality assurance, and documentation became integral parts of the package. This total package approach assured that the software and hardware necessary to perform a given task was available in a timely manner for the performance of the task. In addition to providing guidance through the accomplishment of the task, the document was used to monitor job progress.

The Technical Work Packages consisted of the following sections: Title Sheet, providing administrative control of the work package; Engineering Review and Safety Analysis, documenting the review and evaluation of the work to be performed and any "unreviewed safety questions" as defined in 10CFR50.50; References, Drawings, Sketches, and Specifications, listing documents necessary to perform the work; Description of Work, defining effort included in the package; Materials Lists, ensuring availability of equipment, material, and tools required for the package; Schedule, indicating the time frame for the effort; and Construction and Testing Procedures, including initial conditions, precautions, instructions, and acceptance criteria.

The elements of the Construction and Testing Procedures section, especially the "initial conditions" and "precautions" portions, were useful in attainment of ALARA objectives by specifying radiological controls and ensuring that certain work efforts were completed before others were started. Radiation Safety personnel reviewed each work package from this aspect.

The Technical Work Package approach, by dividing the overall outage effort into tasks that were readily addressed, helped to ensure the performance of the job on schedule, within budget, and within radiation exposure estimates.

STEAM GENERATOR RIGGING AND HANDLING

A well-planned rigging operation, using specialized rigging equipment, was an important prerequisite for handling the steam generator upper shells and replacing the lower shells. The Surry reactor buildings had been designed to allow removal of the steam generators, but the access area had been kept to a minimum due to other considerations (Figure 1). This required that the rigging equipment be compact but capable of handling and maneuvering the 101 MT (111 tons) upper shells and the 200 MT (220 tons) lower shells. The equipment had to be designed to accomplish the rigging operations in minimum time for both radiological and job schedule considerations, with minimal impact on existing equipment and structure.

HANDLING THE UPPER SHELLS

The upper shells of all three steam generators were separated from the lower shells, inverted, and placed in special stands on the operating floor prior to the removal of any of the lower shells. Inverting the upper shells was necessary to provide the access required for the replacement of the feed ring and moisture separation equipment

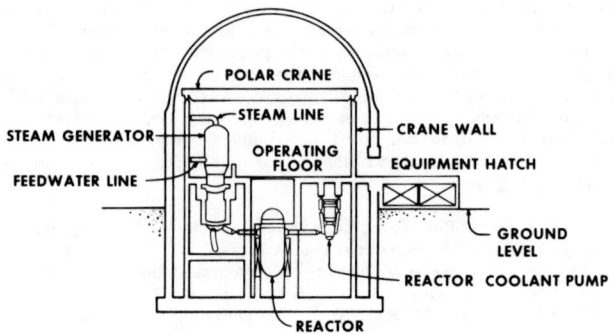

FIGURE 1 Containment Layout

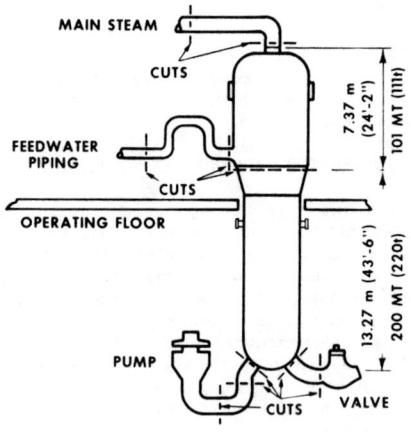

FIGURE 2 Steam Generator Cutting

in the upper shell and for the preparation of the shell end for rewelding to the new lower shells. The method described here was typical for handling the upper shells in all three steam generators.

The attached piping had to be cut and removed prior to removal of an upper shell. The feedwater piping was cut at the steam generator and at a point upstream and the section removed. A section of the main steam piping was removed by cutting at the steam generator and at the crane wall. Miscellaneous small piping and instrumentation were also removed (Figure 2).

After removal of the piping, the steam generator was cut in the transition cone to separate the upper and lower shell. The location for the cut was chosen so that the inside diameter of the upper shell was large enough to allow replacement of the moisture separator equipment and, most important, to ensure that the outside diameter of the lower shell was less than the diameter of the reactor building equipment hatch.

For the removal of the upper shell from the steam generator cubicle, a spreader beam attached to one of the polar crane main hooks was connected to lifting lugs on the top of the upper shell by wire rope. A lift of approximately 3.4 meters (11 feet) was required to clear the swirl vane assembly projecting from the lower shell. When the upper shell was clear of the swirl vane assembly, it was moved by polar crane to a temporary position on the floor for attachment of the rigging equipment required for the inverting operation.

The inverting operation was accomplished by separately rigging the two main hooks of the polar crane. One hook was rigged to the existing trunnions near the bottom of the upper shell and the other to trunnions installed in existing manways near the top. A specially designed inverting saddle was attached to the side of the upper shell. The inverting saddle provided an offset guide for the lower lifting cables. The upper shell was lifted with the main hook attached to the upper trunnions and held suspended above the operating floor. By raising the hook attached to the lower trunnions, outside the center of gravity, and lowering the hook attached to the upper trunnions, the upper shell was inverted. By moving the polar crane, the inverted upper shell was placed in the stand for rework of the internals and weld preparation. The stands were designed to support the upper shell in the inverted position and to distribute the weight to the operating floor (Figure 3).

REMOVAL OF THE LOWER SHELLS

When the three upper shells were placed in their respective stands, the three lower shells were lifted from the steam generator cubicles and removed from the reactor containment structure (Figure 4). They then were transported to a storage facility constructed on the site. Again, the method described here was typical for the removal of the lower shells for all three steam generators. The preliminary work was done simultaneously with the preparation for removal of the upper shells and, for the most part, concurrently on all three steam generators. Each lower shell was rigged and removed from the containment before the next one was started.

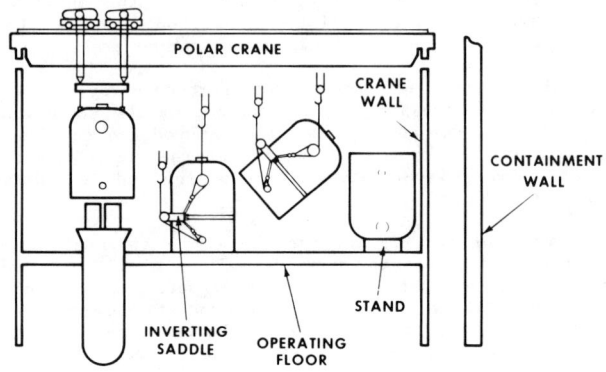

FIGURE 3 Upper Shell Removal

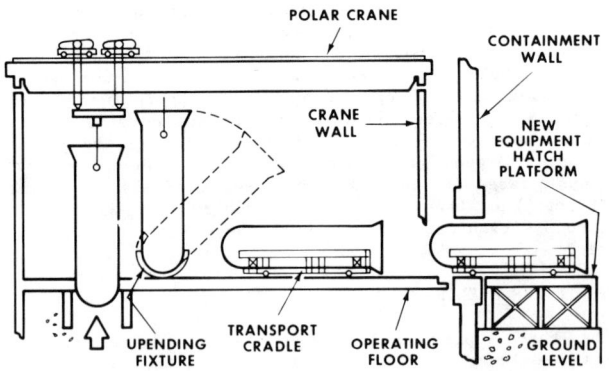

FIGURE 4 Lower Shell Removal

The preliminary work included cutting and removing sections of the hot and cold leg piping (Figure 2). The hot leg piping was cut at the steam generator and on the outlet side of the reactor coolant system isolation valve. The cold leg piping was removed in two pieces by cutting at the steam generator and at the inlet of the reactor coolant pump, plus one intermediate cut. The reasons for the intermediate cut will be discussed later. After the piping was cut, the steam generator was freed from the steam generator support. The pipe cuts and freeing of the supports were done prior to the removal of the upper shells.

To minimize radiation exposure, the cone end of the lower shell was sealed with a 5-cm (2-in) thick steel plate and the nozzles at the channel head were fitted with lead shields. The secondary side of the steam generators were drained after the shielding was installed.

To lift the lower shells, both main hooks of the polar crane were upgraded from 114 MT (125 tons) to 118 MT (130 tons), and the bridge was upgraded from 181 MT (200 tons) to 236 MT (260 tons). Rigging included a special swivel lift beam which allowed rotation of the lower shell while it was suspended from the polar crane. The swivel lift beam used in the original installation was modified to fit between the bridge girders to assure clearance above the operating floor when the lower shells were being lifted from the steam generator cubicle. A lift of approximately 12 meters (39 feet) was required to clear the operating floor.

After removal from the cubicles, the lower shell was moved and lowered to a horizontal position on a transport cradle. A special shoe was attached to the bottom of the lower shell to rotate the shell into a horizontal position and to distribute the load to the floor. It was lowered to a horizontal position by simultaneously lowering the main hooks and traveling with the bridge trolleys to keep the rigging cables vertical at all times. The shoe rotated on a structural steel platform installed in an existing opening on the operating floor. Minor modifications to the existing opening were required to install the platform flush with the operating floor.

Once the lower shell was lowered onto the cradle, a radiological survey was made. The shell was decontaminated as required and secured to the cradle. Using the polar crane, the lower shell and cradle were lifted, rotated end for end placing the cone end towards the equipment hatch, and lowered back onto the track. To center the cone end with the equipment hatch, it was necessary to remove a 0.31 meter (1 foot) deep section of the operating floor in the vicinity of the equipment. To allow clearance for maneuvering the lower shell through the equipment hatch, the crane wall opening was enlarged.

The lower shell was properly aligned to pass through the equipment hatch using special roller assemblies, and was moved on the cradle through the equipment hatch to the outside platform. Two air operated winches, one located inside the containment and one outside, were used to move the cradle. This arrangement allowed movement in either direction.

The transport cradle was designed to transport the lower shell through the equipment hatch. A special track on which the cradle would roll was installed in the operating floor. The cradle was mounted on roller assemblies comprising hydraulic jacking cylinders and low friction surfaces that allowed movement of the top portion of the assembly unit in relation to the rollers. This design of the roller assembly permitted raising, lowering, and lateral movement of the cradle and lower shell for easy alignment with the equipment hatch. The radial clearance between the outside diameter of the equipment hatch was less than 1.27 cm (0.5 in). Two complete sets of roller assemblies were required, one set inside the containment building and one set outside. It was necessary to have two sets because there was not clearance enough to extend the tracks on which the cradle rode from the operating floor through the hatch. This required the cradle to be transferred from the inside set of roller assemblies to the second set on the track outside the containment.

The track system outside the containment equipment hatch was placed on a platform approximately 6 meters (20 feet) above the ground. This platform was constructed specifically for this effort and replaced the original smaller and lighter equipment hatch platform. The original platform was reinstalled upon the completion of the steam generator replacement.

A 363 MT (400 ton) capacity mobil crane was used to lift the lower shell from the cradle and place it on a wheeled transporter for movement to the onsite storage facility. At the storage facility a jacking tower was used to lift the lower shell from the transporter and place it on rollers to be moved into the storage facility.

INSTALLATION OF THE NEW LOWER SHELLS AND REFURBISHED UPPER SHELLS

When all three lower shells were removed, the replacement lower shells were transported into the containment and installed. The same rigging techniques and equipment utilized in the removal of the old lower shells were used to handle the new ones. The installation operation was basically the reverse of the removal operation.

After the lower shells were installed, the refurbished upper shells were righted and reinstalled using the same rigging and techniques utilized for their removal. When the upper shells were in place, the reestablishment of the systems commenced.

STEAM GENERATOR CUTTING, MACHINING, AND WELDING

To support the objectives of maintaining schedule and minimizing radiation exposure, considerable effort was expended on the choice and development of techniques and tools. All work fell under the repair rules of ASME XI since the steam generator replacement was identified as a repair effort. The following techniques were several of those chosen to perform the repair effort at Surry Unit 2.

The activities for each of the three steam generators were performed simultaneously whenever possible. The use of the Technical Work Package approach described previously provided visibility to the critical scheduling of the use of the polar crane. To alleviate this potentially restrictive activity, jib cranes were installed on the crane wall for handling the materials needed for the upper shell refurbishment. The activities described in the following paragraphs are for one steam generator but are typical for all three.

To allow removal of the steam generators it was necessary to remove sections of the main steam, feedwater, and small bore piping. The sections of main steam and feedwater lines were reworked and reused while the small bore piping was replaced with new material.

Sections of the cold leg and hot leg piping were removed to allow removal of the lower shell. The section of the cold leg was cut into two pieces to provide flexibility in the replacement process and ease in rigging the pieces out of the steam generator cubicles. The new lower shell nozzle fabrication tolerances were such that there could have been a considerable deviation from the original locations. By having the cold leg section in two pieces, the refitting could be accomplished without excessive mitering that might have been required with a single piece. The cuts were made at the original weld points by the plasma arc process. The pipe was 78.7 cm (31 in) inside diameter with a wall thickness of 7.6 cm (3 in).

The removed sections of the reactor coolant pipe were decontaminated on the lower level of the containment building. The decontamination was accomplished by an electropolishing process. The decontaminated sections were then removed from the containment building to an onsite hot machine shop for reworking. The heat affected zone was removed and weld buildup was performed using an automatic submerged arc welding machine to compensate for material loss in the cutting operation. It was estimated that the use of the automatic submerged arc process required approximately 45 percent less time than manual buildup and 25 percent less time than automatic gas tungsten arc welding. Weld buildup was required on three pipe ends per generator. The final operation in the hot machine shop was weld end preparation (Figure 5). After the new lower shell assemblies were installed, the reactor coolant piping was replaced. The welding was performed by a manual root and hot pass, then completed using an automatic track mounted gas tungsten arc welding process. The automatic fill activity utilized two welding machines 180° apart to equalize weld shrinkage (Figure 6). All welds were examined and accepted in accordance with ASME III NB-5000 criteria.

The location for the circumferential cut separating the upper and lower shell assemblies was on the transition cone section. As stated earlier, it was necessary to locate the cut so that the inside diameter of the upper shell section was large enough to allow removal and reinstallation of the moisture separator assemblies but still result in the outside diameter of the lower shell being small enough to fit through the equipment hatch. To further complicate locating the cut, the transition cone section had been fabricated by using three roller plates, and therefore was not perfectly round. To ensure that enough metal would be left to machine the weld land for mating with the new lower shell, a specially designed circumferential template was used to determine accurately how much out of round the cone was at the selected cut location.

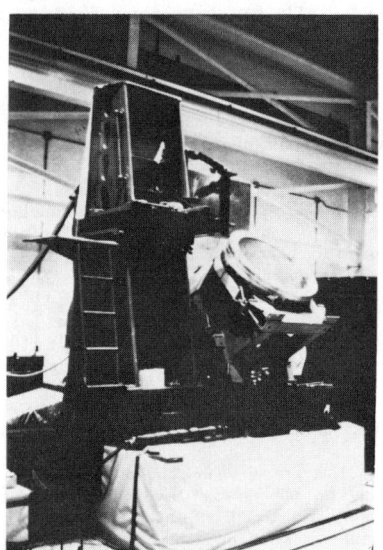

FIGURE 5 Coolant Pipe Section Being Prepared for Machining

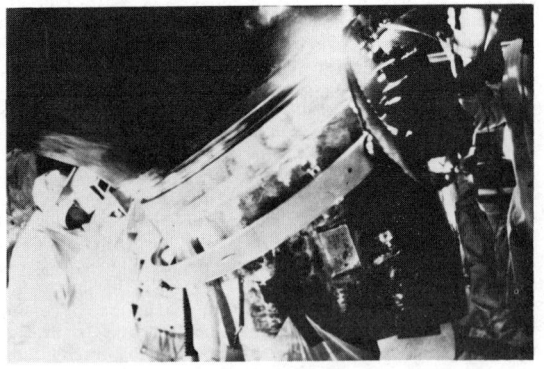

FIGURE 6 Welding of the Coolant Pipe

The outside diameter of the shell at the cut was 4.3 meters (14 feet) and the shell thickness was 10.2 cm (4 in).

During the set-up time to perform the circumferential cut, the cuts for separating the upper and lower internals were made by oxygen-acetylene flamecutting. To accomplish the circumferential cut, a track was magnetically mounted to the generator and track mounted oxygen-acetylene flamecutting equipment was utilized (Figure 7). The flamecut was made leaving four 4.1 cm (2 in) wide bridges 90° apart to support the upper shell until it was rigged for removal. A slight vacuum was maintained in the upper shell during the cutting operation to control potential airborne contamination. An exhaust fan in one of the manholes, discharging through an absolute filter, was used to maintain the vacuum. When the upper shell was rigged, the four bridge sections were flamecut, the upper shell was removed, inverted, and placed on the storage stand for refurbishment.

The track acted as a reference plane throughout the weld end preparation machining process to overcome the difficulty of having the upper shell centerline in a perfectly vertical position. The excess metal was removed by a precision oxygen-acetylene flamecut prior to starting the machining process for the weld land. The machining process was performed in accordance with a detailed procedure, including the transfer of dimensions of the weld land from the new mating lower shell, to ensure the best fit possible. The machining equipment used is shown in Figure 8.

A change in design to improve the performance of the new steam generators required the replacement of the feedring and moisture separation equipment. The existing equipment was removed from the upper shell and the new equipment was installed following the completion of the machining. The feedwater nozzle was modified and a wet lay-up nozzle was added as part of the design change.

Mechanical alignment devices were attached to both the upper and lower shells; then the upper shell was rigged over the lower shell and lowered onto the guides into place. Shims were used to support the upper shell and prevent distortion of the weld land. The weld area was then heated with electrical devices to 176.7°C (350°F) and the root pass made manually. Non-destructive examination was performed to ensure the quality of the root pass. The inside diameter of the weld joint was back-gouged to sound metal, then welded from inside the shell to provide a full penetration butt weld. To provide a satisfactory working environment inside the steam generator while performing the back-gouging and welding operation and still maintaining the preheat temperature, air conditioning and air cooled vests were used (Figure 9). The weld was then completed using manual arc welding, then radiographed and magnetic particle tested. Next the weld was stress-relieved by wrapping the outside diameter with electric heaters and heating to 620°C (1150°F). The feedwater nozzle and wet lay-up nozzle installation were stress-relieved simultaneously. The final radiographic and magnetic particle tests were performed when the stress relief was complete and welds were shown to be free of defects. The reinstallation of the tube bundle wrapper, mainsteam, feedwater, and small bore lines completed the effort.

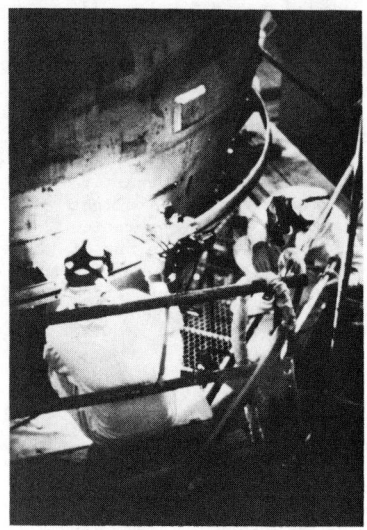

FIGURE 7 Cutting the Steam Generator

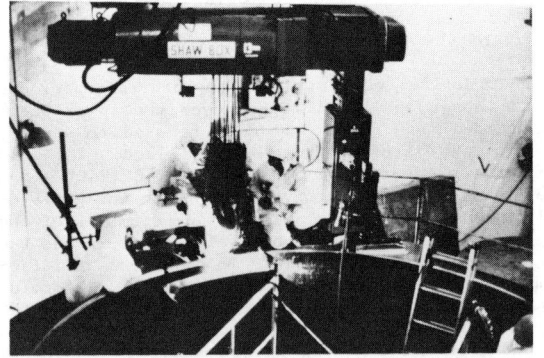

FIGURE 8 Machining the Weld Land on the Inverted Upper Shell

FIGURE 9 Manually Welding the Steam Generator

FIGURE 10 Mock-Up Reactor Coolant Pipe in the Lower Steam Generator Cubicle Area

ALARA CONSIDERATIONS

One of the major considerations in performing work on such a steam generator replacement program was to minimize radiation exposure as low as reasonably achievable (ALARA). A detailed description of the ALARA program was submitted to the Nuclear Regulatory Commission (NRC) for review prior to the approval of the replacement program. The Technical Work Package approach described earlier was beneficial in development of the ALARA program. Each package was preplanned, reviewed, and scheduled to minimize duration, exposure, and number of people in radiation areas. The techniques and equipment were chosen with ALARA in mind.

During the four years prior to the replacement, an average of 1,030 manrem per year, approximately 45 percent of the total station exposure, was caused by inspection and tube plugging of the steam generators. A benefit from the steam generator replacement was that the exposure levels from tube inspection were expected to be between 25 and 100 manrem per year. Table 1 summarizes the occupational exposure history for the steam generator effort.

TABLE 1[1]

OCCUPATIONAL EXPOSURE HISTORY

Year	Steam Generator Work (manrem)	Total Station Exposure (manrem)
1975	638	1,649
1976	1,287	3,163
1977	1,410	2,416
1978	788	1,837
Estimated Post-Repair	25-100	

Steam Generator Replacement

Estimated Repair	2,070 manrem/unit
Actual Repair Unit 2	2,140 manrem/unit

The actual occupational exposure for replacement at Surry Unit 2 steam generators was 2,140 manrems. If the Unit 1 effort results are the same, then 4,280 manrem will be the total exposure from the replacement effort. The resulting exposure then will be recovered in four to five years (4,280 manrem/(1,030-100) manrem/year = 4.6 years). The replacement, therefore, will result in a savings of personnel exposure over the life of the plant, as well as increased efficiency.

EXPOSURE ESTIMATE

As stated, the Technical Work Packages were basic elements in estimating and controlling occupational radiation exposure and radioactive contamination. The personnel radiation exposure was calculated for each task by multiplying the average radiation level in the work area by the manpower required to perform the task. The use of special techniques and equipment was factored into these calculations. The summation of the estimated exposure for the tasks resulted in an estimated 2,070 manrem for the replacement effort for one unit. The actual exposure experienced as the tasks were being performed was compared to the estimated exposure and provided visibility to possible problem areas. As part of the licensing requirements of the steam generator replacement program, reports comparing the exposure being experienced with the original estimates were submitted to the NRC every two months.

CONTAINMENT CLEANUP

One of the first activities performed was an extensive general cleaning of the reactor containment interior. The effect was to remove loose radioactive contamination, thereby reducing the potential of airborne contamination during subsequent activities. This initial cleanup resulted in 23 manrem of exposure. A general work area clean-up effort was continued throughout the program to maintain good radiological working conditions. It was difficult to quantify the benefits of the cleaning effort, but it was significant that respiratory equipment was rarely used except during cutting and grinding of contaminated components. The use of the respiratory devices would have lowered workers' efficiency; therefore, the reduced requirements were beneficial. In addition, there were no instances of significant internal or external contamination which would have been the result of airborne contamination.

SHIELDING

A separate work package was developed for the design and installation of shielding. The use of temporary shielding was an important factor in the reduction of radiation exposure levels. The lower steam generator cubicles were areas where temporary shielding was used extensively. The activities for cutting, removing, and reinstallation of the reactor coolant and associated piping in this area of fairly high radiation levels required a significant amount of manpower. Large pipes and general hot spots were covered with 1/8 in thick sheet lead curtains and blankets. Small bore piping was wrapped with herculite and shielded with molded lead halves. Shield plugs were installed in the pipe openings as the pipe was cut. During rework and reinstallation of the coolant piping, specially fabricated bags containing lead chips were placed in the pipe ends.

The average radiation levels in the lower steam generator cubicles were 75 to 500 mr/hour. The use of the shielding techniques described above reduced these levels to less than 50 mr/hour, a reduction factor of approximately 7. The shielding in the area, based on this factor and compared with actual exposure levels for cutting and removing the reactor coolant piping, gave an indicated savings of 1,280 manrem for this task. The shielding used on the miscellaneous piping in the steam generator cubicles resulted in an additonal indicated savings of 220 manrem, based on a reduction factor of approximately 5.

The installation of the shielding resulted in personnel exposure of approximately 143 manrem. This 143 manrem, compared with the calculated savings of 2,700 manrem, clearly indicated the benefit of the shielding techniques in maintaining personnel exposure ALARA.

Another shielding technique used without incurring any exposure was maintaining water in the secondary system at a level covering the tube bundle during the removal of insulation, piping, steam generator supports, and the upper shells. This resulted in an indicated exposure savings of approximately 576 manrem.

DECONTAMINATION

The decontamination of the reactor coolant piping was essential to reduce exposure during the weld buildup and machining on the removed sections. The method chosen for this effort was electropolishing. In essence, electropolishing is a reverse electroplating process which removed about 5 mils of material, and thereby the surface contamination.

Radiation survey data collected prior to and following the decontamination effort indicated the effectiveness of the process. Average contact readings of 5,000 to 10,000 mr/hour on the inside surface of the pipe were recorded prior to decontamination. These readings corresponded to an average of 1 to 5 mr/hour taken after decontamination. Using the estimated manhours to perform rework of the pipe, a savings of as much as 45,000 manrem was postulated for performing the task. This projected savings emphasized the importance of ALARA consideration as techniques for performing the tasks were developed and planned.

Because of the time and effort expended in decontamination and rework of the coolant piping, a cost/benefit analysis indicated that the replacement of these pipe sections would be more economical. For future work new pipe will be used, and thus the radiation exposure for the rework will be eliminated. Adherence to the ALARA commitment will be the result.

GLOVE BOXES

An NRC imposed license condition of the steam generator replacement program was the use of glove boxes and tents for the cutting and grinding of contaminated piping. Using glove boxes was intended to allow personnel performing tasks on contaminated material to work without respiratory protection. Respiratory equipment decreases efficiency and thus increases exposure time. The installation and removal of the glove boxes in several cases resulted in 50 times more exposure than the operation it was used to shield. A license change has been approved and the necessity for using glove boxes will be evaluated on a case-by-case basis for benefits from an ALARA standpoint.

REST AREA

To avoid unnecessary dose accumulation, rest areas were designated inside the containment. These areas were located where exposure levels were less than 5 mr/hour. Workers were encouraged to spend work breaks, material and equipment delays, work discussion, and other unavoidable inactive time in these areas rather than in the work areas where exposure could be considerably higher. Effective use of this practice contributed to the ALARA program, but was difficult to calculate.

TRAINING

A full size mockup of the reactor coolant piping, steam generator channel head, and steam generator support was constructed to train the work force (Figure 10). The craft personnel received extensive training in the activities that they were to perform by actually doing dry runs on the mockup. Simulated activities were repeated, completion times were reduced, and workers became more familiar with their responsibilities. Again, the savings in exposure was not quantified, but the training did contribute to the ALARA program, as well as to schedule savings.

All personnel involved in replacement activities received radiation training, including instruction in radiation protection practices and applicable federal regulations. Other training sessions were conducted on specific tasks, such as setting up glove boxes and tents and practicing activities to be performed within the enclosures.

During the course of the effort many work activities were photographed or videotaped to record the actual operations. If the activities were to be repeated in the replacement effort, these visual aides were used to familiarize personnel with the actual work prior to its taking place again. These visual aids were used also to evaluate the techniques used and to look for areas of improvement. The replacement work to be performed on Unit 1 will receive the major training benefit of the visual records from Unit 2 and the resultant effects on the ALARA program.

SPECIAL TOOLS

The utilization of special tools and equipment has been described earlier. The ALARA program was a factor in choosing the specific tools or techniques for the various work packages. It was found that, in most cases, the chosen technique satisfied the criteria for schedule, good construction practice, and the ALARA program. Examples of this are plasma-arc cutting of the reactor coolant pipes and the use of automatic welding equipment for the reinstallation of the coolant pipe. Plasma-arc has a high cutting speed and reduced exposure time. The automatic welding equipment resulted in fast, quality welds, and thus lower exposure.

Table 2 summarizes ALARA techniques along with general programs of health physics, training, and contamination control. The result was that the actual exposure was within 3 percent of the estimated exposure.

TABLE 2[1]

SUMMARY OF SPECIFIC ALARA TECHNIQUES

Technique	Exposure "Cost" Incurred (manrem)	Achieved Dose Reduction Factor	Calculated Exposure Savings (manrem)
1. Initial Containment Cleanup	23	Difficult to quantify	
2. Temporary Shielding	142	5-7	2,700
3. Steam Generator Water Level	Negligible	10	576
4. Reactor Coolant Pipe Decontamination	41	1,000	45,000
5. Rest Area	Negligible	Difficult to quantify	
6. Training	Negligible	Difficult to quantify	

An important aspect for an ALARA program that must be constantly considered in performing work in radiation areas is minimizing the manpower in these areas. Special equipment and techniques can reduce the manpower requirements, but a strict management effort is essential to keep manpower to a minimum to further reduce actual exposure and contribute to the success of the ALARA program.

REFERENCE

Surry Steam Generator Repair Program, Rev. 7, Oct., 1978: Surry Power Station - Unit 1, Docket No. 50-280, Surry Power Station - Unit 2, Docket No. 50-281, Virginia Electric and Power Company.

MILLSTONE III CONTAINMENT DOME LIFT

by

James A. Galinsky, P.E.[1]
Member, ASCE

ABSTRACT

It has long been recognized that large field erected subassemblies can often save time and dollars on any construction project. This approach was used on the Millstone III Containment Dome field erection.

INTRODUCTION

In 1974, Graver Energy Systems, Inc. was awarded a contract to construct a Nuclear Containment Liner for the Millstone III Reactor at Niantic, Connecticut. The owner of the facility is Northeast Utilities Service Co. Graver was to act as a subcontractor to Stone & Webster Engineering Corporation.

PROBLEM DESCRIPTION

The Millstone containment liner consists of a cylindrical steel liner with a hemispherical dome top closure. The lower cylindrical portion is 140' in diameter and 130' high. The shell is made of 13 courses of 10' wide by 3/8 thick plate rolled to the 140' diameter. The upper hemispherical portion has a spherical diameter of 140'. It is made of 11 courses and a top circular closure plate, 1/2" thick, hot formed to the 140' spherical diameter.

We began work on the project in July of 1974 at the East Chicago, Indiana, fabrication facility. Field erection began during April, 1975.

ERECTION SCHEMES

The erection scheme called for field assembly of the cylindrical liner on a ring by ring basis. Each rolled shell plate was to be lifted into place, attached to the completed ring below and to the adjacent shell plate with fit up hardware. The assembled ring would then be welded and examined before progressing to the next higher ring.

---

[1] Manager of Engineering, Graver Energy Systems, E. Chicago, IN

The initial erection scheme for the dome involved subassembly of several dome plates on the ground, lifting this configuration to the top of the containment and holding the centilevered assembly with a cable tieback. The tieback tension was resisted by a strongback contilevered from the liner shell below. Inner plates were to be assemblied on a grid type falsework. Early in 1978, a preliminary study of alternate schemes was initiated. Using Graver's past experience with large ground prepared subassemblies, and associated lifts, it was determined that this erection scheme would be cost and schedule effective. Limits of the Manitowoc 4600 ringer crane, however, would dictate the maximum lift and, therefore, the number of subassemblies.

Crane capacity limitations resulted in selecting two large subassembled lifts. The first lift included the second through fifth dome rings. The second lift included the balance of the dome above the sixth ring. The first dome ring was installed on top of the cylindrical containment liner shell without ground sub-assembly.

It was determined that about four months would be trimmed from the schedule using this erection technique with an accompanying sub-contract savings of approximately $1,000,000. Additional cost savings would result since a cherry picker would be used for the ground subassembly as opposed to virtual full time use of a larger crane if plates were assembled plate by plate on top of the containment liner.

Additional benefits realized are:

1. Work delays of other craft working under the dome would be minimized.

2. Dome assembly would be safer since the activity would be conducted at ground elevation.

3. Piping and conduit could be installed with greater ease near the gound.

4. The blasting and painting mess would be minimized.

On May 3, 1979, our customer directed us to proceed with final engineering for this approach. On May 3, 1980, the first lift was successfully made and on May 31, 1980, the second lift was completed.

CONTAINMENT DOME LIFT

To better describe the lifts the following movie is used.

MOVIE SCRIPT

In the course of any major project, there are milestones that mark the path of progress.

In construction of a nuclear power plant, one such progress mark is the completion of the containment or containment liner. It marks the closing of the housing for the reactor. It marks the end of a major phase of construction.

In recent years, closing the containment liner has taken on additional dramatic effect, thanks to an engineering advance pioneered by Graver Energy Systems, Inc. This is the offsite assembly of the containment's dome, and lift of the completed dome as a single subassembly.

Preparations for the Millstone III lift began in the early hours before dawn. This is especially desirable because it is usually the calmest time of day and it allows the maximum daylight hours for the lift and for the fitting and "tacking" of the dome assembly.

Long before this, Graver engineers had determined the number of lift points necessary for the rigging, and had plotted the location of these points to insure minimum distortion of the dome sections.

Earlier in the week, the rigging had been attached to the lift point lugs and the lower dome section, the knuckle section was eased slightly off the supports on which it had been constructed to check the balance of the rigging and the assembly itself. The brief trial lift had gone well and the engineering calculations had proven correct.

The first upward movement is barely perceptable. Then the 460,000 pound knuckle assembly swings clear of the construction support on which it was assembled. After a pause, it begins to move upward twenty feet or so.

Next, it will move horizontally several hundred yards to a position next to the completed side section of the liner.

Then, it will be raised straight up to a point just above the liner shell.

Finally, it will be moved, in horizontal mode, over the liner, where it will be lowered into place.

The idea of lifting and setting the dome as a complete sub-assembly was thought of as a way to speed construction and permit a number of different construction phases to proceed at the same time.

When the roof section was installed in conventional one-piece-at-a-time in-place erection, considerable scaffolding, safety flooring and supports, and extra cranes were required. Each panel was lifted, fitted and welded into place. Because space withing the containment was limited and there was the hazard of people working overhead, other site contractors were necessarily held up. Graver proposed that the entire roof section be assembled nearby, on the ground, as a single unit and then lifted into place as a completed assembly.

The first containment dome lifts in the industry were made in 1970 and '71 by Graver at American Electric Power's Donald C. Cook plant near Bridgman, Michigan. A year later a similar lift was made at Duquesne Light & Power's Beaver Valley Station near Pittsburgh.

Plates for the liner shell and dome were fabricated in Graver's shops in East Chicago, Indiana. Large panels of three-eights inch carbon steel, eight feet by thirty feet, were roll-formed for the sidewall assemblies. Roof panels of one-half inch steel were rough-cut, heated, and then formed to the dome curvature.

The edges of the steel plate were beveled for welding. Then the plate was cleaned, painted with a special epoxy coating, and shipped to the jobsite in special racks.

At the jobsite, the measured and numbered panels were joined. Large wall penetrations were installed. Every inch of the 10,000 feet of welding was inspected, tested and x-rayed. The x-rays become permanent documentation.

While Graver had in the past lifted the entire dome assembly in a single lift, at Millstone III we decided to made the final lift in two sections. The reason was that two smaller assemblies, each weighing about 230 tons could be lifted with a single crane, rather than with two cranes operating in tandem with a lifting beam. One crane offered better control. As the skilled crane operator gently lowers the hugh knuckle section, Graver boilermakers watch expectantly.

Once it is down, they move quickly to secure it to the sidewall assembly.

Later they will complete the task of welding the 440 foot seam.

Soon they will be ready to lift and set the dome section itself.

Innovation has a long history at Graver. The company was founded more than 123 years ago in Allegheny City, Pennsylvania because William Graver thought there had to be a better way to transport the oil from Pennsylvania's newly-discovered fields than in hollowed logs.

It was the beginning of a tank business and a long association with the energy market that continues today.

Graver's association with nuclear technology goes back to 1937 when Graver built a small domed structure for the research cyclotron at Massachusetts Institute of Technology.

When the Atomic Energy Commission built its experimental reactor in 1955 at Argonne National Laboratory to pursue the use of nucler energy, Graver was called to assist the design, fabrication and construction of the housing for this peacetime reactor.

Similarly, when the AEC decided on its second installation, at Idaho Falls, ID, it again called on Graver for fabrication and field construction. Graver has grown up with the nuclear power industry.

The morning of the second lift dawns grey and with the threat of rain. There is a sense of urgency because the decision has been made to try to get the dome lifted and set before the predicted rain. There is no time to waste. Preparations go quickly. The lift crew is confident and sure. The experience of the first lift is still in their minds.

The second lift at Millstone, that of the dome portion, was made four weeks after the knuckle section had been lifted into place. Twenty-one plates had been joined to form the dome in the field and the 468,000 pound assembly was lifted into place in little more than an hour.

At Millstone, the dome section, the roof's actual closure, had been assembled first, before the knuckle section that was raised in the first lift. That assembly, with its larger circumference, had been assemblied around the dome portion.

While Graver field crews worked on the larger knuckle section, other contractors, also working on the ground, had installed the spray piping, header assemblies, insert plates, hangers for electrical raceways and other dome attachments.

The entire dome assembly was sandblasted and then painted, also on the gound. What went up in the final lift was virutally a complete unit, requiring only minimum connections to make the dome systems operable.

Another economy in the Graver system is that the dome sections had been worked and completed at the same time other Graver crews erected the vertical wall panels. Engineering estimates show a savings of four to six months in construction time, and saving tens of thousands of dollars in extra material and equipment as well.

The pace and the intensity of the lift pick up as the dome section seems to slide over the top of the containment's sidewall. Control shifts to the men on top. Gently the assembly is lowered, very gently. Graver supervisors and boilermakers alike watch the progress anxiously. There is a final moment of concern. Then, it is down, matching up all around the 108 foot diameter. Wedges are driven. Buckles are tightened. It is down. It is done.

Today, Graver Energy Systems, Inc. is totally dedicated to the electric power generating field. We continue to participate in the construction of new nuclear power stations, but we are also channeling our capabilities in new directions to serve the changing needs of our customers in the electric utility industry.

Graver has new experience in nuclear retrofit applications in both internal and external torus modifications, and a capability for system modifications, feedwater heaters, PWR steam generators and condensors, and related capital projects.

Graver offers the electric power generating market a single, responsible source. It has the planning experience, construction skills and management ability that building for this specialized market requires.

We are proud of our association with the electric utility industry, with companies like Northeast Utilities, and Stone and Webster, and we are pleased that our innovation and achievement have helped them produce new capacity at a saving of both time and cost.

TECHNICAL DATA

The first lift was 140' diameter by 35' high with a total weight of 460,000 pounds. The second lift was 108' diameter by 25' high with a total weight of 468,600 pounds. Twenty four lift lugs were used for each lift. Each lug was attached to a 1-1/2" diameter cable with a 2" diameter turnbuckle. The cable angle for the first lift was 60 degrees with the horizontal and 45 degrees for the second lift.

FIELD EXPERIENCES

Our first major field concern was to assure a sound weld between the lifting lug and dome shell plate. Our finite element analysis revealed that the concentrated loads at these points quickly dissapated into the liner. These welds were examined after they were made and again on the day prior to lift with Magnetic Particle equipment.

A second concern was fit up of each subassembly to the construction it would attach to. There is very little latitude for adjustment to assemblies of this size. Good controls and constant measurements are helpful in achieving this end.

A third concern was the balance of loading and equalization of tension in the cables. Every attempt must be made during the planning to balance and evenly distribute the loads. Unbalanced loads will cause tilting of the sub-assembly as well as unequal tensions. If unbalanced conditions exists, the erection procedure must plan for the tilting and the stress analysis must provide for the unequal loads.

STEEL FABRICATION IN THE NUCLEAR POWER INDUSTRY

by

Denis Mason[1]

Abstract

An analogy of regulations compiled for presentation to the Conference of Construction of Power Generation Facilities September 16th thru 18th, 1981, Pennsylvania State University. Various codes (e.g.; Code of Federal Regulations 10CFR50. Appendix B, ANSI N45.2, MIL-Q-9858, ASME Section III, AWS D1.1, MIL STD 248) that affect the cost and determine the documented actions required by the Steel Fabricator to the Commercial Nuclear Power Industry. Steps 1 thru 13 set forth the course of action taken by a Steel Fabricator's Client and the Steel Fabricator pointing out where codes overlap, but documents generated must differ. Standardization of Codes, being the goal which, would result in more easily audited systems, clearly identifiable personnel certifications and document requirements that could substantially reduce costs.

Introduction

Prepare to take a tour of the wonderful and sometimes incredulous world of steel fabrication in the Nuclear Power Industry. I use the term "wonderful" advisedly. How else can so much good be accomplished with so little risk to the general public or the environment? Incredulous? Every action taken that MAY affect the operation or safety of a Nuclear Power Plant must be tested, verified and certified. Some of the worlds finest minds coupled to a dedicated personality have undertaken this task, with a history of success, to my mind, unequaled in the annals of our forebearers. Steel fabrication to the Nuclear Industry is steel fabrication of the highest obtainable quality, accomplished and inspected by personnel well trained and certified, with every step of the training documented and scrutinized again and again by many agencies and our many clients representatives.

In order that Thames Valley Steel be allowed to fabricate items and components to the Nuclear Industry, the following sequence of events must successfully take place:

---

[1] Manager, Quality Assurance, Thames Valley Steel Corporation, 18 Eastern Avenue, New London, CT. 06320

Step One: THE CLIENT'S APPROVED VENDOR'S LIST

Prior to being allowed to submit a cost estimate, the client desiring to utilize our services, must accept our written Quality Assurance Manual and determine proper implementation of the manual and subsequent internal procedures. This is done by audits and surveys which generally include a team of individuals and as many as six persons conducting the survey for up to four days. This process is expensive, both to the client and to the prospective vendor, it is time consuming, nerve-wracking and redundant. Every aspect of the program, including the requirements of the codes that will be invoked, is thoroughly examined. Holders of valid certificates of authorization issued by the American Society of Mechanical Engineers (ASME) to Section III may be accepted on the basis of the ASME team audit, however, this had not occured to date at Thames Valley Steel. The prospective client, after receipt of our valid certificate of authorization, still conducts his own full scale audit. All client's audits differ. A Bechtel audit may be very different from a Stone and Webster audit vs. the criteria of a Yankee Atomic Audit. Since January, 1981, Thames Valley Steel has undergone not less than ten independent audits, stimulated by the receipt of a purchase order allowed by the certificate of authorization issued to Thames Valley Steel by the ASME audit.

Step 2: REQUEST FOR QUOTATION

As complicated as a project may be, an appropriate cost estimate can be readily determined provided the following obvious information is included in the Request for Quotation package.

1. Codes & Specifications with revision dates
2. Material specifications with revision dates
3. Submittals of manuals, procedures, drawings and qualifications (including personnel qualifications)
4. Configuration tolerances
5. Inspection requirements with acceptance criteria
6. Hold and witness points
7. Additional tests
8. Coating requirements
9. Packaging and shipping requirements

Step 3: THE AWARDED CONTRACT

No major differences can be invoked within the Purchase Order without the Request for Quote cycle being repeated.

Step 4: PRELIMINARY ENGINEERING ACTIONS

Upon review of the criteria set forth by the client, the Chief Engineer causes a preliminary purchase document to be generated. This document must be transmitted to the Quality Assurance Manager, who must insure that all quality requirements are set forth. The Quality Assurance Manager signs the document as proof of review and transmits it to the Purchase Manager, who in turn, generates the Purchase Order to a Vendor that appears on the Thames Valley Approved Vendor's List.

(See Step One) The formal Purchase Order is transmitted to the Quality Assurance Manager who insures that the quality requirements are set forth. The Quality Assurance Manager signs that document, returns the document to the Purchase Manager for his signature and we begin the next step.

Step 5: RECEIVING OF MATERIALS, ITEMS AND COMPONENTS

Upon receipt of shipments, the receiving agent inspects the physical characteristics of the materials and items and the traceability to the documents received. A Quality Assurance representative verifies the receiving agents actions and inspects and accepts the documents received.

But, let us return for a moment to those thrilling days of yesteryear when a chunk of steel was a chunk of steel, hand forging replaced by drop hammers to automatic computer operated rolling of shapes, 250 feet long from one billet. Not so distant to the swahili iron maker who identifies his hunting spears of farming implements as does the Japanese samurai sword maker, a gesture to invoke the pride of his craftsmanship and the confidence of the user. Today, the state of the art and pride in craftsmanship being what it is, the Certified Mill Test Report (CMTR) traceable to material by actual heat or lot analysis and testing, or an individual material verification testing and analysis report replaces the aboriginal makers mark.

Step 6: CERTIFIED MATERIAL TEST REPORTS (CMTR)

CMTR's are necessary to separate a heat of material that essentially meets the requirements set forth from the material which may, at a later date, prove to be deficient and installed within several Nuclear Power Plants. No apparent better system exists for identification and perhaps subsequent removal of potentially dangerous materials from affected systems. That is the wonderful part. Now to the incredulous. CMTR's will eventually indicate the value of the material we are utilizing and I offer the following criteria as a checklist, because the CMTR, in an acceptable form, has proven to be the most difficult part of the documentation cycle.

1. The CMTR must be lcear and reproducible
2. The CMTR Must include all test criteria
3. The CMTR must show an indication of acceptance by the manufacturer; a signature, a stamped signature or initials is acceptable.
4. The CMTR may require notorization
5. The CMTR cannot be altered
6. The CMTR may be received from a foreign country, in a foreign language, such as Spanish, and a separate translation should accompany this type CMTR

When the above information is not received on the CMTR, the fabricator must take the following steps.

1. The material manufacturer or material supplier is contacted and an appropriate CMTR is received and corrective action accomplished to prevent reoccurance of documentation problem. This might include a full scale audit of the supplier facility or

2. The material may be returned for replacement material and an appropriate CMTR or
3. The material may be sent to an independant testing laboratory for verification testing and analysis, the failure of which would be cause for the material to be returned to the supplier.

(Please remember - Thames Valley Steel is a fabricator and erector of structural steel and ASME Section III items, supports and components. We do not manufacture Nuclear Steam Supply Systems or containment liners). If material manufacturers were made to understand the implication of an unacceptable CMTR and the cost involved where a proper document has not been generated, and subsequent verification analysis and testing must be accomplished to accept the material, our world, yours and mine, would be a more comfortable and less expensive one.

Step 7: QUALIFICATION OF PERSONNEL

We have previously stated that steel fabrication is steel fabrication of the highest quality, and as we are about to actually fabricate, we must be able to verify that all persons involved in the fabrication cycle, not only have the ability to accomplish their part, but that their abilities include training to the program and that they fully understand that training. These actions are documented on an indoctrination and training record. A welder with thirty years experience must still receive training to our procedures and implementation of the procedures by him will be verified by Quality Control individuals, who have also received documented training to the procedures. Let us examine what this document indoctrination and training record entials. Thames Valley Steel works to Military Specifications, American Welding Society Codes, ASME Boiler & Pressure Vessel Code, the American Bureau of Shipbuilders, etc. Each code has its own requirements for acceptable performance levels and what documentation of the performance should be and these are subject to the individual auditors' concept of what they should be. Standardization of codes is clearly indicated. Further, materials that are considered to be installed by the ASME Boiler & Pressure Vessel Code are those that have had their data package recorded at the National Board. Repairs to such systems could require a "mock up" assembly that demonstrates the welders ability to install filler metal to an acceptable quality level. These "mock ups" can cost $100,000 plus. Why is this necessary when the final weld will be accepted by Radiography or other approved nondestructive examinations?

Step 8: FABRICATION CYCLE

The material is brought into the shop from a segregated holding area. The maintenance of indefication/traceability to heat is verified. The identification is setforth on a blast list, the material is blasted and immediately re-identified. The material is sent to an appropriate area. An appropriate area at Thames Valley Steel is one of six attached, but separate shops encompassing 110,000 square feet. The material is identified with a low stress die-stamp so that no possible loss of traceability may occur. Any welding to be accomplished is controlled by a Welding Supervisory Division. The qualified welders are assigned by procedure and position. Closely controlled weld filler metals, traceable to heat or lost are requested by welding production personnel

and issued by Quality Control individuals only. The weld process is
closely monitored and the end result of every inch of deposited material
is inspected and eventually accepted by some means; including everything from radiography, ultrasonics, magnetic particle, liquid pene-
trant and visual and dimensional nondestructive examinations, which
leads us back to the qualification question, the end result of all
welding is verified, accepted and certified, have we, as an industry,
gone over-board regarding the quality requirements and the documentation of the maintenance of the qualifications of a welder? Certain
codes require a welder must pass an eye test annually or sooner, other
codes require he pass a practical test and then demonstrate utilization
of that welding process within a certain time limit. Why? If a welder
passes a practical test and all his welding performed is verified, how
much documentation regarding that welder should be required? Currently,
the qualification records for a welder are extensive. The fact of all
this documentation does not preclude a welder having a fight with his
wife the night before performing unacceptable welds. Which leads us to
bolting sequences. Bolting is no less important to steel fabrication
than welding. Personnel performing bolting operations receive indoctrination and training to a bolting procedure and they may proceed to
accomplish their tasks without further training or any maintenance of
qualification to do bolting and bolting is not as simple as it may
appear. Several acceptable methods, such as turn of the nut, calibrated torque wrench and load indicator washers, may be utilized.
However, each bolt tightened is inspected and accepted. Load Indicator
Washers seem to be the least expensive method for these reasons:
NOTE: It's obvious that a Load Indicator Washer costs money. The
Load Indicator Washer is easy to install, can be torqued rapidly or
slowly, tests have shown the washer can be torqued half way today and
the process completed tomorrow and proper tension will result.
Inspection is a simple matter using a piece of .010 shim stock and a
calibrated torque wrench if thought to be needed will verify the
results. The other methods need either a Skidmore Wilhelm or calibrated
torque wrench to be applied to each bolt.

Step 9: COATINGS WITHIN THE CONTAINMENT STRUCTURE

A coating system that will be installed within the containment is
as specified by the Client and must have passed design basis accident
criteria. Thames Valley Steel's obligation is to insure that the
manufacturer's recommendations are completely complied with by
personnel who have passed written and practical examinations. The
substrate must be developed by an acceptable method of blasting,
cleaned with solvents and coating applied prior to any rust "bloom" or
contamination. The coating must be mixed and allowed to catalyze. In
the case of two part epoxies, the ambient conditions of the substrate
temperature, the atmospheric temperature and relative humidity level
must be acceptable prior to application. The wet film thicknesses
and the dry film thicknesses must be inspected and fall within acceptable parameters.

Step 10: SHIPPING AND HANDLING

Shipping and handling requirements are either set forth by the
Client at the time a Request for Quote is generated or standard

commercial practice that will insure acceptable receipt of materials
and items is instituted. (Standard Commercial Practice generally
follows the guide lines set forth in ANSI N45.2.2-Level D.)

Step 11: NONCONFORMANCE REPORTS

When material or items are found to be descrepant in any manner,
a nonconformance report must be generated. The report must contain
the nature of the nonconformance, the action taken to correct the
nonconformance, the action taken to prevent reoccurrence of the nonconformance, the results of any applicable reinspection and the distribution of the nonconformance report.

Step 12: DOCUMENT CONTROL AND RECORDS

Document Control insures that the information necessary to
accomplish a task is issued to those necessarily interested individuals.
A document control record is generated for each controlled document.
The document control record must reflect the nomenclature of the document, the identification of the persons who have received the document,
the revision level of the document and the disposition of any super-
ceded copies of the document. Records generated in compliance with
controlled documents must be reviewed for completeness and adequacy
and entered into a record file with a controlling identification to
insure retrievability. The records are maintained in separate, remote
locations. The earlier built Nuclear Power Plants have amassed a
notable record for safety and operational longevity, as I previously
stated, unequaled. This has been accomplished without the enormous
collection of verifiable evidence by documentation. Therefore, does
the documentation required for new construction actually result in a
net gain to the owner, the Nuclear Regulatory Commission or the general
public?

Step 13: INTERNAL AUDITS

Every action taken to comply with the program and procedures must
be periodically audited by Thames Valley Steel's personnel utilizing
reference material and checklists and who are free of responsibility
in the area being audited. This process is time-consuming and costly
to the Company and to the persons both conducting audits and being
audited. However, the system is necessary to test the effectiveness
of our program and procedures to determine if any defect trend is
occurring.

In an effort to comply with the intent of this conference, which
is to try to point out overly stringent regulations, Thames Valley
Steel, as a steel fabricator, has learned to so totally comply with
invoked criteria, that we are able to deliver urgently needed items
to the requirements of the Codes within extremely restrictive delivery
schedules. It is difficult for us to criticize a system in which we
routinely conduct our daily business. That is not to say the codes
and regulations cannot be loosened or intensified to certain areas.
The cost of documentation for all aspects of a program appears to be
a major investment. Proper physical performance is judged to be
worthless without it. How much verification by documentation is enough?

So far, it is not for the steel fabricator to say. However, standardization of codes, audit criteria, qualification records and certified material test reports would go a long way to reduce the cost to the owner and the vendor would be better able to comply with all necessary aspects of a single program concept.

NUCLEAR CONSTRUCTION CONCRETE ANCHOR EXPERIENCES

by

Thomas W. Deshefy [1]

ABSTRACT

This paper addresses the experiences and problems Northeast Utilities has had with concrete anchors at the Millstone Nuclear Power Complex. Past and present experiences on Units 1 and 2, now in operation, and on Unit 3, now under construction, will be addressed. It will touch on a few of our past experiences with drilled-in anchors (such as wedge-type and expansion-type bolts) and the installation costs involved in meeting existing federal codes and project specifications. In conjunction with this, the paper will describe a test program that was developed to reduce the installation costs of drilled-in anchors. This test program proved that reinforcing steel would not be damaged sufficiently to impair its strength when drilled with carbide bits. The results provided guidance as to acceptable defects in reinforcing steel which are caused by drilling operations or chipping hammers. Lastly, this paper will relate an alternative method of anchorage system, namely, concrete anchor inserts and their advantages/disadvantages this company has experienced.

The experiences and problems described herein are unique to the nuclear industry due to the strict federal controls and quality assurance criteria integral with design and construction. These problems are not found on conventional construction projects, but it is hoped the information provided will be of use to all readers.

DRILLED-IN CONCRETE ANCHORS

This section of the paper will discuss the use of drilled-in anchors in construction and why they can't be avoided completely. It will relate some specific experiences Northeast Utilities has had with improper installation, rebar interferences and torquing criteria.

It will describe the Nuclear Regulatory Commission's (NRC) concern with proper drilled-in anchor installation which resulted in tighter

---

[1] Civil Engineer, Generation Construction Department - Northeast Utilities Service Company.

controls on design and installation. The effects of the NRC's concerns on project specifications for drilled-in anchors will be discussed, also.

Use of Drilled-In Anchors on Construction Projects

It has been Northeast Utilities' experience on three nuclear projects that design of piping and electrical supports inevitably falls behind the civil/structural designs. Some of the reasons for this are not enough design lead time, design delays/changes due to new codes and criteria, and continual seismic analysis required for piping systems. All of the above have an impact on design schedule.

The results of the design delays of piping and electrical supports do have a profound effect on construction activities. Experience has been that many supports (i.e., embedded plates) are not designed and fabricated by the time the field is ready to place concrete. As a result, concrete has to be placed without these supports in order to not hold up the construction schedule. Later, when the supports are designed and fabricated, they would be installed and supported to the concrete by drilled-in concrete anchors. The drilled-in anchors referred to are the common wedge type or expansion type normally found in construction, such as, Hiltis, Redheads and Thunderstuds.

For non-nuclear projects this kind of thing is not a big problem. However, on nuclear power plants installation of drilled-in anchors can be frustrating. If rebar is struck when drilling holes for the bolts, all work must stop until an Engineering resolution is received.

The delays encountered when rebar is struck is very costly. It has been experienced that rebar interferences occur in approximately 50% of the holes drilled. This drives the average cost of bolt installation up enormously. If the utility had planned to use these bolts exclusively during the project, it is estimated that 248,000 bolts would have to be installed. Direct labor alone would amount to approximately $6.2 million. This is assuming a labor rate of 1.5 manhours/bolt - a rate even lower than we have experienced.

In addition, the engineering cost involved for each re-evaluation due to rebar interferences would amount to approximately $500 for each occurrence or $3.1 million in total. This figure was arrived at by assuming the 248,000 bolts would amount to 62,000 support plates and that an engineering evaluation would have to be made for every 1 out of 10 plates. This figure is also considered conservative. One can see that if drilled-in anchors were used exclusively on Unit 3, the total estimated cost would be approximately $9.3 million.

These figures are not inflated but very real. It can be clearly seen from the above why the utility was searching for another way to anchor support plates to concrete.

Tighter NRC Controls on Concrete Anchor Installation

Results of NRC audits of architect engineering firms showed a wide range of design practices and installation procedures used for concrete expansion anchors. Past practices showed little control over installation activities which probably accounted for the high percentage of bolt failures.

In 1979, the NRC issued IE Bulletin No. 79-02 addressing this problem. This bulletin instructed licensees of nuclear power plants to review designs of base plates with expansion anchors, to assure NRC recommended factors of safety were used in designs, to verify that bolts already installed met design requirements (i.e., the correct size and type of bolt were correctly installed) and that bolts were torqued to their proper value.

For Millstone Units 1 and 2 verification of bolts to meet design requirements was very extensive and costly. Inspection and testing programs were established to verify thousands of bolts for satisfactory installation. This program is still in progress and final costs are not yet known.

NRC Bulletin 79-02 also had a profound effect on the Millstone Unit 3 installation procedures which will now be discussed.

Effect of NRC Concerns on Project Specifications

The effect of NRC Bulletin 79-02 was felt immediately on the construction site. No seismic Catergory 1 bolts had been installed to date. The first issue of the project specification for drilled-in anchor installation reflected the NRC's directives. The intent was to provide enough controls on installation to easily verify by documentation that bolts met the design requirement of size, embedment depth and torque.

However, the initial draft of the project specification was very conservative. Installation and documentation controls were overly restrictive and unnecessarily costly in the eyes of the utility. Of concern, in particular, was the A/E's concern that rebar could be damaged if struck with carbide bits when drilling holes for concrete anchors. The original specifications required that any incident of striking rebar be reported to the lead structural engineer for inspection and evaluation. The utility maintained that rebar would not be damaged sufficiently to impair its strength. If the specification remained "as is", additional lost time would result while awaiting engineering evaluation of the hit rebar.

In a practical sense, this requirement would be nearly impossible to carry out. The utility felt that no one could possibly measure a defect on a rebar drill hole, 4" deep. In addition, how could anyone look down that same hole and make an engineering judgment as to the structural integrity of the rebar?

Although the utility strongly felt that rebar would not be damaged when struck by carbide drills, there was no documented evidence that this was true.

Northeast Utilities decided to run a well documented and carefully controlled testing program to prove their point.

TESTING OF REBAR DRILLED WITH CARBIDE BITS

This section of the paper will discuss and describe the testing program used to conclude that reinforcing steel is not sufficiently damaged to impair its strength when drilled with carbide bits. As stated previously, the reason for doing this was to help produce a concrete anchor installation specification that would provide enough controls to assure proper installation and yet not be overly restrictive and unnecessarily costly.

There was also another area of concern where it was felt that this testing program could help cut down on construction costs; that is, what was considered an acceptable defect in reinforcing steel when hit wil chipping hammers. Occasionally, after stripping of forms, honeycombed areas of concrete are found. Some are large enough to require chipping back to sound concrete, which in some cases is past the first layer of rebar. It is nearly impossible to avoid striking the bars with pneumatic chipping guns. The original reinforcing steel specification for the project provided strict guidelines as to rebar defect/gouge acceptance. In general, bar sizes of N11 and under would have to be cut out and respliced by lapping. Acceptable defects on N18 and N14 bars were only 1/8" diameter by 1/16" deep gouges. Anything larger would have to be ground smooth and the depression filled with weld or the bar cut out and a new piece cad-welded back in. Again the utility felt these restrictions may be too conservative and hoped the testing program would provide enough evidence to loosen these costly guidelines.

Upon doing a little research, it was found that both the ACI code and the ASTM A-615 for reinforcing steel do not address acceptable/unacceptable defects in rebar. The NRC regulatory guides also do not address this area. The reason presumably is that not enough, if any, actual data is available. Theoretical strength loss calculations can be made, but actual field testing provides a much better proof of acceptance criteria for rebar.

Test Procedure

In order to report test results of this program that would be accepted by the utilities' A/E, a formal procedure was developed, reviewed and approved by construction and quality control personnel at the construction site. The utility took the responsibility to develop the test procedure, coordinate the program and write the final report. The A/E's quality control personnel would be responsible for witnessing all tests and measuring all rebar defects. They would also tensile test each drilled specimen and be responsible for the entire documentation of their program.

The approach taken was to use various sizes of carbide drills and reinforcing steel in different combinations. The intent was to try and damage the rebar by drilling and simulate the worst field conditions.

A test jig was fabricated to accept various size rebar from N5 to N18. It was designed to adequately support each rebar without allowing movement during drilling operations to simulate rebar embedded in concrete. Guide tubes slightly larger than each drill bit were an integral part of the test jig. Each tube was approximately 3" long and simulated a hole drilled in concrete prior to striking a rebar.

Drill bit sizes of 3/8", 1/2", 3/4" and 1" were used. All were new Hilti carbide bits, were regularly inspected for wear and tear and replaced on an average of every six samples. Drills were rotary hammer type as used in normal concrete drilling.

Five bar sizes used in the test were No.'s 5, 8, 11, 14 and 18.

Three specimens, each 3' long, of each bar size was placed in the jig and drilled with a carbide bit. This was repeated for each drill bit specified above. The drilled defects were positioned in the body of the reinforcing steel, not on a longitudinal rib or deformation. A control specimen of each bar size was taken from the same bar as the other specimens. The control bars were not drilled but tensile tested for comparison purposes and baseline data.

The driller was instructed to make every effort to drill through each bar. Drilling was done in a downward, vertical postion with the operator applying continuous pressure for not less than 15 seconds. Any other position would detract from the total pressure applied to the reinforcement.

Each specimen was marked and the drilled defect measured for diameter and depth by the quality control inspector. Each specimen was carefully identified so that it could be traceable to all test data. All bars were then tensile tested in a Forney LT1000 Testing Machine. The test results as well as elongation and location of bar break was documented.

Test Results by Bar Sizes

No. 5.

Reduction in ultimate strength due to drilling ranged from 3% to 6%. Each bar broke at the drilled depression when tensile tested as would be expected. It should be noted that even with loss of 6% in ultimate strength, the bar strength exceeded specification requirements due to a "cushion" built into this particular rebar heat.

The defects were typically conical in shape. The deepest penetration into the body of the bar was approxmiately 1/10". The diameter of the defects ranged from 3/8" to 1/2".

No. 8.

Reduction in ultimate strength due to drilling ranged from 0% to 2%. All bars but two broke at the drilled defect but for some reason still had a 2% strength reduction. It should be noted that these two bars had defects with the lowest penetrations into the bar, a fact that may have affected the break location.

The defects again were conical in shape. The deepest penetration into the body was approximately 1/10", while the defect diameter ranged from 1/3" to 1/2".

No. 11.

Reduction in ultimate strength due to drilling ranged from 0% to 2%. All bars except one broke at the drilled defect. The remaining one broke approximately nine inches from the defect but still had a 1% strength loss. The defect size of this bar gives no clues as to why the bar broke in this manner.

All bars broke above the specified ultimate strength. However, 3 bars broke 260 PSI lower than the required yield strength. This was due to the relatively small "cushion" built into this heat.

The deepest penetration into the body of the bar was approximately 1/8", while the defect diameters ranged from 7/16" to 9/16".

No. 14.

Reduction in ultimate strength due to drilling ranged from 1% to 2%. Five of the 12 bars broke at least seven inches from the defect. As was the case with the N11 bars, the defect sizes gave no clues as to why the bars broke in this manner.

All bars broke above specified yield and ultimate strengths due to the large 37% "cushion" of this heat together with the minor reduction in strength.

The deepest penetration into the body of the bar was approximately 1/10", while the defect diamters ranged 7/16" to 9/16".

No. 18.

Reduction in ultimate strength due to drilling ranged from 0% to 1%. Seven of the 12 bars showed no reduction in ultimate strength with three bars breaking away from the defect.

All bars broke above the specified yield and ultimate strengths. The control bar had an ultimate strength 25% greater than was required.

The deepest penetration into the body of the bar was approximately 1/8", while the defect diameters ranged from 7/16" to 9/16".

Conclusions/Recommendations of Testing Program

It is felt that this testing program has conclusively proved that drilling reinforcing steel with carbide bits will not result in significant strength reduction of the bars.

It was found that the reduction in ultimate tensile strength of N8, N11, N14 and N18 rebar ranged from 0% to 2%. However, it was also found that at least one bar of each of these sizes broke away from the defect and still had strength reduction of 1% to 2%. This indicates that the defects had little or no effect of the rebar strength. It also points out that in actuality, the strength of any rebar varies at least 2% along the same bar.

It was also found that the reduction in ultimate tensile strength of N5 rebar ranged from 3% to 6%. These minor strength reductions are relatively insignificant when one considers the area where N5 rebar is used in nuclear construction. In general N5 rebar is used in narrow slabs and walls in non-safety related structures. Many times it is used for temperature steel or for shear bars and a 6% reduction in strength should not be a problem.

In any case, it should be noted that ALL 60 BARS THAT WERE DRILLED STILL EXCEEDED THE REQUIRED ULTIMATE TENSILE STRENGTH -- This is a fact that cannot be over-emphasized.

There are also many other considerations to take into account which add credence to the fact that the strength reductions are insignificants. For instance:

1. There appears to be a built-in "cushion" or factor of safety in the manufacturing of reinforcing steel. Rebar normally has an ultimate strength at least 5% higher than that specified for the Millstone 3 project. This is most likely the case for other projects as well.

2. When designing reinforced concrete, the full capacity of rebar is never completely developed. In design, the engineer picks a rebar size and spacing that will give him more than the required percentage of steel per foot of concrete.

3. Formulas used in reinforced concrete design have conservatisms and factors of safety built into them.

4. During the test the operator drilled for 15 seconds into the reinforcing steel. This is a very extreme test as normally in construction a worker can tell almost immediately that rebar has been struck. Thus, in actual construction the defects will be much smaller.

5. New Hilti carbide bits were used when drilling the rebar and were changed every six bars so that sharp bits were always in use. Normal construction practice would be to use the same bits many times over. Thus the bits will not be as sharp during construction usage resulting in smaller defects.

Based on the test results and all data compiled, the utility recommended that the project specifications reflect the fact that carbide bits will not damage rebar sufficiently to require further evaluation. Any bars struck by carbide bits should be automatically considered acceptable and no work stoppage or reporting to the engineers should be required.

In addition, it was recommended that the accept/reject/repair criteria defined in the project specifications for reinforcing steel be relaxed. The recommendations noted above were accepted by the A/E and the project specifications amended.

Table 1 on the following page summarizes the acceptance criteria of damaged rebar from the original to the present.

The relaxation of the acceptance criteria may seem minor to some, but for each gouge that construction can "Accept As Is" without grinding smooth and rewelding, time and money is saved. This adds up over the course of the entire job.

ALTERNATIVE TO DRILLED-IN ANCHORS

Some of the experiences and problems the utility had with drilled-in anchors on previous units led to a desire to find an alternative method for Millstone Unit 3.

Early in the construction project it became apparent that the electrical and piping designs had fallen significantly behind schedule, while the civil/structural design was pretty much on schedule.

The problem centered mostly on the primary side of the plant (i.e., Auxiliary Building, Containment, ESF). The main concern was that many of the piping and electrical support plates would not be designed (i.e., sized and located on drawings) by the time the field would be ready to place concrete.

In order to resolved this problem, the A/E recommended two alternatives to the utility:

Table 1. Accept/Reject/Repair Criteria on Millstone III Project

Bar Size	Original Criteria 7/78	Utility Recommendation 4/79	Present Criteria 8/80
N14, N18	1) Defects up to 1/8" dia. x 1/16" deep are acceptable. 2) Defects between 1/16"-3/8" dia. x 1/8"-1/4" deep shall be ground smooth and rewelded. 3) Defects larger than 3/8" dia. x 1/4" deep are rejected.	1) 1/2" dia. x 1/8" deep defects should be acceptable.	1) Defects $\leq$ 3/16" deep - acpt. 2) Defects $>$ 3/16" deep, 1/4 the dia. of bar - grind and re weld. 3) Defects $>$ 1/4 of bar dia. rejected.
N10, N11	1) Not addressed in original specification - bar was usually cut out and/or resplicced by lapping.	1) 1/2" dia. x 1/8" deep defects should be acceptable.	1) Defects $\leq$ 3/16" deep - acpt. 2) Defects $>$ 3/16" deep $\leq$ 1/4 the dia. of bar - grind and re weld. 3) Defects $>$ 1/4 of bar dia. rejected.
N6 - N9	1) Not addressed in original specification - bar was usually cut out and/or resplicced by lapping.	1) 1/2" dia. x 1/8" deep defects should be acceptable.	1) Defects $\leq$ 1/8" deep - acpt. 2) Defects $>$ 1/8" deep $\leq$ 1/4 of bar diameter - grind and re weld. 3) Defects $>$ 1/4 of bar dia. should be rejected.

1. Prior to placing concrete, install long strips of embedded plates on a predetermined pattern.

2. Prior to placing concrete, install Richmond screw-type inserts, also on a predetermined pattern.

After reviewing the pros and cons of each alternative, the utility instructed the A/E to have the Richmond inserts installed on all walls and ceilings that did not have definitive piping and electrical support designs. The primary reason for choosing the Richmond inserts was the cost of installation was much lower, and they could be installed to cover a larger area of wall space.

Use on Millstone Unit 3

The Richmond Inserts used at Millstone Unit 3 are 1" diameter, type EC-2W threaded inserts. Each has a flat galvanized anchor base with nail holed to attach the insert to the formwork. The threaded ferrules are tapped with standard NC thread. The 5½" long Richmond Insert has a shear capacity of 8,000 pounds and a tension capacity of 8,270 pounds. These are working loads with a factor of safety of three. This working capacity is comparable to a drilled-in anchor of the same size.

The Richmond Inserts are placed on approximate 6" centers over all available wall areas by nailing or bolting them to the formwork. On Millstone Unit 3, it is estimated that 400,000 inserts will be installed. It is projected that approximately 45% of these will be used for permanent plant supports.

Although the quantity to be installed is enormous, the installation cost is very low. The experienced manhour rate is approximately 0.17 manhours/insert or a total installation cost of $1,020,000.

One can readily see that the direct installation costs of the inserts are much lower than drilled-in anchors. Table 2 on the following page shows a cost comparasion based on this utility's experiences.

The advantages of the inserts are obvious. After the forms are stripped from the walls, the closely spaced pattern of inserts allows construction to position surface mounted plates for picking up the required number of bolts to carry the design load. No drilling of concrete is required; thus there are no problems of rebar interference. These inserts are also used to a great extent as temporary supports during construction, thereby cutting costs in another way. Furthermore, the inserts that are not used during construction will always be available for future backfit and betterment work. Experience has shown that new codes and regulations will always require changes or additions to existing piping and electrical systems. New piping/electrical supports go hand-in-hand with these changes. At Millstone Unit 3 a great deal of future supports will be able to utilize the existing Richmond Inserts, thereby cutting down on backfitting costs.

The Learning Experience with Concrete Inserts

As it is with all new ideas, there was a "learning curve" that had

Table 2. Cost Comparison: Drilled-In Anchors vs. Concrete Inserts*

	Drilled-In Anchors	Concrete Inserts
Material Cost	248,000 x $2.00 = $496,000	400,000 ea @ $2.50 = $1,000,000 plus 68,000 ea @ $2.00 = $136,000 (drilled-in)
Labor	248,000 x 1.5 mhr/bolt x $16.67/hr = $6,200,000	400,000 x .17 mhr/ea x $15/hr = $1,020,000 plus 68,000 x 1.5 mhr/bolt x $16.67/hr = $1,701,000
Engineering Cost for Re-evaluation due to Rebar Interferences.	$500 x 6,200 ea = $3,100,000	$500 x 1,700 ea = $850,000
Est. for Future Insert problems		$500,000
Total Projected Cost	$9,796,000	$5,207,000

*The two comparisons represent estimated costs if the utility used strictly drilled-in anchors vs. the actual plan of using mostly concrete inserts with some drilled-in anchors.

to be dealt with. The use of Richmond Inserts required some experimentation and created some new problems. First, there was a problem of finding the best way to fasten the insert washer securely to the formwork. Initially, the insert was held by two nails driven through the washer holes into the formwork. This allowed for very quick installation, but the insert was not very rigid during concrete placements. If it were struck by a vibrator, one or both nails could be loosened and the insert pulled away from the forms. The insert would set in the concrete at a severe angle with the wall face and thus could not be used for any support plates. The solution to this was to use small bolts instead of nails. It takes a little longer to install the inserts but they are much more rigid and secure during concrete placements.

Another problem was that the insert washer and threads corroded very quickly when exposed to the outdoor environment. In addition, during concrete placement the cement paste would find its way into the threaded ferrule and harden. Each insert had to be cleaned by wire brushing and a threaded plastic-type plug inserted for corrosion protection. The solution to this problem was to buy the inserts with a factory-installed plug and thread lubricant. Various plug materials were tested under Design Basis Accident (DBA) conditions until an acceptable product was found. This has worked out very well. The additional cost of the plug is only half the cost of cleaning the inserts.

Painting of the inserts posed yet another problem. The original painting specification called for two different painting systems for steel and concrete. This would mean the inserts would have to be painted separately from the concrete, thus driving the painting costs up. Because the inserts were in the containment structure, the paint system to be used had to undergo DBA testing. The inserts were painted with the concrete paint system and tested. This painting scheme passed DBA testing and now the concrete and insert washers with plugs are all painted in one system.

A final problem remains to be solved at Millstone concerning the use of Richmond Inserts. When surface mounted plates are located on a wall as shown on the project drawings, usually not enough inserts can be picked up under the area of the plate. It became evident very quickly that engineering had to provide guidelines to expedite installation of surface mounted plates using as-built Richmond Insert patterns.

A generic procedure has been established by the A/E to cover most plate sizes. It has been used on site only a short period of time and has been found to be somewhat cumbersome. The guidelines appear to be too restrictive as far as relocating plates to pick enough inserts to carry the design loads. In order for this procedure to work, more flexibility has to be allowed. This problem is presently being addressed and it is felt that it will be worked out shortly. This is just another phase of the learning curve and patience is important.

CONCLUSION

It is hoped that the experiences at Millstone with concrete anchors will help others in future projects and that this paper has given the reader a better insight into the advantages/disadvantages of both drilled-in anchors and concrete inserts. In closing, it is felt that three key points presented in this paper be summarized:

1. Drilled-in anchors should be used only where it is absolutely necessary. In nuclear power plants the installation costs are not cost effective.
2. Concrete inserts provide a more economical alternative for supporting surface mounted plates. Their advantages outweigh the disadvantages especially when the alternative is to use drilled-in anchors.
3. Reinforcing steel cannot be damaged sufficiently to impair its strength when drilled with carbide bits. The size of an acceptable defect whether done be drilling or chipping hammers was shown to be relatively large.

It is realized that other companies may have had completely different experiences with concrete anchors. Keep in mind that the subject matter contained in this paper is based solely on the experiences of one company. If there are other experiences pertaining to this topic, it would be beneficial to all in the construction industry for these experiences to be shared.

REFERENCES

1. ACI Manual of Concrete Practice, 1980 edition.
2. ASTM-A615. Deformed and Plain Billet Steel Bars for Concrete Reinforcement.
3. Hilti. Anchor and Fastener Design Manual.
4. Nuclear Regulatory Commission. IE Bulletin No. 79-02, dated March 8, 1979.
5. Millstone Unit 3, Project Specifications for Drilled-In Anchors and Reinforcing Steel.

SESSION VI - CONCRETE BATCH PLANT PRODUCTION, TRANSPORTATION AND FIELD ADJUSTMENT PRACTICES

SESSION OBJECTIVES/SESSION CHAIRMAN SUMMARY

by

C. B. Tatum[1] and John M. Fisher, Jr.[2]

Objective of Session

Experiences related to the compliance with codes, standards and regulatory requirements in the production, transportation and field adjustment of concrete on power plant projects are presented in this session. Areas where there is a pressing need for improvement are identified and proposed changes which can be made are noted. In both the presentations and recommendations, achieving a balance between code and standard requirements and project cost and schedule is considered to be an important consideration.

Session Chairman Summary

A description of the impact of codes and standards on the selection and operation of concrete batching equipment was presented by C. B. Tatum and A. P. Demers. B. C. Bennett described general factors to be considered in the referencing of codes and standards in project specifications and experience with the implementation of code and standard requirements regarding field adjustment for pump line effect, determination of adverse weather conditions, and testing of dry pack grout. W. H. Brown discussed factors to be considered in the selection and usage of concrete pumping equipment. The lack of code and standard requirements in this area was highlighted. T. L. Moore related means of determining the intent of codes and standards and the need for increased flexibility in their application. R. M. Esbach reviewed design engineering involvement in the field implementation of design requirements, codes and standards and the necessity for application of individual judgement.

Consensus Recommendations*

Following the paper presentations, the speakers responded to several questions from the session attendees regarding paper topics. The subjects addressed included: verification of concrete adequacy following discovery of batching

---

[1] Construction Superintendent, Ebasco Services Incorporated, Elma, Wa. 98541

[2] Director of Projects, Gilbert Commonwealth International, Inc., Reading, Pa. 19603

*Based upon the panel of Speakers/Audience discussion period at the end of the session.

tolerance errors, means of establishing consistent correlations for pump line effect, implementation of systems for the use of judgement in the field, and means of increasing personnel qualifications.

As a result of the presentations and the panel discussion, the session participants identified the following areas as priority subjects for improvements in codes and standards applicable to concrete production and supply:

1. Placing increased emphasis on the selective referencing of codes and standards in project specifications. This step will assist field personnel in implementation of only the applicable portions necessary to satisfy the design intent and in avoidance of ambiguity and conflict. This action must be taken by individual architect/engineer firms, but could be encouraged by introductory sections in code and standard documents requesting that specific sections be cited in implementing specifications.

2. Develop, in codes, standards and project specifications, improved systems for use of judgement in field implementation. These would include code and standard language allowing field variations and the delegation of authority to field personnel for the use of judgement. This improvement will also be assisted by the development of improved commentaries to explain the full intent of code and standard requirements.

3. Development of industry standards regarding the selection and use of concrete pumping equipment. These standards would provide guidance for application of this equipment in various placing situations and criteria for equipment capacity rating.

CONCRETE PLANT AND PRODUCTION OPERATION STANDARDS

by

Clyde B. Tatum[1] and Albert P. Demers[2]

Abstract

    Supply of concrete meeting precise quality control requirements is a critical factor in power plant construction in the selection and operation of concrete batching equipment, codes, standards, and regulatory requirements must be fully considered along with production demands. Experience in mobilization of facilities and supply of concrete on a large power plant project is presented. The impacts of codes and standards on the various phases of these operations are described. The authors conclude that equipment and techniques are available to satisfy these requirements and that careful planning is essential in plant erection and maintenance of concrete supply.

Introduction

    Concrete production equipment and operations on power plant projects are governed by an extensive set of codes, standards, and regulatory requirements. When combined with the specifications for a particular project, these criteria control concrete production to a precise degree. If the potential impact of these requirements is recognized and planned for, compliance can be achieved in parallel with minimizing adverse consequences. However, if the impact of these requirements is not fully considered, severe production restraints can result.

    The selection, mobilization, qualification and operation of a concrete batching facility, in accordance with these codes, standards, and regulatory requirements, is described in this paper. The basis of this scope and the examples given are for a two unit nuclear project under construction in western Washington state since Spring 1977. The purpose of this experience description is to identify the specific impacts of these requirements, describe the problem areas, and transfer the experience gained in problem solution. With this information, constructive inputs are made for possible improvements in both the requirements of codes and standards and the actions taken for their implementation. In presenting this information, we hope that it will assist in avoiding problem areas on future projects.

---

[1] Construction Superintendent, Ebasco Services, Inc., Elma, WA
[2] Project Manager, Associated Sand & Gravel Co., Inc.

The selection of concrete batching equipment, including criteria for automation and backup facilities, is first reviewed. Experience in the erection and certification of these plants, in accordance with the standards, is then described. The operation of these facilities on a power plant project is next reviewed. In the conclusions and practical applications section, the authors identify significant lessons learned and propose both revisions to the requirements and means of implementation to avoid problem areas.

Criteria for Concrete Plants and Operations

Several industry organizations and regulatory agencies issue codes, standards and other requirements applicable to concrete production operations. The American Society for Testing and Materials (ASTM) has developed a widely used standard (1) covering materials, concrete purchase, batching plant, mixers, mixing and delivery, and inspection. For batch plant certification, the checklist (2) prepared by the National Ready Mixed Concrete Association (NRMCA) is frequently used. This list defines inspection tasks, standards of accuracy for various plant weighing and dispensing systems, and uniformity requirements for qualification. Various portions of the standards issued by the American Concrete Institute (ACI) also define requirements for concrete batching equipment and operations. Examples of topics include storage of raw materials (3), water temperature limitations (4), certification of production facilities (5), and adverse weather conditions (6, 7).

For nuclear projects, the United States Nuclear Regulatory Commission (NRC) has issued general Quality Assurance (QA) Criteria as a part of the Code of Federal Regulations (8) and specific implementing instructions in the form of regulatory guides. The American National Standards Institute (ANSI) has provided a more detailed statement of these requirements in the form of QA standards for various phases of construction (9, 10). See Table 1 for a matrix summary of the relationship of these various documents to the individual steps of concrete production.

Selection of Concrete Supply Equipment

The selection of components and capacities for concrete production requires full consideration of concrete demand, project specification requirements, unique site conditions, and the criteria imposed by codes, standards and regulations.

TABLE I.- Code And Standard Applicability To Concrete Production

CODE OR STANDARD \ CONCRETE OPERATION	BATCHING EQUIPMENT SELECTION	BATCH PLANT ERECTION AND QUALIFICATION	CONCRETE PRODUCTION
ASTM	C-94 C-33	C-94	C-94
NRMCA	PLANT STANDARDS, CHECKLIST	CHECKLISTS	
ACI	ACI-304, 305, 306	ACI-311	ACI-301
ANSI			ANSI-N45.2, N45.2.5
NRC		10CFR50, APP. B	10CFR50, APP. B

Plant Sizing and Selection Factors

Several basic criteria determine plant sizing. The most fundamental, level of site demand, is impacted by the construction schedule, site layout, means of delivery, and means of placing. The forecasted peak period of concrete placement should be evaluated to determine likely placement sizes, means of placement, potential for placing on multiple shifts, and impacts of delay to certain types of placements. In cases where the schedule or productivity for specialty operations not on the critical path, such as cooling tower construction, would be severely impacted by restrictions in concrete supply, larger concrete plant capacity may be justified. A basic capacity of 200 to 300 yd^3/hour has been used on many nuclear project sites. Single unit projects have tended to fall in the low end of this range, and two unit sites toward the high end. The specific requirement is generally selected based on experience at similar projects and consideration of unique factors at the current site.

In evaluating the actual capability of a proposed plant, two key factors must be considered. First, since the output requirement is generally for delivered concrete, the means of both transport and placing must be reviewed. Overall plant capacity, specifically mixer size, must be compatible with the planned transport and placing equipment to achieve the desired output. Second, concrete mixing times must be established to define plant output. These may be defined in the technical specifications but are frequently modified as the result of mixer performance testing.

Concrete Specification Requirements

Batching Facilities. Several requirements in both codes and standards and project specifications directly impact plant selection. Examples include: requirements for separate compartments for storage of different type cements (3), special requirements for automatic or semi-automatic operation, batching tolerances (1, 3), and mixer type. Also, the capability to meet special conditions; such as cold weather, hot weather, or massive placements may require special plant equipment.

Aggregate Processing Facilities. The performance requirements of codes, standards and technical specifications determine the basic aggregate processing plant. The materials specification for concrete aggregates (11), includes basic grading requirements for both sizes and fineness modulus requirements for fine aggregate. Special requirements for washing and rescreening, and storage volume may also be included in the project specifications. Each of these requirements impact plant selection.

Description of Example Batching Facility

It is obvious that both aggregate and batching facilities should be located as close as possible to the point of placement on large power projects. Several factors must be considered. First, standard requirements generally limit the time between mixing and discharge to 1½ hours (1, 3). With fixed time requirements for mixing, testing and preparation for discharging, transit cycle durations are therefore limited. Also, plant location must allow convenient and relatively consistent access to the points of placement.

Site conditions frequently restrict plant location. Space in the immediate proximity of the permanent plant must be assigned to such high priority use as material staging, craft change facilities, offices, and parking. Also, the location of suitable mechanical and electrical utilities must be considered in siting the concrete plant. Figure 1 indicates an example site layout and concrete plant location.

On this example site, two batch plants and support facilities are located on 5.3 acres as shown in Figure 2. The primary 10 cubic yard central mixer designed to produce 200 cubic yards per hour and adjacent 6 cubic yard (100 cubic yard per hour) backup facility are completely independent systems as described in subsequent sections of the paper.

Approximately 3 acres are allocated to aggregate storage. Utilizing a conical stockpile design, and stacking conveyor system, maximum storage is approximately 20,000 tons of 3/4" aggregate and 10,000 tons of sand. The plant includes provisions for automatic discharge of aggregate from stockpiles. Coarse aggregate then crosses a 3 deck screening unit which resizes and rewashes the material. This step was proven to be very important in meeting both gradation (12) and moisture requirements.

Cement storage has the capacity to produce 3300 cubic yards of concrete. Four 1000 barrel silos, with their output capable of being directed to either plant, surround the two batch plants.

Admix storage tanks, computers, and offices are located in a two-story environmentally controlled building. Fully automated controls are located on the second floor with unobstructed views of both plants, aggregate handling systems, and discharge of concrete into transit mixers.

A fleet of 14 transit mixers utilize a washout facility designed to separate the cementious material and aggregates. Settlement weirs and ponds were installed to collect the cement slurry and discharge chemically treated, noncontaminated water to meet the environmental requirements.

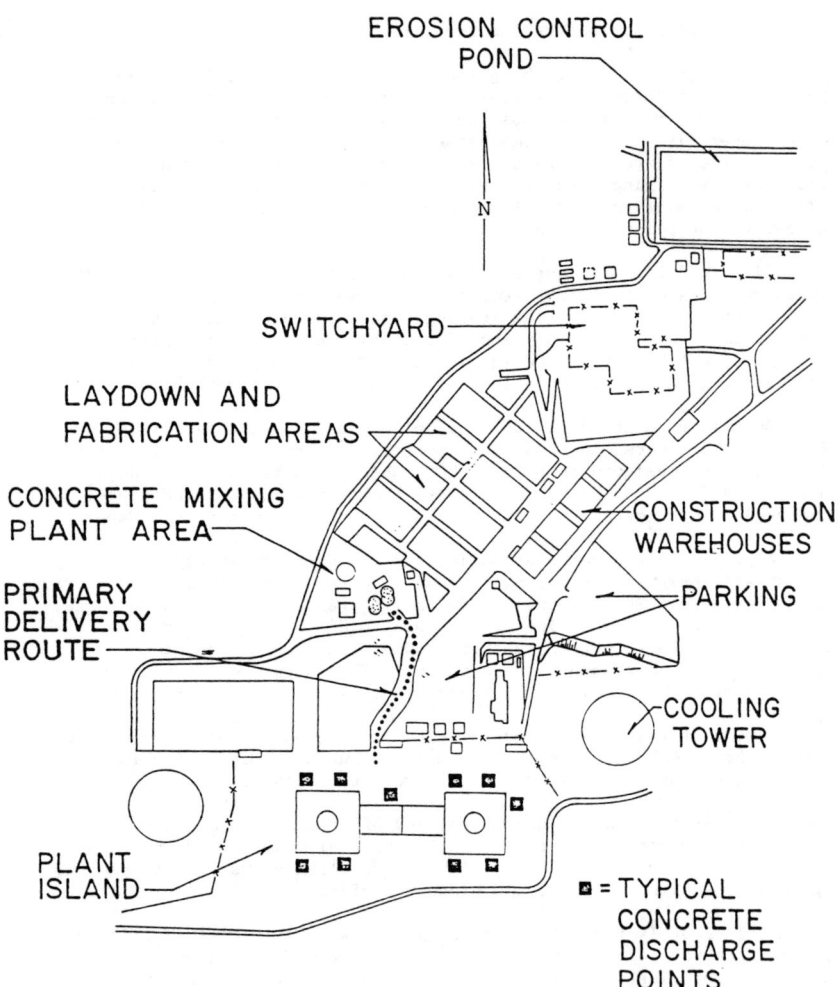

FIGURE I. - Site Layout And Concrete Plant Location

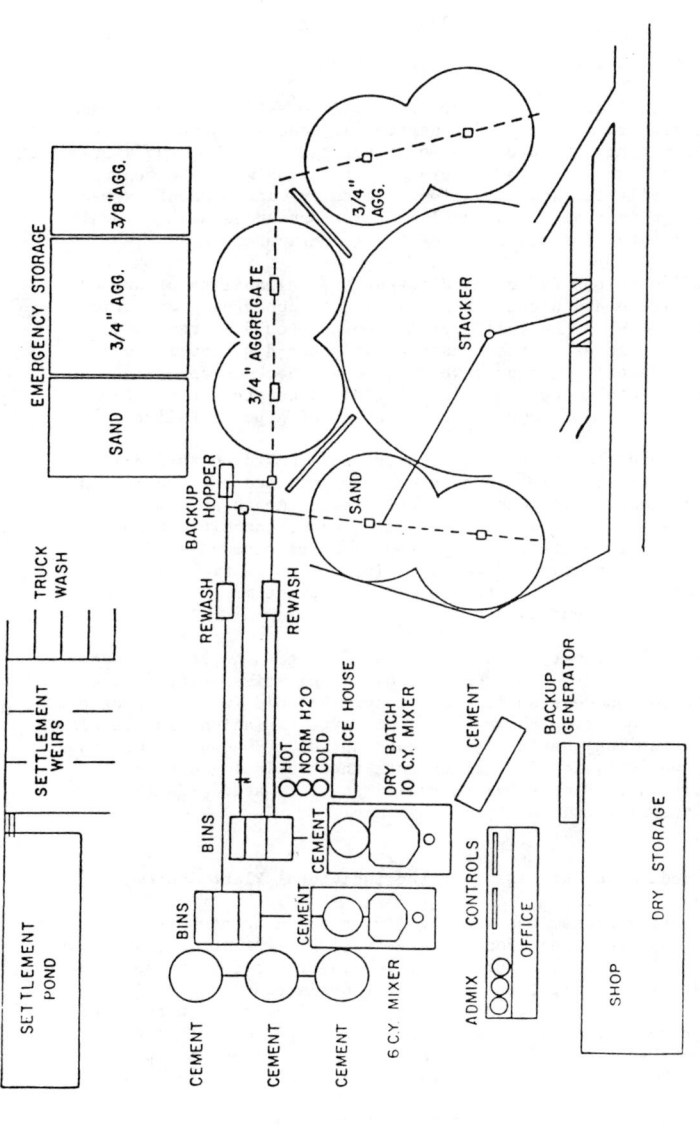

FIGURE 2. - Concrete Plant Area Layout

Plant Back-Up and Redundancy

Back-Up Batching Facilities

Many different approaches have been used to satisfy code, standard and specification requirements for dependable concrete supply. The three main approaches are provision of spare parts and back-up systems within the main plant to allow continued production when the central mixer is not available, and provision of a fully independent back-up plant. As shown in Table 2, several criteria should be used in evaluating these alternates for selection of plant configuration.

In addition to the factors indicated, the reliability of support utilities should be evaluated. Historical data concerning electric power interruptions in the basic supply source should be reviewed. Also, possible failures in the construction power distribution system should be considered. On many power projects a back-up generator has been provided to enable continuation of limited concrete production for the completion of placements in the event of a power failure.

Once an evaluation of back-up requirements based on reliability has been completed, the impacts of production mix and rate variations should be considered. If concrete supply requirements at times will exceed, or fall considerably below the main plant capacity, then a second plant may be merited. Also, if production schedules will call for simultaneous production of various mixes, such as two aggregate sizes along with soil cement, then a separate plant may be merited to satisfy these requirements.

There are, in summary, several types of factors to be considered in evaluating the need for back-up batching facilities. The implications of disruption in concrete supply, and resultant cold joints are of course paramount, but production capacity and mix type variations should also be reviewed. The many variations of back-up and redundant plants found on various power project sites illustrate the continuing difference of opinion within the industry. Detailed industry standards (11) provide additional requirements in this area.

Back-Up and Redundancy Requirements for Individual Plant Systems

In order to increase the reliability of the concrete plant, back-up equipment and numerous redundancies are available for incorporation in batch plant facilities to assure steady, uninterrupted concrete production. At the example site, the contractor designed the main plant with built-in back-up capability. After the start of concrete production operations, requirements for a complete back-up plant were added for reasons of further reliability and capacity. Experience has proven that total back-up systems are beneficial in times of extra capacity or mix diversity, and are fully reliable only if they are used on a scheduled basis to assure their state of readiness.

TABLE 2.- Concrete Plant Redundancy Alternative

BACKUP ALTERNATE / EVALUATION CRITERIA	SPARE PARTS AND REDUNDANT SYSTEMS FOR MAIN PLANT	SECOND BATCHING MODE IN MAIN PLANT	INDEPENDENT BACKUP PLANT
COST	LOWEST	MEDIUM	HIGHEST
PLANT AVAILABILITY	LOWEST	MEDIUM	HIGHEST
DIFFICULTY IN MEETING QUALITY ASSURANCE REQUIREMENTS	LOWEST	HIGHEST	MEDIUM
FLEXIBILITY FOR PRODUCTION RATE VARIATIONS	MEDIUM	LOWEST	HIGHEST
FLEXIBILITY FOR PRODUCTION TYPE VARIATIONS	LOWEST	MEDIUM	HIGHEST

Aggregate Handling. Several factors were included in the design of the aggregate handling system to increase reliability. All head and tail pulley shafts and bearings of the plant conveyors are made one size larger than normal (standard) in order to increase resistance to wear. Conveyor belts and operating speeds have been selected to supply the maximum rate of the batch plant without taxing their capabilities. Since an automated tunnel gate system is utilized, a back-up feed is incorporated to allow the utilization of emergency aggregate storage through a back-up feed hopper. Aggregate bins have two 100% batching gates; either one of which may be operated from the control console.

Cement Handling. The cement silos have two batching gates either one of which may be operated from the control console. Each silo has dual fill pipes for cement tanker use and two pumping facilities are installed. The plant silos have two back-up systems, which utilize independent transfer units, fill, and discharge pipes.

Admixture Dispensing. The system for dispensing concrete admixtures to the primary plant consists of two parallel trains, either one capable of dispensing admixtures independently of the other. The back-up batch plant has the same parallel systems, and all systems are capable of operating on three different sources of power. Admix dispensing systems are necessarily complex since they must be duplicated many times in the vicinity of the storage tanks. Electronic systems were installed to minimize the complexity of this operation.

Water Storage and Handling. To maintain an uninterrupted supply, normal water is stored at the batch plant facility in a 10,000 gallon storage tank. Two additional storage tanks are installed; one for hot water (10,000 gallons) and one for cold water (10,000 gallons). Each of the tanks has its own independent regulator and spare pump.

Concrete Batching. The 10 cubic yard primary central mixer was built with many redundant systems. The concrete mixer is dual driven so if one drive unit fails, production may be continued at approximately 50% of the normal capacity. A facility to dry batch is included with the main plant. This also enables concrete production to continue at a reduced rate, should the main mixer fail.

The main concrete plant has a second complete back-up system consisting of a 6 cubic yard central mixer with independent weigh batchers, aggregate bins, aggregate handling system, power source, and control console. In addition to being a back-up facility, the 6 cubic yard mixer has been utilized as a second plant.

Plant Power Supply. The main and back-up batch plants are supplied independent power sources. Each source has the ability to operate the entire batch plant and aggregate handling facility. In the case of a complete blackout, a 500 KVA back-up diesel generator is provided to allow continued batching at a reduced rate.

Batch Plant Control and Automation

Many concrete supply specifications for power projects require that a fully automated control system be provided. Typical requirements are batching a minimum of three aggregate materials, cement, water, ice and two admixtures. Tolerance interlocks are also required meeting ASTM C-94 weighing requirements to prevent discharge of aggregates, cement, water, and admixtures if the quantities batched are outside the tolerance limits.

Control Unit Description

For the purpose of speed and accuracy, all functions of the batching facility at the example plant were automated, creating a state of the art operation. The control consoles contain two separate functional sections operated by the same computer for aggregate handling, and batching. The operator therefore can control, from one location, all functions needed to assure the continuous supply of concrete.

The unit incorporates a core memory computer mounted in the console, and controls all basic functions of the plant. All functions related to batching are stored in a disk memory system which include mix designs, accounts, variable tolerance limits, self-correcting free-fall compensation, batch material sequences, moisture compensation for fine aggregate, multiple batching, mix water adjustment, material discharge, mixing time, admix controls, water/cement ratios and others. This system is very versatile, providing automatic, semi-automatic and manual control operations, while at the same time allowing the operator to select from a variety of preset formulas.

Since all functions of batching are automatic, the batchman is allowed to edit input, including the formula about to be batched, on a video screen prior to weighing material, and the Engineer receives computer printouts of mix designs, tolerances and water/cement ratios for each formula that may be batched. An automatic printer is provided to allow input commands as well as receiving information, such as completed delivery tickets, mix designs, accounts, and inventories.

Batching Tolerances

Requirements for batching tolerances are detailed in the project specifications and industry standards which specify strict percentages for each ingredient.

Utilizing computer controls with pre-programmed batching tolerances and a positive interlock to prevent a batching sequence from progressing if an ingredient is not weighed within tolerance, the batching operation is precisely controlled to these specification limits. Even though an ingredient is batched within tolerance limits, the automatic controls will continually attempt to cut the variance, though it is acceptable, in half for subsequent batches.

Batch Recording Requirements

Specification requirements for batch recording vary greatly from project to project. The contractor normally requires certain data in addition to that needed by the Owner or Engineer. Automatic plant controls provide an extensive capability in this area.

Information is recorded on each delivery ticket by an automatic printer tied into the plant controls. This level of information, which exceeds normal requirements, allows quick, accurate, and knowledgeable decisions concerning concrete adjustment to be made in the field.

Mixing Time

For a 10 cubic yard mixer, industry standards (1) recommend that the minimum mixing time be $3\frac{1}{4}$ minutes. This requirement may be reduced by successfully performing a mixer uniformity test.

Fifty mixing seconds, along with a good ribboning of ingredients into the central mixer has been found to be sufficient to obtain a uniform mix at the plant described. Uniformity tests are required on each central mixer every six months (10), as well as on a representative mixer truck utilizing the dry batch system and field addition of water.

Moisture Compensation

Precise moisture control is provided by use of an automatic capability of ready adjustment of the weight of water to compensate for the varying moisture content of the aggregate. The percent of moisture in the sand is dialed in by the operator. Using this setting, the computer will automatically draw the proper amount of sand and water, indicating on the delivery ticket the individual batch weights converted to Saturated Surface Dry (SSD), and the amount of water that may be added at the point of placement without exceeding the required water/cement ratio.

Water/Cement Ratio

Each mix formula entered into the computer includes the water/cement ratio selected by the engineer. The computer totals all water by weight and indicates on the delivery ticket the balance remaining in gallons, allowing the technician in the field to easily determine the maximum allowable water if needed for field adjustments. This feature also relieves the batch plant inspector of this responsibility, and eliminates human error in one of the most vital areas related to concrete production.

Batch Plant Erection and Certification

On many projects, erection and certification of the batch plant becomes the critical path item for the start of concrete placement. Durations for plant certification may exceed schedules because of difficulties in meeting the detailed certification requirements. Therefore, this activity merits special planning and prior consideration of the codes, standards and other governing requirements.

Erection and Certification Schedule

A typical schedule for erection and certification of a full concrete production facility is shown in Figure 3. The site preparation activities vary, depending on the location. It is important that schedules provide adequate time for construction of foundations and installation of necessary utilities in addition to site development. The shipping sequence for major components should be based on the preferred erection sequences. Individual sub-components should be checked out as soon as they are operational, to allow resolution of any problem areas prior to certification.

Code and Standard Requirements

The basic requirements for mixing plant certification are defined by the National Ready Mix Concrete Association (NRMCA). This checklist (1) defines all necessary actions for full plant certification. The inspections and witnessing of plant certification operations required under the NRMCA must be performed by a registered professional engineer. It is generally possible to perform portions of the certification once individual plant components are operational; thereby shortening overall the time required.

Batch Plant Erection Experience

Experience in erecting a 200 cubic yard/per hour concrete batching facility is discussed below to give a complete example of the activities and durations required and the potential problem areas. The fully automated plant is described elsewhere in this paper.

Site Development

Site development for the 5.3 acre site for the facility consisted of excavation, back-fill and grading to provide the required surface area and an adequate foundation for both the main plant components and the aggregate handling tunnels. Soil borings were required at specific locations to verify adequacy. For this project site development required approximately 45 days to complete. Upon completion of grading, a full surface topping of 3" Asphalt Treated Base (ATB) was applied to the entire area for stabilization and sealing.

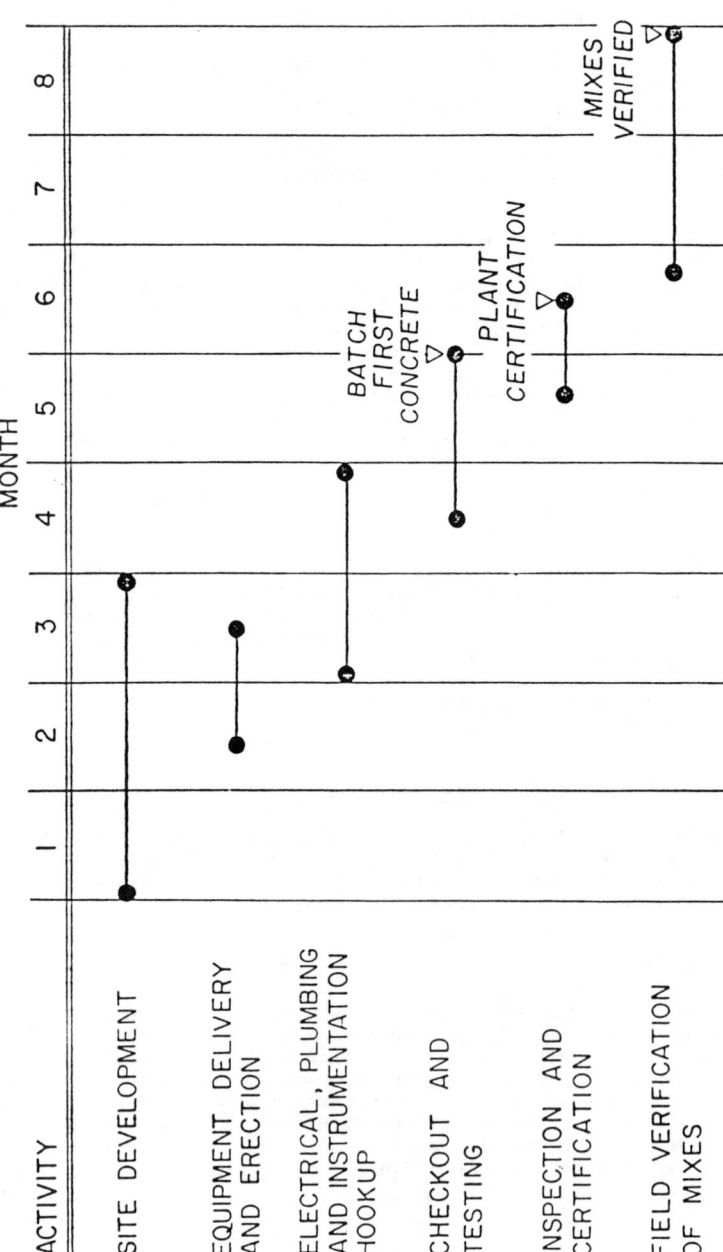

FIGURE 3.-Schedule For Batch Plant Erection And Qualification

Batch Plant Erection

Approximately 20 days were required from the start of deliveries to the completion of setup on the foundations of plant components such as cement silos, the main plant mixer, stacker and conveyor components, and aggregate bins. Electrical, plumbing, and instrumentation hook-up of the various components required approximately 60 days to complete. Checkout and testing of plant components, including isolation and resolution of miscellaneous system problems, required approximately 30 days. Uniformity testing of the mixer system was initiated during this period. Field verification of the concrete mix was completed approximately 90 days following initial concrete batching.

Batch Plant Certification

Completion of the inspection and testing operation necessary for final batch plant certification in accordance with NRMCA requirements proceeded in parallel with testing and mix verification.

Initial inspections of calibrations were initiated during component checkout and full sign-off of all necessary inspections and calibrations was completed within five days. Mixer uniformity testing in accordance with ASTM C-94, appendix 11, was completed in one day.

Other items associated with batch plant certification, including checkout and verification of control systems, required two days. Final NRMCA sign-off was obtained from the testing laboratory serving as an independent inspection agency within four weeks of starting initial inspections. No problems were encountered during the initial NRMCA list check-off and plant systems met or exceeded certification requirements. Subsequent checklist certifications have also been completed without difficulty.

Concrete Production Operations

Requirements for Personnel Training

Quality control personnel must pass an annual physical exam, participate in periodic training programs, and pass a bi-annual performance evaluation. If it is determined that an individual is not qualified to perform a given activity, he must be prevented from performing that activity until such time as he has been trained in the needed skills and has been reevaluated and recertified. Batch plant operators must pass a bi-annual evaluation to demonstrate familiarity with batching tolerances, mixing times, slump, air, and temperature requirements in the field. Each transit mixer driver is issued a manual to familiarize them with equipment and quality control requirements.

Batching Process

After the batch plant operator has dialed a mix number, truck number, account number, quantity, aggregate-free moisture percentages, hot or cold water percentages, and the number of gallons of water per cubic yard he wishes to hold back from the programmed mix design, he initiates the batching sequence. These manual adjustments, made by thumbwheel, can pace the production time frame when the plant is simultaneously supplying to several contractors. An alternate approach to console design would be the use of dials on many items.

After all ingredients are weighed within tolerance, the control system allows the weigh hoppers and admixture dispensers to discharge. The timing of the discharge of each individual material is critical as a ribboning effect is required to attain a uniform mix. The actual seconds of delay, which are programmed into the control console, are a function of the distance from the central mixer and the weighing and measuring of each material.

Delivery

It is important to clearly identify all mixer trucks to eliminate confusion during rapid batching. Truck numbers should be placed at both ends and sides of the vehicles. A communication system should be utilized to minimize the turn-around time of a truck. Also, a delivery ticket pickup should be isolated from the batching area if possible, as this step may become a bottleneck. Air tubes have proven to be an extremely efficient means of providing for ticket pickup.

Impacts of Multiple Concrete Conveyance Equipment

One of the tasks of a concrete supplier is to alter the slump, air, and/or temperature adjustment of concrete to be placed utilizing a pump. Therefore, close coordination between the batching and placing operations is essential for delivery of concrete properly batched to support the placing operation.

Plant Daily Routines

A plant inspection is performed by the plant maintenance personnel, one hour prior to batching. Material handling systems, gates, and scales are checked by the batchman, in accordance with a "State of Readiness" checklist.

Aggregate bins are drained daily, and refilled 30 minutes prior to batching, allowing quality control technicians time to obtain aggregate moisture data. Gradations are taken daily when aggregates are hauled to the batching facilities to assure specification compliance. Batch plant quality control technicians are continuously in the field to observe placements and concrete consistancy. These personnel advise the batchman of requirements for minor mix adjustments.

Conclusions

Codes, standards and regulatory requirements impact concrete supply equipment and operations in several diverse ways. The industry has become familiar with implementing many standards, such as ACI and NRMCA. However, newer requirements, such as ANSI, have required further industry attention.

Plant design, erection and certification, and operation are each governed by various elements of industry codes and standards. Several plants now available, including fully automated units, can meet standard requirements. Experience on the project described in this paper indicates that plant erection/certification schedules and reliable plant operation can be achieved based on careful planning. However, this planning must reflect a full consideration of potential impacts by codes and standards.

Two elements of this plant operation experience indicate a potential need for re-evaluation of technical requirements. First, test frequencies for mixer uniformity appear to be greater (every six months) than necessary for modern plants. Also, the requirement for air-entrainment as a part of the concrete forming a portion of interior structures should be evaluated. This would involve limiting the use of air-entrained concrete to those portions of structures where it is essential. This step would be very beneficial to field operations because of the difficulty in avoiding slight variations in air content. Under the acceptance criteria generally used, these variations can cause concrete rejection and create restraints to placing operations.

Appendix I - References

1. American Society for Testing and Materials, Standard Specification for Ready-Mixed Concrete ANSI/ASTM C-94-78a, 1971 Annual Book of ASTM Standards, Part 14, Concrete and Mineral Aggregates, Philadelphia, PA, 1979, pp.: 54-66.

2. National Ready Mix Concrete Association, Certification of Ready-Mixed Concrete Production Facilities, Second Revision, Silver Spring, MD, January 1, 1972.

3. American Concrete Institute, Recommended Practice for Measuring, Mixing, Transporting, and Placing Concrete (ACI-304-73).

4. American Concrete Institute, Specifications for Structural Concrete for Buildings, (ACI-301-72).

5. American Concrete Institute, Recommended Practices for Concrete Inspection (ACI-311-75).

6. American Concrete Institute, Recommended Practice for Hot Weather Concreting (ACI-305-72).

7. American Concrete Institute, Recommended Practice for Cold Weather Concreting (ACI-306-66).

8. Code of Federal Regulations, Title 10, Part 50, Appendix B, (10CFR50, App. B), Quality Assurance Criteria for Nuclear Power Plants.

9. American Society of Mechanical Engineers, Quality Assurance Program Requirements for Nuclear Power Plants, ANSI N45.2-, New York, 1972.

10. American Society of Mechanical Engineers, Supplementary Quality Assurance Requirements for Installation, Inspection and Testing of Structural Concrete and Structural Steel During the Construction Phase of Nuclear Power Plants (ANSI N45.2.5 - 1974), New York, 1974.

11. Concrete Plant Standards of the Concrete Plant Manufacturer's Bureau, Sixth Revision, Concrete Plant Manufacturer's Bureau, Silver Springs, Maryland, 1977.

12. American Society for Testing and Materials, Standard Specification for Concrete Aggregates, ASTM (33-78).

STANDARDS IMPACT ON MIX
VERIFICATION/ADJUSTMENT

By Bruce C. Bennett,[1] Robert V. Potter,[2] and John R. McCutchen[3]

Abstract

Industry codes, standards, guides, and project specification criteria establish stringent requirements for concrete mix verification and field adjustment. This paper describes actual experience with the implementation of these requirements in the following areas: field adjustment of concrete consistency, field adjustment for pumpline effect, determination of adverse weather concreting conditions, and field adjustment practices for dry-pack grout. In each of these areas, the authors describe techniques used to satisfy requirements, problem areas encountered, and solutions developed. Conclusions regarding industry actions to simplify requirements in each of these areas are also presented.

Introduction

The Nuclear Industry uses and refers to codes and standards which are generally intended for commercial grade or industrial-type construction. In some instances, these guidelines do not sufficiently meet the demands of Nuclear related construction. Documentation and Quality Assurance requirements are generally less stringent for commercial grade or industrial type construction and their guides are not necessarily used in their entirety. Nuclear specifications need to clearly define which guidelines and practices are applicable in order to ensure clarity of intent, inspection criteria, and accept/reject criteria. Care exercised in supplementing the Nuclear Industry codes with this information may have a substantial impact on project cost, schedule, and quality.

In preparation of specifications for nuclear construction, there may be a tendency for restriction beyond the intent of the applicable standards, codes, and guides. This occurs because: 1.) The codes may be broad in their intent and the Nuclear Industries, in their attempt to clarify, may have gone beyond the intent of the code; 2.) The code may not be applicable in its entirety or the code may not directly address the subject matter involved.

The Nuclear Industry will then prepare its own specification to assure that it is conservative enough to meet licensing and NRC requirements. These self-imposed restrictions may set a pattern which is often difficult to change and may be questioned for their overly restrictive

---
[1] Resident Engineer, Ebasco Services Incorporated
[2] Assistant Construction Superintendent, Ebasco Services Incorporated
[3] Lead Construction Engineer - Civil, Ebasco Services Incorporated

criteria.

The application of codes, standards, and guides is approaching a level of complexity which impacts construction progress and cost effectiveness. Some projects have undertaken a program to review specifications, and procedures in an effort to ensure that there are no excessive requirements in the contract document.

The objective of this paper is to illustrate how specifications for mix verification and field adjustments have been impacted by the application of industry codes, standards, and guides. Actual construction practices and experiences will then be discussed covering the following topics:

1. Field Mix Verification and Adjustments
2. Field Adjustment of Concrete Consistency
3. Field Adjustment for Pumpline Effect
4. Determination of Adverse Weather Concreting Conditions
5. Field Adjustment Practices for Dry Pack (Grout)

Field Mix Verification and Adjustments

Industry codes, standards, and guides are evaluated in the development of concrete design mixtures. (1) In many instances, the designs have been completed and qualified in a main laboratory, prior to being submitted to the project for field use.

Codes, standards, guides, and specifications require production of various quantities for field verification of concrete mixtures (2,3). After the master mix designs have been verified and produced in the required quantities, it may become necessary to vary or adjust the master designs for special placing problems or types of equipment: i.e. tremie concrete, pumping, restricted access and/or placing clearances, yield and strength.

Field adjustments for (sand, cement, and slump) proportions should be evaluated by the Engineer and accepted when the evaluation indicates satisfactory results will be obtained. Re-design and re-qualification in the laboratory, and an intensive field verification, are generally not necessary for adequate control. It is often necessary to adjust proportions for yield, w/c ratios, or sand content once a design mixture is being produced from the site concrete facility. This would normally require reverification of the new mix proportion in the field. An abbreviated verification and test program should be an acceptable method. Once the master mix is verified, any further adjustments for field application should not require additional reverification. At the current cost of concrete production, this could have several cost and schedule advantages.

On some projects, it has been necessary to develop additional mix designs in order to document adherence to specification requirements while providing the necessary flexibility to adjust the proportions. A more rapid and wider range of mix adjustment can be accomplished in the field by utilizing this practice.

Field Adjustment of Concrete Consistency

The field addition of mixing water to concrete is an often discussed topic. As indicated in reference (4), it has been addressed in almost as many ways as there are projects. Briefly, we wish to bring the subject up once more with a review of code and standard applicability.

ASTM C-94 (5) states in part, ". . .When a truck mixer or agitator is approved for mixing or delivery of concrete, no water from the truck water system or elsewhere shall be added after the initial introduction of the mixing water for the batch, except when on arrival at the job site, the slump of the concrete is less than that specified. Such additional water to bring the slump within required limits shall be injected into the mixer under such pressure and direction of flow that the requirements for uniformity specified in Appendix XI are met. The drum or blades shall be turned an additional 30 revolutions or more if necessary, at mixing speed until the uniformity of the concrete is within these limits. Water shall not be added to the batch at any later time."

Some projects deem it advantageous to allow addition of mixing water at the point of truck discharge in order to achieve the workability required by placement conditions. Slump loss is a problem experienced on many projects. There are many causes for the loss: scheduling, delays, travel time, distance, false set, premature set, etc. All or any of these items can result in added costs due to lost time, waste, or both. On one project, the addition of water was allowed and controlled as follows:

> Prior to concrete discharge into the receiving hopper or bucket, the load was visually inspected jointly by the Contractor and Engineer's inspector. The Contractor determined the quantity of water desired and the inspector determined whether the quantity was within the maximum allowed w/c ratio, documented the quantity, and verified the mixer revolutions. In the event that an inadequate amount of water was added, another attempt was allowed at that time, provided the maximum w/c ratio was not exceeded.
>
> Water adjustments were only allowed to full batches, in order to document the water cement ratios. Retempering was addressed in association with this practice, and a 30-minute time limit for the addition of water was imposed. This time frame was chosen to satisfy the following concerns: ASTM C-94, states that the "producer shall not be held responsible for the limitation of minimum slump after total waiting period of 30 minutes, at agitating speed or agitating and discharge, and the user shall assume full responsibility for the condition of the concrete thereafter".
>
> It was felt that this 30-minute time frame gave the producer the ability to control the workability for the entire time of his responsibility. Whenever the placing contractor accepted the load within the 30-minute time frame, he then accepted the responsibility. The producer only exercised the option whenever the concrete arrived, at the point of

discharge, at a slump below the minimum specification value. However, the placing contractor more often chose to work closer to the maximum value.

The project was accused of retempering by a placing contractor who wanted to claim lost time due to the addition of water. It was pointed out that he would benefit in that the workability time frame would be extended and a much lesser impact would result from delays. The 30-minute time frame was then applied, thereby placing a new but workable limit, well within requirements of applicable codes, standards, and guides. ACI Publication SP-19 (6) defines retempering as the addition of water and remixing of concrete or mortar which has started to stiffen. ASTM C-94 does not specifically place a time limit on addition of water, but allows addition upon arrival at the job site.

As demonstrated here, one standard (ASTM C-94) allowed addition of water at the site to increase the slump if required, but also it was restricted by the ACI code that addressed retempering with a good but judgemental definition of retempering. The industry must, in this instance, supplement the codes and standards to ensure that both commercial and quality considerations are adequately, but not overly, addressed.

Field Adjustment for Pumpline Effect

Concrete pumps have been used successfully for many years in the placement of concrete. Not only are concrete pumps utilized for difficult placements, i.e.: tunnel linings, congested building locations, etc., but also for construction of typical slabs on grade and other relatively simple placements. Concrete standards employed on most nuclear power plant projects have impacted the smooth conveyance of pumped concrete to the final point of discharge by requiring unreasonable or excessive testing of plastic concrete properties.

Nuclear plants constructed to the requirements of ANSI N45.2.5 - 1974 (7), are required to obtain in-process samples of concrete from the discharge of the last piece of conveying equipment. Therefore, concrete must be sampled from the pumpline discharge unless a deviation is allowed in the site Preliminary Safety Analysis Report (PSAR). ANSI N45.2.5 - 1974 Table B requires in-process tests be performed at specified frequencies; every 50 cubic yards placed for slump, air content and temperature, and every 100 cubic yards for strength.

The economic impact of testing at the pumpline discharge is considerable. The reduced productivity due to lost time in sampling, transportation to vibration-free areas, and the physical difficulty of obtaining representative samples from congested placements are a few of the impacts. The requirement to sample fresh concrete from the pump discharge is excessive if an acceptable pump correlation can be established between the pump inlet and pump outlet thereby allowing sampling and testing for acceptance at the pump inlet.

Considerable experience concerning pumpline effect on concrete has been gained at various nuclear projects. One site has developed a specification that allows for sampling of concrete at the point of leaving the truck agitator, providing that a satisfactory determination of pumpline effect has been established in concrete properties both at the point of leaving the truck agitator and the pumpline discharge. The first 1,000 cubic yards of concrete placed by pumping required correlation testing primarily to assure that the pump had no detrimental effect on the concrete. The initial correlation was established after the initial 1,000 cubic yards had been pumped and tested, and no detrimental effect discovered. Subsequently, tests are performed at the pump inlet and verified periodically.

The following method and guidelines were developed for establishing the initial pump correlation, documentation, and in-process correlation verification. Since correlation testing is dependent upon many parameters including pump type, mix class, line length, weather, and other variables, it was determined to base the correlation for each contractor on the following:

1. Each class of concrete. Class of concrete is designated by the design compressive strength at 28 days (i.e.: 4,000 psi @ 28 days; 5,000 psi @ 28 days, etc.)
2. Pumps of different manufacturers (i.e.: Challenge, Hercules, Thomsen, Schwing, Pecco, etc.).
3. Placements with equivalent horizontal pumpline length in excess of 500 feet and less than 500 feet.

The following guidelines were used for deterimiing equivalent horizontal pumpline length:

1 vertical foot (rise only)	= 6 horizontal feet
$30°$ bend	= 13 horizontal feet
$45°$ bend	= 20 horizontal feet
$90°$ bend	= 40 horizontal feet
1 foot of rubber hose	= 1½ horizontal feet

The initial correlation values were established and continually checked for their validity by tests and observations in the field. The correlation would be revised if needed when tests indicated the properties of fresh concrete varied from the established values by more than the following amounts:

Slump	± ½ inch
Air Content	± 0.5%
Concrete Temperature	± $2°F$

A minimum of one correlation consisting of slump, air content and temperature was required once per week for each established correlation. Additional correlations could be required as conditions warranted.

The application of pump correlation value was done by the Testing Laboratory by applying the specified requirements based on the indicated pumpline length. The following example is intended to indicate how the correlation values were applied to a particular placement.

Example: Placement X   Contractor B, Mix C

                Slump Requirement: 2 - 5"
          Air Content Requirement: 4 - 8%
          Temperature Requirement: 65°F maximum
         Horizontal Pumpline Length: Approximately 630 feet

Established correlation values for Contractor B, Mix C, Line Length greater than 500 feet was:

          Slump   ½ inch loss
          Air Content   0.3% loss
          Temperature   2°F gain

The Testing Laboratory indicated the reflected specified requirements on their field documentation. Consequently, the specified requirements at the mixer truck discharge were the following:

          Slump   2½ - 5½ inches
          Air Content   4.3 - 8.3%
          Temperature   63°F maximum

ANSI N45.2.5 - 1974 has impacted the nuclear industry by not allowing correlation testing of pumped concrete. As stated before, the ANSI committee members realized the requirement was too stringent and therefore it was revised in the 1978 version. It is felt the 1978 revision is a step in the right direction, but still requires excessive verification testing and does not allow sufficient latitude for judgement in in the field.

It is suggested that future changes to ANSI N45.2.5 follow the guidelines of ACI 304-73, (8) Paragraph 9.7 recommends correlation testing of pumped concrete, but is general enough to leave the final correlation/verification program to be determined by each project.

Determination of Adverse Weather Concreting Conditions

Determination of entry into adverse weather conditions on a project requires full evaluation due to the impacts on the concrete placement program. The intent of this section is only to address how the determination is made and not the problems encountered when working within the conditions.

ACI 305 (9) and 306 (10) address and define adverse weather: however, many interpretations and methods of implementation may be developed with the possible result of missing the intent of the standard. The implementation of this requirement on a complex nuclear site can become very cumbersome when dealing with a number of contractors. With each contractor attempting to determine when these weather conditions are to be applied, the concrete supply contractor would be at the mercy of any and all placing contractors. It is not inconceivable that the supplier could be called upon to produce concrete for three concrete weather conditions in a given day or shift. Since precaution and preparation are the keys to handling adverse weather concreting problems, it becomes advisable to have the capability to forecast and establish a plan to be carried out prior to occurence of adverse conditions.

The geographical location of the project is of paramount importance in the chosen approach. Some areas basically have no hot weather as compared to areas, such as the South and Southeast. However, hot weather conditions can and do occur in regions other than those normally thought of as "hot climates".

On one Northwest project, a program was developed to determine weather conditions from data recorded at a weather station located at the batch plant facility. The temperature probe was installed on a louvered weather enclosure at a location that was free from any influence of building and/or equipment. An indoor unit then continuously monitored and graphically recorded the temperature. This unit was declared the official site weather station for the purpose of determining the project's adverse weather conditions. Concrete was supplied and tested based on the recorded weather data.

In isolated cases, some placing contractors were able to establish, by their own quality control group or procedures, that their location experienced conditions differing from the batch plant facility. But in all cases, the supply, delivery, and testing (up to the point of receiving by the placing contractor) was performed under the specified weather condition.

Hot Weather

The project specifications required that hot weather conditions be in effect whenever the dry bulb temperature was $85°F$, and expected to go higher. The decision was made to impose and implement hot weather conditions upon the first occurence of the season, and until further notice, all concrete was batched with type D water reducing set retarding admixture to meet the requirements of the project specifications.

Hot weather conditions were imposed on supply, delivery, and testing upon notification from the batch plant, when the condition existed. Notification of hot weather conditions were noted on all concrete batch tickets of concrete mixed during hot weather conditions.

In this manner, the receiving or placing contractor was made aware that the condition existed. Adherence to the hot weather requirements for placing and curing was the responsibility of the placing contractor. Each contractor was notified (on a seasonal basis) of the system for determining and responding to adverse weather to allow time to prepare for the condition and avoid confusion and delays during placements when the conditions come into effect.

By noting the "hot weather condition" on all batch tickets the site was assured that all concrete during the season was batched with the required hot weather admix (Type D, water-reducer, set-retarding) (11). Other requirements were met only when the actual "condition" existed, which might only have been for a few hours a day on a limited number of days. This generally allowed concreting to proceed under less restrictive requirements.

Cold Weather

Determination and adherence to the cold weather conditions was addressed and implemented in a more restrictive manner than the hot weather requirements. Again, the location of the project is a factor that requires different considerations based on the geographical location.

ACI 306-66 on cold weather concreting defines cold weather as a period when, for more than three (3) successive days, the mean daily temperature drops below $40°F$ ($5°C$). Once again, advance notice and preparation are the keys to a successful program. The project requirement was that cold weather conditions be considered to be in effect whenever the mean daily temperature at the job site was at or below $40°F$ ($5°C$) for two (2) or more consecutive days in a row. The establishment of the cold weather condition could be determined from local forecasts, site records, historical data, current data, and the United States Department of Commerce Climatological Data, which is available for various local areas (12).

The determination of mean daily temperature is often a problem, and various methods of determining the mean are addressed. The project determined the mean from the established weather station at the batch plant. The graphically recorded data was monitored and the mean determined by averaging the values of two-hour intervals from 0200 through 2400 hours. In the fall, the mean (as well as the maximum and minimum temperatures) was plotted daily as advance information to give an indication of the progression of cold weather. When these records and/or local forecasts indicated the mean would be or was expected to be below the specified temperature, a notice was issued to all applicable personnel. This notification provided advance warning as the season approached. A statement was issued declaring cold weather conditions would be in effect as of a given advance date.

ACI 306 states in part, ". . .when temperatures above $50°F$ ($10°C$) occur during more than half of any 24-hour period, the concrete should no longer be regarded as winter concrete. Times and temperatures given for various conditions and situations are not exact values and should not be used as such."

The project site was in a relatively moderate climate, and the method of returning to normal concrete conditions was basically the reverse of going into the cold weather conditions. Temperatures were monitored daily and the long range and local forecasts were utilized; the result being the return to less restrictive conditions when the mean daily temperature was above $40°F$ ($5°C$) for two or more days and the forecasts were favorable for the future. Upon determination to return to less restrictive conditions, written notification was sent with the added note that if conditions other than those forecast were encountered, notification to return to cold weather conditions could occur.

In summary, it was felt that these methods for determining adverse weather conditions were the most appropriate for this type of geographical location. The geographical location should be considered in ACI 305 and 306.

Field Adjustment Practices For Dry Pack (Grout)

Two industry standards govern grout. These are ANSI N45.2.5, Table B, which requires strength tests of grout to ASTM C-109 daily, during production and ACI 210 (13) which establishes a method to determine the correct amount of water in dry pack grout - the handball test.

The first problem that results from the ANSI standard is in the assumption that it is applicable to dry pack. Many sites have applied this assumption. This then results in the second problem - the handball test. This test has frequently been inadequate for determining the proper amount of moisture. The problem is insufficient water content to provide for hydration of the cement, which results in low and inconsistent strengths. The third problem is the preparation of samples. The standard that is usually referenced is ASTM C-109 (14) and this is clearly for material of a plastic or flowable consistency, and is to be performed in the laboratory.

On the site much time was spent on investigation of low strengths of dry pack test specimens. Many items were addressed: bone-dry sand, cold weather, time frame from mix to use, water content, and possible tempering of mixture after the initial mixing or during use. The action resulting from this investigation was to stop drying the sand below the saturated surface dry condition to preclude loss of mixing water due to unknown or inconsistent absorption. Moisture tests were performed to check the accuracy of the handball tests. Over a period of some months, the test results indicated that the handball test was not reliable. A moisture content of approximately 9.5 to 12.5 percent usually resulted in the desired workability and produced adequate moisture for hydration of the cement.

Once it is determined that sufficient water is present for hydration, a method to cast the test specimens is the only remaining item. No industry standard exists for fabrication of dry pack compressive strength specimens. The following are methods that have been used or proposed:

1) Cast specimen in a 6" x 12" rigid metal cylinder mold. Compaction is to be achieved by craftsmen utilizing the same method and tools used in their work.
2) Same method as (1), but cast by laboratory technicians utilizing similar tools and methods used by the Craftsmen.
3) Same method as (1) and (2) above, but using 4" x 8" rigid metal molds for convenience and ease of transportation.
4) Cast 2" x 2" cube specimens.
   a. Dry pack strength specimens shall be cast in 2-inch cube molds conforming to the requirements of ASTM C-109, paragraph 3.5. Uniformly fill each cube compartment loosely and completely with material.
   b. Compact the material using an approximately 1-7/8 inch square hard wood block approximately 4-5 inches in length.

Consolidate the material in each compartment with approximately ten moderately applied blows using a hammer weighing 12-16 ounces. Molds shall be placed on a firm unyielding base during compactive effort.
- c. With the point of a trowel or other suitable instrument, scarify the surface of the compacted material to an approximate depth of ¼ inch to insure good bonding with the next layer.
- d. Fill each cube compartment again and repeat the consolidation process as prescribed in paragraph (b) above.
  Note: The final layer in each cube compartment will need to be overfilled so as to completely fill each compartment after compaction. If the tamping operation results in a subsidence of the dry pack below the top edge of the mold, add additional dry pack to keep an excess above the top of the mold at all times.
- e. Strike off any excess material upon completion of consolidation using the method described in paragraph 9.4.3 of ASTM C-109.
- f. Store and test the specimens in accordance with paragraph 8.5 and 8.6 of ASTM C-109.

5) Cast 2" x 2" x 2" cubes following the basic method described in ASTM C-109, with the exception that the tamping be accomplished by tapping the tamper with an eight-ounce mallot.

6) Cast a 2" x 2" x 10" (slab) mold. This method requires laying the mold horizontal with the Cratsman preparing the specimen with the tools used in the work. Prior to the date of the test, the sample is cut into 2" x 2" x 2" cubes by diamond-sawing. Historically, the cubes are not true and stresses are applied other than in the desired norm.

7) All of the above methods attempt to duplicate some form of field placement practice or method. Laboratory concrete testing employs some standard laboratory method or methods to parallel field placing practices. Therefore if documentation of the acceptability of the material not proof of the placing practice, is the purpose, as in concrete laboratory testing, another method becomes available. The Portland Cement Association (PCA) (15) methods for preparing soil cement strength samples could be adopted as a standard method. Several additional cylindrical specimen sizes are commonly used.

8) The last method could require workmen to be certified both for mixing and installation by an approved qualification process. The program should include a time frame and requirement for re-certification. This subject needs to be addressed, and standards developed, to guide the industry.

As demonstrated here, the parent documents require that grout tests be performed in accordance with ASTM C-109. This standard can not be used by the testing laboratory as explained above; therefore, the Nuclear Industry must determine how it is going to supplement the standard in the specification to provide the testing laboratory the necessary information to perform the test and obtain results that can be analyzed for acceptability. Several primary items must be kept in mind in determining the test procedure to be used when the codes and standards are either not entirely applicable or vague in their intent. The industry must first look at the results to be obtained. Is the laboratory equipment standard, off-the-shelf items? Should the tests be performed in the lab by technicians or in the field by craftsmen? Does the test procedure meet the intent of the parent document and/or codes without being overly restrictive?

Testing of dry pack grout is one area that has remained unaddressed by the codes and standards. Insufficient recommendations or guidelines presently exist for the preparation of dry pack specifications for nuclear projects.

Conclusion

Codes, standards, and guides, while providing the design and installation criteria, recommended practices, and guidelines for the nuclear construction industry, do impact mix verification and field adjustments of concrete more by what they don't say then what they do say. The impacts are caused when the codes are either not applicable in their entirety for the specific use, or are vague in their intent, or are used as an absolute when they are only intended as "Recommended Practices". Other impacts on construction occur when a code is referenced in the specification and applied in total when it was not the original intent.

Addressed herein were just a few of the many areas that have been looked at in the past and should be addressed by the Code Committees, Architect/Engineers, and Contractors in the future to provide the industry with a clear set of criteria for designing and building nuclear plants. As this happens, every related aspect of nuclear construction will begin to feel the benefits whether it be the NRC, Architect/Engineers, Construction Management, Prime Contractor, Owners, Suppliers, or Vendors.

The whole industry needs to look at primarily maintaining quality standards, but we must also look very circumspectly at the problems our industry is facing. These main problems, such as schedule and cost overruns, quality-related deficiencies in hardware and software could be reduced significantly if:

1) The Architect/Engineers would specify exactly what they want or give a range of what they want since the code is broad. Having a vague specification allows a rigid interpretation which may be beyond the intent of the design.
2) If specifics are not required, then the authors should make it very clear in the codes, standards, and specifications that the decisions as to what should be done, as well as how it is to be done, should be left with the construction engineering personnel at the jobsite. These people are usually more familiar with field problems, applications and good installation requirements.

References

1. American Concrete Institute. *Recommended Practice for Selecting Proportions for Normal and Heavyweight Concrete* (ACI 211.1-77), Detroit, Michigan 1978.
2. American Concrete Institute. *Specification for Structural Concrete for Buildings* (ACI 301-72), Detroit, Michigan 1978.
3. American Concrete Institute. *Building Code Requirements for Reinforced Concrete* (ACI 319-77), Detroit, Michigan 1978.
4. ASCE Journal of the Construction Division. *Concrete Production on Nuclear Power Plant Projects*, by H. Randolph Thomas, Jack H. Willenbrock, and James L. Burati, September, 1980.
5. American Society for Testing and Materials. *Standard Specification for Ready-Mixed Concrete* (C-94-71), Annual Book of ASTM Standards, Philadelphia, Pennsylvania 1974.
6. American Concrete Institute. *Cement and Concrete Terminology*, Publication SP-19 (73), reported by ACI Committee 116, Detroit, Michigan 1978.
7. *Supplementary Quality Assurance Requirements for Installation, Inspection and Testing of Structural Concrete and Structural Steel During the Construction Phase of Nuclear Power Plants* (ANSI N45.2.5 - 1974), New York 1974.
8. American Concrete Institute. *Recommended Practice for Measuring, Mixing, Transporting, and Placing Concrete* (ACI 304-73), reported by ACI Committee 304, Detroit, Michigan 1978.
9. American Concrete Institute. *Recommended Practice for Hot Weather Concreting* (ACI 305-72), Detroit, Michigan 1978.
10. American Concrete Institute. *Recommended Practice for Cold Weather Concreting* (ACI 306-66), Detroit, Michigan 1978.
11. American Society for Testing and Materials. *Standard Specification for Chemical Admixtures for Concrete* (C494-71), Annual Book of ASTM Standards, Part 14, Philadelphia, Pennsylvania 1974.
12. United States Department of Commerce. *Local Climatoligical Data Annual Summary With Comparative Data*, Olympia, Washington 1980.
13. American Concrete Institute. *Durability of Concrete in Service* (ACI 201), Detroit, Michigan 1978.
14. American Society for Testing and Materials. *Standard Method of Test for Compressive Strength of Hydraulic Cement Mortars* (using 2-inch or 50 mm cube specimens) (C-109-73), Annual Book of ASTM Standards, Part 13, Philadelphia, Pennsylvania 1974.
15. *Soil Cement Laboratory Handbook*, publication (EB052-065), Skokie, Illinois 1971. Portland Cement Association.

From Batch Plant to Form

by

William H. Brown[1]

Abstract

The major emphasis for quality concrete construction should be in the method of transportation and placing concrete in the formwork. All methods of transportation have some type of drawback or cost. The old time proven methods now have to be weighed against new techniques to result in a structure that provides the best guarantee of quality and performance of concrete design.

Introduction

In this paper we will attempt to explore the perimeters of placing concrete from a practical (job site) point of view, rather than from a theoretical application. We do not say that all the views are in keeping with current A.C.I. or ASTM codes and practices, but in light of better transportation of concrete you should question each of our suggestions in relation to current practices employed by this industry now. It should be noted that the first and foremost aspect of this discussion is the integrity of the structure. We are trying to create an atmosphere where the ability to communicate with those involved will have some impact on all problems with transportation of concrete in its plastic state. We will divide our subject by actual types of handling concrete from batch plant to form. There are many things to be considered in transportation of concrete, such as site terrain, climate, type of structure, and distance from batch plant. Other things might be slump, maximum aggregate size, placement rate, and other outside factors. All of these must be reviewed before deciding on any type of equipment.

Trucks

Some type of mobile vehicle, motor driven, known as ready mix trucks, are the most common used equipment to transport concrete from a batch plant to an intermediate discharge point. There are three primary types of ready mix trucks in use today: revolving drum

---

[1] President, Sparks Equipment Company, Inc.

bodies used for either complete mixing or agitating concrete. Open top agitating bodies in which previously mixed concrete is kept in motion while delivering to job, and open top (dump) non-agitating bodies for hauling concrete without any means of keeping it in motion. These vehicles and the bodies must be kept in good working order and maintained to preset specifications if we are to limit the amount of segregation and any loss of mortar or water from the concrete once it leaves the plant. Most job sites these vehicles will only take the concrete to an intermediate delivery point then it must be re-handled from there, so therefore, insure that this concrete is as close to design after it comes out of the truck as when it went in at the batch plant.

Chutes and Buckets

Drop chutes or elephant trunks are used in cases where the sides of forms or reinforcing steel interfere with the free fall of concrete. Concrete dropped vertically through this equipment segregates very little. Drop chutes are made of steel sections that are round or rectangular in cross section and supplied in two-to three-foot lengths. The lengths can be hooked or chained together. These connections provide some flexibility for the overall chute. Standard diameters for round drop chutes range from about six to twelve inches.

Elephant trunks should be used for placing concrete into deep, narrow forms but steel drop chutes are more suitable for wider forms and less restricted areas. Rectangular or round metal collars are provided at the top of elephant trunks for attachment to collection hoppers.

Open trough chutes used on an incline provide one of the simplest and most economical ways of handling concrete, but they can also be one of the least satisfactory. Special precautions are necessary to ensure that discharge is vertical at the end of the chute to prevent segregation. Open chutes should be as short as possible, preferably less than 20 feet long. The ideal open trough chute is made of metal or metal-lined wood, has a rounded cross section, is rigidly constructed and has sufficient capacity to prevent overflow.

Transporting and handling concrete in buckets is often considered the most flexible method for moving concrete from the point of delivery at the job site to the point of placement in the forms.

Buckets are made in different shapes and in sizes varying from one-third cubic yard up to twelve cubic yards for different applications. Some buckets have rectangular cross sections but most are circular. The concrete is released by opening a gate that forms the bottom of the bucket. For massive work the buckets often have vertical or very steep sides with gates that open to the full area of the bottom. However, for most types of work, buckets having the lower part of the sides sloping to a smaller gate are usually preferred for better control of concrete discharge. On the average job, buckets with gates that can be regulated to control the flow of concrete and closed after only part of the concrete has been deposited are preferred. The gates for buckets should be grout-tight, particularly if material is to be transported for some distance by truck, boat or rail car.

Conveyors

Belt conveyors are used to transfer concrete horizontally and limited distances vertically. Conveyors made expressly for handling concrete are relatively inexpensive and may eliminate the need for other more expensive auxiliary equipment such as cranes. They are particularly useful in areas such as tunnels where space is limited and are widely used on building construction, especially large floor slabs and bridge decks. In massive construction, such as mat foundations, dams and power plants which require large quantities of concrete, conveyors are advantageous because of their placing capacities.

Modern concrete belt conveyors can handle any type of mix but maximum efficiency is achieved with a cohesive mix of approximately two and one-half- to three-inch slump. Low slump concrete is generally best handled by slower moving belts, high slump concrete by faster moving belts. The cement content has very little if any effect on the suitability of concrete for conveyors.

One of the most important considerations for an acceptable concrete conveyor is a properly functioning wiper or scraper to remove mortar and paste from the belt. This device should be located directly behind the discharge pulley in such a position that the mortar and paste scraped off are directed back into the flow of material. Scrapers should be made of abrasion-resistant materials to minimize wear.

Of equal importance for systems using more than one belt are the transfer devices. A properly designed transfer device should be able to receive concrete from the charging belt at any angle at which belts operate (vertically or horizontally) and drop the concrete through the pivot axis onto the next belt in a smooth vertical column. This transfer should take place with no loss in the basic energy and speed of the ribbon of concrete, and with no objectionable rock bounce, segregation or loss of material.

Concrete conveyors running at their correct belt speed and with properly functioning charging hoppers, transfer devices and scrapers do not impair the strength, slump or air content of the concrete that they carry. However, consideration must be given to the possibility of segregation at the discharge point of the last belt. Final discharge should take place through a properly designed discharge hopper, drop chute or elephant trunk to prevent segregation of concrete discharged into forms.

Pumps

The pumping of concrete has become one of the fastest growing methods of transporting and handling at the job site. Pumping is a point-to-point delivery system in which concrete is moved under pressure through rigid pipe, flexible hose or a combination of both to its final destination in the forms with little necessity for further handling of the mix. Hence, with pumping there is less chance for segregation.

There are a number of other advantages in pumping concrete that should be considered when evaluating job site transporting and handling methods. The use of pumps can eliminate or greatly reduce the need for access roads, scaffolding, and other concrete handling equipment. Hoists and cranes are freed to handle other construction materials concurrently with concrete placing. Where tower cranes are being used the use of pumps permits them to be smaller and to speed job completion because other crafts can work relatively unhampered by concrete operations. Furthermore, concrete may be pumped vertically (up or down), horizontally or around corners, and the delivery lines can be run over, under or around obstructions and through windows or other small openings. No other method is as adaptable as pumping for transporting concrete into confined areas or under adverse conditions like inclement weather.

Concrete pumps are rated in terms of cubic yards delivered per hour and in capabilities of horizontal and vertical pumping distances. The concrete pump industry has not embraced a common rating method, so a word of caution is necessary in regard to selecting equipment. The various pump ratings should be based on pipelines of some specific minimal length and maximum diameter just as crane ratings are applicable only to lifts at minimum radius and maximum boom angle. In the final analysis, the limiting factor on capacity usually is the rate at which concrete can be supplied to one or more pumps. Receiving hoppers vary in size from a few cubic feet to several cubic yards. Hoppers should be equipped with remixing blades to maintain concrete uniformity. The pump and primary power equipment are available as truck-, trailer-, or skid-mounted units.

Concrete is pumped through rigid pipe or a combination of rigid pipe and flexible concrete hose. Rigid pipe (also called slick line or hard line) is made of steel and is available in sizes from three inches to eight inches in diameter. Five-inch line is considered a practical maximum without losing the advantage of maneuverability of the pipeline system.

Couplings used to connect sections of rigid and flexible pipeline must be strong enough to withstand wear and deterioration from handling during erection of the system or from misalignment or poor support along the line. Couplings should be capable of withstanding line pressures and surges and should not allow mortar leakage at pipe connections. Also, properly designed couplings should allow replacement of any pipe section in a pipeline without disturbing adjacent sections and should provide a smooth, full internal cross section with no constrictions that would tend to disrupt the flow of concrete.

A wide variety of pipeline accessories is available to tailor a pumping system to fit job site conditions. For example, there are varying lengths of rigid and flexible pipe sections in addition to the standard 10-foot elements. There are curved pipe sections of various radii, swivel joints and rotary distributors, and valves which prevent backflow in the pipeline, direct flow into other lines, or vent air for downgrade pumping. There are connection devices for pumping concrete into forms from the bottom up, transitions for connecting pipe and hose of different diameters, and equipment for line cleanout.

A number of manufacturers offer power-controlled hydraulic pumping booms, some with capabilities for reaching vertically in the neighborhood of 150 feet. This equipment provides rapid and convenient power handling of concrete delivery lines that are permanently incorporated into the booms. These units are especially useful for pumping concrete to columns, walls and scattered small placements. The mobility of concrete pumps and booms mounted on trucks or trailers capable of carrying all necessary pipeline and accessories allows the operator to set up, pump and move on to the next location several times in a normal working day.

Conclusion

Quality is maintained by avoiding equipment and practices that cause segregation, loss of any part of the concrete, or loss of slump. Nevertheless, new, improved, or experimental equipment should be allowed where tests can be made to insure that the concrete will be uniform and equal in quality and economy to concrete handled by usually accepted equipment.

The most common damage to quality during transportation and handling concrete is caused by segregation, that is, a separation of coarse aggregate from the mortar or loss of water from the concrete. Segregation tends to occur during filling and discharging vertically. When discharge is at an angle, the larger aggregate particles are thrown to the far side of the container or forms and the mortar is concentrated on the near side. Any type of equipment that permits part of the concrete to move faster than adjacent concrete will also tend to cause segregation.

One form of segregation is caused by jarring or shaking of concrete during transport in non-agitating equipment. This is more critical over long haul distances but is rarely a problem if the concrete slump is less than four inches. Care in maintaining smooth runways and the use of air entrainment also are helpful in eliminating this type of segregation.

Segregation, if allowed to occur during transporting and handling, may not be corrected during placing of the concrete into forms. In general, segregation can be minimized by some change in the mix design that produces a more cohesive concrete. Lower slumps, adequate cement and air contents, good aggregate gradation, and the proper ratio of fine to coarse aggregate are important factors contributing to cohesion.

Loss of mortar or paste from the concrete during transporting and handling can cause an initially workable concrete to become harsh and difficult to place. Chutes with rectangular cross sections, belt conveyors with improperly functioning belt wipers, and dump trucks with leaky seams are examples of equipment that is most susceptible to loss of mortar or paste. Loss of part of the mortar or paste can affect the strength, density, durability and economy of concrete and cause surface defects.

The most carefully proportioned and mixed batch of concrete can be transformed into a nonhomogeneous mass of unworkable material by careless transporting and handling.

INTENT OF CONCRETE ADJUSTMENT CRITERIA
by
Timothy L. Moore[1]

ABSTRACT

This paper discusses the various requirements for concrete that can be affected by and subsequently adjusted as a result of varying material characteristics and varying field conditions. The intent and proper application of current industry reference requirements are discussed, and field solutions are suggested. Solutions resulting from proper specifying and preconstruction conferences are discussed.

INTRODUCTION

Specifications for materials and construction procedures cannot possibly account for all possible situations that can occur in materials and in the field.

Additionally, there are situations where existing reference specifications and requirements should be selectively waived in the interest of economy and concrete quality. In these situations, rigid enforcement of some requirements is impractical and uneconomical.

In order to successfully evaluate the material or field condition and its effect on the ultimate product, an understanding of the intent of the requirements is necessary. With such an understanding, it is possible to apply good judgement interpreting the specification in the field and in adjusting the concrete mix at the batch plant. Interpreting the specification does not always mean following it precisely.

The origin of reference specifications from ACI and ASTM is examined. This provides a basis for understanding them. The reference documents are not perfect; they are put together by people like you and me. Who puts them together and how they do it are both examined.

Concrete material requirements and construction requirements as related to field and batch plant adjustment are evaluated with the objective of providing the user with a basic knowledge of the requirements and their intent such that adjustments can be made which will result in the best product.

---

[1] Structural Engineer, Gilbert Associates, Inc., Reading, PA 19603

REFERENCE DOCUMENTS

GENERAL

What are the basic requirements for concrete and from where do they come? Some project specifications merely specify a minimum strength. Some project specifications simply refer to ACI 301 "Specifications for Structural Concrete for Buildings." Some project specifications refer to both ACI 301 and ACI 318 "Building Code Requirements for Reinforced Concrete." Some project specifications refer to every document the writer can get his (or her) hands on, just to "be sure we're covered." Because the subject of standards and specifications is covered in another paper, the topic will not be expanded upon here. However, it should be emphasized that project specifications should refer only to pertinent portions of construction standards, such as ACI 301, because such standards often contain inapplicable requirements, such as contractual information. For example, ASTM C 94, "Standard Specification for Ready-Mixed Concrete", contains a procedure to follow in the case of low strength concrete that might not be desirable. Also, documents such as ACI 301 contain options and fill-in type requirements that the specifier must complete. The specifier must be familiar with and determine the applicability of the documents to which he (or she) is referring. The practice of requiring compliance with ACI 301, ACI 318, ASTM C 94, and various other ACI documents relating to hot and cold weather concreting, curing, etc., in their entirety is inconsistent, incorrect, and may impose costly, unnecessary burdens on the Contractor.

DOCUMENT USED THE MOST

No matter what method of specifying concrete is used, almost all specifications rely in one way or another on industry "standards." For concrete construction, the basic documents are ACI 301 and ASTM C 94. ACI 318 is not a construction document. The list at the end of this paper contains other documents to which are often referred. Many of these documents contain similar requirements that conflict. This is a result of the timing in the ACI and ASTM standardization processes, as well as differences of opinion. These documents must be reviewed for applicability before being incorporated by reference.

WHO MAKES UP THE DOCUMENTS

The ACI and ASTM influence is obvious when one checks the reference list at the end of this paper. In order to be able to use these documents effectively, it is very helpful to know who puts them together and what procedures they follow. As ACI states, its purpose is "to gather, correlate, and disseminate information for the improvement of the design, construction, manufacture, use, and maintenance of concrete products and storage." This wordy phrase means that ACI gathers and publishes information relating to concrete use. This is accomplished by technical committees, all volunteer, of which ACI has approximately 90. Each committee deals with one phase of concrete, such as curing, or consolidation. Depending on the work of the technical committee, ACI tries to maintain an equitable balance among designers, educators, and constructors in technical committees. For example, ACI Committee 301 has 26 members. Of the 26 members, 10 are designers, 11 are constructors, and 5 are educators. ASTM has a similar requirement. The main difference between ACI and ASTM is that ACI develops data for concrete construction and design, while ASTM deals primarily with material specifications and the testing of materials.

THE DOCUMENTS ARE NOT PERFECT

The important thing to be considered is that these committees are comprised of people with varying opinions and diverse experiences. The requirements contained in a document are often a result of multiple compromises among committee members. Because of compromises, requirements are often written in very general terms such that they must be supplemented or interpreted by the Engineer or Inspector. For example, ACI 301 contains requirements for mass concrete, but leaves it up to the Engineer to determine just what mass concrete is. ACI 301 states, "Concrete of lower than usual slump may be used provided it is properly placed and consolidated." The meanings of "usual" and "properly" are left for interpretation by those on the job. ACI is working to make requirements more specific, but volunteer committee work is a very slow process. Even though these documents are processed by standardization procedures and voted on by committee members and institute members, they are not always correct for construction and must be implemented with sound engineering judgement. For some situations, sound engineering judgement must supersede the requirements of a standard. One example of this is ACI 301, Section 8.2.2.4. This section says that the slump loss in pumping shall not exceed 2 inches. On one large nuclear power plant project, slump losses of 2-3/4 to 3 inches were encountered. The results of cylinders taken at the pump line discharge were good, and were consistent. The concrete was quite placeable, and no bleeding occurred. No difficulties in pumping were encountered. For the sake of good judgement and experience, the ACI 301 requirement was superseded. Blind obedience to the ACI 301 requirement in this case would have resulted in costly delays and rejected concrete.

The importance of properly using ACI documents cannot be overemphasized. When using ACI 301, for example, all options must be considered and appropriately exercised to fit the particular project conditions. Field personnel should be thoroughly familiar with the document controlling their work. If not, the all-too-often rigid enforcement of the many number limitations (i.e., slump, water/cement ratio, mix proportions, placement times, placement temperatures) at the site, without any allowance for field adjustment based on a knowledge of why the limitations exist, will result in concrete of lower quality and higher cost.

CONCRETE MATERIAL CRITERIA AND ADJUSTMENTS

AGGREGATES

A look at the criteria for concrete materials that can result in field adjustment is appropriate at this point. All of the basic material requirements will not be covered, as they are properly covered in ASTM material standards. However, the properties that affect field adjustments will be discussed.

Aggregates are a source of field adjustment. The fineness modulus (F.M.) of fine aggregate, which is a measure of the fineness of the fine aggregate (the sum of cumulative percentages of the aggregate retained on designated sieve sizes, divided by 100) is regulated by ASTM C 33, "Standard Specification for Concrete Aggregates", by not being allowed to vary by more than 0.20. If the F.M. does vary by more than 0.20 from the F.M. used in the original mix design, ASTM C 33 requires that the material either be rejected or adjustments be made in the mix design to compensate for the change in gradation. An increase in the F.M. will result in a decrease in the amount of coarse aggregate required, an increase in the sand content, and a decrease in the water content. Correspondingly, a decrease in the F.M. will result in an increase in the amount of coarse aggregate required, a decrease in the sand content, and an increase in the water content. Obviously, the higher the F.M., the coarser the sand, and the less fines are available, and more sand must be added to obtain the needed fines, and then less coarse aggregate is required. The ACI proportioning document ACI 211.1, "Recommended Practice for Selecting Proportions for Normal and Heavyweight Concrete", contains the relationship between the F.M. of sand and corresponding increase or decrease in coarse aggregate content. Mixes can thus be adjusted for changes in F.M. greater than 0.20. It is important that judgement be exercised and material not be unnecessarily rejected. In this case, the intent of the ASTM C 33 criteria is that the F.M. be monitored and substantial changes be accounted for by appropriately revising the mix. If this is not done, the water demand will change and the slump will change, resulting in rejected concrete at the point of delivery.

The coarse aggregate gradation should also be monitored. Substantial changes in gradation, outside the limits set by ASTM C 33 from the gradation used in the mix design, could result in a different dry-rodded unit weight due to a changed aggregate void system. For example, if the dry-rodded unit weight increased, then less air voids exist. Therefore, less sand is needed to fill the voids and consequently less water and cement are required (although the minimum required cement content might govern). Similarly, a change in aggregate shape will affect the water demand of the mix. For example, if the mix design is based on using a coarse aggregate which is from a river run (rounded), and a shipment of crushed aggregate is received, it will be necessary to increase the water content in order to maintain the same slump because of the added surface area of aggregate which requires water. The cement demand will increase if the same W/C ratio is maintained. Once again, the intent of the criteria is that the gradation and aggregate shape be monitored, and any substantial changes be counteracted by appropriately revising the mix.

AIR CONTENT

Another adjustment related to aggregates is the air content in the mix. If the aggregate gradation changes appreciably, the required air-entraining agent to achieve a given air content will change. Approximately 9% entrained air in the mortar paste is considered adequate to provide for freezing and thawing protection. For a change in gradation to a larger size aggregate, more of the volume of the concrete will be filled up by aggregate, less by the paste, and the total air content required will be less.

ADJUSTMENTS

Therefore, unnecessary rejection of the aggregates and costly problems with the concrete (varying slump, varying air content, varying water demands, truck delays, placement problems, etc.) can be avoided by properly monitoring the materials and knowledgeably adjusting the proportions of the mix.

For the material variations discussed above, field adjustments to the mix design can be made if a history has been developed for the mix for that batch plant. For example, if an adjustment is made to the laboratory mix design due to fineness modulus changing, the adjustment can be made at the batch plant by a knowledgeable or experienced batchman, as long as the cement content, air content, and slump requirements of the originally approved mix are met. No new mix design should be necessary. Unfortunately, inspection personnel often will require a complete new approved mix design or reject materials that vary as described above. It is true that material consistency should be monitored for both quality and cost purposes, but it is also true that knowledgeable personnel should be able to use materials that might fall slightly out of the limits of the specification by applying methods as described above. ASTM C 33 does provide for using sand with varying F.M. by making adjustments. If the coarse aggregates vary substantially from the gradations of ASTM C 33, re-screening might be necessary. The source of aggregates should be investigated. However, adjustment at the batch plant to account for nominal variations in aggregate gradations should be allowed. Strict adherence to the laboratory mix design proportions, with no regard for manageable material variations, can be very costly, in both time and money.

FIELD ADJUSTMENTS

GENERAL

ACI 301 states, in Section 3.8.1, "Control in the field shall be based upon maintenance of proper cement content, slump, and air content." In other words, when concrete arrives at the job site, cement content must be verified by noting the quantities on the batch ticket, and slump and air content must be determined. If everything meets the requirements, and the materials are basically the same upon which the mix proportions were based, the concrete should be acceptable, assuming it can be properly placed. This leads to the discussion of slump.

SLUMP

Slump is considered to be many things to many people. Some say that slump tells you how much water is in the mix. Some say that slump tells you how much cement is in the mix. Some even say that slump tells you what strength the concrete will be. Slump <u>can</u> indicate all of the above to some degree, but certainly not with absolute accuracy. Slump is a relative indication of the consistency of the concrete mix. More water in the mix increases slump and decreases strength. ACI 301 states, "Unless otherwise permitted or specified, the concrete shall be proportioned and produced to have a slump of 4 inches or less if consolidation is by vibration, and 5 inches or less if consolidation is by methods other than vibration." The 4 inch slump requirement has existed for many years, and is considered to be adequate for most construction. However, with many new designs requiring heavy concentrations of reinforcing steel, particularly in nuclear power plant construction, 4 inch slump concrete can be very difficult to properly place and consolidate. Even though ACI allows a one inch tolerance above the 4 inches provided, "..... the average for all batches or the most recent 10 batches tested, whichever is fewer, does not exceed the maximum limit," it still is not adequate for many placements. The result of forcing a 4 inch slump concrete into some of these placements is often honeycombed concrete and slow placements. This costs money. The slow placements have a domino effect, causing waiting ready-mix trucks to wait longer and lose even more slump.

ACI 301, Section 7.5, does provide for field adjustment of low slump mixes, where the slump is below the <u>specified</u> slump. Increasing the slump in the specifications will be covered in subsequent paragraphs. ACI 301, Section 7.5, allows water to be added to the truck at the site if neither the water/cement ratio nor the maximum slump is exceeded. If the water/cement ratio is exceeded, cement must be added to be within limits. This is a very difficult procedure to properly control and monitor in the field. Due to the fact that about 3% difference in water added can change the slump one inch, it takes very little water (1-2 gallons) per yard to gain one inch slump. It is difficult to accurately measure the water added because of the motion of the truck and the up-down motion of the water in the water sight gauge. Eight to sixteen gallons added for an 8 yard truck can be very hard to measure. Additionally, most sites do not have bagged cement at the placement site or weight scales to measure the amount of cement to be added. The biggest problem with this procedure is knowing how much water can be added before the water/cement ratio is exceeded. To know this, the moisture content of the aggregates before batching <u>and</u> the total water added to the mix at the batch plant must be known. This is necessary because most mix designs are in terms of aggregates in the surface-saturated-dry (SSD) condition. However, aggregates at the batch plant are not always in the SSD condition, thus necessitating that the aggregate moisture content be verified prior to batching. Some plants check the aggregate moisture content twice a day, some once a day. Some do not check it at all, but rely on the batch plant operator's experience and expertise. The intent of the ACI 301 provision to add water at the site is that the water content of the aggregate plus the total amount of water added at the plant be known. This requires close coordination between the plant and the site, but is very useful in using one of ACI 301's most practical field adjustment procedures.

If the slump is too high, which can happen because too much water was added at the batch plant, because the aggregates were too wet and the moisture not compensated for, or because too much water was added at the site, four things can happen: (1) cement can be added, (2) the truck can sit and revolve the drum until the slump decreases, (3) the concrete can be rejected, and (4) judgement can be applied. Number 1, adding cement is a possibility but not practical for most job sites. Number 2, revolving the drum can take too much time and delay the job. Number 3 is not in anyone's interest. Let us consider Number 4, judgement. If the slump is one inch too high, that means that about 3% too much water has been added. The water/cement ratio will change by about 1.3% for the average mix. This will not affect strength to a measurable degree. If one is worried about shrinkage because of the excess water, ACI SP27, "Designing for the Effects of Creep, Shrinkage, Temperature in Concrete Structures", indicates that the effects of a slump increase from 4 inches to 5 inches can normally be neglected. A similar statement is made for the effect on creep. If the member is massive, then shrinkage would be a consideration only if it was not reinforced. In most cases, the 5 inch slump would be acceptable. However, variations in slump indicate something is wrong in the concrete making and delivering process, and should be checked. Unfortunately, such concrete is often rejected because of the ACI 301 limitation. I disagree with the ACI 301 limitation as it is written. ACI Committee 301 is now in the process of changing the slump requirement to indicate a more workable tolerance, and to differentiate between heavily and lightly reinforced sections.

For concrete which is to be heavily reinforced, the specifier should specify higher than a 4 inch slump. In the supplemental requirements to ACI 301, it states, "Slump, if other than 4 inch maximum for vibrated concrete or 5 inch maximum for other methods of consolidation should be designated if applicable." Thus, ACI 301 does recognize that higher or lower than 4 inch slump might be necessary, and provides for it to be specified. <u>There is nothing sacred about 4 inches</u>. As ACI 318 and ACI 349, "Code Requirements for Nuclear Safety-Related Concrete Structures", both state, "Proportions of materials for concrete shall be established to provide:

A) Adequate workability and proper consistency to permit concrete to be worked readily into the forms and around reinforcement under conditions of placement to be employed, without excessive segregation or bleeding.

B) Resistance to freezing and thawing and other aggressive actions.

C) Conformance with strength test requirements."

If a 5 inch slump concrete satisfies these requirements, use it. The slightly higher cost will pay for itself in case of placement in congested areas. For other areas, the more economical 4 inch slump mix should be used. Where practical, from a placement standpoint, use even a lower slump concrete, as it will be less expensive from a materials standpoint.

As mentioned before, slump is a measure of consistency. It is also a measure of the variation in the concrete from one batch to the next. Big variations mean something is not consistent in either the materials, the batching and mixing, or the delivery. Aggregate moisture content could be changing, water could be inaccurately batched, truck mixer blades could be worn, or delays might be occurring. Because variations in slump can act as a warning to other problems, it is often prudent to specify a minimum slump, as well as a maximum. The variation in slump and, hence, the associated causes are thus controlled to some degree. Once again, however, judgement must be used. For example, if the specifications require a minimum 2 inch slump and a 1-1/2 inch slump arrives at the site for a very lightly reinforced flat area, and is placeable, it is wiser to place the 1-1/2 inch slump concrete in lieu of adding water and mixing for a while, thus causing delays. Once again, the field personnel must be aware of the intent of the requirement and not be totally concerned with narrowly comparing numbers only.

MASS CONCRETE

Of particular consternation to many in the concrete industry is the ACI 301 requirement for mass concrete. The ACI 301 maximum slump for "mass concrete" is 2 inches. Although ACI 301 does not specifically define mass concrete, it suggests in a footnote that 2-1/2 feet in the least dimension might define the critical size for mass concrete. For many structures, elements 2-1/2 feet or greater are heavily reinforced. Placing 2 inch slump concrete in such areas is not practical, and the consequences are voids, honeycombing, and truck delays. The intent of ACI 301 is that mass concrete be defined as not reinforced or only nominally reinforced, not heavily reinforced, and that very large aggregate (up to 6") can be used. Dam construction is a typical application. ACI Committee 301 is now revising this requirement in order to be more practical, with heavily reinforced areas being taken into account. For "mass concrete", which has more than nominal reinforcing steel, 2 inch slump is not adequate.

TEMPERATURES

Temperature limitation for concrete is also established by ACI 301. For mass concrete, the maximum placing temperature of the concrete is 70°F. The intent here is obviously to minimize total temperature build-up by limiting the starting temperature. However, judgement should once again be exercised. If "mass concrete" is delivered at 75°, but is cooled by water spraying metal forms (formed on both sides), it will be better off than mass concrete delivered at 70° and formed on one side only (cast against existing concrete) with wood forms. An inspector going merely by the numbers would probably reject the 75° concrete. Additionally, if the "mass concrete" is more than nominally reinforced, any cracking that might be induced by thermal shrinkage is minimized by the presence of reinforcement, and the problem is less than that for traditional mass concrete, as in unreinforced dam sections. Once again, field personnel should be familiar with the intent of the requirement. Additionally, 90°F is often specified as the maximum allowable concrete temperature in hot weather for non-mass concrete because several documents refer to 90°F as an upper limit. This must be used with judgement, because concrete higher than 90° is perfectly acceptable under certain conditions, such as high humidity and where water curing will be applied. However, under dry and hot windy conditions, the 90°F limit should, as a rule, be used.

TIME LIMITS

ASTM C 94 imposes a time limit for when the concrete must be discharged into forms. It states, "Discharge of the concrete shall be completed within 1-1/2 hours, or before the drum has revolved 300 revolutions, whichever comes first ....." Many inspectors apply this rule as hard and fast. If concrete is not too hot and the slump is placeable, there is no reason to reject concrete that is outside these time limits. ASTM C 94 also states, "These limitations may be waived by the purchaser if the concrete is of such slump after the 1-1/2 hours time or 300 revolution limit has been reached that it can be placed, without the addition of water to the batch." The inspector should use judgement and apply this last statement if at all possible.

SOLUTIONS

Thus far, it appears as though I am advocating disregard for specification and contract requirements and obligations in the field. This is not my intent. Possible solutions will be investigated at this point.

In order to properly apply the principles I have described herein, there must be close cooperation between the specifier (engineer), the producer, and the user. Several things can occur on any given project where concrete requirements are specified:

1. The engineer specifying the requirements is so familiar with the field conditions and materials to be used in the

project that he anticipates all possible problems and writes a specification that everyone is happy with throughout the life of the project.

2. The engineer specifying the requirements is not familiar with the field conditions and materials to be used in the project and writes a specification which refers to all of the standard ACI and ASTM References in order to cover himself.

3. The engineer specifying the requirements is not familiar with the field conditions and materials to be used in the project and, therefore, writes a specification which refers correctly to the standard ACI and ASTM requirements and provides the logistics for deviation from these requirements should field or material conditions require it.

Number 1 above is a theoretical case. Such a specification would be very long, unless the project is very limited in scope. I have not seen such a specification. It is virtually impossible to anticipate all possible field problems that can occur. On large scope projects, you will never see such a specification.

Number 2 above occurs quite often. It is usually the result of the engineer not being familiar with the ACI and ASTM requirements and relying solely on them because they are the "standards" of the industry. If there is no preconstruction meeting and the specifications are applied in the field exactly as written, there will be problems. Such problems have been described in previous paragraphs. If there is a preconstruction meeting between the engineer, the producer, and the constructor (and the Owner), some of the potential problems can be avoided by appropriately revising some of the "standard" requirements, based upon fruitful discussions between all parties. The Owner should be aware of such changes and how they affect the inspection of the work. Unfortunately, all ACI and ASTM documents are not "standards" in that some of them are not in mandatory language, contain alternatives, and merely suggest means to meet certain requirements. Even if such a preconstruction meeting does result in changes to avoid some of the problems, other problems will undoubtedly arise in the course of the work.

Number 3 above should occur more often. It is the result of an engineer being familiar enough with the ACI and ASTM references to recognize that they are not all in mandatory language, they contain alternatives, and they require supplemental information from the engineer or specifier, and that only relevant portions should be referred to. This type of specification will use the ACI and ASTM references properly, and will exercise properly the options therein contained. Furthermore, this specification will take into account the items pointed out in previous paragraphs of this paper, such as slump and time limits. How such items are taken into account is very important. There must be a communication network set up such that changing materials and field conditions can be accounted for. This means that the logistics for continuous cooperation between the engineer, constructor, concrete supplier, and inspection agency must be established. This all must begin with a preconstruction meeting including the above parties. Any problems on interpreting the requirements should be addressed and mutually resolved. An understanding of how the requirements are to be met should be worked out. All of this must be documented.

This type of specification must specify which parties are responsible for implementing and interpreting the various requirement. It must describe the process by which changing field conditions or materials can be compensated. For example, the specification can provide for the possibility of high slump concrete arriving at the site occasionally by requiring approval by the Owner or his agent. At the preconstruction meeting, it should be discussed under what conditions the Owner, through his inspection agency, will permit the higher slump concrete. Similarly, the specification should require approval by the Owner or his agent, in accepting for use concrete exceeding the ASTM C 94 time limit requirement of 90 minutes from initial batching to placing concrete. The preconstruction meeting should result in an understanding of under what conditions the deviation will be considered.

The preconstruction meeting cannot cover everything. Problems will arise in the field which may not be covered by the specification and for which "approval by the engineer" would take too long to obtain. For example, if the specifications require that concrete temperature not be higher than $90°F$ when placed, and it arrives at the site at $92°F$, no one really knows whether or not the concrete will be workable in the forms or not, and it would take too long to consult with the engineer. Such concrete should be rejected if no precedent has been established for this case (either at the preconstruction meeting or by a previous case in the field). However, the site (constructor and inspectors) should determine if the concrete is workable enough to be placeable (by possibly taking a small sample and testing it). This case should be documented and referred to the engineer who can, after consulting with the constructor and inspector, permit the inspector to allow such concrete to be placed under certain conditions (such as only in nominally reinforced areas) after determining its effect, if any, on the structural integrity. It is obvious here that continuous interaction must take place between the constructor, engineer, Owner (inspector), and concrete supplier.

The specification, if the project format requires it, may have to be revised periodically to take into account such cases, or the project might require that field procedures be established to supplement the specification. The preconstruction meeting is the time to clarify the logistics through which these cases are handled. Whatever procedure is established, it is essential that allowed deviations be documented.

GENERAL

I am advocating the proper use of ACI and ASTM requirements.

They must be understood by those applying them, and good common sense combined with an understanding of the requirements must be applied early in the project in the specification writing stage and preconstruction conference stage. The specifier must understand the requirements and must consider field conditions and ease of construction when writing the specifications. Even though ACI 301 is the basic document in the concrete industry for construction specifications, it does not cover all possibilities. It must be supplemented by experience and judgement. Although it is basically a sound, workable document, ACI 301 is not a Bible, and only the relevant portions of it should be referred to. The variations and adjustments described earlier in this paper are essential considerations to be addressed in the specification and at the preconstruction meeting.

REFERENCE LIST

1. ACI 211.1, "Recommended Practice for Selecting Proportions for Normal and Heavyweight Concrete."

2. ACI 301, "Specifications for Structural Concrete for Buildings."

3. ACI 301, "Recommended Practice for Measuring, Mixing, Transporting, and Placing Concrete."

4. ACI Title No. 74-33, "Hot Weather Concreting."

5. ACI Title No. 75-18, "Cold Weather Concreting."

6. ACI 308, "Recommended Practice for Curing Concrete."

7. ACI 309, "Recommended Practice for Consolidation of Concrete."

8. ACI 318, "Building Code Requirements for Reinforced Concrete."

9. ASTM C 33, "Standard Specification for Concrete Aggregates."

10. ASTM C 94, "Standard Specification for Ready-Mixed Concrete."

Is Criteria as Written Enough?

by

Robert M. Eshbach[1]

Abstract

The degree to which quality assurance programs, and through them the formal adherence to codes, standards, and regulations, are able to actually assure the real quality of concrete structures is discussed.

Introduction

During the past fourteen years, I have been associated with construction of power facilities in the capacity of design engineer, quality assurance engineer, field engineer, and specifications writer and interpreter. That association has resulted in the formulation of a number of ideas relevant to the assurance of quality in concrete construction. Some of these ideas are totally supportive of traditional concepts of assuring quality. These are not emphasized here. Others, however, are less supportive of the methods normally employed today and it is these ideas which will be discussed. In doing so, I should like to emphasize that the following may appear to be critical of traditional quality programs and of the way they are implemented. That criticism is intended to be constructive. In no way do I suggest that quality programs are not necessary or that the concept of quality assurance is less than totally valid. Indeed, no other attribute of a structure is so important as its quality. However, at the same time, I should stress that my primary focus is on the structure itself and, secondarily, on the programs governing its construction. The existence of mountains of records and documentation is not an assurance of quality construction.

To consider construction quality, one must necessarily consider the procedures, codes, standards and regulations relating to the construction of power facilities. The collection of such documents,

---

[1] Structural Engineer, Gilbert Associates, Inc., Reading, Pennsylvania, 19603

together with the drawings and specifications, might be collectively labeled as written criteria. These documents establish what is to be done, how it is to be done, how well it is to be done and who is to do it. The written criteria are essential to the successful completion of the project. Nevertheless, one might question if the adoption and implementation of written criteria, while necessary, is also sufficient to assure the required level of quality. If the sufficiency condition is not satisfied, then it is important to recognize that under some circumstances, the adoption of a quality program could result not so much in high quality concrete structures as in high quality documentation which may not necessarily be representative of reality.

Realistically, then it is appropriate to recognize that a one-to-one, cause and effect relationship between a quality program and the actual realization of quality construction may not exist. Where such is the case, a breakdown in the system which ensures the attainment of quality has obviously occurred. Since such breakdowns continue to occur in spite of the ever expanding QC and QA programs, it is appropriate to ask why.

The Problem

In that regard it is my opinion that the primary shortcoming of many quality programs is their lack of recognition of the human element, the failure to utilize the very considerable capacities of trained and qualified people. Further, the attitudes, feelings and needs of the people who affect or are affected by the program are not given ample consideration. This basic shortcoming would seem to lead to the development of a variety of conditions the existence of which tend to limit the effectiveness of efforts to maintain high standards of quality. These conditions include:

1. Total dependence on the program
2. Wasted resources
3. Excessive deference to the expertise of the design engineer
4. Apathy
5. Rigidly defined responsibilities
6. Need for self-preservation
7. Tunnel vision
8. Imposition of inappropriate requirements

<u>Total Dependence on the Program</u> When individuals associated with the project become totally dependent on the quality program they tend to become effectively less competent people. This is the case because the utilization of rational thought, judgment, intuition, and common sense, experience and all the other factors which determine an individual's overall ability, are placed in a secondary position. They are subordinated to the words in the program. People become robot-like and learn to depend on the accuracy, completeness, and effectiveness of the written criteria, rather than on themselves.

This situation can be critical because it can affect virtually anyone on the project. The worth of a craftsman with many years of experience setting forms or finishing concrete might be diminished to the point where he is no more valuable than the recent trade school graduate who has memorized all the right procedures. No longer does he feel as if he contributes to the success of a project through the application of the expertise he has developed over the years. Rather, he is told to unquestioningly accept the dictates of the program. He begins to feel that he is regarded more as a tool than as a human being. The challenge of continuously turning out high quality work becomes overshadowed by a tendency to let the program tell him when his work is not good enough. He avoids responsibility for a procedure that is not his own. His morale suffers and so does the quality of his work. Parodoxically, the program which was to assure quality instead encourages mediocrity. How ridiculous it is for those of us who design structures or write specifications to complain about pride of workmanship when the programs to which we've contributed have taken much of the pride out of workmanship.

<u>Wasted Resources</u> Consider that the organizations responsible for the policies and criteria for the concrete construction effort are generally the design engineers and the regulatory personnel. While such personnel are usually well qualified in terms of degrees and licenses, they are, in general, probably less well informed regarding practical considerations than those of other organizations associated with the project. This is obviously not an infallible truism. It would seem realistic to assume however, that those with the most practical knowledge of concrete are those with experience which relates closely to the actual production, transportation and placement of concrete. Certainly we must recognize that experience alone is not enough. Concrete is a complex material, and there are those who may not understand the many influences on and determinants of the various properties of hardened concrete. But what about, as an example, QA and QC engineers who choose to work in the field. Many such individuals have appropriate formal qualifications, a genuine interest and real knowledge backed up by valuable experience. Those persons are valuable resources, and they are utilized, although unfortunately to a rather limited extent, within the framework of most existing quality programs. However, their expertise is totally ignored when specifications or industry standards need to be interpreted. They are not consulted and certainly not relied upon when concrete fails to achieve its specified properties. And they are not counted upon to point out potential construction difficulties to the extent possible. In general, their expertise is wasted.

<u>Excessive Deference to the Expertise of the Design Engineer</u> Traditional practice dictates that the resolution of construction problems is an extension of the design process and, therefore, the responsibility of the design engineer. This is not unreasonable until one recognizes that the engineer responsible for a particular design may not

be available when a decision is needed. If he is available, he
may not be able to appreciate the situation simply because he
may be hundreds of miles away. Most significantly, in matters of
concrete technology he usually has little or no expertise. His
only qualification for addressing questions on concrete
technology is his familiarity with the design, a qualification
which is often meaningless since his concern for the material
properties of concrete is, in most cases, limited to a concern
that the design strength is obtained. QA and QC people, on the
other hand, are totally qualified to address issues which are
not dependent on the details of the design and should be permitted
to do so.

Apathy Any perusal of newspapers and magazines indicates that
apathy is prevalent in our society as a whole, and quality programs
do little to prevent it on the construction site. For instance,
detailed construction procedures, while serving many useful functions, are supportive of those individuals who prefer not to get
involved, to let someone else exercise initiative and assume
responsibility. For these people, detailed procedures provide a
substitute for their personal built-in barometers of quality.
A worker may be inclined to let the program tell him when what he
is doing is incorrect or inappropriate or when the result is of
unacceptably low quality. This, of course, is no problem so
long as someone is watching everything he does. To require that,
however, places an unrealistically strenuous burden on the efforts
of QC personnel. As a result, one could logically question
everything which is not specifically inspected. And, unfortunately,
that question cannot be eliminated with paper work, documentation
and additional criteria.

Rigidly Defined Responsibilities Industry organizations such
as ACI and ASTM, design engineers, suppliers, contractors, inspectors, auditors, regulatory personnel, consultants, and even
members of the academic community all interrelate in the development and utilization of the written criteria. Unfortunately,
that process has some hitches which relate to the way each group
perceives its responsibilities, and those hitches very definitely
affect quality and cost. Consider the following scenario.
Various industry groups prepare the codes and standards, but the
language, in order to simultaneously ensure technical accuracy and
general applicability, is sometimes not clear and in need of
interpretation. Unfortunately, such groups do not normally consider, and perhaps, for good reason, that their responsibility
includes that of making official interpretations. Nevertheless,
the engineering community utilizes these standards, providing their
own interpretations, in the specifications. These interpretations
may or may not be conservative; they may or may not be correct.
But they do become gospel for the constructors who are responsible
for preparing the procedures to satisfy specification provisions
with which they may not be in agreement. As a result, their viewpoint or the way a job should be done, and the way they have done
it for years, may differ from what appears in their own procedures.

Next, the QA and QC personnel get involved, not so much to control or assure quality, but to ensure compliance with the specification and procedures. The intent is to guarantee that the engineer gets what he wants, even if they think that some requirement is inappropriate or that the engineer does not know what he wants. It is not their responsibility to make such judgments, and it would be a conflict of interest to do so. Eventually, regulatory personnel enter the scene to determine the extent to which the criteria is being satisfied. And, of course, they do discover some findings. They then, indirectly, make further contributions to the written criteria through their evaluations of the response to the findings. Unfortunately, the evaluation may consist of comparing the proposal of the project in question with the most conservative practice on any other project in their territory. And then, finally, there is the owner who attempts to resolve these findings, but whose primary responsibility in the scenerio is to suffer. So he opens his wallet and pays the engineer and the contractors to resolve the problems. And he opens it still wider when he commits to whatever the regulatory personnel want, even if he is 100% opposed, because to fight it would simply cost too much, especially in terms of time. Unfortunately, his agreement becomes a precedent for the next project on which a similar finding occurs.

<u>Desire for Self Preservation</u> The desire for self preservation is equivalent to the desire to avoid blame. It can be manifested in a wide variety of situations. For the sake of illustration, consider a situation in which the rigid enforcement of a quality program can actually be detrimental to the real quality of the structure. Suppose that concrete is being placed on a hot summer day, and for some reason a delay in the supply occurs. The exposed fresh concrete starts to harden. A cold joint becomes a good possibility. Finally, a truck arrives in time to prevent the cold joint, but a test indicates that its temperature is in excess of the maximum allowed by the specifications. The person in charge of accepting or rejecting the concrete now has to make a decision. He can recognize that the benefit to be realized by preventing the cold joint exceeds by far the potentially deleterious effects of a single batch of high temperature concrete. He may, therefore, choose to accept the concrete. In doing so, he may also prudently choose to make extra test cylinders to demonstrate that the concrete, as received, is acceptable. Such cylinders, together with an explanation of the situation, would provide the primary justification for his actions to his supervisors and to the organizations to whom those persons are dependent for approval. On the other hand, if those organizations had been known in the past for rigid adherence to procedures, with little allowance for the application of educated common sense, or if the cylinders for any reason, fail to develop the required strength then he can anticipate that acceptance of the concrete may have been inadvisable. Any such development would make him personally responsible for the "inferior" concrete in the structure. However, by rejecting the concrete, even if doing so did offend his conscience and even if he did recognize that it was a detriment to the structure, to the owner's budget, and to the project schedule, he is free of blame and liability.

Tunnel Vision   One inviolable requirement of quality programs is that QA and QC personnel must be concerned primarily about quality and have little concern with such practical matters as cost and schedule. This is obviously to prevent a conflict of interest, to guarantee that their only interest is quality. But, being a design engineer, I find myself disturbed with the philosophy because it seems to indicate that concern for quality is the only indicator of overall design acceptability.

To the contrary, cost is of equal significance to the design profession. It is an engineer's ethical responsibility to consider both economy and quality. Why should quality personnel not be similarly inclined? A conflict of interest between quality and cost is no more necessary on the construction site than in the design office so long as the ultimate goal is to assure quality. Unfortunately, that requires consideration of what degree of quality is sufficient, and that degree of subjectivity is generally not permitted.

Imposition of Inappropriate Requirements   It seems likely that a variety of papers presented at this conference would draw attention to the contention that the provisions of the written criteria are frequently invoked with little consideration for specific circumstances and with little indication of any understanding for the purpose or intent of specific provisions. Fortunately, this sort of occurrence is not indicative of the quality engineers who may be involved. But is is indicative of their conscientious desire to conform to the provisions of the program, a program that permits only literal interpretations of written criteria, permits no exercise of judgment or common sense, and which is based on the basic tenet that the design engineer who prepares the specifications is responsible for delineating any and all exceptions. That philosophy does not recognize, of course, that no specification dealing with either the supply or the placement of concrete could possibly anticipate every unusual circumstance or contingency and, therefore, that every exception cannot be identified. Unfortunately, the practice of enforcing inappropriate requirements, while usually having little effect on quality, positive or negative, does add substantially to the cost and to the schedule. Probably of greater significance, however, is the possibility that the enforcement of seemingly inappropriate requirements could result in a loss of respect among workers for the quality program in general, and that is a severe loss which should not be tolerated.

Summary

The preceding discussion was not intended to be a revelation, certainly not to the group of people attending this conference. It is intended to emphasize that more can be done to assure a high level of quality in the construction process than is presently being done. And that extra effort involves the recognition of the

of the contribution which can and must be made by competent
people, regardless of the formal organizational role they are
assigned. That contribution and only that contribution can
eliminate the conditions just discussed.

It also implies that concrete construction activity, by its
very nature is less predictable than most manufacturing or fabricating processes for which the quality programs are based, to a
large extent on statistical methods and on easily applied acceptance criteria. Perhaps, the application of rigidly enforced rules
and regulations is appropriate there. Concrete construction is,
however, a complex interaction of related operations, no single
one of which can be treated on a go:no go basis without consideration of the affects of such decisions elsewhere. For that reason,
rules governing concrete construction must be supplemented with
the judgment and general competence of qualified personnel.

The August, 1981 issue of Concrete International contained an
introductory memo from ACI President Mr. T.Z. Chastain in which
he indicated that ". . . knowledge alone will not result in
quality structures; there are other factors involved in producing
good concrete. Following knowledge - the know-how to do the job -
there comes pride - the willingness to do the job not just correctly but with an added incentive to do it as near perfectly as
possible . . . All of us need to . . . do our best to eliminate
attitudes which lead to poor results. There can be 'progress
through knowledge' but there can never be real progress without a
desire for quality and pride in the job."

Mr. Chastain seems to be indicating that the focus for assuring
quality must be on people and on the challenge to instill in those
people the desire to conscientiously perform their functions with
pride and to the best of their ability. Specifications, procedures,
quality programs in general, must be supplemented by a much more
far-reaching application of qualified human expertise and rational
human judgment. Doing so will not solve all the problems on any
project, but is should help to achieve the required level of
quality at a reasonable cost and in a reasonable time. Ultimately,
our only faith in the overall quality of the work is the faith
we have in the people who batched, transported and placed the
concrete and it is there that efforts for improving quality ought
to be directed.

The most valuable resource available to the effort of constructing power facilities is the overall ability of people.
Codes, standards, procedures and regulations, as valuable as they
are, do not comprise as valuable an asset.

SESSION VII - CONCRETE FORMWORK, REINFORCEMENT AND
PLACEMENT TOLERANCES AND PRACTICES

SESSION OBJECTIVES/SESSION CHAIRMAN SUMMARY

by

Glen A. Chauvin[1] M, ASCE and Aldo Palmeri[2] M, ASCE

Objective of Session

The object of the session is to indicate the reasons and need for ACI Standard 117: Tolerances for Concrete Construction and Materials. How restrictive specification tolerances multiply safety factors that are in codes and standard used in design and how such specifications restrain Construction Management from using cost effective judgment are indicated. A construction viewpoint about whether the tolerances in use should be strict, practical, economical or whether they should be more stringent is provided. The field and fabrication problems involving reinforcing steel congestion, restrictive tolerances and the resultant fabrication requirements are noted. The advantages of using large scale engineering - check models produced by design engineers (with input from construction personnel) for complex reinforcing steel placements is also presented.

Session Chairman Summary

One of the papers described the difficulty that the ACI Standard 117 committee had in writing the standard. Most concrete standards contain tolerance requirements and it was found that there were conflicts and inconsistencies between standards. In addition, some tolerances specified are not achievable or practical. The committee made judgments to eliminate differences and close the gaps for open ended requirements. The commieteee realizes that the standard is controversial and that it will be revised to reflect practicality and clarity. The intent of the committee is that it should be used as a guide and not be the basis for rejection.

Three of the papers dealt with the subject of "Tolerances"; views were provided by a design engineer, construction engineer and supplier.

The design engineer wondered why we complain today about tolerances that have been in existence for years. He noted that the reason is that in the past, we were not faced with the increased demands on the performance of the structure and the enforcement of strict quality assurance and quality control programs.

---

[1] Associate & Head Structural Project Engineering Division, Sargent & Lundy, Chicago, Ill.

[2] Supervisor Civil Engineering, Ebasco Services, Inc., Princeton, New Jersey

The construction engineer described things that can go wrong on a construction site. Most of these are due to intangible things that are difficult to quantify such as: project location, craft training, union versus non-union labor, personalities, type of contract, management practices, alcohol, drugs and communication.

The supplier indicated that tolerances for reinforcing steel fabrication to any degree can be met. This is done through sophisticated equipment and a strict quality control program in conjunction with well-trained craftsmen. All of these factors increase the cost of the product.

The "Modeling" paper described the value of modeling complex reinforcement placements. This can only be accomplished if the model is to a sufficient scale, complete to include all embeds and is constructed by a team of design engineers in conjunction with field personnel. In addition, it should be transported to the field where it can be studied and used during actual construction.

Consensus Recommendations*

The speaker panel and audience agreed on the following five recommendations:

1. ACI Standard 117 Tolerances for Concrete Construction and Materials has now been issued. The engineer must realize that this is a guideline and that it must be used with discretion. The engineer must specify the tolerances he requires for the structure he has designed. Since the committee for ACI-117 has used their judgment in eliminating tolerance conflicts from other standards, they are asking for comments from the profession with regard to the actual guidelines.

2. Some ACI standards have commentaries. It was a consensus that all ACI standards should have a commentary. The commentary should be more definitive than presently given in order to provide justification for the requirements in the standard.

3. Engineers should not blindly accept tolerances in the standards. They should only specify what is achievable in the shop and field and what is reasonable for the structure design.

4. Engineers and owners should pool their resources when testing and justification is required for methods of design, material and installation. Examples of this are the recent work which has been completed on concrete block walls and expansion anchors in the nuclear power plant industry.

5. Something must be done to instill in field craftsmen a pride in the work that they do. They should be made aware that careless mistakes cause excessive project costs and on occasion jeopardize the safety of the structure. It was a consensus that two factors: communications and recognition, would greatly help in this endeavor.

---

*Based upon the Panel of Speakers/Audience discussion period at the end of the session.

ACI STANDARD 117: CONCRETE TOLERANCES

by

J. Doug Sykes[1]

Abstract

An overview of ACI Standard 117 - Tolerances for Concrete Construction and Materials which has recently been published by the American Concrete Institute as the standard on tolerances is presented. There are many standards which affect the design and construction of power plants. Most of these standards appropriately address quality of design and construction. In general, all of these standards presently also address tolerance concerns to a minor extent. It is noted that the new ACI 117 standard will replace the others.

Introduction

With the past developments in the construction industry and with the anticipated future construction procedures becoming more complex, more expensive and thoroughly inspected; the previous tolerances need careful review and improvement. We have all experienced projects wherein tolerances created problems. At every gathering of concerned designers and builders, we all hear "horror" stories of tolerance problems. ACI Committee 117 hopes that the publication of a standard on tolerances will be the *first* step toward minimizing these tolerance difficulties.

During the development of the present standard, it became obvious that the legalistic interpretation of codes and standards presents a unique challenge toward developing appropriate tolerances. This was particularly recognized during the development of the standard and resulted in removal from this standard any reference to the Code for Concrete Reactor Vessels, ACI 359.

History

It is useful to look at the history of the development of the ACI Tolerance Standard. The committee was constituted in 1963. The original mission of the committee was to review the existing ACI standards and to develop an overview for the practicality of the tolerances. The committee was charged with the responsibility to achieve uniformity and practicality of tolerances through coordination with all other committees.

---

[1] Chief Civil Engineer, Ebasco Services Incorporated, Norcross, GA

For many years, the committee received much input regarding inconsistencies of tolerances and determined that most of the tolerances that existed were not realistic or enforceable. Papers were written and seminars were conducted to discuss these inconsistencies and difficulties. Everyone seemed to know where problems existed, but no one knew the best way to resolve the problems.

Based on extensive discussions, the Tolerance Committee decided, about four years ago, that the best approach was to develop a single ACI tolerance standard; this standard is now available. It consists of a compilation of tolerances currently shown in the ACI documents which are referenced in the standard. The committee's aim was not to redefine existing standards nor to create new standards, only to collect all of the tolerances into one document.

During development of the present standard, however, it became obvious that the same tolerances had different values in different documents, and that there are certain open items not addressed in any of the existing documents. As a result, certain judgments were made by the committee to eliminate differences and close the gaps.

The present tolerance document resulted from extensive discussions with other ACI committees and with concerned members.

The committee feels that this document represents the present state of the standards of all ACI committees. We do not believe that the document is complete nor final. It will be revised to reflect the concerns of useability and clarity as experience with its application determines. This standard should be used as a guide and "should be used with judgment as a range for acceptability and not a limit for rejection."

Philosophy

During the development of the Tolerance Standard, the committee had extensive discussions regarding the philosophy of tolerances and the philosophy of the standard. The printed introduction to the standard addresses this. Several excerpts from the introduction are important.

"The designer must specify and clearly identify those items which require either closer or more lenient tolerances as the needs of the project dictate." This implies that the designer of the specific structure must take into consideration tolerances, during the conceptual development of the structure, and must reflect these tolerances in the project documents, so that the constructor can understand the requirements and can achieve the results.

"Necessity rather than desirability should be the basis for selecting tolerances." This implies that every individual would like the best possible results, but that high accuracy is not required in every circumstance. Good engineering judgment is always required to achieve cost-effective construction and inspection.

"No structure is <u>exactly</u> level, plumb, straight, and true." If this fact can be fully appreciated by the designer, constructor and inspector, many conflicts will be avoided.

Overview of Standard

The standard has been divided into six sections, each of which addresses a related area. The first section is devoted to Definitions and Principles, and is necessary to establish the basis for all of the data contained in the other sections. As is normal, the principles of a document, as described in the opening paragraphs, are seldom understood by the people who attempt to interpret the document.

The second section deals with General Building construction which accounts for the largest quantity of concrete construction in the building industry. This is the section that deals with the majority of tolerances with which most people are involved.

The third section deals with Special Structures. These structures are the exceptions to normal buildings and require special considerations due to their use or method of construction.

Section 4 covers Pre-cast Concrete. This is a major segment of the concrete construction industry. Due to the usually better controlled conditions for construction, tolerances can be different.

The fifth section covers Masonry Construction. This addresses "units joined together with mortar" for which construction techniques are considerably different from normal <u>placement</u> of concrete.

The sixth and final section deals with Materials. This section addresses concrete and reinforcing steel as would normally be expected. It also addresses other embedded items.

Illustration

Rather than trying to review each paragraph and each tolerance, selected illustrations will be used to clarify the content and the intent of Committee 117 in producing the Tolerance Standard. The normal tolerance on vertical control is stated that the line should be plumb within 1/4" in any 10 feet. It is obvious that this is unlimited as far as total height of the structure is concerned and, therefore, requires additional clarification. The limitation in the standard is one inch for the total height of the structure.

This also produces a question regarding what is the possibility of height of a structure. Therefore, a footnote was added which states that the total height is limited to 100 feet. Further, the footnote requires that if the structure is more than 100 feet high, a special tolerance must be determined.

In this tolerance for plumb, there is no plus or minus designation. This omission has been questioned many times. The reason that positive and negative is omitted from plumb is the difficulty in determining which way is plus - right or left.

With regard to horizontal dimensions, a limitation of plus or minus 1/2" in any bay for planned dimensions is stated. The tolerance for floor finish is divided into several classes, varying from a tolerance of 1/8" to 1/2" below a 10 foot straight edge. This is a uni-directional tolerance measured at any location by use of a flat tool of specific length.

The standard then adds a comment that the floor tolerance measurement must be made within 24 hours and prior to removing any form work. This illustrates one of the major oversights in tolerance standards. The time of measurement, the method of measurement, and the conditions of the structure during measurement can significantly influence the results.

The sensitivity of owners to the flatness of floors can obviously produce some interesting results. If the owner requires compliance with a flatness tolerance at the time of occupancy, this becomes a problem of considerable magnitude. Producing a flat floor which will remain flat forever obviously dictates construction technique and cost. We must remember that any load or temperature change produces deformations.

In the section on concrete dimensions, tolerances are specified for overall dimensions of the finished concrete. In the section on Materials, tolerances on the location of the encased reinforcing steel are based on the design depth of the member. This is one illustration of a difference in criteria within the present standard.

A final illustration of the contents of the standard concerns material, in which tolerances are given regarding proportions of the components of concrete. These deal with such things as air entraining additives for which tolerances are frequently ignored. The standard also addresses tolerance in slump which is a property of the concrete frequently measured during placement. It is important to note that the standard comments that this tolerance on slump only applies when the project specification does not address the slump requirements. The intent is that the project specification will require a range of values for slump rather than a single value. Actually specifying the designer's limitation of range is preferable.

Implementation

Implementation of the above illustrated standards for tolerances is an item which the committee intends to address during its continuing deliberations. The General Building tolerances applying to most cast-in-place concrete deals with completed dimensions and locations. These dimensions are controlled during form construction, installation of encased items, and placement of concrete. Thus, the resulting actual location, which is measured after placing concrete, is determined before the concrete is placed in the forms. Recognizing this difficulty of securing proper in-place control through measurements is one of the main concerns of the committee.

Since these present tolerances came from existing ACI standards, such as 301 - Specifications, 304 - Mixing, 318 - Building Code, and 347 - Form Work, the committee must consider the intent of these previous committees, but must also consider the practicality of achieving the specified results. During the past development of tolerances by the many ACI committees, the prime concern of the individuals on that committee was their standard and not tolerances as specifically related to the inspection aspects. Committee 117 is particularly concerned with developing data and concepts that will provide a reasonable basis for implementation of tolerance standards.

Future

Committee 117 is now embarking on the integration of tolerance requirements with actual construction experience. The goal is to improve the numerous values for the tolerances to provide both safe and economical construction. At the present, we are involved in resolving the difference in the basis for member size and rebar placement measurements. Design depth is important for safety, but difficult to interpret in the field. Overall dimensions can be easily measured but misleading as to strength. The engineer who designs the structure must take this concept into account and apply both the design and construction standards appropriately. The project specification must convey his requirement to the contractor for use during quality control.

Committee 117 will attempt to resolve differences in depth-design and depth-overall, in rebar-fabrication and rebar-placement, and in any other items which are brought to our attention. Please help us by conveying your concerns, in a specific manner, with recommendations that can be applied in an overall tolerance standard. The entire standard is included herein as Appendix A.

Our work has just begun by publishing one tolerance standard. All other committees must remove tolerances from their standards and refer to ACI 117. Field data must be obtained. Then an economical, enforceable and safe tolerance can be specified.

ACI STANDARD 117

APPENDIX A

Authorized Reprint From
Concrete International,
Vol. 2, No. 8, Page 38

Proposed ACI Standard: Tolerances for Concrete Construction and Materials

Reported by ACI Committee 117

This proposed standard covers the tolerances currently shown in the referenced ACI documents. In this initial attempt, the committee's direction was to collect the existing tolerances into one tolerance document. The situation where the same tolerance had different values in different documents was resolved in several ways; the attempt being to find some tolerance acceptable to both parties. A few tolerances were added to "complete" or "close" existing tolerances.

It is intended that this document will be used for tolerances rather than several other committee standards. This process will take some time since each affected committee will have to incorporate a reference to this document and delete specific tolerances.

Keywords: bending (reinforced steels); building codes; concrete construction; concretes; formwork (construction); masonry; mass concrete; piles; precast concrete; prestressed concrete; reinforcing steels; splicing; **standards; tolerances (mechanics).**

Contents

Introduction

1.0 — Definitions and Principles, page 39
 1.1 — Definitions
 1.2 — Principles

2.0 — General Building — Cast-in-place, page 40
 2.1 — Tolerance applying to concrete dimensions and locations only
 2.2 — Tolerances for finished slab surfaces

3.0 — Special Structures, page 41
 3.1 — Tolerances for mass concrete structures
 3.2 — Tolerances for concrete canal lining
 3.3 — Tolerances for monolithic siphons and culverts
 3.4 — Tolerances for bridges, checks, overchutes, drops, turnouts, inlets, chutes and similar structures
 3.5 — Slipformed construction
 3.6 — Pavements

4.0 — Precast Concrete, page 42
 4.1 — Tolerances for precast nonprestressed elements
 4.2 — Tolerances for precast prestressed elements
 4.3 — Precast panels
 4.4 — Precast piles

5.0 — Masonry Construction, page 45
 5.1 — Tolerances in joints
 5.2 — Tolerances for walls

6.0 — Materials, page 45
 6.1 — Reinforcing steel
 6.2 — Concrete materials
 6.3 — Concrete
 6.4 — Embedded materials

Discussion closes Nov. 1, 1980.
Copyright © 1980 American Concrete Institute. All rights reserved including rights of reproduction and use in any form or by any means, including the making of copies by any photo process, or by any electronic or mechanical device, printed or written or oral, or recording for sound or visual reproduction or for use in any knowledge or retrieval system or device, unless permission in writing is obtained from the copyright proprietors.

APPENDIX A

Introduction

No structure is exactly level, plumb, straight, and true. Fortunately, such perfection is not necessary. Tolerances are a means to establish permissible variation in dimension and location; giving both the designer and the contractor parameters within which the work is to be performed. They are the means by which the designer conveys to the contractor the performance expectations upon which the design is based or the use of the project requires. As such, specified tolerances should reflect design assumptions and project needs; being neither overly restrictive nor lenient. Necessity rather than desirability should be the basis of selecting tolerances.

The required degree of accuracy of performance depends on the interrelationship of several factors:

Structural strength and function requirements.
The structure must be safe and strong reflecting the design assumptions, and accurate enough in size and shape to do the job for which is was built.

Esthetics.
The structure must satisfy the appearance needs or wishes of the owner and designer.

Economic feasibility.
The specified degree of accuracy has a direct impact on the cost of production and for the construction method. In general the higher degree of accuracy required, the higher the cost of obtaining it.

Relationship of all components.
The required degree of accuracy of individual parts can be influenced by adjacent units and materials, joint and connection details, and the possibility of the accumulation of tolerances in critical dimensions.

Construction techniques.
The feasible tolerance depends on available craftsmanship, technology, and materials.

Properties of materials.
Specified degree of accuracy for shrinkage and prestressed camber should recognize the difficulty of predetermining deflection due to shrinkage and prestressed camber.

Compatibility.
Designers are cautioned to use finish and architectural details which are compatible with the type and anticipated method of construction. Finish and architectural details used should be compatible with the concrete tolerances which are achievable.

Job conditions.
Unique job situations and conditions must be considered. The designer must specify and clearly identify those items which require either closer or more lenient tolerances as the needs of the project dictate.

Measurement.
Mutually agreed upon control points and bench marks must be provided as reference points for measurements to establish the degree of accuracy of items produced and for verifying the tolerances of the item produced. Control points and bench marks should be established and maintained in an undisturbed condition until final completion and acceptance of the project.

The tolerances stated in this document have been taken from numerous ACI standards and reports. In some cases the tolerances in these documents conflict and may conflict with the tolerances shown in this document. In the event of conflict it is the *responsibility* of the *specifier* to *clearly state which document governs.*

The stated tolerances in this document were, in many cases, derived by consensus opinion due to the lack of definitive data; *and should be used with judgement* as a range for acceptability and not a limit for rejection.

1.0 — Definitions and principles

1.1 — Definitions

Arris — The ridge formed by the meeting of two surfaces.

Basic Dimension — The dimensions shown on the contract drawings or called for in the specifications. The basic dimension applies to size, location, and relative location.

Bowing — The curvature of an element in one plane. The curvature may be concave (negative) or convex (positive).

Camber — The maximum change in elevation from a straight line through the end points of an element. A deflection that is intentionally built into a structural element or form to improve appearance or to nullify the deflection of the element under the effects of loads, shrinkage, and creep.

Clearance — Interface space between two items. Normally specified for concrete coverage of reinforcing steel or to allow for tolerance and for structural movement caused by volume change or elastic deflection.

Variation — The difference between the actual dimension and the basic dimension. Variations may be either negative (less) or positive (greater).

Level — A line or plane perpendicular to plumb.

Plumb — A vertical direction radiating from the center of the earth commonly determined by a suspended weight.

Skew — An out-of-square variation from a rectangular shape. This is normally measured by comparing the length of the diagonals.

Tolerance (T) —
1. The permitted variation from a basic dimension or quantity.

APPENDIX A

2. The range of variation permitted in maintaining a basic dimension.
3. A permitted variation from location or alignment.

1.2 — Principles

1.2.1 Tolerances are generally expressed in terms of their negative and positive components. Where variations are symmetrical, the tolerance (T) should be expressed as ±V.

1.2.2 Tolerances should be reviewed prior to specifying so that they reflect design assumptions and requirements, use of the element, accumulation of variations allowed for various elements, functional need, appearance and practical ability, and cost to achieve. If the conditions warrant, the *specifier should indicate the age and temperature criteria at which tolerances apply.*

1.2.3 Where no tolerances are stated in the specifications or drawings for any individual element, structure or feature, recommended permissible variations are contained herein. No tolerance indicated shall be construed to permit encroachment of the structure beyond legal property boundaries.

1.2.4 Preconstruction meetings should be held for the purpose of reviewing critical tolerances, methods of making measurements, and the basis for acceptance/rejection of completed work to avoid misunderstandings at the time of final inspection.

2.0 — General building — Cast-in-place (Refer ACI-301, 302, 347)

2.1 — Tolerance applying to concrete dimensions and locations only.

2.1.1 *Plumb: (allowable variation)*

2.1.1.1 *In the lines and surfaces of columns, piers, walls, and in arrises:*
In any 10 ft .... ¼ in.
Maximum for the total height of the structure* .... 1 in.

2.1.1.2 *For exposed corner columns, control-joint grooves and other conspicuous lines:*
In any 20 ft .... ¼ in.
Maximum for the total height of the structure* .... ½ in.

2.1.1.3 *For slipformed walls or columns with respect to a reference point at the base of the structure, including both translational and rotational components:*
In any 5 ft of height .... 1/8 in.
In any 50 ft of height .... 1 in.
Maximum in total height of structure (up to 600 ft) .... 3 in.

2.1.2 *Level or from the grades and elevations specified in the contract documents:*

2.1.2.1 *In slab soffits, ceilings, beam soffits and in arrises, measured before removal of supporting shores:*
In any 10 ft .... ±¼ in.
In any bay or in any 20 ft .... ±3/8 in.
Maximum for the total length of the structure .... ±¾ in.

2.1.2.2 *In exposed lintels, sills, parapets, horizontal grooves, and other conspicuous lines:*
In any bay or in 20 ft .... ±¼ in.
Maximum for the total length of the structure .... ±½ in.

2.1.3 *Linear building lines from the basic dimension in plan and related position of columns, walls, beams, and partitions:*
In any bay .... ±½ in.
In any 20 ft .... ±½ in.
Maximum for the structure .... ±1 in.

2.1.4 *Size of sleeves, floor openings, and wall openings* .... ±¼ in.
Location of the center lines of sleeves, floor openings, and wall openings .... ±½ in.

2.1.5 *Cross-sectional dimensions of columns, beams, walls, and slab thickness. (Including walls and columns constructed using slipforms.)*
Up to 12 in. .... +3/8 in.
.... −¼ in.
More than 12 in. .... +½ in.
.... −3/8 in.

2.1.6 *Footings:*

2.1.6.1 *Horizontal dimensions*
(formed) .... +2 in.
.... −½ in.
(unformed excavation) .... +3 in.

2.1.6.2 *Misplacement or eccentricity:*
2 percent of the footing width in the direction of misplacement but not more than .... ±2 in.

2.1.6.3 *Cross-sectional thickness* .... +no limit
.... −5 percent

2.1.6.4 *To receive masonry construction:*
Alignment in 10 ft .... ±¼ in.
Maximum for entire length 50 ft .... ±½ in.
Level in 10 ft. .... ±¼ in.
Maximum for entire length 50 ft. .... ±½ in.

2.1.6.5 *Level — Footings for construction other than that shown in 2.1.6.4 (but not to exceed limits of 2.1.6.3)*
+½ in.
−2 in.

2.1.7 *Stairs:*

2.1.7.1 *For an individual step:*
Riser .... ±1/8 in.
Tread .... ±¼ in.

2.1.7.2 *In a flight of stairs:*
Rise .... ±1/8 in.
Run .... ±¼ in.

2.2 — Tolerances for finished slab surfaces

2.2.1 *Class of tolerance.* Specifier shall designate class to be used.

---

*Total height is taken to be less than 100 ft. Structures with heights in excess of these values are to be considered special cases and other overall tolerances should be considered and/or specified.

APPENDIX A

2.2.1.1 *Class AA Surface Finish Tolerance*
Depressions in floors between high spots shall not be greater than 1/8 in. below a 10 ft long straightedge.*

2.2.1.2 *Class AX Surface Finish Tolerance*
Depressions in floors between high spots shall not be greater than 3/16 in. below a 10 ft long straightedge.*

2.2.1.3 *Class BX Surface Finish Tolerance*
Depressions in floors between high spots shall not be greater than 5/16 in. below a 10 ft long straightedge.*

2.2.1.4 *Class CX Surface Finish Tolerance*
Depressions in floors between high spots shall not be greater than ½ in. below a 10 ft long straightedge.*

2.2.2 *Floor tolerance measurements should be made the day after a concrete floor is finished and before the shoring is removed, in order to eliminate any effects of shrinkage, curling, and deflection.*

2.2.3 *Cost of achievability factors*

2.2.3.1 *Class AA or closer finish tolerances are extremely difficult and expensive to achieve on large areas. They should be specified only for critical areas where such tolerances are vital for the operations that will take place in the areas. Specifications for bidding contractors should thoroughly cover:*

(a) the importance of achieving the tolerance specified.
(b) the exact areas involved.
(c) the minimum joint spacing permitted.
(d) the precise method of measurement using a 10 ft long straightedge that will be used to approve or reject the floors involved.

Full clarification of such information before bidding will enable the contractor to price the work realistically, avoid misunderstandings, and provide greater assurance of obtaining the desired results.

2.2.3.2 *Class BX finish tolerances are generally practical for floors over metal decking or precast beams if proper compensation has been made for deflection. The lack of planeness usual in the decking or precast beams makes closer tolerances quite difficult.*

Class BX finish tolerances are generally suitable for Class 1, 2, and 3 floors. They may also be suitable for Class 4, 5, 6, and 7 floors where traffic and other use considerations do not require closer surface tolerances.
See ACI 302 for definition of class of floors.

2.2.3.3 *Tolerances for floors cast on metal decks or other easily deflected material shall not be less than the calculable deflection anticipated.*

3.0 — Special structures (Refer ACI-316, 347)

3.1 — Tolerances for mass concrete structures

3.1.1 *Linear outline:*
  3.1.1.1 *In any 20 ft of length*.....................±½ in.
  3.1.1.2 *In any 40 ft of length*.....................±¾ in.
  3.1.1.3 *In any 80 ft of length or greater*..±1¼ in.

3.1.2 *Plumb, the specified batter, or from the curved surfaces of all structures including the lines and surfaces of columns, walls, piers, buttresses, arch sections, vertical joint grooves, and visible arrises:*
  3.1.2.1 *In 10 ft of height*...........................±½ in.
  3.1.2.2 *In 20 ft of height*..........................±¾ in.
  3.1.2.3 *In 40 ft of height or greater†* ....±1¼ in.

3.1.3 *Level or from grades specified in the contract drawings for slabs, beams, soffits, horizontal grooves, and visible arrises:*
  3.1.3.1 *In 10 ft of length*..........................±¼ in.
  3.1.3.2 *In 30 ft of length or greater*..........±½ in.

3.1.4 *For buried construction the tolerance allowable is twice the above amounts.*

3.2 — Tolerances for concrete canal lining

3.2.1 *Established alignment:*
  Tangents...±2 in.
  Curves ....±4 in.

3.2.2 *Established profile grade*......................±1 in.

3.2.3 *Thickness of lining cross section — 10 percent of specified thickness provided average thickness is maintained as determined by daily batch volumes.*

3.2.4 *Width of section at any height:*
  ±0.0025 times specified width plus 1 in.

3.2.5 *Height of lining:*
  ±0.005 times established height plus 1 in.

3.2.6 *Surfaces:*
  3.2.6.1 *Invert-in any 10 ft length*...............±¼ in.
  3.2.6.2 *Slope of side-in any 10-ft length*.....±½ in.

3.3 — Tolerances for monolithic siphons and culverts

3.3.1 *Alignment* ....±1 in.
3.3.2 *Profile grade* ....±1 in.
3.3.3 *Cross section thickness at any point*
  + greater of 0.05 times thickness or ....½ in.
  − greater of 0.025 times thickness or ......¼ in.
3.3.4 *Inside dimensions* ±0.005 times inside dimension.
3.3.5 *Surfaces:*
  3.3.5.1 *Inverts in any 10 ft length*............±¼ in.
  3.3.5.2 *Slopes of side-in any 10 ft length*...±½ in.

3.4 — Tolerances for bridges, checks, overchutes, drops, turnouts, inlets, chutes, and similar structures

3.4.1 *Alignment* ....±1 in.
3.4.2 *Grades*..±1 in.
3.4.3 *Plumb or specified batter in the lines and surfaces of columns, piers, walls and arrises:*

---

*Compliance with the designated limits in four of five consecutive measurements should generally be satisfactory unless obvious faults are observed.
†Total height is taken to be less than 100 ft. Structures with heights in excess of these values are to be considered special cases and other overall tolerances should be considered and/or specified.

APPENDIX A

3.4.3.1 *Exposed*
In any 10 ft length..................................±½ in.
Maximum for entire length..................±¾ in.
3.4.3.2 *Backfilled*
In any 10 ft of length ....±1 in.
Maximum for entire length ....±1½ in.
3.4.4 *Level or grades for slabs, beams, horizontal grooves and railing offsets:*
3.4.4.1 *Exposed:*
In any 10 ft of length ....±½ in.
Maximum for entire length ....±¾ in.
3.4.4.2 *Backfilled:*
In any 10 ft of length ....±1 in.
Maximum for entire length ....±1½ in.
3.4.4.5 *Cross-sectional dimensions of columns, beams and piers, and thickness of slabs and walls*......+½ in.
..........−¼ in.
3.4.6 *Bridge slabs (thickness)*........................+¼ in.
........................−1/8 in.
3.4.7 *Footings*
See Section 2.1
3.4.8 *Size and location of slab and wall openings*..±½ in.
3.4.9 *Plumb or level in sills and side walls for radial gates and similar watertight joints:*
3.4.9.1 *In any 10 ft of length*....................±1/8 in.
3.4.9.2 *Maximum for entire length*..........±3/16 in.

3.5 — Slipformed construction
3.5.1 *Variation from prescribed inside dimensions for noncircular structures between opposite walls shall not exceed:*
Per 10 ft ....±½ in.
Maximum..±2 in.

3.6 — Pavements
3.6.1 *Mainline pavements (from specified slope)*
3.6.1.1 *Longitudinal direction as measured with 10 ft straightedge* ....1/8 in.
3.6.1.2 *Transverse direction as measured with 10 ft straightedge*..¼ in.
3.6.2 *Ramps and Intersections (from specified slope)*
As measured with 10 ft straightedge........¼ in.
3.6.3 *Dowels*
3.6.3.1 *Placement (from specified location)*..±1 in.
Alignment for 18 in. dowel......................±¼ in.

4.0 — Precast concrete
(Refer ACI-347, 533)

4.1 — Tolerances for precast nonprestressed elements
4.1.1 *Length of element*
Per 10 ft of length ....±1/8 in.
Maximum for entire length....................±¾ in.
4.1.2 *Cross-sectional dimensions*
Sections less than 6 in.........................±1/8 in.
Sections 6 in. and less than 18 in. ....±3/16 in.
Sections 18 in. to 36 in. ....±¼ in.
Sections over 36 in. ....±3/8 in.

4.1.3 *Variation from straight line*
In any 10 ft of length...........................±1/8 in.
Maximum for the entire length...............±¾ in.
4.1.4 *Camber (Variation from specified)*
Per 10 ft of span..................................±1/8 in.
But not greater than ....±½ in.
4.1.5 *Differential in camber between adjacent units in erected position*
Per 10 ft of span..................................±1/8 in.
But not greater than ....±½ in.

4.2 — Tolerances for precast prestressed elements*
4.2.1 *Length of element*
Per 10 ft of length ....±1/8 in.
Maximum for entire length....................±¾ in.
4.2.2 *Cross-sectional dimensions*
Sections less than 6 in.........................±1/8 in.
Sections 6 in. and less than 18 in. ......±3/16 in.
Sections 18 in. to 36 in. ....±¼ in.
Sections over 36 in. ....±3/8 in.
4.2.3 *Variation from straight line*
In any 10 ft of length............................±1/8 in.
Maximum for the entire length...............±¾ in.
4.2.4 *Camber — variation from specified*
Per 10 ft of span..................................±1/8 in.
But no greater than ....±1 in.
4.2.5 *Differential in camber between adjacent units in erected position*
Per 10 ft of span..................................±1/8 in.
But not greater than ....±1 in.

4.3 — Precast panels
4.3.1 *Casting tolerances*
4.3.1.1 *Height and width of panel*
Basic Dimension:
Under 10 ft ....±1/8 in.
10 ft to 20 ft.......................................+1/8 in.
....−3/16 in.
Over 20 ft to 30 ft...............................+1/8 in.
....−¼ in.
Each additional 10 ft increment in excess of 30 ft..±1/16 in.
4.3.1.2 *Thickness*...+¼ in.
....−1/8 in.
4.3.1.3 *Skew-measured by the difference in length of the two diagonals*
Per 6 ft of diagonal length ....−1/8 in.
....+¼ in.
4.3.1.4 *Openings cast into panels*
Size of opening ....±¼ in.
Location of center line of opening..........±¼ in.
4.3.1.5 *Location of embedded items*
Inserts, bolts, pipe sleeves, etc............±3/8 in.
Flashing reglets at panel edge..............±¼ in.
Reglets for glazing gaskets ....±1/8 in.
Electrical outlets, hose bibs, etc. ....±½ in.
4.3.2 *After casting tolerances (monolithically cast panels)*

---
*For further information see Manufactured Product Standards.

APPENDIX A

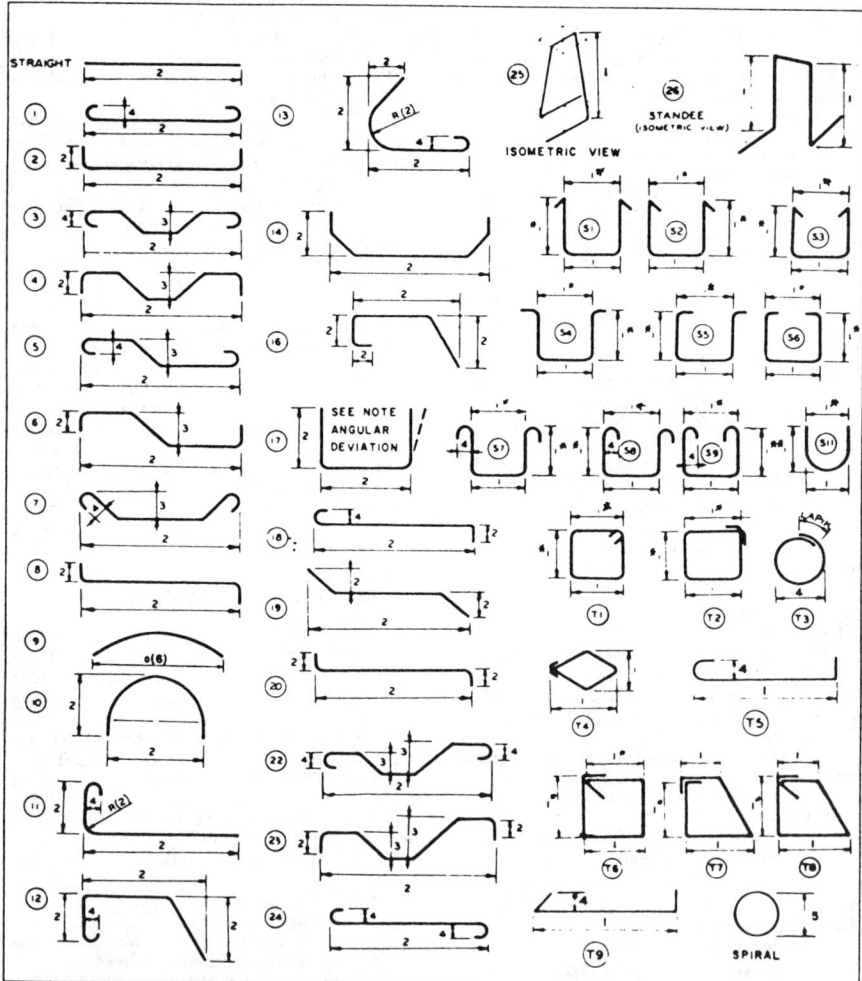

Note: 1. Entire shearing and bending tolerances are customarily absorbed in the extension past the last bend in a bar.
2. Tolerances for Types S1-S9, S11, T1-T9 apply to bar sizes #3-#8 inclusive only.
*Dimensions on this line are to be within tolerance shown but are not to differ from the opposite parallel dimension more than 1/2 in.

Fig. 6.1.1a — Standard fabricating tolerances for bar sizes #3 through #11

ACI STANDARD 117

APPENDIX A

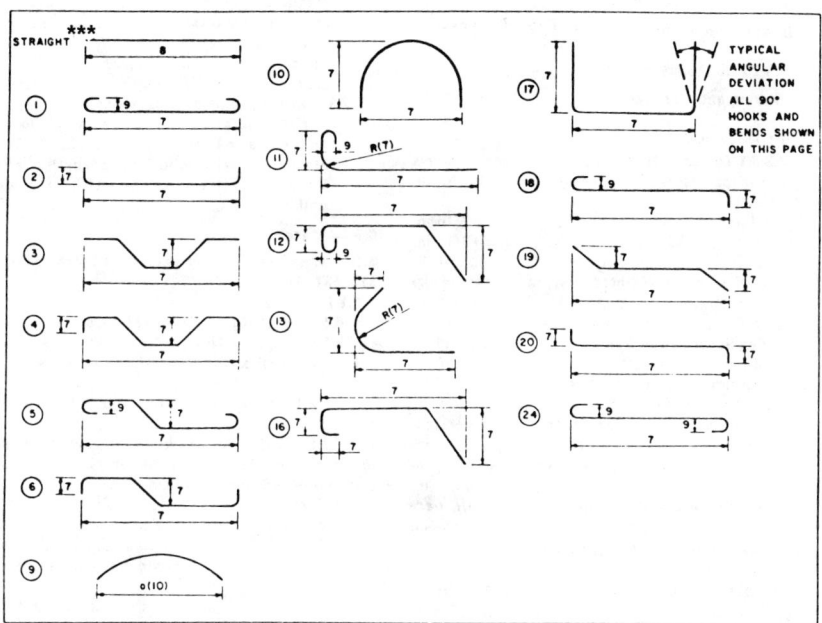

Tolerance Symbols for bar sizes #3 through #11

1. Bar sizes #3, 4, 5 = Plus or Minus ½" (gross length < 12'-0")
   Bar sizes #3, 4, 5 = Plus or Minus 1" (gross length ≥ 12'-0")
   Bar sizes #6, 7, 8 = Plus or Minus 1"
2. Plus or Minus 1"
3. Plus 0", Minus ½"
4. Plus or Minus ½"
5. Diameter ≤ 30" = Plus or Minus ½"
   Diameter > 30" = Plus or Minus 1"
6. Plus or Minus 1.5% × "0" dimension ≥ ± 2" Min.**

Tolerance Symbols for bar sizes #14 & #18

	#14	#18
7 = Plus or Minus	2½"	3½"
8 = Plus or Minus	2"	2"
9 = Plus or Minus	1½"	2"
10 = Plus or Minus 2% × "0" Dim. ≥	± 2½" Min.	± 3½"** Min.

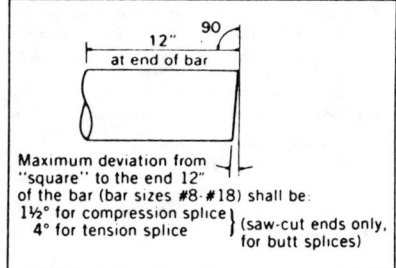

Maximum deviation from "square" to the end 12" of the bar (bar sizes #8-#18) shall be: 1½° for compression splice, 4° for tension splice (saw-cut ends only, for butt splices)

**If application of positive tolerance to Type 9 results in a chord length equal to or greater than the arc or bar length, the bar may be shipped straight.

Angular Deviation — maximum ± 2½° or ± ½"/ft on all 90° hooks and bends.
***Saw cut both ends — overall length ± ½".
All tolerances single plane and as shown.

Fig. 6.1.1b — Standard fabricating tolerances for #14 and #18

APPENDIX A

4.3.2.1 *With intermediate support*
Bowing and warpage........± $\dfrac{\text{Panel Dimension}}{360}$

4.3.2.2 *Without intermediate support*
Bowing and warpage........± $\dfrac{\text{Panel Dimension}}{240}$

4.3.3 *Erection tolerance*
  4.3.3.1 *Distance between panels at face of panels* For panels with dimensions (normal to the joint)
  Of under 10 ft ....±3/16 in.
  Of 10 ft to 20 ft ....+3/16 in.
  .... −1/4 in.
  For each 10 ft increment in excess of 20 ft..±1/16 in.
  4.3.3.2 *Joint taper (panel edges not parallel)*
  Per lineal foot of joint.............................1/40 in.
  Minimum allowable ....1/16 in.
  Maximum for entire length....................3/8 in.
  4.3.3.3 *Panel alignment*
  Alignment of horizontal and vertical joints..1/4 in.
  Offset in exterior face of adjacent panels...1/4 in.
  4.3.3.5 *Location of openings in wall panels*..±1/4 in.

4.4 — Precast piles (Refer ACI 543)
  4.4.1 *Length*
  4.4.1.1 *Manufactured length per 10 ft of length*..±3/8 in.
  4.4.1.2 *Cut-off length*.....................................±2 in.
  4.4.2 *Cross-sectional dimensions*.....................+1/2 in.
  ....−1/4 in.
  4.4.3 *Wall thickness of hollow sections* ....+3/8 in.
  ....−1/4 in.
  4.4.4 *Departure from a straight line parallel to the center line of the pile*
  Per any 10 ft length..................................1/8 in.
  4.4.5 *Departure of internal core or void from position indicated by plans or specification*...........±3/8 in.
  4.4.6 *Departure of pile head from plane at right angles to longitudinal axis of pile*
  Per foot of head dimension ....±1/4 in.
  4.4.7 *Pile head surface irregularities*...........±1/8 in.

5.0 — Masonry construction (Refer ACI-531)
5.1 — Tolerances in Joints
  5.1.1 *Head and bed joint thickness*..............±1/8 in.
  5.1.2 *Head joint vertical alignment and bed joint level:*
  In any 10 ft of length ....±1/4 in.
  Maximum for the entire length..............±1/2 in.

5.2 — Tolerances for walls
  5.2.1 *Cross-sectional thickness, multiple wythes*...±1/4 in.
  5.2.2 *Wall alignment:*
  Plumb
    In any 10 ft ....±1/4 in.

    Maximum per floor..................................±1/2 in.
    Maximum for total height........................±1 in.
  Horizontal
    In any 10 ft..±1/4 in.
    Maximum for entire length.....................±1/2 in.
  5.2.3 *Top surface of bearing walls*
  Variation in grade between adjacent floor elements in any 10 ft ....±1/8 in.
  Maximum variation for entire length ....±1/4 in.
  Variation in level within width of a single unit...±1/16 in.

6.0 — Materials
6.1 — Reinforcing steel (Refer ACI 318, 349, 301, 315, 531, 543)
  6.1.1 *Fabrication*
    6.1.1.1 *Fig. 6.1.1a for #3-#11 bars*
    6.1.1.2 *Fig. 6.1.1b for #14 and #18 bars*
  6.1.2 *Placement for flexural members, walls, and compression members*
    6.1.2.1 *Tolerance in clear distance to formed soffit*..−1/4 in.
    6.1.2.2 *Tolerance in depth "d" (where d is the distance from the extreme compression fiber to the centroid of tension reinforcement).*
    depth "d"
    8 in. or less ....±3/8 in.
    More than 8 in.......................................±1/2 in.
    6.1.2.3 *Tolerance on minimum concrete cover* depth "d"
    8 in. or less..−3/8 in.*
    More than 8 in. ....−1/2 in.*
    6.1.2.4 *Tolerance on minimum distance between bars*...−1/4 in.
    6.1.2.5 *Tolerance in uniform spacing of reinforcement from theoretical location*..................±2 in.
    6.1.2.6 *Tolerance in uniform spacing of stirrups and ties from theoretical location*.......................±1 in.
    6.1.2.7 *Tolerance in longitudinal location of bends and ends of bars.*
    General...±2 in.
    Discontinuous ends of members..............±1/2 in.
    6.1.2.8 *Tolerance in length of bar laps*.....−1 1/2 in.
    6.1.2.9 *Tolerance in embedded length*
    #3-#11..−1 in.
    #14 and #18...−2 in.
  6.1.3 *Placement of prestressing steel and prestressing steel ducts in precast element (Refer ACI 301, 318, 349, 543)*
    6.1.3.1 *Tolerance on depth "d".*
    See Section 6.1.2.2
    6.1.3.2 *Slabs*
    Horizontal tolerance in any 15 ft of tendon length...±1
    6.1.3.3 *Tolerance for bearing plate concentricity and perpendicularity with tendons and concrete*..±1 deg

---

*But not to exceed 1/3 specified concrete cover.

APPENDIX A

6.2 — Concrete materials (Refer ACI-301, 302, and 304)

6.2.1 Tolerance in measurement of quantity of materials for batching concrete.

	Batch weights greater than 30 percent of scale capacity		Batch weights less than 30 percent of scale capacity	
	Individual Batching	Cumulative Batching	Individual Batching	Cumulative Batching
Cement and other cementitious materials	±1% or ±0.3% of scale capacity whichever is greater		Not less than required weight or more than 4% required weight.	
Water (by volume or weight) percent	±1%		±1%	
Aggregates, percent	±2%	±1%	±2%	±0.3% of scale capacity
				±3% of required cumulative weight whichever is less
Admixtures (by volume or weight), percent	±3%		±3%	

6.3 — Concrete (Refer ACI-301, 302, and 304)

6.3.1 Tolerance in slump (where no range specified)*...±1 in.

6.3.2 Tolerance in air content, by volume 4% minimum (where no range specified)*..........±1½ percent

6.4 — Embedded materials

6.4.1 Tolerance from specified clearance relative to reinforcing..±1 in.†

6.4.2 Tolerance from specified location.........±¼ in.

Section 2.1.2.3
Elevation control points for slabs on grade
In any bay or 20 ft ....±3/8 in.
Maximum for total length of structure..±3/4 in.

ACI Committee 117

Reference documents

ACI 301-72, (Revised 1975) Specifications for Structural Concrete for Buildings
ACI 302-69, Recommended Practice for Floor and Slab Construction
ACI 304-73, Recommended Practice for Measuring, Mixing, Transporting and Placing Concrete
ACI 315-74, Manual of Standard Practice for Detailing Reinforced Concrete Structures
ACI 316-74, Recommended Practice for Construction of Concrete Pavements and Concrete Bases
ACI 318-71, Building Code Requirements for Reinforced Concrete
ACI 345-74, Recommended Practice for Concrete Highway Bridge Deck Construction
ACI 347-78, Recommended Practice for Concrete Formwork
ACI 349-76, Code Requirements for Nuclear Safety Related Concrete Structures

ACI 531-79, Concrete Masonry Structures — Design and Construction
ACI 533 Report, Part 3 of the Manual of Concrete Practice, Design of Precast Concrete Wall Panels
ACI 543 Report, Part 3 of the Manual of Concrete Practice, Recommendations for Design, Manufacture and Installation of Concrete Piles
CRSI Manual of Standard Practice (22nd Edition) MSP 2:77

---

*Where range is specified there is no tolerance.
†But not less than diameter of the reinforcing bar.

This report was submitted to letter ballot of the committee which now consists of 15 members; 11 members returned affirmative ballots, 4 ballots were not returned. It has been processed in accordance with the Institute standardization procedure and is approved for publication and discussion. Participation by Federal agency representatives in the work of the American Concrete Institute and in the development of Institute standards does not constitute Government endorsement of ACI or the standards which it develops.

Section 4.5 Erection Tolerances
Refer to Section 2.1 for in-place tolerances of elements covered in Sections 4.1, 4.2, and 4.3.

PROPOSED ACI STANDARD:
TOLERANCES FOR CONCRETE CONSTRUCTION AND MATERIALS

A. Ernest Fisher, III
Chairman

Robert C. Bates
Robert D. Bay
Philip W. Birkeland
Norman K. Brown
Thomas C. Heist
Kai Holbek
Richard A. Kaden
Roy P. Keslin

Rex I. Lancaster
W. Robert Little
Michael A. Lombard
Kurtz L. Paulson
Dean E. Stephan
J. Doug Sykes, Jr.
Leonard J. Westhoff
Eduard Witta

RELEVANCE OF ENFORCED CONSTRUCTION TOLERANCES

By

Allen J. Hulshizer[1] - F. ASCE

ABSTRACT

Specifications are necessary to establish standards to insure that constructed facilities fully comply with parameters, factors and assumptions utilized in the basic design. Over-restrictive (beyond actual design requirements) specification tolerances multiply safety factors already incorporated in codes and standards used in the design work and restrain construction management from exercising cost effective judgment in non-design controlled areas. The strict enforcement of specification tolerances in today's quality control atmosphere puts further demands on accuracy in construction that can unnecessarily bind schedules, be very costly and often requires preciseness well beyond what standard fabrication and installation tolerances and experience allows for. The role of the Code Maker, Designer, Constructor and Quality Control Bodies are discussed with respect to achieving a balance between necessary design requirements and unnecessary costs.

INTRODUCTION

As the cost of power plant construction continues to increase (especially nuclear facilities), the search for construction economies is strengthened. Worker productivity is examined (1), methods and equipment are improved, better materials are utilized but claims to slowed progress still remain to be that of over-restrictive or unrealistic tolerances.

It was the initial intent of the Author to review some of the current practices and standards relative to nuclear concrete construction work with the high hopes of making certain tolerance panacea recommendations. In this vein, the proposed tolerances of ACI Committe 117 (2) would have been critiqued with the aid of construction personnel. However, in considering the types and variableness of tolerance difficulties encountered over the years, it would be imaginative to think that one more set of numbers would remove the problem that so many have sought for years to solve. However, there is much that can be done with this claimed problem of tolerances, and it is to that end that this paper will seek to evaluate the role of the Constructor, Designer and Code (Standard) Maker in mitigating tolerance associated cost, particularly for nuclear facilities.

---

[1] Supervising Structural Engineer, United Engineers and Constructors Inc., Philadelphia, Penna.

DEFINING THE BASIC PROBLEM

Why, in recent times, has the cry grown louder and louder with regards to "intolerable tolerances" when the same requirements have existed in specifications, codes and standards for many years? Two prime factors can be attributed to this dilemma:

o Increased demands on the designed performance of structures

o Enforced requirements through strict Quality Assurance and Quality Control programs

Increased Demands on Design

Essentially, the structural design of today's nuclear facilities have become even more difficult due to ever increasing loads from increased seismic resistance requirements, pressures and loads from postulated energy system failures and sometimes restrictive allowable design parameters. The net result is that concrete sections are often designed to their structural limits by utilizing maximum reinforcement quantities in complex arrangements. Couple this with numerous anchor plates and bolts and openings for piping, ducts, cable trays, equipment, etc. and too many things begin to occupy the same space. Then come the problems, what stays and what goes, what moves and by how much. Meantime, progress is slowed or stopped waiting for decisions, and there to make sure everything is as specified or drawn are the "enforcers" (Q/A).

Enforced Requirements

Why are yesterday's specifications so difficult to live with today apart from the effect of increased design requirements? The basic reason appears to be that of today's rigidly enforced rules and regulations in contrast to previous eras where "good judgment" prevailed. In other words, not everything worked to the specification tolerances in the past, but things were "straightened out" in the field, leaving the designer and Code Makers with the general impression that things could be built to the tolerances given. As it turns out, when all the facts are in, this cannot always be done.

First, there is nothing wrong with a good Quality Assurance program to assure that the design assumptions utilized were in fact correct and that the structure should perform as planned. Since the responsibility of the inspector is limited to checking the work done against the specified requirements, the problem of unreasonable preciseness does not lie with what the inspectors' find and report but with the criteria they are given to evaluate by (that is when it is unreasonable). If no tolerances are given, they are often taken as zero, further compounding the problem.

The problem with promulgated tolerances is in the fact that no one set of numbers or descriptions can accurately cover each and every situation or condition, and sound judgment is required to adapt these standards to carry out the design or fit intent. With respect to Quality Control inspectors, no judgment is allowed to alter the specification requirements. In fact, as will be discussed later, even the designer's judgment is often not permitted in allowing other than the "official" code tolerances.

However, the code committees recognize the limitations of establishing the "only" set of correct tolerances, as an example from ACI Committee 117 proposed standard (2) which states:

"The stated tolerances in this document were, in many cases, derived by consensus opinion due to the lack of definitive data; and should be used with judgment as a range for acceptability and not a limit for rejection."

To make tolerances workable, either more extensive values must be developed to meet the various cases that can arise and/or the means established whereby on-the-spot judgment decisions can be made by persons qualified to see that the specification intent is met.

BASIC TOLERANCE DEFINITION

The essential reason for tolerances is the need to convey the degree of accuracy needed to construct components so that they:

o Perform structurally as intended under various conditions

o Permit required items to fit together without major alterations or affecting basic design

o Promote the desired appearance

For the purposes of this paper, discussion will be limited to the first two items.

Design Performance

Tolerances related to design performance in concrete structures can be broken down into two groups:

1. As they relate directly to the inherent properties of a component to resist load and deformation with the desired margin of safety, generally classified as strength and stiffness characteristics.

2. Those tolerances that have been determined to ensure that the components will remain functional under conditions that would tend to reduce the structural aspects of the members such as corrosion, fire and chemical attack.

These two groups are not necessarily mutually exclusive.

Installation Fit

Tolerances generally related to dimensional aspects, particularly those required to assure the proper fit of future interfacing elements, would include such items as location of anchor bolts to receive columns; beams and equipment; insert plates and anchorages for component attachment; sleeves for through passage of pipes and conduit; and dimensions to ensure the fit of other prefabricated units such as temporary hatch way plugs.

UNDERLYING TOLERANCE PROBLEMS

The design of the basically heterogeneous concrete section has been satisfactorily defined by theory and testing to the point that variations in properties (member size, concrete strengths, reinforcing areas, strengths, locations, etc.) can be reasonably evaluated to determine the effect on performance and safety margins. However, since many of the design parameters were developed by tests utilizing protective covers and other limiting factors, the designer is usually unable to determine how much variation can be permitted without affecting the basic design. For example, if a cast-in-place beam tie cage shifted such that there was only one quarter inch cover over the ties, which were side lapped "U" bars installed for constructibility, would the provided cover be sufficient to develop the lap splice if fire and corrosion were not involved?

During the course of design, numerous factors are used or are built into the codes to provide adequate margins of safety, some of which are to cover the things that were never thought about in the first place. These factors all become cumulative (as they should) and tend to further increase the safety factors originally deemed sufficient. Some of these accumulations occur within the following items:

o   Load Factors

o   Strength Reduction Factors

o   Design vs. Actual Concrete and Reinforcing Strengths

o   Design Loads

The difficulty is not in being aware that some additional conservatism may be inherent in the basic design but to what extent and on which items could one begin to pare out "margins" for the sake of easier construction and still achieve basic structural performance with adequate reserve to cover the undefined.

Another source of difficulty comes from the definition of how and where measurements are to be taken. As previously alluded to, inspectors must have a definite base against which they can factually (by defined values) determine if the item in question "complies" with the specification or drawings, without the flexibility of judging whether the intent of the design was fulfilled.

An example of this lack of code definition, yet implied measurement, can be seen from ACI 318-77(3), Section 11.7.9, which states:

"For the purpose of Section 11.7 when concrete is placed against previously hardened concrete, the interface for shear transfer shall be clean, free of laitance and intentionally roughened to a full amplitude of approximately ¼ inch."

When put to the test determining on what basis an inspector can accept a prepared joint on other than visual appearance (which is a subjective evaluation), the request comes: "Provide a measurement criteria". Apart from the first problem of determining if there is any variation permitted in the statement, "a full amplitude of approximately ¼ inch", one must try to determine how to define a means of measuring the surface, before even trying to provide a tolerance which is not thoroughly defined in the code. Consider, in this case, that the method of surface preparation and size of aggregate will produce varying surface textures, some of which will have difficulty in achieving any kind of ¼ inch amplitude (for example bush hammering with a small coarse aggregate mix). If a measurement scheme and tolerances are derived for this application, how practical and necessary is it to provide a satisfactory joint?

Measurements, and therefore tolerances to those measurements, are necessary to establish a means of acceptibility of material and construction but often the strict, un-evaluated rejection on the basis of just missing these values becomes costly. Consider how many truck loads of good quality concrete are rejected because the slump (which is really only an <u>indicator</u> of performability and quality) was outside the specification by ¼ inch, yet its condition would perfectly satisfy design and consolidation requirements. This is also true of time limits imposed on batching to placing. Certainly, there are more items, such as measurement of lift heights, how much contact is required to qualify a splice as a contact splice, etc., which could be added to this list where absolutes can be yielded to qualified judgment.

Numerous questions are raised to the designer by Owners and Constructors alike regarding the costly establishment or adherence to "seemingly" over-restrictive tolerances while, on the other hand, regulatory bodies continue to challenge the designer to justify each and every variation from the "established" code values. This often results in costly tests to defend the designer's acceptance of work that has been built outside the code given measurements and tolerances. In other words, many times even the designer's judgment is not considered acceptable to permit variations in the <u>generalized code requirements</u>.

With respect to tolerances as they relate to "Installation Fit", much more tolerance latitude is available if the cost of making things fit after the fact can be balanced against the precise installation of the accepting item. This area is essentially divorced from code restrictions and does not, therefore, handcuff the designer with respect to establishing appropriate installation tolerances. Although the designer is not frustrated by these installation tolerances, construction is, and perhaps some of the highest unnecessary construction cost can be attributed to tedious positioning and re-positioning of embedded items whose final location could have tolerated more liberal placement requirements with proper pre-planning.

MITIGATION OF TOLERANCE PROBLEMS

If any headway is to be made in bringing realism into nuclear concrete construction tolerances, there has to be an acceptance of individual responsibility by each of the groups - code maker, design and construction - to examine their role in developing necesssary and workable criteria and in seeing that these resulting tolerances are actually adhered to. In other words, the solution to this problem cannot be shifted to the other party. Finally, the outside regulatory and quality assurance bodies must examine their criteria and standards to determine in which areas qualified judgment decisions would provide a satisfactory and economical remedy to certain tolerance requirements.

Code Maker

In perusing the various concrete codes (and standards), there is no doubt, from their language, that the majority of the stipulations and generalized guidelines are meant to be applied or varied to suit individual design and construction requirements. However, in the climate of imposed exactness to the "letter of the law", variations are not easily justified by the designer who is not party to, or knowledgeable of, the reasonings behind the criteria eventually written into the code documents.

In the absence of an impractical and probably impossible revision to the codes to clear up these problems, one solution would be to provide a supplement to each code (or standard) which would provide the designer with at least the following information for stated criteria or tolerance thereto:

1. Basis for criteria (strength, stiffness, corrosion, etc.)

2. Reasoning behind final values

3. Intent of Criteria (rigid, guideline, indicator, etc.)

4. Type and range of criteria variation that:

    a. should be handled by designer

    b. could be judged by qualified field inspectors

    c. could be assigned to construction personnel

Armed with this type of information, the designer would be in position to develop job specifications and guidelines that would establish the proper realm of tolerance and judgment responsibility, which would eliminate an area of unnecessary construction exactness while at the same time, satisfying regulatory and quality assurance concerns.

Designer

Although certain help will be necessary from the code makers (and regulatory bodies), as previously discussed, to enable the designer to effect broader and more meaningful tolerance values in specifications, there are other areas which are within the designer's control which would provide for greater economies without sacrificing quality (and, in some cases, improving it).

First of all, there is an area relating to design values where it will become necessary to do testing work on a generic and specific job basis to establish tolerances which, for the most part, will probably never be addressed by code committees. An example of this would be the establishment of criteria for the acceptance of reinforcing bars and anchor bolts which have been damaged (dinged, notched, dented, chain marked, etc.) during normal installation. The proof of acceptance becomes involved with stress concentrations and energy capacity loss which can best be evaluated through test work. It would be most helpful to the engineering community if a forum for collecting and disseminating this data were developed to prevent "re-inventing the wheel on different projects".

Further, due to the complex nature of nuclear concrete construction, it is incumbent on the engineering profession to become more knowledgeable of construction practices, techniques, problems and fabrication tolerances so that drawings and specifications will contain only relevant requirements and tolerances so as not to unnecessarily hamper construction by over-precise requirements and impractical details. Sometimes the specification, code or detail requirements are more restrictive than that to which the shop fabricated item has been made. Drawings and specifications should not impose restrictions that would restrain construction management from exercising cost-effective judgment in non-design controlled areas.

In this vein the use of scale models of complex reinforced concrete structures or sections can be very helpful in bringing the construction problems home to the designer and eliminating "too many things in the same spot" before they become costly field delays and corrections (4)(5).

There remains one area of unnecessary preciseness that, with a little early design pre-planning, could eliminate considerable tedious field installation and surveying. That is the location of most anchor bolts and insert plates used for future welded attachments.

Since almost without exception, anchor bolts are not (and cannot without a reamed, bound, full contact fit) used to transmit shear, the size of base plate and equipment base holes should be provided with reasonably large holes to allow for anchor bolt location tolerances of $\pm$ 1/8 inch without sleeves and $\pm$ 1/4 inch with sleeves without sacrificing member installation accuracy. If closer tolerances are required, templates should be furnished by the equipment manufacturer that are identical to the bolt pattern and dimensions for the item involved. Sizing of holes in equipment requires early job coordination with a specifying discipline to be certain that the equipment is purchased and manufactured with appropriately sized holes. Holes for building component base plates (columns, beams, etc.) usually fall within the domain of the construction-oriented designer and, therefore, hole sizes can easily be established to suit best needs and practices.

Insert plates are most often used to allow for flexibility in welding future attachments which already have certain provisions for installation variations. However, the installation tolerances on these plates do not allow for a reasonable fit when they are required to be placed in heavily reinforced sections. Insert plates should, of necessity, be provided oversized to allow for a minimum of two reinforcing bar diameters movement and at least nominally three inches in any direction.

Where possible the designer should evaluate increasing member sizes and, subsequently, the tolerances to be able to more reasonably and economically accommodate reinforcing, anchorability and insert plates. Special tolerances should only be given to reflect special requirements.

Construction

Initially, construction must be aware that certain tolerances will have to be maintained - and plan and instruct the installation accordingly. Often, after working out special limiting tolerances to accommodate tight or as-installed field conditions, the designer is asked, "And what is the tolerance of this?"

Sometimes there comes a limit to which the designer can make further adjustments for design, fit or justification reasons.

Construction bears the pressures of trying to economically build to the tolerances specified. However, once construction has built a "non-conforming" structure or member, it is the designer who receives the final pressure to accept, justify or reject the item in question as the cry comes back, "You mean we have to rip this down, you (the designer) are holding up the job and wasting money, etc."

If the code makers and designers are successful in establishing a better set of tolerances, the constructor must see their role in carrying out work within the specified limits.

CONCLUSIONS

Are specified construction tolerances relevant to design requirements and the high cost of perfection? The answer is yes and no.

Where the requirements are unnecessarily restrictive the code makers, designers, regulatory and quality assurance bodies must work together to ferret out the unnecessary from the necessary and develop some means of restoring on-the-spot judgment decisions to qualified construction personnel. There is no doubt that not everything built in the past complied with specification requirements and sometimes problems resulted. But neither can it be overlooked that the majority of work survived very successfully with less than perfect installation tolerances. Some reasonable balance must be brought back into the nuclear concrete construction to reduce unnecessary costs.

On the other hand, it will take more than loose tolerances to optimize the cost of concrete construction, and constructors must work at perfecting their work when and where it is required.

There is room for considerable and worthwhile improvements in the area of construction tolerances. Extended cooperation and joint efforts will be required on the part of all involved to know their role - and to identify and resolve issues if any benefits and cost reductions are to be realized in the concrete construction of nuclear facilities.

REFERENCES

1. Borcherding, John D., Sebastian, Scott J., Samelson, Nancy M., "Improving Motivation and Productivity on Large Projects", Journal of the Construction Division, ASCE Volume 106, No. Co 1., Proc. Paper 15272, March 1980, pp 73-89.

2. ACI Committe, "Proposed ACI Standard: Tolerances for Concrete Construction and Materials", Concrete International, August 1980. pp 38-46.

3. ACI Standard, "Building Code Requirements for Reinforced Concrete (ACI 318-77), American Concrete Institute, Detroit, Michigan.

4. Hulshizer, Allen J., Mobley, R.M., "The Use of Scale Reinforcing Bar Models for Heavily Reinforced, Complex Concrete Structures", Proceedings, Seminar 80, American Engineering Model Society, May 5-8, 1980, Pittsburgh, PA.

5. Hulshizer, Allen J., Mobley, R.M., "Increased Productivity in Placing Reinforcing in Reactor Containment Structures Through the Use of Scale Reinforcing Models", Proceedings No. 80-662, 1980 ASCE Annual Convention and Exposition, October 27-31, 1980, Hollywood, Florida.

ENGINEERING TOLERANCES:
A CONSTRUCTION VIEWPOINT

By

Frank J. Freiseis[1]

Abstract

Observations selected from construction practices and conditions which adversely affect tolerance are presented. Typical examples of tolerance variations in reinforced concrete and formwork and the causes of variations are also described. These findings are the result of an ASCE sponsored specialty conference on experience with the implementation of construction practices, codes, standards, and regulations in the construction of power generation facilities held at Pennsylvania State University.

Introduction

Other papers in this session address tolerance standards and their relationship to the designer, the supplier, the modeler, and the standards. This paper will examine tolerance and its relationship to the construction engineer. Construction may be viewed from several positions. There is the picture of construction as seen by the utility-owner, the architect-engineer, the contractor or constructor, the public, and the regulatory agency, particularly when nuclear facilities are considered. This paper will, for the most part, present a view of construction as seen by a construction manager employed by an architect-engineer to represent the utility-owner and work directly with the contractor on the project. So the view can best be described as a composite of those three interests, but especially that of the A-E and the utility-owner. The observations incorporated in this paper are drawn from personal experiences of this writer on both nuclear and fossil projects but primarily reflect fossil experience. The paper will focus on practices which adversely affect tolerance. An examination of intangible factors affecting tolerance will first be made, followed by a review of observed tolerance problems, conclusions, basic recommendations, and summary. Observed tolerance problems have been grouped about three areas of interest: formwork, reinforcing steel and concrete placement.

---

[1] Senior Construction Engineer, Sargent & Lundy Engrs., 55 East Monroe, Chicago, IL 60603

Intangible Factors Affecting Tolerance

Several broad factors establish constraints on a construction team and limit its ability to effectively adhere to any reasonable grouping of tolerances. These factors are in general beyond the control of the project management. They do not lend themselves to change and thus become the conditions under which all work is performed.

Geography

Accuracy, like productivity, seems to vary geographically. The exact reasons are difficult to discern. National origin may play a part. Perhaps causes can be found in work heritage, weather, more stable ways of living, or a fortuitous collection of traits among the factors discussed below. Whatever the reasons, some areas produce more conscientious and careful craftsmen than other areas.

Craft Training

Some unions require little training, others require formal training for several years during apprenticeship. The practice of purchasing a card that automatically qualifies the holder for the same job as others is still prevalent. Some union BA's will require such holders to become travelers and work out of different districts. Such individuals are likely to appear on projects where qualified labor is in short supply. They frequently do less than quality work. Qualified men also travel to find work.

Labor

Projects utilizing non-union labor tend to have a greater number of quality and tolerance problems than projects using union labor. Although it is not uncommon for union personnel to show up on non-union projects (such was the case on a project in Texas where Unit 1 was union and Units 2 & 3 were non-union), a greater number of less qualified, less trained craftsmen do show up on non-union projects. And, for instance, an ironworker may be asked to dig trenches for footings and more efficiently utilize his time. Cost efficiency is increased at the expense of craft pride. Thus the training received by union personnel in the apprentice system plays a major role in preventing tolerance problems.

Conflicts

Disputes between individuals occasionally go underground and result in sabotage. What might have been work within tolerance can become unacceptable work.

Level of Inspection

Project contracts which call for a high level of inspection quite naturally tend to experience a lower incidence of tolerance variation. Craftsmen soon learn, foremen in particular, that correcting unacceptable work is costly. In this regard the most important element is follow-up, having once discovered a variation and requiring a correction. The tendency of some foremen to argue must be met with firmness, tempered with judgement.

Type of Contract

Cost plus contracts tend to produce fewer unacceptable situations than hard money contracts. There is, of course, greater willingness by the contractor to make corrections on cost plus contracts. There is also a tendency to miss schedules with such contracts. In certain areas also the differences between cost plus and hard money contracts regarding tolerance adherence is not significant. Apparently sincere but inattentive craftsmen will make errors regardless of how their employers are compensated.

Management Practices

The tone of any project will be set by the utility-owner, with or without his awareness. A utility which insists on quality control from the beginning and follows up will set a standard which construction managers will follow. It can be said conclusively that a project with significant quality or tolerance problems in the finished project requires at least the tacit approval of the owner. Most utilities want a quality product within cost limits.

Social Habits

Alcohol has been a problem for some construction workers just as it is for society in general. However, for a large segment the beer consumed at lunch may show up as poor judgement in the afternoon. The size of the segment can be estimated by observing the size of the herd headed toward the parking lot at lunch time. Drugs, particularly marijuana smoking, have become a definite problem on jobsites. Drugs are available on jobsites and dealers among craftsmen are generally known. Their increased use probably account for the decrease in general quality of work and increase in carelessness that this writer has observed over twenty (20) years of field work.

Communications

Tolerance variation can result from an incomplete or inaccurate transfer of information. Failure to communicate intentions clearly by designers, managers, superintendents or foremen results in guess work and assumptions by the individual charged with performing the task. Complete failure to transmit instructions can be symptomatic with certain responsible individuals or the result of occasional oversight or excessive pressures. Clearly it is in the interest of all that instructions be conveyed in a timely and accurate manner. How best to motivate responsible persons to this level of communication remains a problem.

Observed Tolerance Problems

Formwork

Excessive variations show up in all systems of formwork. Many contractors use one of the prefabricated systems. Other contractors, particularly in parts of the south or where labor is less expensive, build forms in place with plywood and lumber. Whichever system is used, differences between actual dimensions and basic dimensions show up.

During Assembly. During this stage when using a prefabricated system, the supplier may not have in stock the exact size section required or may ship the wrong size by error. The detailer may designate a piece in error. A needed piece may be tied up in another panel which is being saved for a later placement. A piece of warped plywood may be used when forming in place.

While Setting. A chalkline may have been accidentally deflected by an object when marking a wall line on the foundation. A string baseline from which measurements are taken is removed by one worker because it interferes with his work, then later reattached to the wrong nail by another worker who needs the baseline. A strong wind may blow a long plumb line when plumbing wall forms or the line may be deflected by an object without the layout engineer's awareness. The layout engineer can be unaware that he is working with a 10' offset line set by surveyors instead of a baseline, and thus lay out an entire substation with a 10' variation and discover the error after foundations are partially placed. A layout engineer can be unaware that his assistant at the other end of the tape is holding the 1'0" mark because the 0'0" mark is too worn on the tape to read and result in a coal yard junction house being built 1'0" in variance.

After Setting of Forms. After formwork is complete and while waiting for other work to complete or waiting for weather to clear, new plywood may be warped from wet weather. Temperatures may change the lengths of formed elements. Earth may cave in or wash into excavated work and press against marginally braced work. The wedge to which a foundation brace is attached may move in the wet soft soil it is driven into as winds blow against the formwork. Form ties may not be all checked after bad weather delayed a pour, particularly if they were once checked earlier. Rechecking would be costly and delay pour. Form ties wrapped around resteel may be moved as craftsmen work to install embedded items at same dimensions as resteel or ties.

During Concrete Placement. On thinner walls the rate of placement may exceed the capacity of the formwork to resist the outward pressure of wet fresh concrete. Foremen anxious to complete the placement as soon as possible ignore the specified rate of vertical advancement of the pour. Placement crews improperly supervised will place concrete in one area until it is much higher than adjacent areas, so they won't have to move, thereby placing undo strain on the formwork in that area. In large foundations the tendency to chuting as much concrete as possible sometimes encourages the hasty creation of earth ramps for concrete truck access to all parts of foundation. The pressure from the earth and the concrete trucks can cause forms to move. Vibrators which concentrate too long in one area or too long next to forms can cause a failure of the forms in that area. Failure of the contractor to leave sufficient carpenters on watch to correct loose ties can result in adjacent ties developing excessive stresses and possibly breaking. A bucket load may nudge formwork accidentally causing movement or damage.

During Initial Curing. Inadequate earth foundation design or preparation can cause differential settlement while concrete is being placed. In this case formwork variance may be the least of one's problems.

Reinforcing

Variations in dimensions, angle of bends, radius of curve, condition of steel, show up at all stages of progress in the project. Even acceptable work left sitting for a considerable time is subject to considerable abuse at the hands of mechanical and electrical crafts. Below is a summary of the common problems.

Shop Fabricated Steel. The most common variation is found in length where long pieces will vary 3" or so. Ability to deliver correct numbers of bars varies with fabricators. It is not uncommon on every project to encounter several instances where the angle of bend has been completely misinterpreted. Frequently, column or beam tie dimensions will be several inches off. Curved members are also a common problem, particularly in nuclear work where it occurs more frequently. The expense of replacing can be high and affect on schedule can be significant since field fabrication of curved members is generally not possible.

Field Fabricated Steel. Correcting shop fabrication errors is a routine phenomenon and results in the need for stock reinforcing steel on site. The most common problem encountered is the tendency for the contractor to make bends using heat from a cutting torch rather than setting up a barbender. Ironworkers who have not been taught otherwise may even compound the problem by using a cutting torch without rosebud attachment. The obvious problem is radius of curvature which can become, at its worst, an almost perfect right angle.

Storage of Steel. Storage of steel is normally not a major problem. Damage can occur in yard storage or after shake-out and locating near placement area, however, as project activity picks up. A cherry picker working in the area is a common cause. Concrete trucks maneuvering for a dump are another major cause. It is tempting to use these damaged bars, particularly if the level of detailed inspection is low or perfunctory. Often the contractor is unaware of damage until setting when the pressure of meeting a schedule is heaviest.

Setting of Steel. As pointed out above, there is a vested interest by many parties to use damaged reinforcing, schedule and cost factors overriding quality control concerns. Due to the high level of inspection of nuclear projects, such damaged material is not likely to be set. However, on fossil plants the tendency remains strong. During continuous slip forming one could not ignore the probability that curved bars with major variations would be encountered. The pressure to use such bars at a time of low visability (such as night shifts) would be strong. Continuous slip forming of reactor buildings, unless steel is completely set and inspected, is therefore not possible. The setting

of dowels, particularly smaller dowels, after concrete placement is
common practice on fossil plants. While embedment length is chalked
off normally, this is not always the case, particularly if it is late
or near quitting time. Mislocating the position naturally occurs on
occasion. Often it is not known for certain if correct embedment
length was obtained. Spacing of additional bars at openings is often
improper. Indeed it is often impossible. One encounters conflicts
with embedded frames and their anchors. Dowels set in concrete may be
set after initial set of concrete has occurred, therefore creating a
void around the bar or leaving much doubt as to the quality of bond
development. Exactly what to do with all the hooks at ends of beams is
usually a problem, similar to placing ten pounds of soybeans in a five-
pound bag. The design showing very long dowels for which there is no
adequate way to brace still occurs and is always followed by a request
for field modification. Rebar density, particularly on nuclear plants,
is still a major problem. This writer still has bad memories about an
incident where design added seismic #11 bars into floors in auxiliary
and fuel handling buildings where they joined the reactor building.
Threading in #11 bars in an already extremely congested zone does not
leave much room for concrete aggregate. Turbine foundations on fossil
plants present the greatest challenge to rebar congestion. Penetration
and openings with additional steel requirements are frequent sources of
congestion.

After Setting & Before Placement. As other craft work to install
embedded anchor bolts, conduit and pipe sleeves, etc., it is quite common to encounter conflicts with reinforcing. These conflicts are usually
resolved by relocating reinforcing. They may be resolved by the craft
jacking open a space, using come-alongs, or sometimes untying and leaving loose steel to make room. Cutting torches have been applied without regard to design requirements and without consulting with engineers.
Chairs or bar supports may collapse under construction loads where inadequate support is provided.

During Concrete Placement. At this stage the most common source of
variation is loosening ties and moving rebar to make room for hoppers or
buckets placed strategically to receive the concrete and convey it to
tremies. In the hustle of placing concrete, particularly as topping
out occurs, it is easy to overlook replacing and tying of relocated rebar. Improperly tied steel can be vibrated out of location. Cranes
swinging in buckets of concrete will become entangled with dowels dis-
locating them. The matter of setting dowels during placement has been
covered above. But intending to set them and then forgetting is an
additional hazard. Chuting of concrete also requires frequently the
separation of reinforcing to provide openings for dumping which may not
be replaced. Pumping of concrete can also dislodge reinforcing as hose
is dragged. Conveyor systems have designated discharge points again
requiring the movement of reinforcing to provide clear openings for the
discharge of concrete. Conveyors have the additional problem of requiring relocation of sections. Dragging conveyors to the new position can
sometimes disturb marginally tied steel. Vertical dowels very often
turn out to be in the path of movement. Relocation of pipe for pumping
can have similar problems. Some conveyor systems provide for a swinging
or rotating section which may also interfere with vertical dowels.

These dowels may have been left out intentionally to be installed later, or may be removed on the spot. The quality of their replacement again must be questioned because of the hurried nature of the work effort, the generally tired condition of workers on a large placement, and accompanying possibility of oversite or sloppy work when reinstalling. Wheel barrow or Georgia buggy placements generally result in little or no disturbance on floor pours. They roll on pathways of boards or plywood laid on top of reinforcing and are versatile enough to go anywhere, thus usually not causing disturbance to dowels.

Another common occurence is the spattering of concrete on the rebar. Such spatter quickly dries on the steel. If not removed prior to covering steel with concrete, the bond between the steel and concrete is adversely affected. This material is easily removed with hammers and/or wire brushes.

After Placing Concrete. Here the problems of maintaining tolerances are limited. Carpenters may find horizontal dowels extending from a wall the ideal platform from which to stand to check a cathead above. Laborers may utilize them for access to reach some remote work area. Aside from possible damage to bonding, dowels thus disturbed and bent can escape correction when steel is later tied to them, particularly if relatively inaccessible.

Concrete Placement

Concrete and its additives present unique variation problems to field personnel. Its production is subject to many variables. It does not allow for the same degree of uniformity as factory controlled products such as steel. Ideal control may be achieved best by onsite batch plants. Methods of contracting also play an important role. Batch plant operators are reluctant to discuss problems they may encounter with material measuring equipment, short delivery, unauthorized substitution of additive brands, field instructions, etc. The usual method of including the supplying of concrete with the contract for forming and placing simplifies his management requirements but places the utility-owner in a disadvantaged position. Greater control over the quality of concrete may perhaps be achieved when the owner contracts directly with the concrete supplier, thereby improving his ability to obtain information. In any case, surveillance remains the only real solution.

Rebar Spacing. Honeycombing sometimes results from incorrect aggregate size being used. The problem occurs in larger placements where a localized section of the placement contains rebar densities much greater than the typical density. Instead of using an approved mix containing small aggregate, (or possibly grout in extreme situations) in the small localized dense section, there is a tendency to stay with the approved design mix using the larger aggregate for 95% of the placement. The tendency is to vibrate (rather than consolidate) to move sufficient material into the small section. The dense rebar screens the larger aggregate and blocks the flow of concrete. Honeycombing results, and in extreme situations, voids may develop. Qualified and experienced inspectors is one answer to this problem. Placement plans developed prior to approval to commence placement should evaluate the need for supplementary mixes.

Water Content and Slump. The ever present problem on almost every concrete pour: how to keep slump within tolerance. Sources of problems are many and their solutions are frequently hard to achieve. Water content in the sand and gravel pile varies. No sooner is the concrete arriving at proper slump, then suddenly it changes. The cause is frequently back at the batch plant when the front end loader operator suddenly decides to dig in a different part of the pile, where the content of water is different. Also, the absorption values of one source or delivery of aggregate may differ from another. Improperly stored aggregate, inhibiting drainage, is a common problem. But properly stored aggregate is also a problem. Wetting agent or air admixture measuring equipment may clog or malfunction or run out of the admixture. Cooperation in correcting malfunctioning scales and euqipment may be nonexistent. This is especially true when dealing with the only qualified supplier in a 25-mile radius. On some projects the supplier will manually add admixtures with the additional complication of human error.

At the pour site the addition of water to the truck is taken matter of fact. Some foreman will routinely add water to a truck without even first discharging. Thus water is added to a truck that is possibly too wet already. On floor or large area foundations some contractors place finishers in charge of placing concrete. Finishers love to add water. When chuting concrete, the addition of water become irresistable. It will make the concrete flow to remote areas of the pour, reducing the amount of work to be completed with crane and bucket.

Mistakes on slump tests are another way of increasing water content and maintaining the record or appearance of staying within tolerance. Clever foremen will take the test out of the first chute full or front of truck load. Then after test is made, add water. Another favorite way to pad the reading is to record measurement at the top of the slump, rather than on the concentric centerline of the test.

Truckloads which have been rejected for being too old have been known to take on additional water back at the batch plant and return to pour as a fresh delivery.

When trowel finishing of floors is not undertaken at the correct time and becomes too set, then the addition of water by brush shaking is necessary to obtain a proper finish. Such addition, if excessive, can add much water to the top 1" or so of floor and reduce the durability of the surface. Cracks may develop above rebar. Concrete color will develop a whitish tone.

Recommendations

No attempt can be made here to recommend solutions to such an extensive listing of pitfalls in the pursuit of quality workmanship within code tolerance. To do so here would be presumptuous. Nevertheless, it is possible to reach a basic conclusion and make obvious broad recommendations.

The majority of work performed is generally within tolerance and the above situations, while always and ever present, still represent the

exception rather than the rule. One should not assume, as might be the
tendency after reading a long list of shortcomings, that all field work
is shoddy. This is definitely not the case. Rather, some of the possibilities have been listed here, not the probabilities. Nevertheless,
the pitfalls are numerous, can occur at any time and any place.

General Recommendations

Utility Management Practices. The No. 1 responsibility for quality
work within code tolerance must lie with the utility-owner. He should
make known to his construction management team at the beginning of the
project exactly what his expectations are with reference to quality and
tolerance in various types of work. He should also make known exactly
what price he is willing to pay to achieve those goals. A management
team of sufficient size and experience should be assembled or contracted
to give adequate coverage of all work. Policy for use of outside consultants for special situations as vertical and horizontal control
checking, underwater inspection, and materials testing should be established.

Management Materials. Normal survey equipment in working condition
should be available. Copies of recommended practices and standards
governing the work to be performed should be on file. On fossil plants,
everyone assumes such simple materials are available. Too often they
are overlooked. Copies of specifications should be made available to
those personnel working with a particular contract as well as office
copy.

Training. In addition to staffing the project with trained, qualified, and experienced managers, supervisors, and technicians, consideration should be given to training the craftsmen. Occasional training
sessions held onsite may be beneficial to eliminating repetitive type
variations. Such instruction could be given to the craftsmen following
a safety meeting to minimize unproductive time. The added expense is a
recognized factor. But the reduction in rework will offset its cost.
A recent study indicated rework was the third largest cause of unproductive time. A before-after onsite trial might show interesting results.
Some craftsmen take pride in their work and many respond favorably to a
company interest in their career as opposed to the usual and expected
interest in their productivity.

Auditing of Construction Activities. Inspection procedures should
be established to cover all activities. These procedures should be set
down on paper to be more meaningful to the inspecting engineer. Particular attention should be paid to follow-up of any discrepancies which
are to be corrected.

Good Judgement. Many structural codes and standards apply equally
to nuclear and fossil construction. The interpretation of these documents by construction management is the same. However, the ability to
enforce or verify compliance differs significantly due to public and
regulatory requirements on nuclear projects. Field management staff
size is typically 10 to 20 on fossil projects, with typically 7 to 10
persons making field judgements on tolerance variation as well as other
responsibilities. Given the lower level of public pressure, governmen-

tal involvement, and smaller staff, the use of good judgement becomes an unavoidable necessity. Considerations such as structural soundness, affect on interfacing work, frequency of the variation, affect on schedule, and cost of rework must all be evaluated.

Where structural soundness is a concern, normally the design engineer must be consulted and an engineering change will be prepared and issued. The engineering cost of drawing changes is normally borne by the owner, although caused by construction deficiency. This situation is usually due to contract limitations. Where such limitations do not exist, engineering cost codes for daily computer entry can aid proper billing.

Where affect on interfacing work is a concern, frequently it is more expedient and less expensive to change the interfacing work rather than correct the actual tolerance variation. Good judgement may require the correction of minor variances where the frequency of the variances becomes a factor. Affect on schedule is a major consideration when evaluating a variation. Thus good judgement should lead the manager to the most expedient solution which is also consistent with sound engineering. It is in everone's best interest that the cost of rework be considered, so that solutions to tolerance problems be the most economical solution available to meet the engineering requirement.

A somewhat different approach is taken on nuclear projects. For reasons mentioned earlier, it has been determined that the price to approach the pure state of perfection will be paid. Field staff sizes and paper procedures have escalated to theoretically carry out this mission. The judgement of one individual, as applied on fossil projects, has been replaced by a system to process tolerance variations utilizing the wisdom of many and several levels of inspection.

Summary

An examination of certain construction practices and conditions which adversely affect tolerances has been made. The point of view has been that of a construction manager employed by an A-E. Intangible factors which establish constraints on the project team and act to limit the level of quality include geography, craft training, labor, conflicts, level of inspection, type of contract, management practices, social habits and communication. Observed tolerance problems typically found on projects were reviewed for formwork, reinforcing steel and concrete placement. Finally, five basic recommendations were made to encourage project management to implement tolerance requirements. These recommendations include the management practices of definition of tolerance policy and acquiring qualified management team, providing basic tools and codes, auditing of construction activities with emphasis on follow-up and recognition of the judgement factor, if any, on the project.

Acknowledgements

1. ACI Committee 117, Tolerances for Concrete Construction and Materials, American Concrete Institute.

2. Concrete Reinforcing Steel Institute, Recommended Practice for Placing Reinforcing Bars.

3. ACI 318 Building Code, Requirements for Reinforced Concrete.

4. ACI 304 Recommended Practice for Measuring, Mixing, Transporting, and Placing Concrete.

5. ACI 347 Recommended Practice for Concrete Formwork.

6. SP-4 Formwork for Concrete (ACI)

7. Journal of the Construction Division (ASCE), September, <u>Work Force Motivation and Productivity on Large Jobs</u>, (1981).

NUCLEAR TOLERANCE REQUIREMENT
IMPACT ON SUPPLIER

by

Robert A. Yockin[1]

ABSTRACT

This paper deals with special tolerance and bending requests for reinforcing bars in nuclear power plant construction. It outlines the background of non-standard bending, restrictive tolerance requirements, improved fabricating techniques and the impact on the supplier to meet project requirements. It also deals with the supplier's viewpoint in minimizing field and fabricating problems.

INTRODUCTION

Much has been written regarding the potential conflicts among components of construction that are fabricated and/or constructed to standard tolerances. Since the advent of nuclear power plant construction there has been a progressive increase in the focus of attention on those conflicts and their related standard tolerances.

ACI Committee 117 has collected standard construction tolerances from all code publications and created one tolerance document. The Committee also makes recommendations of approach for those critical areas where standard tolerances for individual elements of construction such as reinforcing bars, placing, formwork and concrete may conflict. Section 1.2.4 of the Committee's Proposed Draft Report states, "Preconstruction meetings should be held for the purpose of reviewing critical tolerances, methods of making measurements and the basis for acceptance/rejection of completed work to avoid misunderstandings at the time of final inspection."

From a reinforcing bar supplier's viewpoint it is not standard fabricating tolerances that create an impact for us but rather those "critical tolerances" that are incumbent with most nuclear power plant construction. This is what we will examine today. In retrospect we will attempt to provide a better understanding of what the restrictive tolerance impact has meant to reinforcing bar fabrication.

BACKGROUND

In the early 60's when Bethlehem Steel started to supply reinforcing bars for nuclear power plants very little progress had occurred during the previous decades in terms of fabricating tech-

---

[1] Assistant Chief Engineer, Reinforcing Bars, Piling and Construction Specialty Sales, Bethlehem Steel Corporation, Bethlehem, Pennsylvania

niques and equipment. If there was one type of construction that can take credit for stimulating new techniques, improved equipment and more restrictive tolerances in fabrication than ever imagined, it is nuclear power plant construction.

Each new project brought increasingly more unique and difficult demands relative to fabricating tolerances and bar configurations. Initially, #14 and #18 bars were supplied straight or with one 90° bend. Then, one daring design group decided they wanted to offset #18 bars around obstructions to maintain continuity in the reinforcing. Soon after came more requirements for non-standard bar types and tolerances that were unique to the reinforcing bar industry. Experience gained in fabricating the varied configurations of #14 and #18 bars was increasing. In lieu of any published standards, fabricating tolerances for #14 and #18 bars had to be mutually agreed upon with consideration given to cost, prior experience and project requirements. By 1976, CRSI published the first industry-accepted bar types and tolerances for #14 and #18 bars (Figure 1) all due to past construction needs of nuclear power plants.

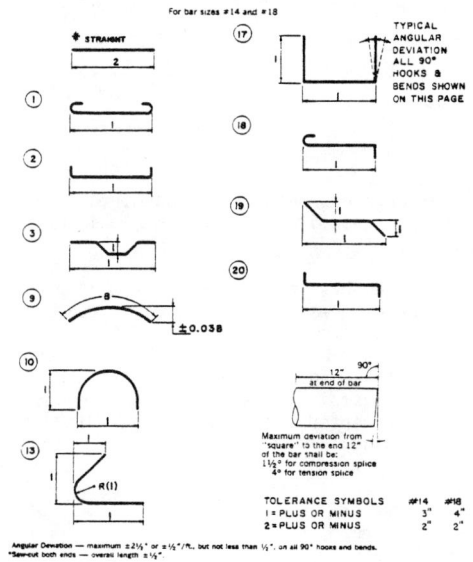

FIGURE 1

Today there is a different and much improved fabricating technology to depend on for nuclear power plant groups. A look at the CRSI Manual of Standard Practice 1-80 page 7-5 (Figure 2) indicates bar type and tolerance standards that would have been very costly special bending a scant five years ago. Yet, we had been asked to fabricate beyond the bounds of these types and tolerances prior to their existence as standards. It is this kind of fabricating requirement, where no

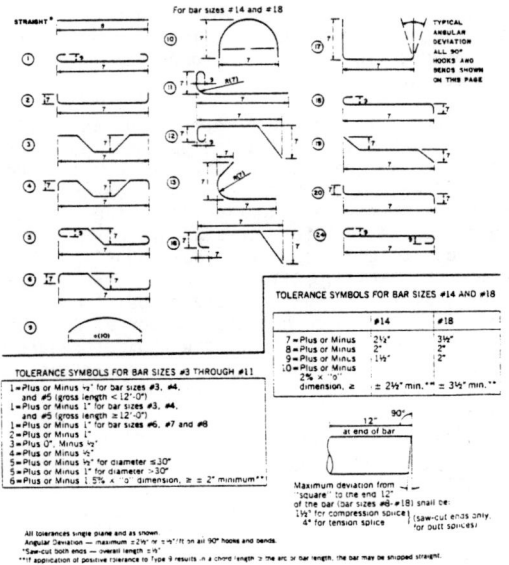

FIGURE 2

procedure or prior experience has been established, that has created an impact on the supplier. It has literally forced him into realms of fabrication never before ventured.

We feel it important to emphasize at this point that publishing bar types and tolerance standards does not automatically guarantee their total acceptance by all suppliers.

Tolerances of $(\pm)\frac{1}{2}$" or $(\pm)1$" for #18 bar fabrication became and still are fairly common requests. The tolerance that was usually agreed upon and generally accepted was ± one (1) bar diameter ($2\frac{1}{2}$") for single and multi-plane bending. In some instances congestion was so great and the problem of placement so costly in terms of construction time that we agreed to $(\pm)\frac{1}{2}$" tolerance for #18 bar bending ... at no trivial cost to the client. One example of that single plane bending is shown in Figure 3. We agreed to (-)0" (+)1" tolerance but would not accept responsibility for the template construction. The customer furnished the template and was responsible for accepting or rejecting each piece at the shop after fabrication and template check.

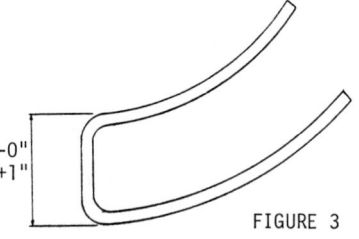

FIGURE 3

EQUIPMENT

To better understand the impact of restrictive tolerances on the supplier one must know something about the equipment used to produce the finished product. Let us quickly examine some basic fabrication. In Figure 4 we see a Coast Bender rolling a radius into a #18 bar. The radius dimension is controlled by moving one of the adjustable back-up rolls forward or backward relative to the drive wheel prior to the bar being rolled through. Predetermined positions of the back-up roll support sleeve vary the radius dimensions in increments of approximately ten feet. The required radius setting within any ten foot increment must be fine tuned by trial and error.

FIGURE 4

Some designers of nuclear power plants require #18 bars bent to a minimum of 32 bar diameters. This is a radius bend by ACI definition, that is, as compared to "standard pin" bending. In the initial stages of supplying reinforcing bars to nuclear power plants we rolled that radius requirement with a great deal of difficulty to try to prevent the bar from warping. Figure 5 shows an innovative piece of equipment that converted the rolling of that tight radius to a simple bend. We found that the investment for this bending wheel more than paid for itself by improving our efficiency for that kind of bending requirement.

SPRINGBACK

With every simple angular bend there is an incumbent springback of the bent segment. To obtain a 90-degree bend the segment might have to be bent 105 degrees or more. When the pressure is relieved the bar segment will spring back to approximately 90 degrees.

Springback varies with bar size, grade, heat and angle of bend. There is no known method to determine the precise amount of springback.

FIGURE 5

Many tests were made to arrive at averages so we could approximate the amount of overbend required for the various bends and bar sizes. The practical approach for bending has been for the bender operator to chalk line the required angle of bend on the bend table top and also the approximated overbend location that allows the bar segment to spring back to the required angle of bend. This has been sufficient for standard tolerance bending. Not so for special tolerances.

Probably the most valuable piece of equipment in producing a bent bar or any configuration to a restrictive tolerance is a hydraulic jack. This jack (Figure 6) was designed and constructed for that very purpose. It is extensively, if not exclusively, used in all multi-plane, multi-radius bending. Without it, the required restrictive tolerances would seldom be achieved by the standard bending equipment.

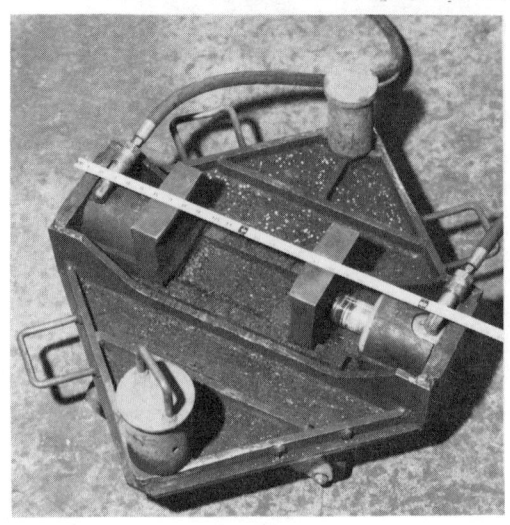

FIGURE 6

TEMPLATES

Simple, one-plane special tolerance bending can be achieved as long as the number of bends are few. The more bends in a bar, the more difficult it is to maintain a special tolerance on the overall dimension. As long as special requirements are known and discussed well in advance of their need, most requirements can be met.

FIGURE 7

We have had excellent success with special tolerances for single plane bending. Chalk templates are usually drawn to scale on the floor or (Figure 7) on a scored flat metal template. Once bent, the particular bar is laid on the template and if necessary the hydraulic jack is used to bring any segment within a required tolerance, as long as there is some reference point or bench mark. Multi-plane bending to special tolerances poses a more difficult problem. Too often one or more segments of the bent bar rise to some theoretical point in space. When this occurs, measuring for a special tolerance is sometimes impossible without the use of a formed template. The larger the template the greater the impact.

One reason for this is the tremendous amount of space taken by a full sized formed template to check fabrication. Another, is the scheduling of manpower to construct the template. Dedication of manpower, equipment and space to one or more projects creates significant additional costs which must be passed on to the client. Careful consideration should be given by design/construct groups to the question, "Is the investment justified by the resultant savings in construction time?"

In the background of Figure 8 we see a full sized segment of a template for a BWR containment structure. This template was used primarily to check fabricated bundled horizontal #18 bars that had to be offset around penetrations and obstructions (Figure 9). This same shop template was also used to check and adjust the tolerance of the diagonal seismic reinforcing. This was especially critical for the seismic reinforcing bars in the transition area from the cylindrical base to the conical upper section of the containment structure. The length of this shop template was 95 feet. It was also used to fabricate and check the multi-layer ring bars around the equipment hatch (Figure 10). The ring bars generally were fabricated in three segments to complete a 360° ring. The agreed upon tolerance was (±) one bar diameter laterally from the theoretical bar location and (+) one inch off the template surface. The pre-assembly of the ring bar cage (Figure 11) was made in the field on a poured concrete template of the equipment hatch area to assure the easiest pre-assembly possible.

Perhaps the greatest challenge of fabricating to special tolerances was for offset seismic reinforcing in the conical segment for a BWR Containment Structure. The design requirement was for the seismic

FIGURE 8

FIGURE 9

FIGURE 10

FIGURE 11

FIGURE 12

reinforcing to climb at 45°, perpendicular to the surface at any given point. Creating the mathematics for computer programming was no trivial chore. In effect we had to develop for the cone what spherical trigonometry is for the sphere. Then came the conversion of complex results to simple measurements for shop personnel.

Because of the sensitivity in fabricating to the geometry of a cone, a new concrete slab was poured around the bending equipment to provide a level base for measurements from floor to various segments of the bar.

Figure 12 shows one of these bars being bent. Nothing was left to trial and error. Every movement and measurement had been predetermined. A set of 21 separate computerized bending instructions for each diagonal offset bar, guided the bending operator to produce a bar that would require a minimum of jacking to within the (±) one bar diameter laterally and (+) one inch off the template surface.

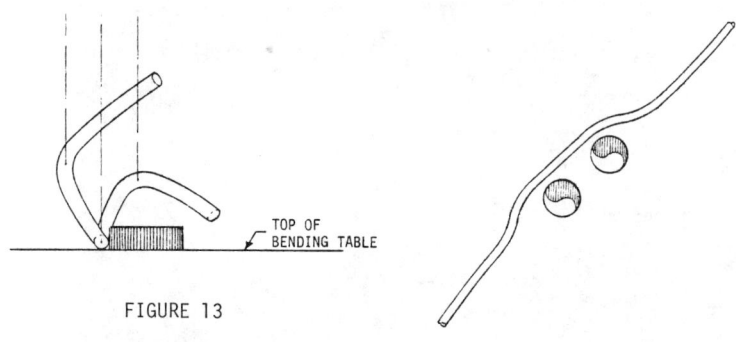

FIGURE 13

FIGURE 14

Figures 13 and 14 are sketches to show what the fabricated bar on the bender looks like from an end view and its theoretical position in the structure.

The position of the legs (Figure 13) at each end of the bar will vary, up or down depending on whether the offset is above or below the opening and/or the direction of climb is diagonal, right or left.

FIGURE 15

In the far background of Figure 8 there is another template under construction for a containment structure of a PWR for yet another nuclear power plant. (Figure 15) The size of this template is 80 foot square. This template, as with the other, was designed to be adjustable to accommodate the varied radii requirements from inner to outerface multi-layered bars. Every bar in every direction around the equipment hatch had to be plotted and painted on the template. This required engineers and painters to plot and simulate the bar locations on the template. As each layer was fabricated complete, the surface of the template was painted, the template adjusted to a new radius setting, and a new direction of bars for another layer was plotted and painted.

As we can see, the amount of area dedicated to the two projects was quite substantial. This entire area had once stored all the #14 and #18 bars for the entire fabricating operation. The storage area had to be relocated outside the building, and serviced with a new overhead crane. Shear lines and the saw cutting operation had to be relocated close to the end of the building. A new conveyor system had to be installed, running from the stock area outside the building, through a new opening in the wall, to the shear and saw. Additional bending equipment was purchased and set up in the template area and dedicated totally to special bending requirements so that the other bending equipment could maintain their normal volume of production. All of this resulted in a major capital expenditure ... to meet project requirements of configuration and special tolerance.

We feel it pertinent to point out at this time that the fabricator generally knows how long it takes to fabricate standard bar types to industry tolerances and prices his fabrication accordingly. It is the deviation from those standard types and tolerances that creates an impact on the fabricator especially if he is supplying reinforcing bars to many nuclear projects at the same time. Fabricating to restrictive tolerances or to bar types that are not standard is upsetting to a supplier's normal service and could produce a slow down in the volume of production and shipment.

PLANNING

One example of examining the constructability of a design philosophy is illustrated in Figure 16. Here we see a model of a containment structure with multi-layers of simulated horizontal, vertical and diagonal reinforcing bars continuous around the equipment hatch. Many meetings were held very early in the design stage of this project relative to bar configuration, tolerance requirements and fabrication capabilities. Extensive modeling was used to illustrate potential problems in reinforcing bar fabrication, scheduling, field storage and constructability.

The simulated bar layout in Figure 17 was modeled from geometric output from our computer program and was also a check for the design group against the validity of our calculations.

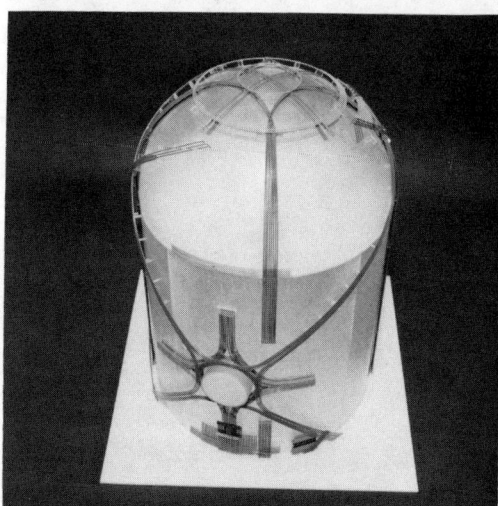

FIGURE 16

Our problem as a fabricator was rather complex. The equipment hatch was 28 foot in diameter. All #18 reinforcing bars were to offset around the equipment hatch as well as small penetrations or obstructions. Half of the sixty-four bars in each layer closest to the equipment hatch were between 71 to 86 feet long. Too long to safely handle. The design group instructed that mechanical splices (Figure 17A) were in order. Those particular bars were fabricated in two pieces, template checked and jacked within tolerance. Once accepted by the customer's agent, an alignment stripe (Figure 18) was painted at the splice location to facilitate field placement.

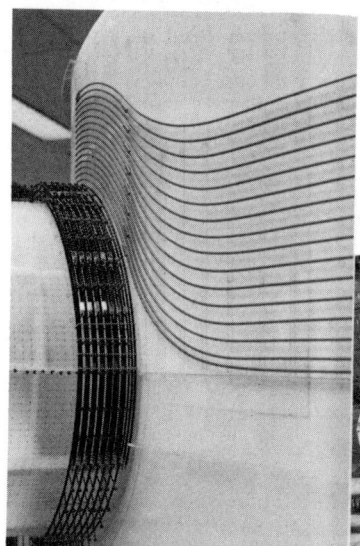

FIGURE 17A

FIGURE 18

TOLERANCE REQUIREMENT IMPACT   519

The layout of the bank of openings shown in Figure 19 posed very little problem for bars in the vertical and horizontal positions. However, the diagonal bars were a nightmare! Not only because they were radius bent to the cylindrical shape of the containment, but also offset to miss the penetrations, while maintaining the code required distance between bars. To manually calculate, detail and document the geometry of each bar would have resulted in a horrendous time factor and monumental costs. To handle that problem we developed computer programs which took over one year to complete.

Included in our computer program output were coordinates for template plotting and painting the horizontal, vertical, diagonal and ring bar locations around the equipment hatch (Figure 15) and step by step fabricating instructions to the bender operator for each bar. The more precise the initial bending, the less time spent jacking to the required tolerance.

As the template checks and jacking adjustments were made for each bar, the customer's agent or inspector would accept the end results on the template when the restrictive tolerance was met. Once accepted the bar would be placed on special shipping frames (Figure 20) designed and constructed for those unique bars. These special frames served a two-fold purpose. One, to maximize the number of bars that could be shipped on one load; and two, to minimize the relaxing of the bends by securely strapping the load to avoid movement over potentially rough roads for a long distance. Not all of these bars were shipped on special frames. Only those whose offset height exceeded five feet were

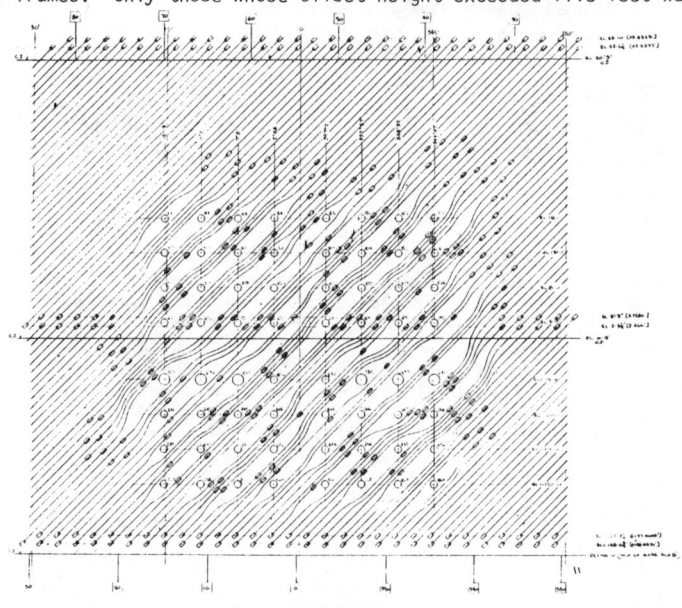

FIGURE 19

found to require the special frames. Those bars whose offset height was less than five feet could be blocked and strapped well enough to prevent any potential relaxation of bend. But, this again, was the customer's decision.

FIGURE 20

SUMMARY

As a supplier, our current position for reinforcing bar fabrication is that each special requirement must be examined individually to evaluate costs and the degree of success in obtaining the requested tolerance. Although not addressed previously this also holds true for special tolerance requests in bar sizes #3 to #11. In many cases the cost to achieve extremely restrictive tolerances may become prohibitive and other alternatives must be explored.

The experience gained in supplying reinforcing bars to over sixty nuclear power plants has taught us to be openly aggressive in requesting meetings very early in the contract and, sometimes in the precontract stage, to discuss real and potential tolerance problems. We feel this is mandatory not only for a smoother running project but selfishly, to minimize any impact in our standard fabricating routine. We have also learned that in this type of meeting, the field personnel who are responsible for the construction are definitely an asset in helping to resolve congestion and tolerance problems.

Another reason for the "early" meetings is to better plan and schedule special tolerance fabrication well in advance of the project requirement. Assumptions should not be made by the design group of fabricating capability and time. This becomes especially true when templates must be designed and constructed. Our policy has been and still is, to have the design group and interested field personnel witness test bending of special requirements, this allows for better understanding of capability, costs and mutually agreeing on an end result.

In closing, we hope what we've presented today will provide you with a better insight in dealing with those tolerance problems tomorrow.

ACKNOWLEDGEMENTS

1. Manual of Standard Practice, MSP-1-76, 22nd Edition, Concrete Reinforcing Steel Institute, Chicago, 1976.

2. Manual of Standard Practice, MSP-1-80, 23rd Edition, Concrete Reinforcing Steel Institute, Chicago, 1980.

3. ACI Committee 117, Tolerances for Concrete Construction and Materials, American Concrete Institute, Detroit.

4. Public Service Company of New Hampshire, Models - Seabrook Nuclear Power Plant.

5. United Engineers and Constructors, Philadelphia, Pennsylvania Models - Seabrook Nuclear Power Plant.

Modeling of Complex Reinforcing Steel Placements

by

Alan J. Boos[1]

Abstract

This paper provides an example of how engineering design and construction field forces cooperated in building an engineering model for a complex reinforcing steel placement. The benefits from this modeling effort included design detail enhancement, construction cost and schedule savings, and a better quality product. Although modeling of reinforcing steel placements is rarely required, this paper attempts to provide insight into how modeling can be used to anticipate and resolve problems with complex placements.

Introduction

The construction of a nuclear power plant is noted for its massive, heavily reinforced concrete sections which are provided to withstand postulated accident and severe environmental conditions. These conditions are generally attributed to pipe rupture and seismic events, respectively. Thick concrete sections are also provided as shielding against radiation. During construction of the internal concrete structure for a pressurized water reactor containment building, field construction personnel expressed concern over the constructability of two such massive, heavily reinforced structures. These structures, the two side walls of a refueling pool located over the reactor, are approximately 29 feet high and 58 feet long and vary in thickness from 4 feet-6 inches to 7 feet-8 inches. Figure 1, which illustrates the completed reinforcing steel model for one of the walls, gives a view of the wall looking toward the inside face. This figure illustrates the structural complexity of the walls. The noteworthy structural feature of these walls is the numerous embedment plates (with anchorage lugs) on the outside face of each wall. These plates, which are as large as 5 feet wide, 5 feet long, and 3-1/2 inches thick, were included to provide anchorage for the steel framing or hydraulic snubbers attached to them. Design tension and compression loads on these plates may reach 2,400 kips, creating the need for extensive local reinforcing. Furthermore, some of these embedment plates were to be set in skewed haunches or pockets. Figures 3 and 4, which also show the completed model for one wall, include these embedment plates.

---

[1]Assistant Project Manager, Midland Nuclear Project, Bechtel Power Corporation, Ann Arbor, Michigan

FIG. 1 Inside Face of South Refueling Canal Wall
(Looking Southeast)

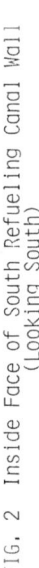

FIG. 2 Inside Face of South Refueling Canal Wall (Looking South)

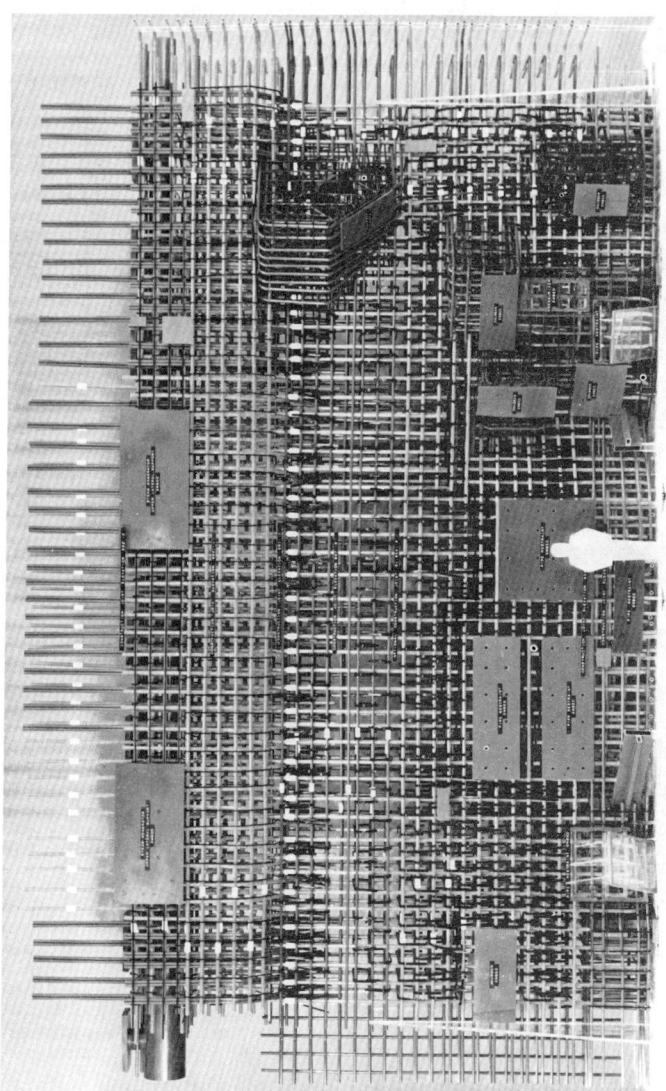

FIG. 3 Outside Face of South Refueling Canal Wall (Looking North)

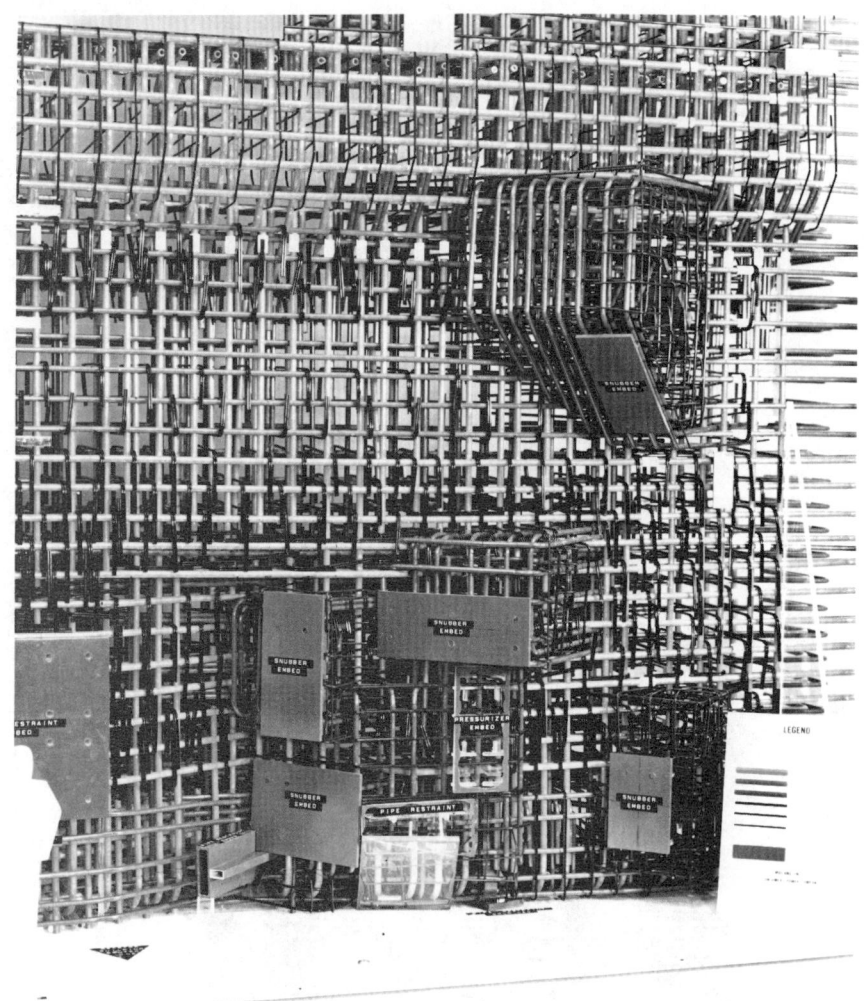

FIG. 4 Outside Face of South Refueling Canal Wall (Looking Northeast)

The design office furnished preliminary wall elevation reinforcing steel design drawings (at 1/4-inch scale) and detail sections (at 1/2-inch scale) to the field. These drawings raised concerns by construction personnel about the constructability of the walls. The main concerns included:

a. The high density of reinforcing steel (preliminary estimates of 600 to 700 pounds per cubic yard) and complex bar configurations, coupled with embed anchorages, would yield numerous interferences not identifiable on two-dimensional drawings. If not corrected, this would likely cause the following:

    1) Increased difficulties, delays, and costs in producing detailed fabrication and erection drawings

    2) Problems in locating construction joints and setting sequences for bar placement

    3) Delays in construction, with attendant cost and schedule effects while the interferences (which were identified during actual placement) were resolved with the design office

b. Placement tolerances provided by the installation specification would be inadequate to construct this complex placement.

Meetings were held between design office representatives and construction field personnel to review these concerns. Alternatives were discussed and a preferential plan of action was agreed upon. The essential elements of this plan were:

a. The design office would produce an engineering model of each wall at a scale of 1-1/2 inches equal to 1 foot. All reinforcing bars would be depicted.

b. The design office would produce detailed fabrication and erection drawings including bill of material sheets.

c. The model would be built in the design office to maximize input from and provide feedback to the design engineers and detailers.

d. Construction personnel would periodically visit the design office to provide input on preferred construction joint placement and bar installation sequencing.

e. Upon completion, the model would be shipped to the jobsite to be used as a reference during construction. The design office would provide personnel, if required, to keep the model current.

The design office already had a professionally staffed model shop. This is not a prerequisite to constructing a model, however, because many professional modeling agencies exist.

The modelers' review of the existing design drawings identified the need for additional drawings, particularly in congested areas, to support model construction. These drawings were produced by the design group. Following normal procedures, differently colored plastic materials were used to represent different bar sizes. These bar sizes included No. 6, 7, 9, 11, 14, and 18. Materials and dimensions were kept as close to scale as possible. The model makers initially built the model in 6 weeks. Although disagreements arose between the modelers and the design engineers during the early stages of model construction, this is not unusual and is generally ascribed to the designer's pride of authorship as design interferences are identified. This situation quickly remedied itself and the modelers and design engineers became a cooperative team. The modelers used a system of tagging to identify interferences. Approximately 35 interferences were noted during the initial construction of the model. These were located primarily in wall-to-wall and wall-to-slab intersection areas and in the region of the skewed embedment plates. Review and input by construction field personnel, including craft supervisory personnel, proved valuable in setting bar lengths, cadweld locations, and sequences of erection.

Review of the completed model by the design engineers resolved the noted interferences and improved other details of the overall design. As a result of this review, the model was essentially disassembled and reconstructed. This second phase was completed in 4 weeks by 6 model makers. The completed model was shipped to the jobsite, where final adjustments were made after transit.

The completed model was placed in a construction field office as close to the actual construction as possible. During bar fabrication, field engineers, quality control inspectors, schedulers, craft supervisory personnel, and craftsmen all became familiar with the model. This improved detailed work scheduling. During installation, the model was referred to frequently, enabling quick resolution of placement questions. Delays caused by design changes were virtually eliminated, and minimal problems were encountered during inspection by quality control personnel. Aggressive schedule milestones set for the placements were met. These results would have been unlikely if the model had not been constructed.

Cost is always a concern in deciding whether an engineering model should be constructed. When constructed in 1977, the cost of this model was approximately $36,000. Although it is difficult to quantify the exact cost savings realized with the use of the model, they easily exceed the cost of constructing it.

The success of this effort had its roots in the early anticipation and concurrence between the design office and field personnel that a model was needed. This concurrence is fundamental to the successful use of any engineering model for reinforcing steel placements.

SESSION VIII - CONCRETE PLACEMENT AND POST PLACEMENT PRACTICES

SESSION OBJECTIVES/SESSION CHAIRMAN SUMMARY

by

Alice Gannon[1] and Jack H. Willenbrock[2] M, ASCE

Objective of Session

Concrete placement practices with a specific emphasis on hot weather and cold weather conditions, mass concrete and post placement control are discussed in this session. Case histories which relate experiences on specific power plant projects are presented.

Session Chairman Summary

Experience at the Limerick Nuclear Generating Plant indicated that many changes were made to concrete placement and post placement practices over the seven years of structural concrete activity, including several PSAR (Preliminary Safety Analysis Report) amendments, thirteen major revisions to the concrete specification, twenty-two revisions to the QC instructions, and seventeen revisions to the job procedure for concrete construction. Many of these changes were required because of a trend toward literal interpretation of codes, standards, and regulations, rather than a continuance of the past practice of using them as guidelines with the project specification taking precedence in areas of conflict.

Changes were made to:

- Make PSAR commitments consistent with design intent.
- Make the specifications compatible with the amended PSAR.
- Provide for field engineering interpretation of specifications.
- Provide clarification as problems occurred.
- Improve economy without reducing quality.

The state of the art of hot-weather, mass concrete placement and post placement was presented. It was noted that this general area is one of the least coded, least regulated, and least standardized in the power plant construction industry today. Unlimited latitude for the specifier results from indefinite

---

[1] Engineering Management Specialist, Gilbert Commonwealth Companies, Reading, Pa. 19603

[2] Professor, Department of Civil Engineering, The Pennsylvania State University, University Park, Pa. 16802

and very limited information with regard to:

   a. Definition of the hot-weather variables: temperature, humidity, high wind velocities, etc.

   b. Definition of mass category: "generally" members over 2'6" thick, etc.

   c. Effects of hot weather on concrete with admixtures.

   d. Usefulness of admixtures in alleviating hot-weather, mass concrete problems.

It was also noted that only general "pointers" are provided for practical matters such as: use of ice, shading/sprinkling, white paint, temperature limitations, curing schedule, etc. It was felt that a consensus must be reached among specifier/designers, technical committees, and users in order to achieve uniformity in hot-weather mass concrete specifications.

The state of the art of cold-weather, mass concrete placement and post placement was also presented. It was noted that cold weather concreting requires several precautions to assure that concrete is not damaged by freezing and that it attains adequate strength. Precautions regarding characteristics of concrete at low temperatures were discussed. It was noted that curing and protection of the concrete is important at any temperature (high or low). The post placement concerns of form removal were also mentioned. It was felt that good quality concrete requires:

   a. Quality materials.

   b. Properly designed concrete mixes.

   c. Properly batched and mixed materials.

   d. Proper placement and consolidation techniques.

   e. Adequate curing and protection.

The final paper was presented in the form of a case history of various inspections and tests which were made on selected concrete placements on a nuclear power plant project to determine as-constructed conformance to the design requirements. This experience has proven once again that nondestructive testing methods can be reliably used to evaluate the quality of in situ concrete. Nondestructive methods must be correlated to destructive testing methods (i.e., compressive strength test of concrete cores). This is because the codes do not allow acceptance based only on nondestructive test methods due to the significant variations of test conditions and concrete constituents. These nondestructive test methods should use calibration techniques similar to the ASME and ASTM nondestructive examination requirements for metals. The test data should be statistically evaluated to provide a comparative measure of reliability used by the concrete industry.

Consensus Recommendations*

The Panel/Audience discussion period focused on:

1. Technical aspects of the concrete specification development, the need for a minimum 5" to 6" pumpline diameter when large aggregates are used, membrane curing on construction joints, testing for complete membrane removal, location of internal anomolies in concrete through ultrasonic testing, usability of frozen concrete, cold-weather protection of concrete and of aggregates, and avoiding voids in concrete placement.

2. Practical aspects of the problem of public safety information including the publicizing of the successful containment of radiation at TMI-2, safety and quality of concrete at other plants, boundaries against overtesting, introduction of routine and economic public reassurance measures (such as inclusion of strength gauges and other economical NDE techniques into concrete structures to assure structural integrity, redirecting employee and public safety questions to industry individuals capable of answering these questions, etc.).

3. When questions or negative observations by workers come up, management should immediately investigate or resolve them, since recent industry history has shown they will reappear (often larger than life) later and the lack of follow-through will appear to have been an indifferent or cover-up behavior by management.

---

*Based upon the Panel of Speakers/Audience discussion period at the end of the session.

Concrete Practice Changes At The Limerick Project

by

Alex C. McLean[1]

Abstract

Concrete practices underwent many changes during 7 years of construction as the Project 1) changed from the old concept that considered codes, standards, and specifications as guidelines to the new concept that codes, standards, and specifications must be literally interpreted and implemented - this required extensive changes to the PSAR and the specification because the PSAR over-committed and the specification did not fully reflect the PSAR commitments; and 2) benefitted from changes based on in-progress experiences to improve economy without compromising quality. Changes in concrete practices and changes to the project documents occurred together - changes to the documents reflect changes to the practices. Changes to the PSAR and the original specification are reviewed to show the scope, nature, and effect of the concrete practice changes. Conclusions are that 1) the PSAR should not over-commit; 2) the construction documents should fully reflect the PSAR commitments, provide for the exercise of field engineering judgment, and be written with full participation of the field groups; and 3) development of effective change procedures and cooperation of the involved groups to identify and implement changes should be encouraged.

Project Description

Limerick is a two unit, boiling water reactor, electric generating plant located 21 miles northwest of Philadelphia, Pennsylvania. Each unit is rated 1100 MW electrical. It was designed and is presently 50% constructed by the Bechtel Power Corporation (Designer/Constructor) for the Philadelphia Electric Company (Owner). The total concrete requirement is 560,000 cy. This includes 5,000 cy of high density concrete and grout for shielding walls, 45,000 cy of concrete for the two 500 foot high cooling towers and 200,000 cy of lean concrete fill.

3,000, 4,000, and 5,000 psi mixes are used for the structures; 2000 psi for backfill and work slabs under structures; and 1-1/2 sack, 80 psi "fillcrete" for backfill. Precast exterior panels for the buildings were cast off-site with 5000 psi concrete.

The concrete is produced by a 200 cy per hour capacity on-site batch plant. Placing equipment included conveyors and swingers for the turbine pedestal; pumps for walls, slabs, and containment structures; a tower crane for cooling tower shells; and a creter-crane for placements from grade.

---
[1] Engineer, Philadelphia Electric Company, Construction Division

Introduction

Concrete placement and post-placement practices employed at Limerick to implement codes, standards, and regulatory requirements have undergone many changes during the seven years since the first structural concrete was placed in July of 1974. The extent of the changes to concrete practices is indicated by the number of changes made to the documents which state the project requirements. For example, the desire to change a concrete practice usually required a specification change, and conversely a specification change usually produced a change to a concrete practice. This paper will indicate the scope, nature, and effect of the concrete practice changes by reviewing changes to the Preliminary Safety Analysis Report and changes to the original specification.

The concrete related portion of the Preliminary Safety Analysis Report (PSAR) which was submitted to the AEC (now the NRC) as part of the application for the construction permit, was changed to incorporate 24 amendments during the first three years of construction. 13 major revisions have been made to the concrete specification, 22 to the quality control instructions, and 17 to the job procedure for concrete construction.

Changes made to concrete practices are presented in two categories. First, changes that were made because of changing industry-held concepts during the late 1960's and early '70's regarding the purpose of codes, standards, and specifications, and second, changes which paralleled on-going construction to improve cost, productivity, and constructability without reducing quality.

First Category Changes

Changes in the first category resulted from over-committment by the original PSAR and failure of the original specification to fully reflect the PSAR commitments.

The PSAR committed concrete construction to the following codes and standards without taking exception to any of the requirements which they contain:

    ACI 301 - Specification for Structural Concrete for Buildings

    ACI 318 - Building Code Requirements for Reinforced Concrete

    ACI 347 - Recommended Practice for Concrete Formwork

    ACI 306 - Recommended Practice for Cold Weather Concreting

    ACI 605 - Recommended Practice for Hot Weather Concreting

    ACI 614 - Recommended Practice for Measuring, Mixing and

Placing Concrete

ACI Manual of Concrete Inspection, Publication SP - 2

The PSAR committed too much and the specification did not incorporate all the commitments. The Designer and the Owner did not consider this a problem when the PSAR and the specification were written. They were working on the concept that the codes and standards committed by the PSAR could be used as guidelines and that the specification requirements would govern when they conflicted with requirements of the codes and standards.

That concept began to change soon after the start of construction in July, 1974 as the result of interaction with the AEC.

In September, the AEC inspected placement of the Unit I containment base mat. The inspector wrote an apparent violation for performing slump, air, and temperature sampling at the truck discharge rather than at the point of placement as required by the ACI SP-2 "Manual of Concrete Inspection". During the same inspection the AEC identified several specification requirements that deviated from the ACI codes and standards, and several instances where the specification failed to incorporate requirements of the codes and standards.

The AEC advised that deviations from the codes and standards referenced in the PSAR would be construed as violations unless documented by formal PSAR changes. Times had changed - guideline interpretation was out and literal interpretation was in.

Placement of concrete in safety-related structures was suspended for a week while an interim procedure was written to implement the most stringent interpretation of the codes and standards for deviations identified by the AEC. The interim procedure remained in effect for 5 months until exceptions were documented by amending the PSAR.

The interim procedure required sampling for slump, air, and temperature from concrete placed in the forms; depositing concrete in 18" instead of 24" layers; ensuring that air temperature immediately adjacent to mass concrete did not fall more than 3° in any one hour during and at the conclusion of the cure period; curing mass concrete 14 days instead of 7 days; keeping the forms continuously wet for 48 hours after placement; and defining mass concrete as 30" rather than 36" thick.

After the PSAR was amended, placing 24" layers, 7 day cure without continuous form wetting, and 36" thickness designation of mass concrete were resumed. The PSAR amendment eliminated the ACI 301 requirement for not more than a 3° drop in air temperature during the cure period and the ACI SP-2 requirement to sample from the forms for air, slump, and temperature. Sampling was permitted at the truck discharge or pump line discharge provided that daily air, slump, and temperature correlation testing is performed between batch plant and field discharges.

The foregoing relates only the first PSAR amendment experience.

Seven pages of concrete related amendments accompanied by continual specification changes were made over a three year period to state exceptions to the codes and standards, to provide clarification in instances where disparities among the codes and standards were identified, and to permit implementation of some cost-effective construction practices. A better specification resulted, but the atmosphere of continual change disrupted the smooth development of concrete practices, especially during early construction when the job was on its learning curve.

When it was realized that the project was really living in a world of literal interpretation and implementation another big change was needed. The following example indicates how it came about: The original specification required that construction joints be thoroughly wetted, free of water, and in a saturated, surface-dry condition prior to placement. Throughly wet, free of water, saturated surface-dry. The resulting placement delaying discussions involved quality control, field engineering, supervision, and Owner's representatives. These delays, and an AEC violation written when concrete was placed on a construction joint where a small quantity of water had not been removed, focused attention on a generic specification problem.

The specification needed either to provide tolerances to eliminate the need for judgment, or it needed to specify who can make judgments upon specification items requiring interpretation.

The resolution was to delegate the specification writer's ultimate responsibility for specification interpretation to the Project Field Engineer. He was specified responsible to determine how wet a wet construction joint should be. The specification was completely revised to make the Project Field Engineer responsible for interpretable items, and to incorporate tolerances where possible. The Project Field Engineer in turn issued a field directive to conditionally delegate many of his new judgment responsibilities to the field engineering groups. This was a significant change for job harmony as well as for cost-effectiveness.

In cases when quality control disputes field engineering judgment, prompt resolution is usually obtained from the Project Resident Engineer who reports directly to the Designer.

Establishment of the on-site resident engineering group was another early change made to expedite resolution of field problems relating to design and interpretation of project engineering documents.

Second Category Changes

Changes in the second category were initiated by continual project effort to improve concrete practices and economy.

Field tests demonstrated that normal vibration of concrete during placement does not cause intermixing of sawdust, wood chips, and organic debris left in the trough portions of metal decking. The specification was revised to permit up to 1/2" of organic material in the troughs, or up to 50% of the trough depth when approved by the

Project Field Engineer. Other changes permitted leaving tie wire and nails on construction joints; and provided direction that the temperature of construction joints, forms, and embedded items be above 40° during cold weather.

Placing of a 6" layer of starter mix on construction joints was given as an alternative to placing grout which has to be thoroughly boomed in or flowed ahead by the concrete. Starter mix is the same mix used for the remainder of the placement but it has a 3" to 5" slump instead of a 3" or lower slump. This option proves beneficial at congested and inaccessible locations where placing grout and controlling its depth is a problem.

The 8" thick cooling tower shells, placed in 5' lifts, originally required 1/2" of grout on horizontal construction joints and placement of concrete in maximum 24" layers. The specification was changed to permit pouring the 5' lift with two 30" layers of starter mix thereby eliminating the 1/2" grout and one placing pass.

Deletion of the ACI 318 requirement to trowel grout on vertical construction joints eliminated the need for cement finishers working inside of wall forms to apply grout in 24" increments as placement layers progressed.

A requirement that vibrators not be used to move or spread concrete was clarified to permit the flow of concrete up to 5 feet from the point of deposition if segregation does not occur and the flow results from effort to consolidate - not from effort to cause the concrete to flow. The requirement that vibrators be inserted in a near-vertical position was modified to permit using them in positions necessary to consolidate concrete under large penetrations and embedded items.

A costly requirement to top out wall placements 1" above finish elevation and strike off to finish elevation just prior to initial set was eliminated. This requirement had no apparent codes or standards basis.

Specification wording which defined that a cold joint occurs when an operating vibrator will no longer sink into the concrete of its own weight was too vague. The specification was changed to follow the wording of ACI 614, and now states that a cold joint has occurred when vibration will not cause intermixing of the concrete layers. The specification states further that if a cold joint does occur, concreting is to be stopped and the cold joint treated as a construction joint by green-cutting or sandblasting.

Early specification revisions made to comply with the most stringent PSAR referenced code requirements established that form deflection could not exceed 1/240 of the span per ACI 301. This requirement and the use of a factor-of-safety of 2 per ACI 347-68 for form ties resulted in an allowable pressure of 1125 psf for gang forms instead of the manufacturer's 1500 psf rating. Additionally, concrete pressure on the compressible plastic foam material left in place to form gaps for seismic separation of structures (e.g. primary containment from the reactor building) was required to be held below 700 psf.

For several years the rate of rise of concrete in the forms and against the gap material was controlled to prevent exceeding the allowable pressures by using an empirical "liquid head" procedure which involved continual probing with a 3/4" rod during palcement to control the depth of fresh concrete above the elevation of initial set. Liquid head is the depth of fresh concrete above the elevation where initial set has occurred. The liquid head multiplied by 150 pcf, the weight of concrete, gives the form pressure. Maximum liquid heads were given by field engineering for placements against the gap material and for the several forming systems in use. When the maximum head was reached, placing was delayed until the elevation of initial set as determined by rod probing rose enough to permit placing the next 24" layer. The validity of this procedure was confirmed by instrumenting the form ties of several gang-formed placements with load cells.

More recently, a modification of the formula for maximum lateral pressure of concrete from ACI-347 has been used. The rise in feet per hour of concrete in the forms correlates well between the two control methods, but the use of a formula-based nomograph which gives maximum allowable pour rates per hour for gap material and the various form systems permits continuous placing at a uniform rate and eliminates the labor of rod probing.

Bullfloating and finishing of slabs with bleed water present were problems during the first slab placements. The specification was revised to incorporate the recommendations of ACI 302 that screeding and bullfloating be performed before bleeding occurs and that no slab finishing shall be performed in the presence of free water.

Revisions also provided for making sample slabs to ensure consistency in the appearances of the several slab finishes and incorporated a broom finish for slabs to receive special coatings.

All concrete sampling and testing is performed at the batch plant with the following exceptions:

- Slump, air, and temperature sampling for Class I structural concrete is performed at the truck or pump line discharge.

- Slump sampling for cooling tower shell concrete is performed at the truck discharge.

- Once-a-day sampling for slump, air, and temperature is performed at the pump line discharge for Class I structures for correlation with the batch plant test results.

This sampling program reflects a change to discontinue field sampling for slump, air, temperature, and compressive strength except as indicated above. The change greatly reduced QC receiving inspection manpower and eliminated field cylinder care and transport.

Construction joint keys are not used. The design premise is that properly prepared joints will provide the equivalent of monolithic concrete. The methods specified for joint preparation are waterblasting, green-cutting, sandblasting, or bush hammering. The use of a set retarder with the green-cutting process is optional.

Sandblasting and green-cutting have been used almost exclusively; bush hammering is utilized mostly for small areas. Prepared surfaces are considered satisfactory if all laitance is removed and aggregate surfaces are uniformly exposed. Acceptable surface textures range from exposed sand to exposed coarse aggregate without appreciable undercutting of the coarse aggregate.

An early economy change to use liquid membrane cure as an alternate to water curing all construction joints for seven days was adopted after field tests were performed to demonstrate that normal sandblasting is adequate to remove the membrane.

Green-cutting is used in preference to sandblasting at congested locations, on joints which will be water cured, when sandblasting operations would affect other work activities, and when it is desired to strip and raise forms rapidly.

Joint cutting requirements were eliminated at the locations where masonry walls contact concrete slabs and walls.

The basic curing requiremets are that mass concrete (thicker than 36") and thin-section concrete (36" and thinner) be cured for 7 days, and maintaining temperatures between 50°F and 70°F during cold weather. Enclosures are to remain in place at least 24 hours after the 7 day cure period and the concrete temperature prevented from dropping more than 30°F at the end of the cure period.

By original specification requirements, mass concrete and all construction joints required water cure, with liquid membrane acceptable only for thin-sections. Mass concrete forms had to be removed within 7 days and thin-section forms within 4 days.

As mentioned before, an early change permitted membrane curing of all construction joints. Other changes provided that forms left in place for 7 days satisfy the cure requirements, that field engineering can decide when to remove forms, and that mass concrete can be membrane cured when below freezing temperatures are forecast. These changes allowed flexibility in utilization of forms and eliminated cold weather water curing difficulties and concerns that forms would have to be stripped within heated enclosures to meet the 4 or 7 day stripping requirements, and the gradual drop-in temperature after 7 day cure requirement. Leaving forms in place to receive successive form lifts was also permitted.

Instructional changes included those that concrete surfaces are to be maintained in a moist condition until final cure method is started, that the concrete surface is to be damp at the time liquid membrane is applied, and that burlap shall be kept wet and in direct contact with the walls where practicable. Provision was made that wet burlap inadvertently not in direct contact with the concrete shall be considered sufficient to provide the necessary cure.

The ACI 306-66 recommendation that unvented combustion heaters not be used during and before 24 hours after concrete placement is followed by the specification. This requirement has limited the use of combustion heaters to heating of formed spaces prior to cold weather placement.

Except for the cooling towers, very little poured-in-place exterior concrete is exposed to view. For unexposed and interior concrete an early decision was made to avoid expenditures for cosmetic repairs and surface treatments that do not contribute to structural integrity. Sack rubbing was deleted from the specification leaving only requirements for "rough" and "natural" finish walls. No surface work is performed on rough finish walls except for removal of loose splatter and run-down grout from succeeding placements. The few natural finish locations require grinding of fins and protrusions. Form offsets are not treated in either case. For walls to receive special coating, consultation with the coating supplier and trial coating eliminated requirements to blend all form offsets by grinding and to flush grind all fins - this saved hundreds of miles of grinding.

Other significant cost reducing changes permitted eliminating patching of all form tie spacer and bolt holes except those at exterior exposed or earth backfilled locations; leaving tight tie spacer cones in place; flush cutting form ties; and flush cutting protruding nails, wire, and construction aid bolts instead of cutting them 1-1/2" below the concrete surface.

The specification was revised to provide criteria to define major concrete defects. The Project Field Engineer was given the responsibility to disposition major repairs and to identify which minor defects require chipping to ensure that they are not major. An original requirement that repairs be started no later than 4 days after form stripping was changed to 28 days, and a requirement for 1/2" saw cutting around areas to be replaced was deleted.

Changes to Quality Control and Job Work Procedures

The Quality Control Instructions (the QCIs) for concrete placement and concrete post-placement and the job procedure for concrete construction (the Job Rule) have been revised as the specification has changed. Additionally, they have undergone several major changes and continual minor changes to prevent recurrence of problems identified during construction progress.

Prior to their issue for construction, QCIs are reviewed with the Constructor's field QC group and are approved on-site by the Owner's engineering, field QA, and field construction groups. Early experience taught that the cooperation and input of the field groups is necessary to develop good Quality Control Instructions.

The Job Rule for concrete construction is a field written document which describes "The methods and procedures by which the project ensures adequate review and control of concrete construction". The Job Rule includes: definition of the responsibilities of the field construction and engineering groups; instruction for the use of concrete placement sign-off cards, concrete order sheets, post-placement stripping and curing forms, and the procedure for release of concrete from the batch plant to the placement.

Two early experiences served to emphasize the project commitment to quality and set the stage for continual upgrading of the QCIs and the Job Rule. The first experience involved stopping placement of the first structural foundation pour because of low slump and placing problems. The concrete was ultimately removed because of rock pockets at the bottom rebar mat which were evident at the face where placing was stopped. The second experience occurred several weeks later when a wall pour was stopped because the wrong mix was being used.

Concrete activity was suspended until major changes were made to the QCI and to the Job Rule. The changes included incorporation in the Job Rule of a drawing to show how emergency construction joints are to be formed and sloped in the event a placement has to be stopped, and addition of an ACI 301 reference to provide QCI inspection criteria for ensuring that placing equipment is properly used to prevent segregation. As an example, segregation problems occurred when creter-crane trunks were held off-vertical, and concrete was permitted to impinge on forms or rebar. Another change was inclusion in the Job Rule and the QCI of a step-by-step procedure, independently verified by QC, to ensure that the correct concrete mix is placed. The procedure covers the entire process from drawing take-off through ordering, batching, transporting, receiving, conveying, and placing.

The early problems benefited future construction - the Constructor and Owner groups rapidly got to know each other and learned to cooperate to resolve problems. Greater emphasis was thenceforth put on pre-placement planning and the groups continue to attend pre-placement meetings for major and unique placements.

A later problem involving voids in the outside face of a primary containment placement reiterated the value of pre-placement planning and the need for increased placement supervision and inspection within heavily congested forms. Extensive analysis of the placement assuming similar voids against the inside steel containment liner indicated a negligible effect on the rebar and concrete stresses.

These problems are related for their experience value. They comprise the major exceptions to Limerick's excellent overall quality record - there have been few major concrete repairs, no forms have been lost, and no major deficiency report has been written for concrete.

The quality and cost of the concrete at Limerick attest the spirit and willingness of the QC and concreting organizations who climb into the forms, meet the challenges of congestion, access, and change - and get the job done.

Summary and Conclusion

"Learn" and "change" summarize Limerick's experiences with the implementation of construction practices, codes, standards, and regulations.

During its early years, Limerick and the nuclear construction industry were undergoing a major learning experience - learning about expanding and changing regulatory requirements, and learning how to adapt previous experience to comply with them. The old concept that codes, standards, and specifications could be used as guidelines was changing to the new concept that requires literal interpretation and implementation of the codes, standards, and specifications.

The Limerick Project accepted the new concept and made many changes as it learned that:

- The PSAR must satisfy the regulatory requirements <u>and</u> the requirements of the Designer/Constructor and the Owner. It should not commit construction to requirements more stringent that those needed to provide a structure which meets the intent of the design and the requirements of the regulatory agencies.

- The specification and the quality control inspection plan must fully reflect the PSAR commitments.

- The specification must be written to be followed exactly by Quality Control and executed completely by the Constructor. To be workable, it must provide flexibility in its application by specifying tolerances and by delegating the Designer's ultimate responsibility for interpretation to the Field Engineer where possible.

- The Constructor's and the Owner's field organizations should contribute as fully as possible during the development of the specification and the quality programs.

In-progress changes for economy and preventing recurrence of problems will and should occur. Encouraging cooperation from the outset of construction among the Designer/Constructor, the Owner, and their Quality Organizations to identify the needs for those changes; and having effective procedures for implementing changes, will pay off as they did at Limerick to expedite construction and reduce costs.

HOT WEATHER TEMPERATURE PLACEMENT

OF MASS CONCRETE

BY

James P. Lonergan [1]

Abstract

Specifiers, producers and consumers of hot-weather mass concrete must become more involved in the specification and use of this product. An overall lack of definitive information and current up-to-date research on the subject has led to inconsistent job specifications, vague regulations and codes, and over-restrictive guidelines for the production and placement of hot-weather mass concrete. Current codes, regulations and specification practices are examined, along with two recent experiences in dealing with mass concrete, placed during periods of high ambient air temperatures.

INTRODUCTION

Hot-weather mass concrete is one of the most uncoded, unregulated and unstandardized materials in use in the power industry today. It is the most single, highly opinionated subject area in the field of concrete. The specification of temperatures, use of admixtures and methods of curing mass concrete are subjects of lengthy debates and considerable disagreements. From a constructors point of view it is a relatively easy product to work with when specified thoroughly, reasonably, and handled with care. It is also a product he fears.

The complications of placing mass concrete during high ambient temperatures are not what the builders are afraid of. Adverse hot weather effects are very well described by the concrete associations, and understood by others. Committees have been established and hundreds of reports have been published. Regulations, codes, standards and guidelines have been issued. All temperature-related subject

---

1 Sr. Construction Engineer, Bechtel Power Corporation, 15740 Shady Grove Rd., Gaithersburg, Maryland 20877

areas have been covered: specification, production, transport, placement and curing. So, what is the problem that constructors and utility owners face today?

An examination of current industry codes, standards and regulations concerning the problems of hot-weather mass concrete is needed in order to answer this question, and is presented in the manuscript that follows. Actual construction practices and experiences dealing with the implementation of these industry requirements are reviewed, along with their subsequent effect on quality, cost, and schedule.

EXISTING HOT WEATHER CODES
REGULATIONS AND STANDARDS

Several associations, societies and governmental bodies have written and published requirements governing the specification and use of concrete. By far, the largest treatise is taken on concrete design, with product (component) specification following second. Of third importance appears to be the production/batching aspects. Transport and placement practices seem to rank a meager fourth place.

Treatment of specific subject areas, includng hot-weather mass concreting are also covered by most of these agencies. However, actual regulations, codes or standards on this topic are not usually presented in a formal text. "Recommendations" by committee are the normal form of expression, which leaves the anticipated code user with an unlimited latitude for the specification of his project. Further examination into any of these publications will generally yield the following limited information about hot-weather mass concrete:

a.) that hot weather can be construed to be almost "any combination" of hot temperature, humidity and high wind velocities;

b.) that over one-half of the publications subject matter is devoted to the detrimental effects of hot weather on plain concrete (ie. concrete without admixtures);

c.) that actual recommendations to solve the adverse effects of hot weather are generally summarized by merely observing the following "pointers":

    1. use of cold water or ice,
    2. shading and sprinkling of aggregate piles,
    3. paint all equipment with white paint,
    4. concrete temperatures should "probably" be

limited to somewhere between $75°F$ and $100°F$, and

5. to immediately commence with the curing of concrete, as soon as practical, for varying lengths of time.

d.) that concrete members over 2'-6" thick in the least dimension should generally start being considered to fall into the "mass" category;

e.) that one paragraph, maybe two, addresses the use of admixtures, and notes that they have been shown to alleviate several of the common hot-weather mass concrete problems, but nothing more is added or said about it.

Information on the use of hot-weather mass concreting has remained in this format for well over twenty years. It has changed little since its initial publication. Outward appearances would have it that the current "state-of-the-art" in this field is afflicted by the incurrable effects of hot weather, and that existing problems cannot be overcome. Modern technology has finally come to a stand-still.

Realistically speaking though, this is not the case. There has been a lot of advancement in hot-weather mass concrete technology over the years. The problem revolves around the fact that a lot of people are just reluctant to recognize it and use it. Project specifications are written every day using the information covered above; and, as many of you are well aware of, a lot of disagreement exists on how it should be used. Typical project specification practices are examined in the following section.

CURRENT PROJECT SPECIFICATION PRACTICES

Project specifications covering the production and placement of hot-weather mass concrete are fairly typical throughout the industry today. Most have been formated into the company standard, routed through word-processing/computer printouts, and have found themselves ready-for-issue to any part of the world. Generally, they begin with a definition of hot weather, a definition of mass concrete, a statement on how to produce cool concrete (ie. chipped ice or flaked ice, etc.), what admixtures may be allowed, temperature limitations as a function of structural size, whether temperature is measured at the point of placement or discharge, time limitations between mixing and discharge, and finally, a lengthy write-up on special provisions for curing mass concrete. But here, the similarity ends.

MASS CONCRETE PLACEMENT 545

REQUIREMENT	PROJECT SPECIFICATION				
	A	B	C	D	E
Definition of hot weather.	90° or higher	90° or higher regardless of humidity 80°-89° w/humidity less than 60% 70°-79°F w/humidity less than 60%	None Specified	None Specified	None Specified
Definition of Mass Concrete.	4' thick in least dimension.	Varies (see temperature limitations).	None Specified	3' thick in least dimension.	Varies (see temperature limitations).
Cooling Procedure	Ice to 90% or by Liquid Nitrogen injection system; spray/shade aggregates.	Chilled water or Ice or both to 100%; Spray Aggregates.	None Specified	Ice to 100%; Spray Aggregates.	Ice to 100%; Spray Aggregates.
CEMENT TYPE					
Admixture Selection	TYPE II All types OK with Approvals	TYPE I Air-entraining OK with approval. Others upon demonstration and with approval.	TYPE I OR II Air-entraining OK. Water-reducing upon demonstration and approval. Retardents NOT ALLOWED.	TYPE II All types OK with approval.	TYPE II All types OK with approval.
Slump	3"	Not Specified	Not Specified	2"	2"
Concrete Temperature Limitations	80°F for Mass, 90°F All Others.	Up to 2'-6" thick 90°F. Over 2'-6" to 4'-0"-80°F. Over 4'-0" - 70°F.	100°F All Sizes.	70° for Mass, 90° all others.	Up to 3'-0" thick - 90°F Over 3'-0" to 6'-0"-75°F Over 6'-0" - 65°F.
Place of Temperature Measurement.	Point of delivery, except at pump lines, use end of pump line.	Same as "A".	Not Specified	Same as "A".	Same as "A".
Discharge Time Limit	1 1/2 hr - normal; 45 min. hot weather.	60min - all conditions.	None Specified.	1 1/2 max or 300 Rev.	1 1/2 max or 300 Rev.
Curing Requirements	Moist for 48 hrs. then membrane 2 weeks, except if air over 90°F resume moisture loosen forms ASAP.	7 days min. moist or use membrane If moist is not practical.	Membrane cure all. Form cure OK.	7 Day min. moist loosen forms after 12 hours.	7 Day min. moist. Loosen forms after 12 hours.

Figure 1.0 Comparison of Hot-Weather Mass Concrete Data For Five Power Generation Facilities

Recent experiences in dealing with various industry counterparts have shown considerable disparity in the way hot-weather mass concrete is specified. Comparison of existing codes and regulations to the job-issued specifications yield an even greater difference. Figure 1.0 was developed from projects experienced during the past 12 months, and attempts to highlight some of these differences. Through inquiry, it was learned that each of these specification requirements are of the company/corporate "standard" type, and are generally considered inflexible from project to project. The reasoning behind this stand is unknown.

It is known though, that the often one-sided approach taken in the specification of hot-weather mass concrete is not at all appropriate when considering the complexity of this product. For example, nearly all of the specifications dealt with during the past year exhibited the following characteristics:

a.) unreasonably low maximum temperature limits for hot-weather mass concrete pours, with no recognition of the structures reinforcement, the structures shape, size or intended use;

b.) arbitrary definitions of "hot-weather," especially without regard to humidity;

c.) arbitrary definition of "mass concrete," again with no recognition of the structures reinforcement, the structures shape, size or intended use,

d.) lack of specification for the use of admixtures (Air-entrainment admixture appeared to be one exception.)

A failure is reached when the "company standards" are issued, irrespective to project locations, conditions and design. Contractors are forced to use a specification that does only one thing--lower the specifiers overhead costs by eliminating the need for an "engineered" structure. Use of innovative products or approaches are being ignored or shunned, unless contractors provide independent, on-site testing and demonstrations. The years of research and issuance of proven data that product manufacturers (and societies and associations) have to offer are infrequently used on a power plant project. The contractor and/or owner must pay the bill. Oftentimes, the bill is paid by using the standard specifications, by trying to cope with low effectiveness, high purchase and placement costs, and continuously being behind schedule.

The brief items discussed in this and the foregoing section can easily shape the basis on how projects will perform on concrete placements. With hot-weather mass concrete placements being even more specialized and subject to greater scrutiny, the more this shaping is affected. The sections to follow examine two actual cases of hot-weather mass concrete placemets on current power projects. The inter-action of codes, standards and specifications on these projects will be viewed along with the construction techniques and experiences used to make the job successful.

HOT-WEATHER, MASS CONCRETE-A NUCLEAR EXPERIENCE

Project Characteristics
  Project Type: Nuclear
  Location: Lower South Central U. S.
  Contract Type: Complete engineering, procurement & construction services (cost plus basis.)
  Summer Temperatures: Very hot and humid
  Winter Temperatures: Frequent freezing
  Concrete Batching Facilities: Twin 150 c. y./hr. central mix plants, within 3,000 feet of most placement areas; cooling by chilled water and/or chipped ice; no facilities to shade aggregates; facilities for sprinkling
  Concrete Specifications: See Figure 1.0, Project "D".
  Placement Equipment:
    - Two-four stationary 4" pumps (W/2 backup units);
    - One conveyor belt mounted on all-terrain hydraulic boom crane
    - One or two latice-boom crawler cranes each W/2 c.y. buckets as backup.
  Pour Sizes: Between 500 c.y.'s to 5,000 c.y.'s (avg. pour about 1,500 c.y.)
  Placement Seasons - All year.

Project Experiences
  Production and placement of concrete at this site is a classical case of technology "by-the-books." Craft superintendants, field engineers, quality control and quality assurance personnel all had some role in the production and placement of concrete. Every step was monitored thoroughly from the selections of aggregates to the use of curing logs to monitor the final completion of each pour. Essentially, the entire process went as follows:

  While batch plant erection was in progress, a thorough testing of proposed aggregates, admixtures and preliminary trial mix designs was underway. Petrographic examinations, Los Angeles abrasion test, water analysis, unit weight samples, etc. were performed until all components were found to be acceptable. Upon completion of erection, the batch plant also was tested, personnel were trained, mixer uniformity tests run, and finally a certification was issued for production.

While this was underway, planning began for the placement of mass concrete. Each mass placement was subjected to the jurisdiction of a general contingency plan, that was developed specifically to deal with any unplanned interruption of concrete flow. The plan concentrated primarily on how to treat the in-place concrete, and dealt with the prospect of cold joints, reinforcement, bulkheading and the resumption of placements. Interruptions may have included such difficulties as mechanical failure of placement equipment, breakdown of the batch-plant itself, possibly the loss of slump control or, uncontrolled temperature rises in the concrete itself.

On the largest of mass pours (ie. turbine pedestals, reactor foundations, etc.), a special plan was developed. This dealt with the assignment of personnel to various shifts, the chain-of-command, communications and access control concrete truck routing, equipment assignments, and the needs of special interest groups such as the testing laboratory or quality control personnel.

Fortunately, a majority of the largest pours were scheduled during the winter months, when high ambient air temperatures were not a problem. However, it was necessary to place several other pours during the heat of the year. In this case, pours normally began in the early morning hours, as close to 6 a.m. as possible, and continued throughout the day until completion. Normal rates of pour were targeted at the 125-150 c.y./hr. range. Actual pour rates were generally much less than this.

In order to successfully monitor such rates, and be able to cope with the ever present danger of mechanical breakdowns, good communications were a must. All placements were assigned a pour supervisor, who maintained continuous contact with the batch plant by use of an FM hand-held radio. Batching of trucks generally proceeded through this meadium, and once loaded, they rarely had problems discharging within the $1\frac{1}{2}$ hour/300 revolution time limits. The technical aspects (slump, tem., air) of concrete flow was also coordinated by radio, and communications between test lab personnel and quality control engineers were constant. Correlation of tests between the batch plant and point-of-discharge was a simple operation.

Coordination and control of this degree is an expensive proposition though, as the manpower requirements are not light. Craft superintendents, field engineers, test lab personnel, quality control engineers, quality assurance people, all with their own procedures, plans and courses of action to follow-to meet the "nuclear" committment. Even so, with such an extensive set-up, technical difficulties still tend to arise on the hot-weather, mass concrete pour. The best men, machines and money cannot always overcome the handicaps involved with the specification of low slump, low temperature mass concrete, placed by "conventional" methods.

Depositing 2" slump concrete, with $1\frac{1}{2}$" aggregate, by a 4" diameter pump line 150' above grade, while trying to maintain a temperature of less than $70°$ when the ambient air is near $100°F$ is certainly considered a handicap. By today's standards, it is also not considered unconventional. A desirable situation to be in? No, certainly not. But when you must, well, the outsome is inevitable. On more than one occasion, pours had to be shut down, and the general contingency plan was put into effect. The reasons for such shut-downs were many, and they often included:

1. temperature consistently above $70°$ F. The use of $71°F$ or $72°$ F concrete was "out-of-spec." and was rejected

2. slumps could not be held consistently at or below the 2" level. The use of too many loads of 3" slump concrete was "out-of-spec." and was rejected.

3. the use of low-slump concrete with $1\frac{1}{2}$" aggregates in a 4" diameter pump line was prone to line plugging, especially in vertical lines and at bends in the line; (and unfortunately, placement by other means was impractical.)

4. mechancial breakdowns, causing lengthy delays in the flow of concrete, contributed to cold joints, especially when such low-slump concrete is utilized.

Aside from these difficulties though, all else proved to be "normal." Each summer was approached with the same set of specifications, the same low slumps, the same large aggregates, the same low temperatures--the cure-all to the adverse effects of hot-weather mass concrete pours.

It can be said, without a doubt, that the adverse effects of hot-weather mass concrete (from a product point of view) were avoided on this project. The quality of concrete at this site is excellent. The cost of producting it however, in the late 1970's, ran over $50 per cubic yard. Add to it, the cost of placement, quality monitoring and testing, and you have some of the most expensive concrete produced today. Add to it the delays caused by shutting down a pour for so called "out-of-spec." concrete and the loss in schedule that can never be regained. This is the nuclear experience.

HOT-WEATHER MASS CONCRETE-A FOSSIL EXPERIENCE

Project Characteristics
  Project Type: Fossil (coal)
  Location: Lower Southeast U. S.
  Contract Type: Construction only, lump sum ("hard-money") pricing.
  Summer Temperatures: Very hot and humid
  Winter Temperatures: Rarely freezes

Concrete Batching Facilities:
  -1st half of project: Single 90 c.y./hr. truck-mixed plant, in metropolitan area, approximately 7 miles from placement area (w/several active train crossings); plant shared committments w/local community and other contractors; no cooling available; no shading of aggregates available; limited sprinkling of aggregates available;

  -2nd half of project: Single 90 c.y./hr. truck-mixed plant, approximately 4 miles from placement area (w/one train crossing); total committment; cooling by liquid nitrogen injection or chilled water; 2nd plant at same site, @ 40 c.y./hr., shared committment, cooling with chipped ice or chilled water.

Concrete Specifications: See Figure 1.0, Project "A".
Placement Equipment:
  -One conveyor belt mounted on all-terrain hydraulic boom crane
  -Two latice-boom crawler cranes each W/2 c. y. buckets.
  -Two truck mounted swing-boom pumps-5" (used on one 1,700 c. y. pour only).
Pour Sizes: Between 500 c. y. to 2,100 c. y. (avg. pour about 1,000 c. y.)
Placement Seasons: Late spring, summer, early fall.

Project Experiences
  Production and placement of concrete at this site started out with all the flavor and excitement of a back-woods mom and pop operation. When contracted to do the job, it was found that aggregate tests, as specified, could not be complied with; sand, as specified did not meet gradation requirements; cooling, as specified, could not be provided; water, as tested, did not meet the specifications, and slump control was a dirty word. As typically found in many rural areas, a lot of small local concrete producers have never dealt with the larger engineering firms especially on power plant projects. The "standard" concrete specification most often issued is very different from exist-

ing local and state agency specifications, and are rarely tailored to suit specific local conditions.

In this particular case, an attempt was made to change this. Through discussion and reference to other existing local projects, the concrete producer was able to justify the ingredients of his product, proving that the strength, durability and appearance of his concrete was equal to that originally specified. Concessions were made on all aggregate selections, water and temperature requirements. None were made on air content or slump. Without these changes, the project never would have started on-schedule.

A majority of concrete to be placed on this contract was mass concrete. Pour sizes of 60' x 125' and 6'-10' in depth were average. Schedule requirements dictated that most of it be placed between May through September, when hot-weather extremes were greatest. In preparation for the mass placements a general contingency plan was drawn up, very similar to that mentioned in the nuclear experience. Placements were planned for starting in the early morning, and working through the heat of the day. Normal rates of pour were targeted at 120-140 c. y. /hr. range. Actual pour rates averaged somewhat below that figure.

Mass placement rates were normally higher in the morning than in afternoons. This was made possible by driving most delivery trucks right onto the existing foundations, up next to the ongoing pour. The base mat design incorporated the use of a 1'-0" fill slab (to be placed later upon completion of all foundation and structural steel work) and "removable" column anchor bolts. During the first 2 to 4 hours, placement of concrete by direct truck discharge normally averaged 120 to 150 c. y. /hr. This method had a distinct advantage over others, as critical equipment, particularly cranes, could be freed to work elsewhere. During this period, little communications with the batch plant was necessary.

During the remainder of each pour, however, communications between the placement area and batch plant became more critical and difficult. As temperatures began to rise and placement methods were changed (from truck discharge to crane/buckets), the concrete changed also. Trucks were not needed so fast, back-ups tended to occur, and slump was easily lost. With a batch plant over 4 miles away and a 20 minute batch and trip time in between, four trucks were on the road before more could be halted. Site-added water became a common ingredient by mid-afternoon.

Afternoon ambient air temperatures were very often above $95°F$. With chilled water being the only cooling ingredient available, a lot of reliance was placed on water-reducing agents (so that water could easily be made up on site, without affecting strength), and retardants. Temperatures at the point of placement very often exceeded 80°F., and in some cases, exceeded 90° F. Shocking by many power plant "standards". However, prompt attention to the curing of concrete by ponding and flooding with water alleviated the most common concern of high ambient air temperatures-cracking of the concrete surface. Cylinder test results also confirmed its strength acceptability, with average breaks at 7 days meeting the 28 day requirements. Curing was continued on the mass pours for at least the first 48 hours. Constant use was made of the ordinary lawn sprinkler, both rotary and linear spraying models. Although not very durable, their use is very adaptable to moist curing of concrete.

Subsequent to the production of nearly 20,000 c. y. of uncooled concrete, a new plant was brought on line. The remainder of concrete was placed with the benefit of cooling, by the liquid nitrogen injection process. In this process, water is passed through a charging tube, aimed at the rear of each truck, while a lance emits liquid nitrogen gas. This forms a mixture of slush. Problems beset the plant almost immediately though. Concrete temperatures below 80°F were hard to obtain with ambient air temperatures above 95°. Later it was found that the lance/injection angle was critical to performance, and new positions were tried. Additionally, a second lance was added, to enhance the cooling effect. Unfortunately though, full perforance of this new cooling system was not achieved during the active portion of this contract.

Overall, the requirements and procedures used on this project were flexible and fairly easy to work with. Recognition was made of the individuality of a project site, its design, and its intended use. A "by-th-book" approach was often used, but not as a golden rule. Concrete production costs were kept at under $45 a cubic yard. Schedule was maintained through the job, and quality did not suffer. This was a fossil experience.

SUMMARY

It would be somewhat inappropriate to draw conclusions, and summarize the events and experiences shared in this manuscript. They stand on their own, on behalf of the industry we serve, and the technical organizations that serve us in return. What is needed, instead, is to ask ourselves where it all leads to and what is around the corner for us in the future.

The future of concrete technology hopefully holds more than its past. Of all the publications, documents, journals and literature published during the last ten years, how many have been devoted to originals, innovative techniques in the hot-weather mass concrete industry? How many are really being considered for incorporation into existing codes, regulations, standards, or project specifications? Before this can happen though, a much stronger uniformity must exist in the current specification of hot-weather mass concrete. A concession must be reached whereas technical committees, specifiers of the product and the users themselves are in total agreement as to what hot-weather, mass concrete really is. The challenge is there for us to accept.

COLD WEATHER PLACEMENT
AND POST PLACEMENT CONTROL
by
Donald E. Dixon, P.E.[1]

ABSTRACT

Cold weather concreting requires several precautions to assure the concrete is not damaged by freezing and attains adequate strength. This paper presents and discusses the precautions regarding the characteristics of the concrete at low temperatures. Curing and protection of the concrete is very important irregardless of the temperature (high or low!). This paper also discusses the affects of these post placement concerns including form removal.

COLD WEATHER CONCRETE PLACEMENT

Many active construction areas experience low ambient temperatures for a relatively long period of time. While there are precautions to be taken when placing concrete during cold weather, cold weather does not present as many problems as does hot weather concreting. For the purposes of this paper, cold weather concreting involves ambient temperatures below 40°F (5°C). Above this temperature, only minimal additional precautions are necessary, the most important of which is the in-place strength needed for form or shoring removal. With many of the concrete placements for power plant constructions being massive and on soil for foundation support even this precaution may be of low priority.

Ready mix concrete producers in those high volume construction areas with extended cold months have installed heating equipment in their plants to meet the demands of the projects. These facilities include

---

[1]Corporate Materials Consultant
Law Engineering Testing Company
Atlanta, Georgia 30324

hot water heaters and means to heat aggregates. Insulated forms and bats of insulation material are available to help maintain much of the heat of hydration within the concrete. The major concern during sub-freezing weather is to prevent the plastic concrete from freezing. When the free moisture in the freshly mixed concrete freezes to form ice crystals (about 30°F (-1°C)), the concrete is considered frozen. The concrete must be protected until the cement hydration has progressed sufficiently to reduce the amount of free moisture and create enough strength to withstand the stresses of freezing.

Requirements

Most codes and specifications[1,5,6] have some restrictions and requirements for cold weather concreting. The general theme of these requirements is to prevent the concrete from freezing. Typical general statements (such as ACI 318 Building Code[1]) would be - "Adequate equipment shall be provided for heating concrete materials and protecting concrete during freezing or near-freezing weather." Also, regarding post placement requirements, statements such as - "All concrete materials and all reinforcement, forms, fillers, and ground with which concrete is to come in contact shall be free from frost" and - "frozen materials or materials containing ice shall not be used" are made. Similar requirements can be found in ACI 349[5].

The recommended concrete temperatures shown in Table I are typical for concreting under various conditions. When the mean daily temperatures are below 40°F (5°C) the temperature of the concrete as mixed, placed and maintained should follow these requirements. Also the length of time that the concrete should be maintained at that temperature is dependent upon the cement factor, type of cement, use of an accelerating admixture and type service category for that member. Typical requirements are shown in Table II. It is important to remember that concrete can be protected for a long enough period of time to prevent freezing damage but the concrete may not necessarily have developed enough load carrying capacity during this period.

Effect of Low Temperatures on Compressive Strength

The rate of cement hydration is affected by temperature - low temperatures retard the hydration which slows down the concrete's hardening and strength gain, while high temperatures accelerate these properties. Figure 1 shows the affect of low temperatures (40° and 55°F (4° and 13°C)) on compressive strength as compared to the standard moist cured environment (73°F (23°C) and 100% RH). As can be seen, strengths are lower than standard for the lower temperatures at early ages. Standard 6x12 inch concrete cylinders were used as the test specimens and were moist cured at the temperatures indicated up to 28 days. It is noted that the specimens cured at lower temperatures attain a higher level of strength than the standard beyond 28 days age. It may be inferred that concrete placed at low temperatures, but above freezing, may develop higher ultimate strengths. This is true provided the moist curing is maintained for a longer period of time, since strength gain essentially stops when moisture required for cement hydration is no longer available.

TABLE I
TYPICAL RECOMMENDED CONCRETE TEMPERATURES

Air Temperature °F (°C)	Section Size, Min. dimension, in. (mm)			
	less than 12 (300)	12-36 (300-900)	36-72 (900-1800)	more than 72 (1800)
Min. concrete temp. as placed and maintained, °F (°C)				
---	55 (13)	50 (10)	45 (7)	40 (5)
Min. concrete temp. as mixed, °F (°C)				
Above 30 (-1)	60 (16)	55 (13)	50 (10)	45 (7)
0 to 30 (-18 to -1)	65 (18)	60 (16)	55 (13)	50 (10)
Below 0 (-18)	70 (21)	65 (18)	60 (16)	53 (13)
Max. allowable gradual temp. drop in first 24 hours after end of protection °F (°C)				
---	50 (28)	40 (22)	30 (17)	20 (11)

TABLE II
PROTECTION FOR CONCRETE PLACED IN COLD WEATHER

No. of days protection to be maintained at temp. in Table I

Type Cement	Protection from damage by freezing		Protection for attaining safe strengths	
	Type I or II Cement	Type III Cement*	Type I or II Cement	Type III Cement*
Service Category				
No Load, no exposure	2	1	2	1
No Load, exposed	3	2	3	2
Partial load, exposed	3	2	6	4
Full Load	3	2	(See Chapter 7, ACI 306R)	

* or with accelerator, or 100lbs/yd^3 extra cement

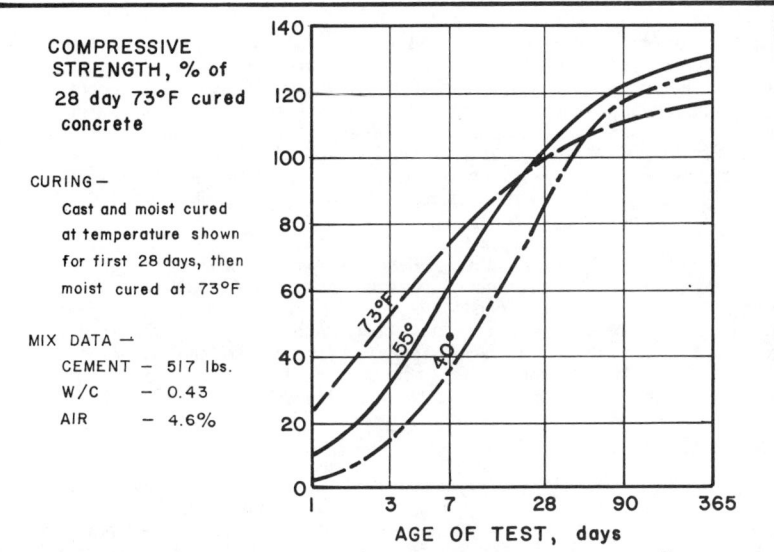

Fig. I – Effect of low temperature on concrete strength[3].

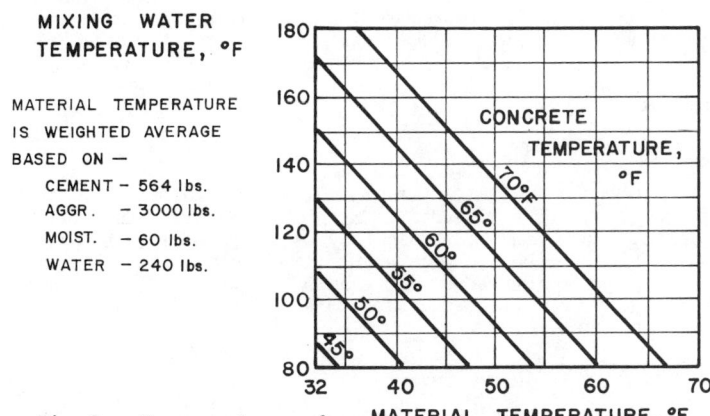

Fig. 2 – Temperature of mix water needed for concrete at req'd temperature[3]

Concreting Procedures

Several procedures are typically used during cold weather concreting. Many areas use heated water to raise the concrete temperature. Figure 2 shows the water temperature necessary for several concrete temperatures needed and the weighted average material temperature (cement, fine and coarse aggregate). The weighted average material temperature is basically determined by cummulatively adding the numbers found by multiplying the temperature of each material (cement, aggregates, etc.) by the ratio of its weight to the total weight of materials (including mix water).

Precaution needs to be given when adding hot water to the mix (or for that matter when adding aggregates heated above 100°F (38°C)) since a flash setting of the cement can occur. The heated water should first be added to the aggregates (which may also be heated) and mixed prior to adding the cement. The temperature of the water and aggregates should not be heated to the extent that the combination is over 80°F (29°C). The water alone should not be heated over 180°F (82°C).

Accelerators

Small amounts of accelerators, ASTM C 494, Types C and E[7], are commonly used to speed the rate of hardening of the concrete during cold weather. Precautions are necessary since many of the available accelerators contain high levels of chlorides. Probably, the most commonly used accelerator is Calcium Chloride (in solution and added at a rate up to 2% by weight of cement). Calcium Chloride, which does not meet Type E requirements, does not change the general chemical reaction (hydration of cement), but acts as a catalyst to free calcium oxide and permit earlier hydration of the cement. Accelerating admixtures, containing chlorides, therefore, present a much higher potential for corrosion of embedded steel. For this reason and others, admixtures containing chlorides should not be used in prestressed concrete, concrete with embedded aluminum, concrete in contact with galvanized steel, and concrete subject to alkali-aggregate reactions.

High-Early-Strength Concrete

Higher early strengths are desirable in cold weather to shorten the length of time the concrete needs to be protected (see Table II). These higher strengths can be obtained by using High-Early-Strength Cement (Type III or IIIA) and/or curing the concrete at higher temperatures (steam curing or heated forms). Figure 3 shows the advantage of using Type III cement for higher early strengths when cured at low temperatures.

Summary of Precautions

ACI Committee 306 in their report[2] give a thorough discussion of cold weather concreting. The reader is recommended to be fully familiar with this document before becoming involved with any aspect of cold weather concrete. The principal recommendations are paraphrased as follows:

1. Use air entrained concrete
2. Minimum concrete temperatures as placed (generally 50°F (10°C) or higher - See Table I) should follow appropriate guidelines regarding protection against freezing and attaining sufficient strength for the loading and exposure to be expected.
3. Protect all concrete from freezing during its early life (24 hrs or more)
4. Recognize that the lower the ambient temperature the longer protection is required.
5. The use of accelerators (with proper note that some concretes should not contain chlorides in excess of a stated limit), high early cement (type III), or more cement, could reduce the time that protection is required.
6. In certain very cold climates, the use of insulated forms may be advantageous and cost effective.
7. At the end of the protection period the concrete's temperature should not be allowed to drop more than about 40°F (4°C) (See Table I) in the first 24 hours. This temperature drop should be gradual during the "cool down" period.

Advantages

The main advantages of cold weather concreting can be summarized as follows:

1. With increasing amount of wintertime construction being accomplished advantages are realized by the labor force, contractors and, therefore, the economy.
2. Projects can be completed earlier when construction continues into the winter months.

As an added benefit, properly protected and cured concrete placed during cold weather could ultimately gain higher strengths than concrete placed and cured at the standard 73°F (23°C) temperature. The optimum curing temperature for highest ultimate strength appears to be near 55°F (13°C).

Disadvantages

The main disadvantages of concreting during cold weater include among others:

1. Lower early age concrete strengths delays form removal. The tragic construction collapses of recent times were, at least partially, caused by the concrete strength being less than anticipated (reference is made to the Skyline Plaza collapse at Baileys Crossroads, Virginia, and the cooling tower form collapse at Willow Island, West Virginia).

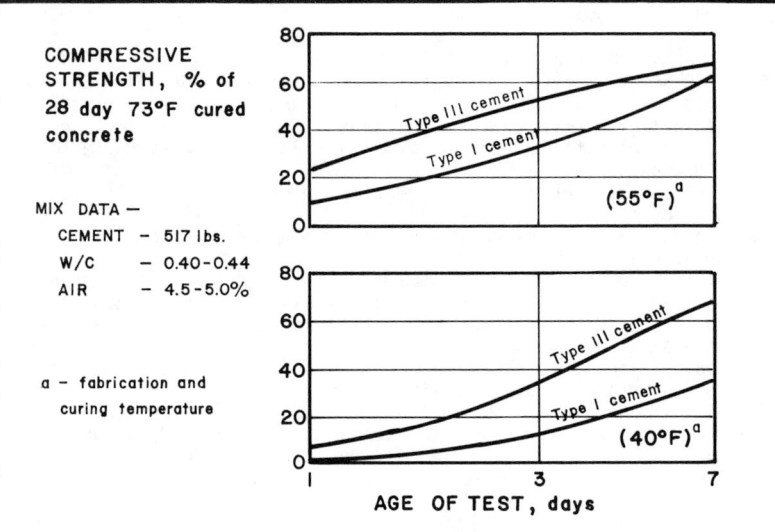

Fig. 3 - Early compressive strength relationships[3].

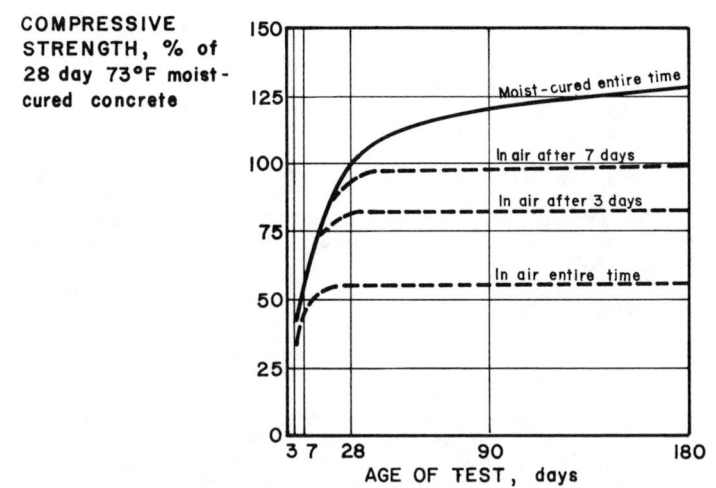

Fig. 4 - Compressive strength of air dried concrete[2].

2. Extra equipment in the form of heaters, water boilers, steamlines, insulating material, etc. is required - which means additional cost
3. Labor force is less efficent and extra personnel are needed during the night to monitor heating equipment.

Perhaps the most significant disadvantage involves the in-place concrete strength at early ages. There are several nondestructive tests available for estimating the concrete's strength through in-place measurements. Calibrated instruments used by properly trained and experienced people, and correlated to compressive strength measurement of the concrete used can give results with which one can have a high degree of confidence. These procedures are discussed in detail in the ACI Monograph on Nondestructive testing.[8]

POST PLACEMENT CONTROL

After placing concrete, in whatever weather constraint, post placement procedures are required. These involve curing and protection, and form removal. Curing of the concrete, which generally includes protection, involves maintaining a satisfactory moisture content and suitable temperature so adequate hydration of the cement can occur.

Curing

Curing is an exceedingly important step in the manufacture of concrete since the hydration of cement requires continued moisture and favorable temperatures. These conditions must be maintained for a period of time normally called the curing period. Most general specifications set this interval, in terms of preventing moisture loss, at 7 days for concretes with Type I cement and 3 days with type III cement.

Curing Methods

Forms provide a satisfactory protection against the loss of moisture from the concrete provided in exposed surfaces of the concrete are kept wet. For concrete surfaces not in contact with forms, the following methods are generally specified as being suitable to prevent loss of moisture from the exposed surfaces:

1. Ponding or continuous sprinkling
2. Absorptive mats or fabric kept continuously wet
3. Sand kept continuously wet
4. Continuous steam (not over 150°F (66°C)) or mist spray over the surface
5. Waterproof sheet materials (conforming to ASTM C171 "Waterproof Sheet Materials for Curing Concrete")
6. Acceptable membrane forming curing compounds (conforming to ASTM C309, "Liquid Membrane-Forming Compounds for Curing Concrete")
7. Other approved moisture-retaining system or covering

Curing Temperatures

Curing temperatures in the range of 60 to 90°F (16 to 32°C) appear to be favorable for most concretes. Massive concrete placements, however, generally need somewhat lower curing temperatures. The lower the curing temperature, relatively speaking, the lower the early strength of the concrete. Cooling of some concretes placed during hot weather is sometimes necessary, particularly for massive placements such as the large foundation mats commonly used at many power plant constructions.

Curing of concrete in hot weather is more critical than in cooler weather. In order to prevent the loss of moisture from exposed surfaces of the freshly placed concrete, curing should start as soon as the surface is finished. The first few hours are the most critical. Continuous moist curing is desirable. As discussed by others, cooling of the freshly mixed conrete may be necessary for massive concrete placement. Detailed discussions of curing during hot or cold weather may be found in ACI 305[9] and 306[2] respectively.

Curing Period

Concrete should be protected against loss of moisture for periods of time which is dependent upon the type cement (I as opposed to III), mix proportions, required strength, size and shape of the member cast (thin sections to massive placements), weather, and future exposure condition. This period could be only a few days to months in length. Since desirable properties of the concrete are enhanced by curing (kept continuously moist at favorable temperatures), the curing period should be for as long as economically practicable. Most specifications require 7 days moist curing (when ambient temperature is above 40°F (4°C)). Massive placements may require a minimum of two weeks. At the end of the curing period rapid drying should not be allowed.

The compressive strength of the concrete is affected by the length of time moist curing is provided. As previously mentioned, strength gain essentially stops when moisture is no longer available for cement hydration. Figure 4 shows this relationship as compared to standard cured (73°F (23°C) and 100% RH) specimens. Note that this relationship is made at a constant temperature, namely 73°F (23°C). The compounded effect of both low temperatures and length of moist curing will determine the strength attained by the concrete in place.

Protection

Practically speaking, protection of the concrete goes hand in hand with curing. That is, curing and protection are both required to attain the desirable properties of the concrete. Sometimes the terms are used interchangeably. Protection involves curing but also includes:

1. Protection against temperature, wind, and humidity; and
2. Protection from mechanical injury

Temperature, Wind, and Humidity

Protection from the elements includes considerations for both hot and cold weather concreting. Figure 5, however, shows the effect of the various combinations of temperature, wind, and humidity on the rate of evaporation of surface moisture on the freshly placed concrete. When this rate exceeds about 0.2 pound per square foot per hour[9], precautions against the occurrence of plastic shrinkage cracks are required. The precautions include wind breaks, facilities for shading the area, fog sprays and timely curing.

Mechanical Injury

During the initial curing period concrete is subject to damage by load stresses, heavy shock, and excessive vibration. Impact by heavy objects on the new concrete may result in irreversible damage. Finished concrete surfaces should be protected from damage by construction equipment. Self supporting members should not be loaded to produce excessive stresses within the new concrete resulting in permanent damage.

Form/Shoring Removal

The last subject discussed by this paper may be one of the most critical steps in post placement control. Forms/shoring should not be removed before such time that the member can sustain its own weight plus any live loads to which it may be subjected. Several construction collapses have occured due (at least in part) to the premature removal of forms (Skyline Plaza, Bailey's Crossroads, Virginia and Harbour Cay Condominums, Cocoa Beach, Florida). Undoubtedly, many others have occurred. Form removal time is directly related to the concrete's strength which in turn is directly related to curing and protection. The real problem lies in the determination of the in-place strength of the concrete. Most specifications now rely on the field cured cylinder (ASTM C31) or minimum curing and protection requirements (prevent loss of moisture and maintain certain temperatures). It would seem desirable to know the actual in-place strength rather than rely on estimates from similarly cured specimens (that are not of similar dimensions of the member) or assumptions that curing and protection has in fact produced sufficient in-place strength. Many contractors, among others, are now using and advocating the use of the several nondestructive tests[8] available to make predictions with more confidence. These include penetration tests, pull-out tests and maturity determinations.

Summary of Controls

After placing, consolidating and finishing the concrete, several important actions are needed to produce good quality concrete. The

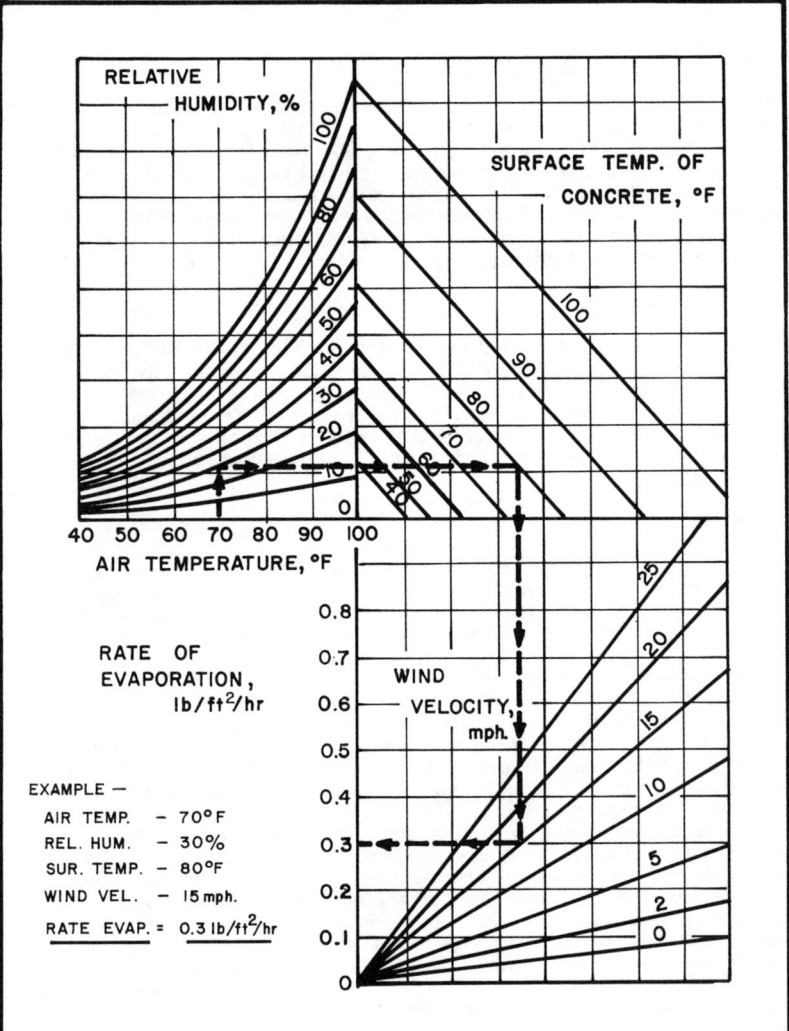

Fig. 5 - Effect of ambient conditions on rate of evaporation of surface moisture[3].

concrete must be cured and protected to assure no loss of moisture during its early life, to prevent damage from hot or cold weather and/or mechanical injury. The curing and protection needs to be maintained for a period long enough to assure adequate strengths are obtained prior to form or shoring removal and to assure that the ultimate strength of the in-place concrete attains the desirable level. Good quality concrete in the project involves :

1. Quality materials,
2. Properly designed concrete mixes,
3. Properly batched and mixed materials,
4. Proper placing and consolidation techniques, and
5. Adequate curing and protection.

All of these procedures need proper attention in order to obtain the quality of concrete intended for the project.

REFERENCES

1. ACI Committee 318, "Building Code Requirements for Reinforced Concrete (ACI 318-77)," American Concrete Institute, Detroit, 1977

2. ACI Committe 306, "Cold Weather Concreting (ACI 306R-78)," American Concrete Institute, Detroit, 1978

3. Engineering Bulletin, "Design and Control of Concrete Mixtures," Eleventh Edition, Portland Cement Association, Skokie, 1968

4. Troxell, Davis, and Kelly, "Composition and Properties of Concrete," Second Edition, McGraw-Hill Book Company, New York, 1968

5. ACI Committee 349, "Code Requirements for Nuclear Safety Related Concrete Structures (ACI 349-76)," American Concrete Institute, Detroit, 1976

6. ACI Committee 359, "Code for Concrete Reactor Vessels and Containments (ACI 359-80)," American Concrete Institute, Detroit, 1980 (Also, ANSI/ASME BPV-III-2, "ASME Boiler and Pressure Vessel Code", Section III, Division 2, Code for Concrete Reactor Vessels and Containments, The Amreican Society of Mechanical Engineers, New York, 1980)

7. ASTM C494-80, "Chemical Admixtures for Concrete", 1980 Book of Standards, Part 14, American Society for Testing and Materials, Philadelphia, 1980

8. Malhotra, V.M., "Testing Hardened Concrete: Nondestructive Methods", ACI Monograph No. 9, the Iowa State University Press, Ames, and American Concrete Institute, Detroit, 1976

9. ACI Committee 305, "Hot Weather Concreting (ACI 305R-77)," American Concrete Institute, Detroit, 1977

QUALITY EVALUATION OF NUCLEAR CONCRETE STRUCTURES

by

Joseph F. Artuso[1], M. ASCE

ABSTRACT

The use of nondestructive methods of determining the quality of in-situ concrete are presented. The use of commonly accepted methods contained in A.S.T.M. methods are emphasized. The limitations of these methods as well as the reliability to determine the properities of uniformity, consolidation and strength are discussed. The paper also contains correlations of nondestructive properties versus actual concrete compressive strengths of cores.

INTRODUCTION

This paper describes the inspections and tests made on selected concrete placements of safety related concrete structures of a given Nuclear Power Plant to determine whether the structures "as constructed" conform to the design requirements. A combination of visual inspections, measurements, documentation review, nondestructive tests and compressive strength tests were utilized to accomplish the evaluation.

A description of each inspection, measurements and test is given. The accumulation and evaluation of data is presented. The inspections and test methods used in the study are those that have been utilized in the concrete industry with a sufficient verification of accuracy and applicability. A discussion of the validity and reliability of each test and inspection is made to provide assurance that the methods enabled a comprehensive evaluation which can accurately assess the quality of in place concrete.

---

[1]President, Construction Engineering Consultants, Inc. Laughlintown, PA

DOCUMENTATION REVIEW

The documentation review was made to determine whether all requirements of construction were properly verified and documented. This review also enabled the clarification of design requirements including all approved revisions. The information revealed the specified location and dimensions of embedments, penetrations, anchor bolts and construction joints.

Drawings, specifications, engineering revisions and construction procedures were studied to determine the project requirements. The specific documents which were required to verify compliance included: change notices, documented nonconformances and each respective resolution, inspection reports, test reports, pour cards and checklists. This study enabled a check of conformance with the Project Quality Control procedures and regulatory requirements for the specific project. The information developed provided the data and requirements of each placement that was used in the visual inspection of each placement.

VISUAL INSPECTION OF CONCRETE PLACEMENTS

The common method of visual examination of the placement was performed to evaluate consolidation, signs of distress and reactivity, alignments and adequacy of repair. The following items were addressed on each inspection:

1. General appearance of the surface
2. The nature and extent of cracking
3. Evidence of volume change
4. Evidence of cement/aggregate reactions
5. Secondary deposits on surface (efflorescence, exudation, incrustation)
6. Secondary deposits in cracks or voids (efflorescence, leaching, incrustation)
7. Construction joint alignment
8. Construction joint cleanliness
9. Control joints
    a. Expansion
    b. Contraction
10. The nature and extent of deflections
11. The nature and extent of dislocations resulting in joint movement, tilting, shearing or misalignment of structural elements
12. Apparent effectiveness of curing
13. The extent and significance of surface characteristics:
    a. Scaling
    b. Spalling
    c. Peeling
    d. Popouts
    e. Pitting or construction scarring

    f. Dusting
    g. Staining or discoloration
    h. Cold joints
    i. Pour lines
    j. Corrosion of reinforcement
    k. Soft spots
    l. Sand streaks or pockets
    m. Honeycomb
    n. Air/water voids
    o. Segregation or stratification
14. Indications of adequate consolidation in general
15. Indications of adequate consolidation behind embedments
16. The adequacy of repairs based on soundness and appearance:
    a. Tie Holes
    b. Cosmetic
    c. Structural
17. Satisfactory embedment of penetrations based on appearance and sounding as applicable
18. Dislocation or misalignment of embedded plates
19. Satisfactory embedment of plates based on appearance and sounding as applicable
20. Apparent consolidation surrounding anchor bolts
21. General appearance of seismic joints
22. Evidence of grout leakage
23. Drilled Cores and Bore Holes

    Each placement was inspected on the basis of a rating system for each item given above. The ratings system consisted of a numbering system of 1 to 4. 1 excellent, 2 good, 3 fair and 4 poor. This rating was applied by judgement of the evaluator based on experience in the concrete industry. Visual examinations were supplemented by manual sounding with a hammer to detect integrity near embedments and anchor bolts. The items were distinguished by those considered substantial or minor. Substantial items are those with structural significance and minor are those of an architectural or esthetic nature. Typical results are shown in Table 1. Evaluation of this data consisted of determining the average quality rating of each placement and the average of all placements inspected. A review of the data of each placemer. necessary to determine whether there are any deficiencies requir.. ⸗ additional testing or repair. A compilation of data developed by this inspection is shown in Table 2. The average rating and statistical evaluation can be obtained from this data.

SURVEY OF AS BUILT DIMENSIONS

    Field measurements of the selected placements were made by a survey party using survey instruments to determine conformance to the drawing and specification requirements. This is considered as As Built Verification which was primarily a check of dimensions for alignment, thickness and plumbness. Location and dimensions of

Table 1

POUR NO. _____ DATE OF POUR _____ SHEET __1__ OF __
DATE OF EVALUATION _____

VISUAL INSPECTION

ITEM NUMBER	ITEM	SUBSTANTIAL	MINOR	QUALITY RATING				COSMETIC OR STRUCTURAL REPAIR REQUIRED	COMMENTS	SEE SHEET NUMBER
				1	2	3	4			
1.	General appearance of the surface	S								
2.	Nature and extent of cracking	S								
3.	Evidence of volume change	S								
4.	Evidence of cement/aggregate reactions	S								
5.	Secondary deposits on surfaces		M							
6.	Secondary deposits in cracks or voids		M							
7.	Construction Joint Alignment		M							
8.	Construction Joint Cleanliness	S								
9.	Control Joints									
a.	Expansion		M							
b.	Contraction		M							
10.	Nature and extent of deflections	S								

Table 2

* * * * * VISUAL INSPECTION QUALITY RATING * * * * *

1=EXCELLENT 2=GOOD 3=FAIR 4=POOR 0=NOT APPLICABLE

BLDG	GENERIC TYPE	DATA TYPE	POUR NO	1	2	3	4	5	6	7	8	9 A	9 B	10	11	12	13 A	13 B	13 C	13 D	13 E	13 F	13 G	13 H	13 I	13 J	13 K	13 L	13 M	13 N	13 O	14	15	16 A	16 B	16 C	17	18	19	20	21	22	23
SUBSTANTIAL/MINOR				S	S	S	S	M	M	M	S	M	M	S	S	S	M	M	M	M	M	M	M	S	M	S	M	M	S	M	S	S	S	M	M	S	S	S	S	S	S	M	S
C 1	NW	VI	01	2	1	1	1	1	1	3	1	0	0	0	1	1	1	1	1	1	1	1	1	1	1	1	1	1	1	1	1	1	1	2	2	0	1	1	1	0	0	1	1
C 1	KW	VI	02	1	1	1	1	1	1	1	1	0	0	0	1	1	1	1	1	1	1	1	1	1	1	1	1	1	1	1	1	1	1	1	1	0	1	1	1	0	0	1	1
C 1	HW	VI	03	2	1	1	1	1	1	1	1	0	0	0	1	1	1	1	2	1	2	2	1	1	1	1	1	2	1	2	1	2	1	1	1	0	1	1	1	0	0	1	1
C 1	KS	VI	04	2	1	1	1	1	1	1	1	0	0	1	1	1	1	1	1	1	2	1	1	1	0	1	1	1	1	1	1	1	0	0	1	0	0	0	0	1	0	1	1
C 1	NS	VI	05	2	1	1	1	1	1	2	1	0	0	1	1	1	1	1	1	1	2	1	1	1	0	1	1	1	2	1	1	2	1	0	2	0	0	0	0	0	1	2	1
C 2	NW	VI	06	2	1	1	1	1	1	3	1	0	0	0	1	1	1	1	1	1	1	1	1	1	1	1	1	2	2	2	1	1	1	1	0	0	1	1	1	0	0	1	1
C 2	KW	VI	07	1	1	1	1	1	1	1	3	0	0	0	1	1	1	1	1	1	1	1	1	1	1	1	1	1	1	1	1	1	1	1	0	0	1	1	1	0	0	2	1
C 2	HW	VI	08	2	1	1	1	1	1	2	1	0	0	0	1	1	1	1	1	1	1	1	1	1	1	1	1	1	1	1	1	1	1	1	2	0	1	1	1	0	0	2	1
C 2	KS	VI	09	1	1	1	1	1	1	1	1	0	0	1	1	1	1	1	1	1	1	1	1	1	1	1	1	1	1	1	1	1	0	0	0	0	0	0	0	1	0	1	1
C 2	NS	VI	10	1	1	1	1	1	1	0	0	0	0	1	1	1	1	1	1	1	1	1	1	1	1	1	1	1	1	1	1	1	1	0	0	1	1	1	1	1	0	1	1
M 1	NW	VI	11	2	1	1	1	1	1	2	1	0	0	0	1	1	1	1	1	1	1	1	1	1	1	1	1	1	1	1	1	2	1	1	1	1	1	2	1	0	0	2	1
M 1	KW	VI	12	2	1	1	1	1	1	2	1	0	0	0	1	1	1	1	1	1	1	1	1	1	1	1	2	2	1	2	1	1	1	1	1	0	1	2	1	0	0	2	2
M 1	HW	VI	13	2	2	1	1	1	1	3	2	0	0	0	1	1	1	1	1	1	1	1	1	1	1	1	1	2	1	2	1	1	1	1	1	0	1	2	1	0	0	2	1
M 1	KS	VI	14	2	2	1	1	2	2	1	1	0	0	1	1	1	1	1	1	1	1	1	1	1	1	1	1	1	1	1	1	1	1	0	0	0	1	2	1	0	0	2	2
M 1	NS	VI	15	2	3	2	1	1	1	1	1	0	0	1	1	1	1	1	1	1	1	1	1	1	1	1	1	1	1	1	1	1	1	0	0	0	1	2	1	0	0	2	2
M 2	NW	VI	16	1	1	1	1	1	1	2	1	0	0	0	1	1	1	1	1	1	1	1	1	1	1	1	1	1	1	1	1	1	1	1	0	0	1	2	2	0	0	2	1
M 2	KW	VI	17	2	2	1	1	1	2	2	2	0	0	0	2	2	1	1	1	1	2	2	1	1	1	1	2	2	2	2	1	1	1	1	1	0	1	2	2	0	0	2	1
M 2	HW	VI	18	2	2	1	1	1	2	2	2	0	0	0	2	1	1	1	1	1	3	1	1	1	2	1	2	2	2	2	1	1	1	1	2	0	1	2	2	0	0	2	1
M 2	KS	VI	19	2	2	1	1	1	2	2	2	0	0	1	2	2	1	1	1	1	2	1	1	1	1	1	1	2	1	2	1	1	1	0	0	0	1	2	1	0	0	2	1
M 2	NS	VI	20	2	2	1	1	1	2	2	2	0	0	1	1	1	1	1	1	1	2	1	1	1	1	1	1	1	1	2	1	1	1	0	1	0	1	2	2	0	0	2	1
F 1	NW	VI	21	2	1	1	1	1	1	2	1	0	0	1	1	1	1	1	1	1	1	1	1	1	1	1	1	2	1	2	1	1	1	1	1	0	1	2	1	0	0	1	1
F 1	KW	VI	22	2	1	1	1	1	1	2	1	0	0	0	1	1	1	1	1	1	1	1	1	1	1	1	1	3	1	2	1	1	1	1	1	0	1	2	1	0	0	1	1
F 1	HW	VI	23	2	1	1	1	1	1	2	1	0	0	0	1	1	1	1	2	1	1	1	1	1	1	1	1	2	1	1	1	1	1	1	1	0	1	2	1	0	1	1	1
F 1	KS	VI	24	1	1	1	1	1	1	1	1	0	0	1	1	1	1	1	1	1	1	1	1	1	1	1	1	1	1	1	1	1	1	0	1	0	1	2	2	0	0	1	1
F 1	NS	VI	25	2	1	1	1	1	1	1	1	0	0	1	1	1	1	1	1	1	3	1	1	1	1	1	1	1	1	1	1	1	1	0	0	0	1	0	0	1	0	1	1
F 2	NW	VI	26	2	2	1	1	1	1	2	1	0	0	0	2	2	1	1	2	1	2	1	1	1	1	1	1	2	1	2	1	1	1	1	1	0	1	2	1	0	0	2	1
F 2	KW	VI	27	1	1	1	1	1	2	2	1	0	0	1	1	1	1	1	1	1	1	1	1	1	1	1	1	1	1	1	1	1	0	2	2	1	1	2	1	0	0	1	1
F 2	HW	VI	28	1	1	1	1	1	1	1	1	0	0	1	1	1	1	1	1	1	2	1	1	1	1	1	1	1	1	2	1	1	1	1	1	1	1	2	2	0	0	1	2
F 2	KS	VI	29	2	3	1	1	1	1	2	1	0	0	3	1	1	1	1	1	1	2	1	1	1	1	1	1	1	1	1	1	1	1	0	0	1	0	2	1	1	0	0	1
F 2	NS	VI	30	1	2	1	1	1	1	1	1	0	0	1	1	1	1	1	1	1	3	2	1	1	1	1	1	1	1	1	1	1	0	0	0	0	0	0	0	0	0	1	1

embedments, penetrations and blockouts were also verified. In areas of the specific placements that contained dowels and reinforcements of placements not made, the reinforcement was checked for size, cover projection and lap splice. The identification number for any embedded item which was on the embedded item was recorded. The survey parties used existing reference points from which additional horizontal and vertical control lines were developed on each side of the specific wall or slab.

In addition to checking control on individual placements, a closed traverse was run from the existing baseline at the plant. Control points within the building were then tied to the closed traverse. This was closed within a first order accuracy. In order to show the relationship between buildings and location of the points with respect to the plant site coordinate system, a comparison was made between completed coordinates and the construction coordinates after the control was checked, a comprehensive survey was made of each placement. Drawings were developed to record all measurements.

Thickness of walls were determined on a 4' grid which was placed on each side. The measurements at each grid intersection were calculated to determine actual as constructed thickness. A 20' grid was used for the slabs and thickness determined using elevation surveys. The plumbness and alignments were developed and recorded on the as built drawings. Deviations from specified tolerances were highlighted and reported.

NONDESTRUCTIVE TESTING

The most common and accepted types of nondestructive tests were performed to determine the structural integrity of the selected concrete placements. The nondestructive test methods used were: Pulse velocity (Pulse through method), Magnetic detection of Rebars, and Penetration Resistance (Windsor Probe). These methods have been extensively used in evaluation of concrete and have been established in National Standards of Testing.

Pulse Velocity Testing (ASTMC597)--This method consists of the determination of the pulse velocity of the propagation of compressional (sound) waves through concrete. Time measurements are made of the travel of a pulse or train of waves through a measured path length using a sending transducer on one side of the concrete surface and a receiving transducer on the other side (Pulse through Method). The James Electronics V. Meter was used for this testing. Measurements were made on a one foot grid of test sample areas within each selected placement. This was a stratified sampling procedure and enabled a statistical analysis of test results. This method of testing utilizes the established criteria of sound transmissions through concrete. It is recognized that high velocities in the range of 13,000 to 16,000 or regular weight concrete is an indication of dense high quality concrete. Lower velocities will indicate lower strength, porosity, voids, honeycomb or cracks. Since the surface and subsurface porosity will impair the sound transmission, additional testing is required to verify the specific condition. Therefore when a low velocity is encountered at any grid point, additional readings were taken within a 6 inch radius. If high velocities were determined in this proximity, the high value is used for the specific grid point. If a higher velocity is not determined, a hole is bored into the concrete at this location and then visually examined for soundness. Readings on surface plates embedded in the concrete were taken. However, when the bond between concrete and plate is not complete, misleading velocities are developed, therefore, the velocities taken on plates are of limited value. In addition to the time measurements obtained with the sonic meter, a Tektronics oscilloscope was used to display the sonic curves. This enabled a visual examination of the curve depicted on the screen for each pulse velocity determination. A smooth curve results when the sound wave passes through monolithic and sound concrete. The first wave of the curve (shown in the photographs) is the significant one for evaluation purposes. An irregular curve, flat curve or a straight line is produced on the oscilloscope when a defect is encountered. Therefore the combination of velocity measurements and curves were used to evaluate the integrity at each grid point. The oscilloscope used contains a memory feature so that two curves can be viewed simultaneously. In this study, the curve of a known quality of concrete was displayed as the reference curve. The reference was selected near a core or bore hole that indicated sound and dense concrete. The reference point curve was then used for comparison with curves developed at various grid points. Documentation of the curves were made by photographing the reference point with various grid points. This provided documentation of the testing and also enabled a check of the velocities developed at the grid points. (See Figure 1 for a typical photograph)

Calibration (Verification) Blocks--Two concrete test blocks with known internal voids were used to provide correlation of velocities and curves for known conditions. This was also a means of verifying the validity of the test method to detect voids and thus provided the criteria for the pulse velocity testing of selected concrete placements.

Generally, the pulse velocity was reduced approximately 30 to 40 percent (to less than 10,000 ft/sec) when a void was encountered. When a 16" void was intercepted in a 4 foot thick test block, the velocities were reduced from 12,300 to 7,300 ft/sec. When a 12" void was intercepted in a 2 1/2 foot thick test block, velocities were reduced from 13,800 to 9,800 ft/sec. The curve produced generally changed from a well defined sine curve to an elongated curve or horizontal line when the void was intercepted in the 2 1/2 foot chick test block. This type of curve void relationship was not

CONCRETE STRUCTURES EVALUATION

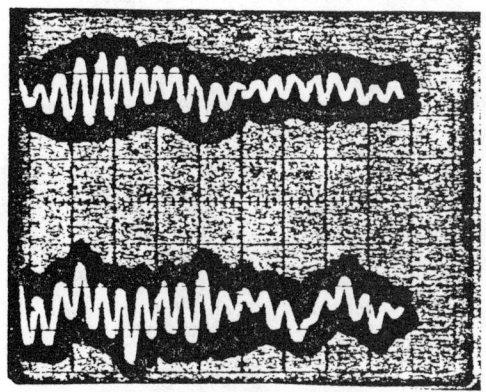

Grid 1-A
(Sound Concrete)

Reference Grid

Photo No. 11-1

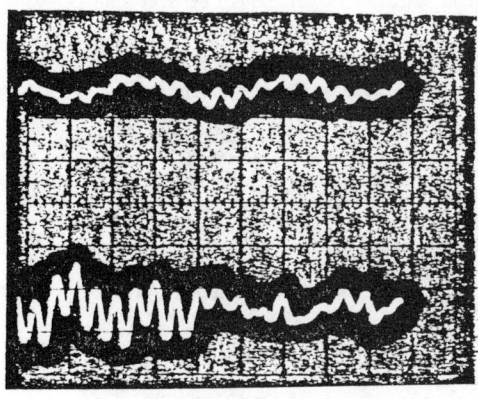

Grid 1-C
(Cntr. Ln. 12" Void)

Reference Grid

Photo No. 11-4

FIGURE 1
2'-6" THICK TEST BLOCK

as pronounced for the 4 foot test block. These velocity reductions established a relationship of void size to wall thickness tested. Smaller voids would be detected in thinner walls or slabs utilizing the same velocity reduction criteria.

The voids of the 4 foot test block were filled with coarse aggregate to simulate honeycomb and resulted in the same relationship of reduction in pulse velocities.

Figures 1 and 2 represent the data and photos of curves taken on the calibration blocks:

2-1/2 FOOT THICK TEST BLOCK (SEE FIGURE 1)

GRID	LOCATION	VELOCITY FT/SEC.	PHOTO NO.
1-A	Grid Point (sound concrete)	13,800	11-1
1-C	Center line of 12" void	9,800	11-4

4 FOOT THICK TEST BLOCK (SEE FIGURE 2)

1-A	Grid Point (sound concrete)	12,300	11-15
1-C	Center line of 16" void	7,300	11-18

The bottom curve in each photo represents the sound concrete at the reference grid and the top curve represents the curve developed at the designated grid point. The top curve should either be flat or have wider peaks and valleys if the grid point is directly over the void.

This calibration proved the criteria that approximately 30 to 40 percent reductions in pulse velocities (lower than 10,000 ft/sec.) and the absence of a sine curve (i.e. the presence of a flat curve) could indicate the presence of a significant void. As can be seen by the photographs, the presence of a possible void cannot be based on the curves alone, but is primarily related to significant reductions in velocities in thick reinforced concrete sections. Uniform and high pulse velocities in the range of 12,000 to 16,000 ft. per second indicated high quality dense concrete. Generally, the maximum pulse velocity that can be obtained in dense high quality reinforced normal weight concrete is about 16,000 ft/sec. Reductions to the 13,000 ft/sec. range may indicate either a surface or near surface irregularity such as entrapped air voids or micro-cracking. Velocities below 10,000 ft/sec. indicate the possibility of voids or cracks or poor surfaces. The surface condition of the block indicated that the sonic determinations were significantly influenced by surface (coupling) conditions. A rough surface caused significant reduction in velocities and therefore, did not truly represent the condition of the internal concrete. This condition was confirmed by retesting a smooth surface within about 3" of the rough surface with a corresponding change to high velocities.

CONCRETE STRUCTURES EVALUATION 575

Grid 1-A
(Sound Concrete)

Reference Grid

Photo No. 11-15

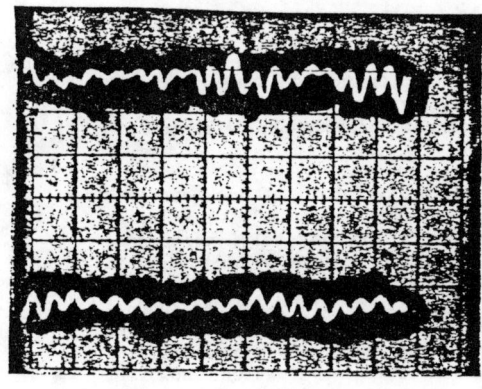

Grid 1-C
(Cntr. Ln. 16" Void)

Reference Grid

Photo No. 11-18

FIGURE 2
4' THICK TEST BLOCK

Surface Influence on Sonic Testing. - A study was made on the effects of surface condition by testing rough or questionable areas and then retesting the same area after chipping about 1 inch deep and replacing with dry packed mortar. This is shown by the following comparison:

LOCATION	SURFACE CONDITION	VELOCITY FT./SEC.
C11-W41A Test Sec. 6; 20', El. 26'	as is	10,400
C11-W41A Test Sec. 6; 20', El. 26'	after 1" chipping & grouting	11,900

As stated previously, rough or questionable surfaces may cause an inadequate coupling. An attempt was made to obtain a good coupling by retesting or by isolating the low velocity with higher velocities in the immediate vicinity. If this was the case a replacement reading was assigned. This is shown by the following comparison:

LOCATION	SURFACE CONDITION	VELOCITY FT./SEC.
C11-W83, Test Sec. 8; 7', El. 81'	as-is (original)	11,006
C11-W83, Test Sec. 8; 7', El. 81'	as-is (replacement)	15,284

When low readings (below 12,000 ft./sec.) were obtained (approximately 1% of all readings) bore holes were made and the concrete was visually inspected using fiberoptics to determine its acceptability.

General Discussion of Pulse Velocity Data. - A one foot grid was applied to all randomly selected test areas of the selected placements. There was a potential for approximately one hundred readings per test section. This number was reduced by omitting grid points that fell on embedded steel plates, and other obstructions such as penetrations or intersecting walls or slabs. Since surface condition was so critical that readings at grid points which fell on rough surfaces were either replaced with sonic values obtained near the grid points or not used in the statistical evaluation. However, the grid points of unknown sonic values were bored and visually inspected with fiberoptics to determine internal quality.

Micro-cracking, presence of bleed water pockets, and/or reinforcing steel conjestion will adversely affect pulse velocities, it is expected that thicker sections and sections of high reinforcing steel congestion would exhibit greater variations in sonic values. This would also be the case of coated surfaces, because the coating may mask surface irregularities and affect sonic transmissions from the transducer to the concrete. A review of the data and comparisons of thick wall values to thin wall values show this condition.

Sonic Data Evaluation. - ACI 214 "Recommended Practice for Evaluation of Strength Test Results of Concrete" (a statistical procedure for interpretation) contains standards of concrete control. An excellent rating is given for a coefficient of variation of 10% for a design strength of 4000 psi. Extending this criteria to the pulse velocity data, the quality rating of the concrete placed can be made.

A typical tabulation of all data obtained by the sonic testing program is shown in Figure 3. The final evaluation is made on the valid data which does not contain readings at grid points that were located on steel plates or grid points that were physically inspected by visual examination of bore holes made into the concrete.

The three (3) test sample areas on each placement are combined so that each placement is presented and average values can be determined.

The data obtained are summarized as individual readings, mean, range, standard deviations and coefficients of variation.

The test data range of a typical study of a number of placements is shown below:

Mean Velocity	14305 to 15608	ft./sec.
Standard Deviation	565 to 152	ft./sec.
Coefficient of Variation	3.9 to 1.0	percent
97.7% of all tests above	13350	ft./sec.

This data indicates superior concrete quality of all concrete placements tested. The low standard deviation is in the excellent range when the ACI evaluation criteria is applied. The pulse velocity data was reduced and presented on histograms (frequency distribution curves) to graphically depict the uniformity of all placements. See Figure 4. A review of the histograms can indicate graphically the quality of concrete as determined by sonic testing.

Penetration Resistance of Hardened Concrete (ASTM C-803) (Windsor Probes) -- This method uses a steel probe energized by a driving unit delivering a uniform amount of energy. The penetration into the concrete is proportional to the concrete strength.

The Windsor probes were effective in concrete with strength below 8000 p.s.i. A limitation of this method is the dependence only upon surface conditions. Therefore, in older concrete that may

POWER GENERATION FACILITIES

+26'00"	E_{PLATE}	A^{14583} A^{14463} A^{14583} A^{14583}	N_{BLOCK} N_{BLOCK}	A^{14344}				
	E_{PLATE}	A^{14706} A^{14463} A^{14706} A^{14831}	N_{BLOCK} N_{BLOCK}	A^{14857}				
	E_{PLATE}	A^{14463} A^{14463} A^{14344} A^{14463}	E_{PLATE} E_{PLATE}	A^{14831}				
	E_{PLATE}	A^{147C6} A^{14583} A^{14113} A^{14706}	E_{PLATE}	A^{14857} A^{15086}				
	E_{PLATE}	A^{14583} A^{14463} A^{14463} A^{14706}	E_{PLATE}	A^{14583} A^{14857}				
	E_{PLATE}	A^{14706} A^{14583} A^{14463} A^{14344}	E_{PLATE}	A^{147C6} A^{14706}				
	E_{PLATE}	A^{14583} A^{14463} A^{147C6} A^{14583}	E_{PLATE}	A^{14463} A^{14831}				
	E_{PLATE}	A^{14583} A^{147C6} A^{14344} A^{14344}	E_{PLATE}	A^{14463} A^{14857}				
	E_{PLATE}	A^{14583} A^{14228} A^{14344} A^{14344}	E_{PLATE}	A^{14463} A^{14583}				
	E_{PLATE}	A^{14583} A^{14113} A^{14463} A^{14344}	E_{PLATE}	A^{14228} A^{14583}				
	E_{PLATE}	A^{14228} A^{14000} A^{14463} A^{14000}	E_{PLATE}	A^{14228} A^{14706}				
	E_{PLATE}	A^{14463} A^{14228} A^{14463} A^{14228}	E_{PLATE}	A^{14000} A^{14831}				
	E_{PLATE}	A^{14583} A^{14583} A^{14583} A^{14228}	E_{PLATE}	A^{14228} A^{14583}				
	E_{PLATE}	A^{14706} A^{14463} A^{14463} A^{14113}	E_{PLATE}	A^{14113} A^{14228}				
	E_{PLATE}	A^{14706} A^{14463} A^{14463} A^{14463}	E_{PLATE}	A^{14113} A^{14857}				
+11'00"	A^{14706} A^{14463} A^{14463} A^{14463} A^{14463} A^{14228} A^{14113} A^{14691}							
	08'00"W						01'00"W	

PLOT OF TEST SECTIONS	TEST AREA 35	SCALE:
Figure 3	PLACEMENT 12	1 IN = 2 FT
		SHEET 1 OF 1

Figure 4

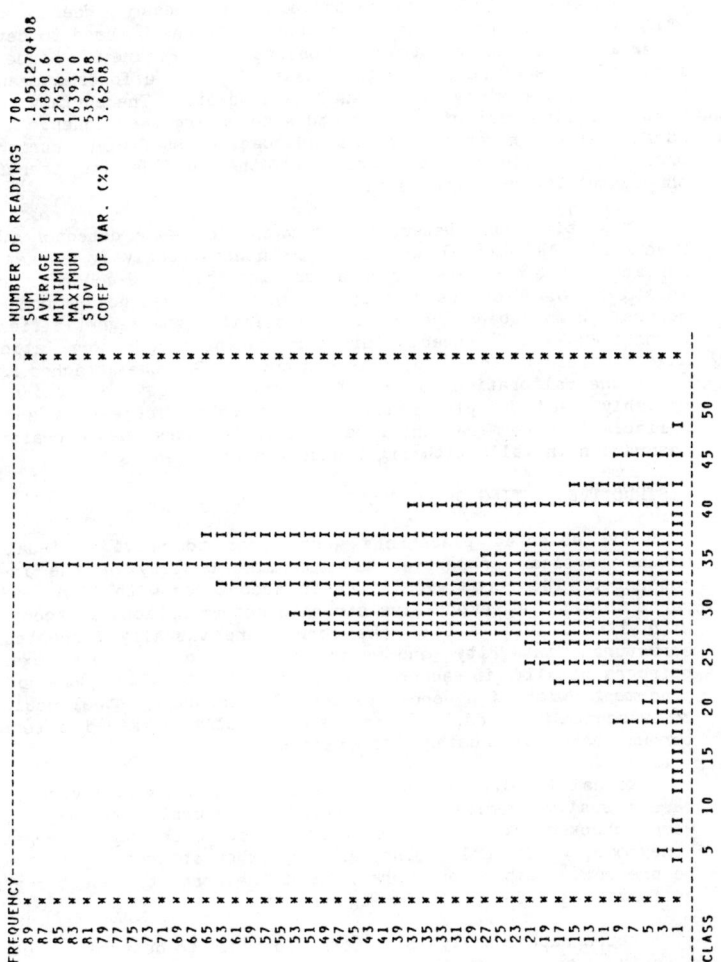

be carbonated, spurious strength values may occur. When properly utilized it can supplement other tests. It can be used to develop general comparisons of strength results. See Figure 5. Since this method is a surface penetration test, it also affords a means of verifying the quality of the surface concrete. The surface is the most sensitive part of the structure to curing techniques. A high surface strength can confirm the adequacy of sufficient curing and proper hydration of the cement at the surface and therefore, throughout the concrete section.

Magnetic Rebar Detector--The magnetic rebar detector (James Electronic "R" meter) was used to nondestructively survey the surface of the concrete for rebar spacing, size and concrete cover. This type of a device is significantly influenced by conjested rebars. When rebars are in close proximity, the magnetic field is significantly influenced. This can prevent proper correlation of magetic field to location, size and cover. See the attached Figure 6 for the calibration curve. Therefore, this device could not be reliably used in placements of high rebar conjestion and not reliable in determination of bar sizes. It does enable reasonable assurances in walls with light widely spaced rebars.

DESTRUCTIVE TESTING OF CONCRETE

The selected placements were cored to provide visual and actual physical properties to determine quality of the in-situ concrete. The cores were taken in accordance with ASTM C-42 and parts tested for compressive strength determinations in accordance with ASTM C-39. All of the cores were visually inspected for structural integrity and some of the cores were examined petrographically in accordance with ASTM C-295. Some of the placements with high concentration of reinforcing steel could not be adequately cored. These were visually examined internally through bore holes using a fiberscope.

Visual Examination of Cores--After securing the cores, they were visually examined to determine structural integrity. They were checked for specific conditions, such as segregation, honeycomb, voids, cold joints and any other abnormality that might be present. Comparisons were made of the cores to detect any non-uniformities. See photo in Figure 7.

Petrographic Analysis of Core--In order to determine whether a reactive condition existed and durability characteristics, representative cores were examined petrographically. The petrographic examination was performed using optical microscopy. The determinations included classifications and identification of the constituents of the sample. The primary purpose was to determine uniformity and whether the characteristics conformed to the design requirements. Additional determinations included reactivity potential and presence of any contaminations. This also

Figure 5

COMPARISON OF WINDSOR PROBE TESTS AND
CORE COMPRESSIVE STRENGTH TESTS

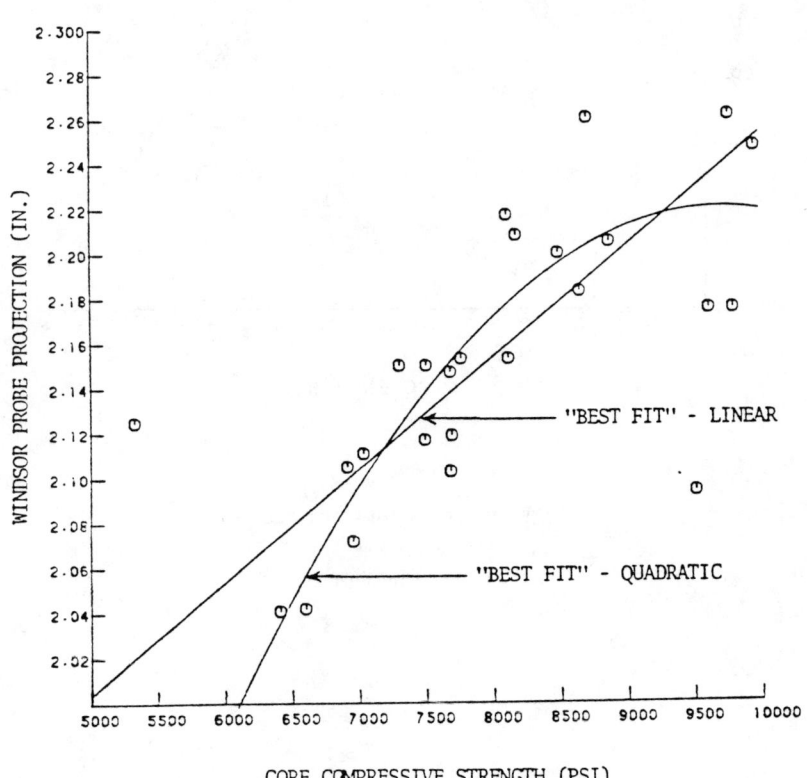

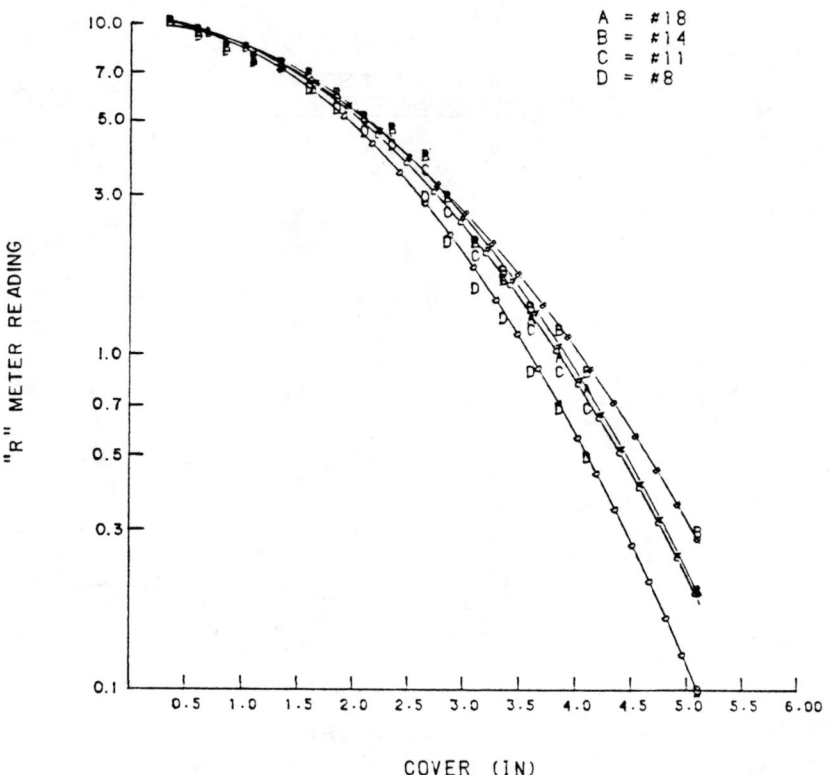

"R" METER CALIBRATION TEST FOR
#8, #11, #14 & #18 BARS

Figure 6

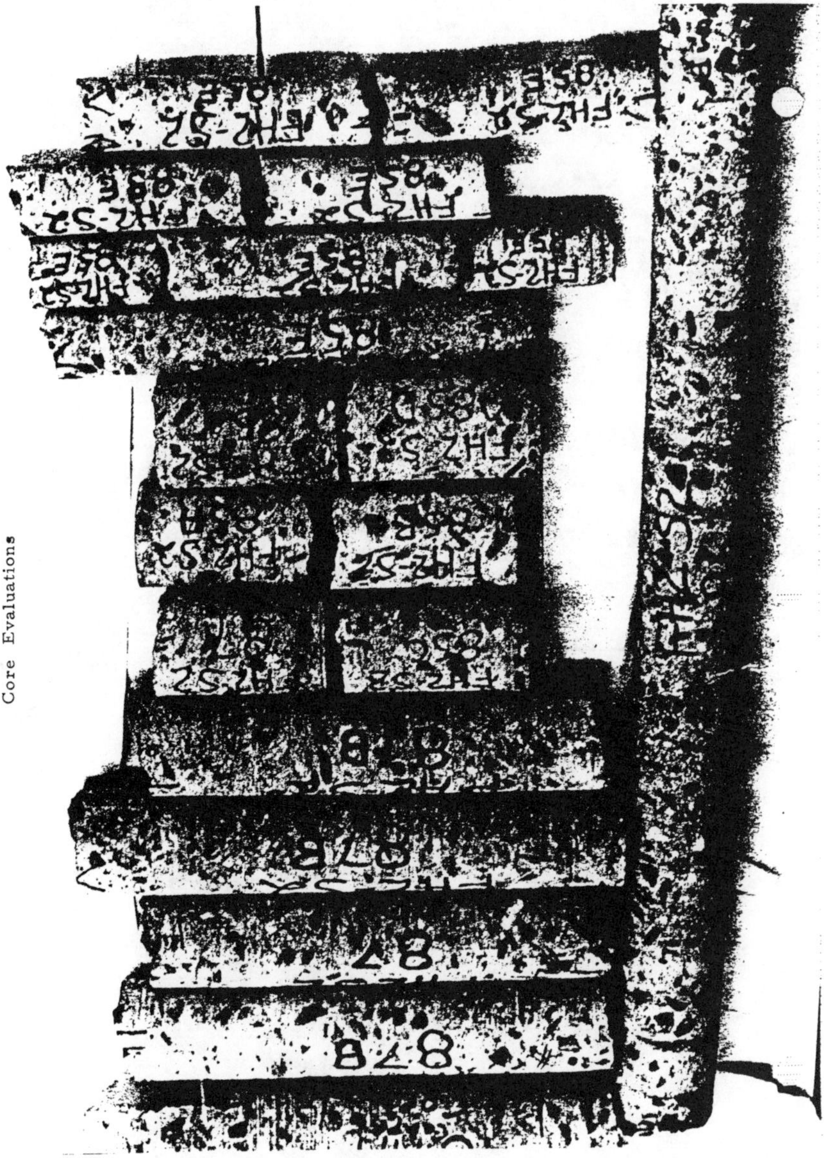

Figure 7

Core Evaluations

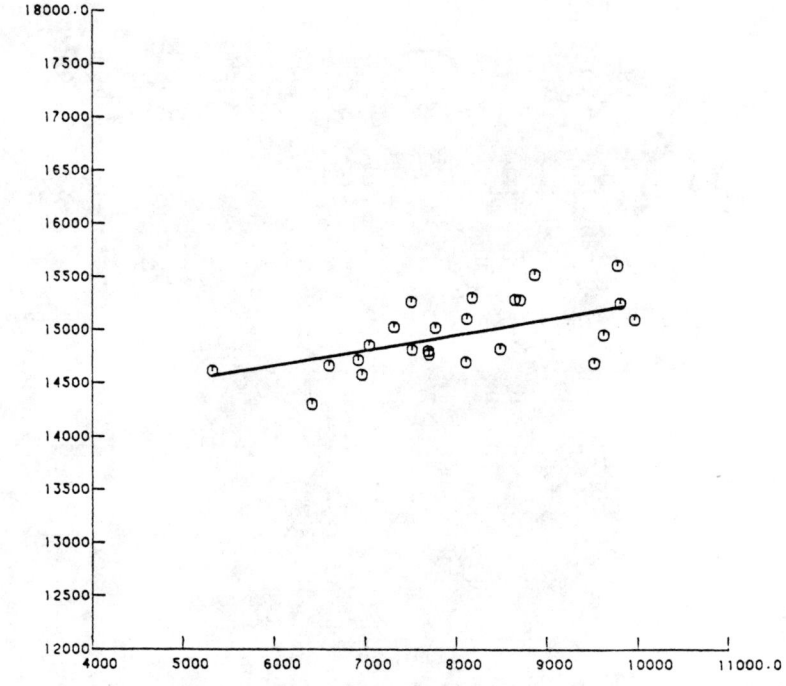

COMPARISON OF AVERAGE PLACEMENT ULTRA SONIC
VELOCITY AND AVERAGE CORE COMPRESIVE STRENGTH

FIGURE 8

Table 3

SUMMARY OF TEST DATA

Placement No.	Generic Type	Pour No.	Concrete Mix	Date Placed	Cu-Yds Placed	Ave. Compressive Strength (psi)	Test Area	Ave. Velocity (fps)	Std. Deviation	Coeff. of Variation	Windsor Probe Reading (in)	Core No.	Total Length of Cores (in)	Average Core Strength (psi)	Ave. Unit Weight of Core (pcf)	Age of Core (days)
29	KS	FH2-S	B-1-3-11	1-19-79	1290	5530	—	14691	394	2.7	2.094		117	9518	147.5	660
							85	14744	87	0.6	2.158	85	4	10490	147.2	660
							86	14417	681	4.7	2.050	85B	4	8430	145.3	660
							87	14821	261	1.8	2.075	86	45	8900	148.0	660
												87A	14	10070	150.2	660
												87B	50	9700	146.8	660
30	NS	FH2-S1	B-1-3-11	6-22-78	108	5350	—	14665	30	2.3	2.042		82	6600	143.9	871
							88	14672	260	1.8	2.042	88	36	6030	143.3	871
							89	14731	320	2.2	1.967	89	24	7010	141.1	871
							90	14550	393	2.7	2.117	90	22	6760	147.4	871

provided another determination of uniformity of the concrete placement.

Compressive Strength of Cores--ACI Codes and specifications require that concrete of questionable strength be evaluated by the Compressive Strength Tests of cores as the primary evaluation method. This is a destructive test and final resolution requires this information. Nondestructive testing can serve as an adjunct to the compressive strength testing of cores. Correlations of compressive strengths to the nondestructive testing can be used to provide a more extensive and comprehensive analysis. The correlations obtained in this study are shown in Figure 8.

Pulse velocities and penetration resistances are correlated to compressive strengths. A typical data presentation to develop this correlation is shown in Table 3.

The compressive strength of cores were analyzed using ACI statistical methods for both compliance and uniformity. The analysis in this specific study for all placements tested are as follows:

Mean Strength	7960 p.s.i.
Standard Deviations	610 p.s.i.
Coefficient of Variation	7.7 percent

In this study using ACI-214, the control of concrete is considered excellent (below 10%). Based on the statistical evaluation, 99 percent of all core strengths would be greater than 6530 p.s.i. Since ACI-318 and ACI-359 codes permit drilled cores to be 85 percent of design strength (.85 x 4000 = 3400 p.s.i.) and coupled with the 1% allowable deviation of ACI codes, the actual quality greatly exceed specification requirements.

CONCLUSIONS

Nondestructive Testing Methods can reliably be used to evaluate quality of in-situ concrete. The nondestructive methods must be reliably correlated to destructive type testing methods (compressive strength test of concrete cores). The codes will not allow acceptance based on only nondestructive test methods because of the significant variations of test conditions and concrete constituents. The nondestructive test methods should use calibration techniques which should be similar to the standards of NDE of metals required by A.S.M.E. Codes and A.S.T.M. Test Methods. The test data should be statistically evaluated to provide a comparative measure of reliability used by the Concrete Industry.comparative measure of reliability used by the Concrete Industry.

FINAL WRAP-UP SESSION: RESULTS OBTAINED ON THE U.S.D.O.E. SPONSORED RESEARCH PROJECT

SESSION OBJECTIVES/SESSION CHAIRMAN SUMMARY

by

Jack H. Willenbrock[1], M. ASCE

Objective of Session

The results of a U.S.D.O.E. sponsored research project entitled, "Structural Concrete Quality Assurance Practices on 9 Nuclear and 3 Fossil Fuel Power Plant Projects," are discussed by the Civil Engineering faculty members who participated in the study.

Session Chairman Summary

The review of the current practices as well as suggestions for improvements were made in the areas of:

   a. Organization and management of Quality Assurance
   b. Procurement and testing of concrete components.
   c. Production and evaluation of fresh concrete.
   d. Concrete preplacement, placement, and postplacement practices.

It was noted that when the results of the study are considered in conjunction with the information which has been provided by all of the speakers in the Opening Keynote Session and the Eight Concurrent Sessions, it should be recognized that a satisfactory bridge between what has occurred in the past and the improvements that are needed in the future has been established.

It was felt that the conference will be considered successful if this information is instrumental in causing fundamental changes to be made to industry codes and standards and governmental regulatory and quality assurance requirements where they are required. Those changes will certainly influence how effectively electric power generation facilities are built from both a time and a cost standpoint during and beyond the 1980's.

Consensus Recommendations

There was insufficient time at the end of the conference to conduct a Speakers/Audience discussion period.

---

[1] Professor, Department of Civil Engineering, The Pennsylvania State University, University Park, PA 16802.

CONCRETE Q.A. PRACTICES ON NUCLEAR POWER PROJECTS

by

Jack H. Willenbrock[1] and H. Randolph Thomas, Jr.[2]
Members, ASCE

Abstract

The major observations related to nuclear power plants that resulted from a U.S.D.O.E. sponsored research project entitled "A Comparitive Analysis of Structural Concrete Quality Assurance Practices on Nine Nuclear and Three Fossil Fuel Power Plant Construction Projects" are presented. The observations present both the current practices as well as suggestions for improvements in the area of: (1) Organization and Management of Quality Assurance, (2) Procurement and Testing of Concrete Components, (3) Production and Evaluation of Fresh Concrete and (4) Concrete Preplacement, Placement, and Post Placement Practices.

Introduction

The period from 1967 (when the A.E.C. released for comment a General Criteria for Light Water Cooled Nuclear Power Plants as Appendix A to Title 10, Part 50, Code of Federal Regulations (i.e., 10 CFR Part 50, Appendix A)) to 1977 was a period of rapid growth within the nuclear power plant industry, both in terms of the number of Construction Permits and Operating Licenses that were issued as well as in terms of the number of regulatory requirements and industry standards related to Quality Assurance that were developed. On each of the nuclear power plant projects that were in the process of construction during that period the parties involved (i.e. the Utility, the Engineer and the Constructor, etc.) implemented Quality Assurance programs which were largely based on their own interpretation of what the regulations required. Under the environment of changing requirements it might be expected that the degree and complexity of these Quality Assurance programs differed from project to project. Some Utilities, Engineers and Constructors, when interpreting the rules and regulations, probably implemented unnecessarily severe Quality Assurance programs. Others may have been able to meet the same goal of quality with more efficient and simpler programs. These different levels of reaction to perceived requirements undoubtedly influenced the cost and duration of the nuclear power plant projects that were being built.

---

[1] Professor, Department of Civil Engineering, The Pennsylvania State University, University Park, Pa., 16802

[2] Associate Professor, Department of Civil Engineering, The Pennsylvania State University, University Park, Pa., 16802

One of the areas of power plant construction that has a tremendous impact on the project in terms of both cost and duration is structural concrete. Approximately one-third to one-half of the field controllable manhours and about three to five years of construction time are consumed in this phase of work on a typical project. It is also one of the areas that is most directly influenced by field related Quality Assurance practices since concrete is essentially produced and placed by combining raw materials at the project site. Structural concrete that is used in the construction of the safety related (i.e. Category I, Q listed, etc.) structures of the power plant must be of very high quality. The achievement of this level of quality has proven to be more costly to the Utility than comparable concrete produced for conventional construction. Some of the high cost is probably related to the Quality Assurance practices which are required to insure this necessary quality.

Because of this important role of structural concrete, a U.S.D.O.E. sponsored research project was begun at The Pennsylvania State University in early 1977 which had the objective of performing a comparative analysis of the Quality Assurance practices related to the structural concrete phase on nine nuclear and three fossil fuel power plant projects that were (or had been) under construction in the United States during the period from 1967 to 1977. James L. Burati, Jr., (presently an Assistant Professor of Civil Engineering at Clemsen University), and graduate students Robert Fron and Patrick Wilson (presently with Bechtel Power Corporation) and James Rouland (presently with the Pennsylvania Power and Light Co.) assisted the writers in the research effort. Assistance was also obtained from members of the Power Plant Construction Advisory Group to The Pennsylvania State University. The results of the study, which appear in several research reports: (1), (2), (3) and in several papers which have been published in the Construction Journal of the American Society of Civil Engineers: (4), (5), (6), (7), represent, to a large degree, an interpretation of the data which was collected as a result of many hours of personal on-site interviewing and both on-site and off-site analysis of project records. The majority of the data collection effort took place from April 1977 to December 1977. The observations which were made do not therefore reflect the changes in concrete practice that have occurred since that time period.

An attempt was made to select the nine nuclear power plant projects using the following criteria: a) The Construction Permit dates should be reasonably spread over the entire ten-year time frame from 1967 to 1977, b) They were constructed under the jurisdiction of several different NRC regional offices and c) They were constructed using different management modes: (1) C. M. Mode - Management by a Construction Management Firm, (2) E.P.C. Mode - Engineering, Procurement and Construction by one firm retained by the Utility and, (3) Owner Mode - Engineering, Design and Construction by the "Owner Utility". The projects that were selected are shown in Table 1.

This paper will highlight the major observations that resulted from the research effort in the areas of: (1) Organization and Management of Quality Assurance, (2) Procurement and Testing of Concrete

Table 1.1--Nuclear Projects in Research Study

Project	A	B	C	D	E	F	G	H	I
Description[a]	2 UNIT PWR	UNIT 2 PWR	2 UNIT BWR	2 UNIT BWR	2 UNIT PWR	2 UNIT PWR	2 UNIT BWR	2 UNIT BWR	1 UNIT PWR
Management Mode	C.M.	C.M.	E.P.C.	E.P.C.	Owner Design Construct	C.M.	E.P.C.	C.M.	E.P.C.

[a] Description refers to Unit for which data was collected.
2 Unit--project consisted of 2 Units
Unit 2--data only collected on second unit of 2-unit facility.

Components, (3) Production and Evaluation of Fresh Concrete and (4) Concrete Preplacement, Placement and Post Placement Practices. The interested reader is referred to the previously noted research reports and papers for a presentation of the background material which supports these observations.

Organization and Management of Quality Assurance

Nuclear Regulatory and Quality Assurance Environment

The major observations that were made with regard to the environment under which the nuclear plants were being built are:

1. The ten year period which began with the issuance of Appendix A of 10 CFR Part 50 in 1967 was one of rapid growth, both in terms of the number of nuclear power plants under construction as well as the number of regulatory requirements and industry standards that were developed related to Quality Assurance. The efforts of the AEC (and later the NRC), ANSI, ASME and ACI during this period all attest to this fact.

2. The Quality Assurance practices related to structural concrete which were implemented on each nuclear power plant construction project were influenced by a number of factors, some external and some internal to the project. Some of the major factors appeared to be: (a) The development of nuclear construction "Standards", (b) The difficulties associated with the interpretation of requirements, (c) The increasing role of the NRC, (d) The need for comprehensive documentation, (e) The cost of the Quality Assurance effort, and (f) The recognition that there were additional benefits that could be derived from a Quality Assurance program.

3. The difficulty which the nuclear industry faced with regard to an interpretation of the sometimes conflicting regulatory requirements and standards (particularly since some of these were in a state of flux during this period) is probably one of the reasons why the parties on

some of the nuclear projects adopted what appeared to be overly restrictive Quality Assurance practices related to the structural concrete phase.

4. It was noted that a clear direction for future standards development was needed and that the possible consolidation of the many existing standards into one reference source was felt to be desirable by some industry personnel. In this regard it should be noted that the joint efforts of ASME and ACI which led in 1974 to Section III, Division 2 of the ASME Boiler and Pressure Vessel Code (which dealt with Concrete Reactor Vessels and Containments)(also called ACI 359-74) and the related ACI 349-76 code (entitled "Code Requirements for Nuclear Safety Related Concrete Structures") were developed too late in the period to greatly affect many of the nuclear plants which were examined. Most of these are being, or were, built under the ACI Codes, "Recommended Practices", etc., which applied to "General Construction" rather than specifically to "Nuclear Construction".

Organizational Aspects

As noted earlier the nine nuclear projects were divided into three groups according to the "Management Mode" which was used on each project. Some of the other observations that were made with regard to organizational aspects are:

1. All projects appeared to reflect, both from an organizational as well as from a procedural viewpoint, a distinction between Construction, Quality Control and Quality Assurance responsibilities. This relationship can be reflected by considering the "Hierarchy of Documents" model which was typical for most projects.

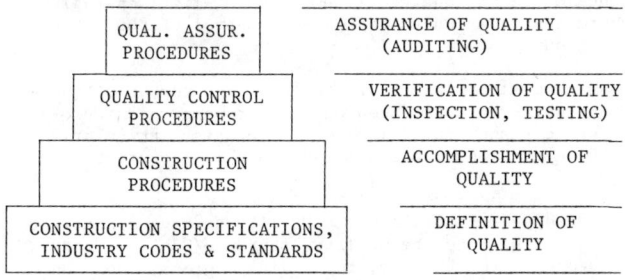

Figure 1--Relationship of Quality Related Documents

2. There appeared to be a number of organizational alternatives to the "Independence" requirement for Quality Assurance and Quality Control personnel which is cited in the criteria of 10 CFR Part 50 Appendix B. All of the Utilities and Constructors established a separate Quality Assurance group within their organizations. With regard to the Quality Control group, in some cases it was directly associated with the Quality Assurance group, while in others it was organizationally related to the Construction group. There appear to be advantages and disadvantages to both approaches in terms of the level of documentation required,

the overall coordination of quality related activities, etc. It was felt that further study should be made in this area to determine how the "Independence" criterion can be most effectively met by the firms within the industry.

3. It was felt that the E.P.C. projects exhibited the least direct influence of the Utility on the overall project, particularly from a Quality Assurance/Quality Control standpoint. One of the points of commonality on all of the E.P.C. projects was that the Utility was primarily placed in a Third Level Auditing role. The primary responsibility for the First Level Quality Control and Inspection function was assumed by the E.P.C. Firm.

There appeared to be a certain amount of duplication of effort with regard to the <u>Audit</u> function since, in general, the E.P.C. Firm QA group was also involved in this activity. The only apparent justification for such overlap would be an increase in the quality of construction which would have resulted under such a scheme. The <u>Surveillance</u> function was also in need of further definition on the projects in this group since it was often assumed by both the Utility QA group, the E.P.C. Firm QA group, and sometimes also by the E.P.C. Firm QC group. Criterion X of 10 CFR Part 50 Appendix B indicates that <u>either</u> inspection <u>or</u> process monitoring (i.e. Surveillance) <u>or both</u> are viable verification activities which may be used. If it is determined that Surveillance is necessary for a particular activity which is a part of the concrete phase, then it should be assigned to only one group in order to avoid duplication of effort.

With regard to the Quality Control groups of each of the E.P.C. Firms, it was noted that on only one of the four projects were they involved in the Balance of Plant Concrete (i.e., non "Q"-list) operations. These placements were generally under the direct control of the E.P.C. firm's Construction group.

A point was also made with regard to the large number of quality related documents which were required on nuclear power plant projects in order to define responsibilities, etc. This is particularly important because there are many revisions made to these documents over the life of a project.

4. The C.M. Mode appeared to allow the Utility to have a somewhat greater influence with respect to the Quality Assurance/Quality Control functions. On all projects the Utility, to varying degrees, appeared to be directly involved in the Third Level <u>Auditing</u> phase <u>as well as</u> the Second Level <u>Surveillance</u> and First Level <u>Quality Control and Inspection</u> phases. On one of the projects in this group, for instance, the Utility was in direct charge of all of the Quality Assurance/Quality Control activities on the project. This was accomplished without utilizing many permanent Utility employees.

It appeared that there was less overlap of quality related activities on the projects in this group. This was probably due to the fact that the Utility had been the prime mover in assuring that each party was involved in only a portion of the total spectrum of quality related activities.

An important point was made with regard to the Work Package approach which was implemented on one of the projects in the group. It was felt that since the concrete work was divided into three work packages awarded to three independent contractors, it may have been to the Utility's benefit to develop a standardized set of construction procedures, instructions, etc., which would apply to all of these contractors. The task of reviewing a vast number of procedures and tracking the changes to them could perhaps lead to problems which could be avoided if the Utility adopted a standardization approach.

5. Only one project in the research study, was being constructed under the Owner Mode. Several unique organizational solutions to the 10CFR Part 50 Appendix B Independence criterion for the Quality Assurance and Quality Control groups occurred on this project. There appeared to be very little overlap of quality related activities. This is understandable, since only one party, the Utility, was involved. It seems that as more organizations become involved, the need for increased Auditing and Surveillance also arises. It is certainly not feasible for every Utility to adopt the role of the Utility in this group. Some of the practices which occur under such an arrangement however, deserve serious consideration.

6. On most of the projects there seemed to be an overlap of responsibility for quality between the Quality Control and the Construction groups of the firms performing the actual construction work. There appears to be a need to develop a clearer definition of the interface between these two groups by the industry in order to avoid the situation where the Construction group abdicates its traditional responsibility for the quality of the final product to the Quality Control group.

It is recommended that during the preplanning stages of a project the Utility and the other parties involved mutually develop a quality system which considers, among other things, the above areas of potential overlap and duplication. The savings which result over the life of a project, both in terms of human and financial resources as well as documentation efficiency, will probably justify the effort.

On the "Owner Mode" project a strong attempt was made to leave the prime responsibility for quality in the hands of the Construction group, a place where it traditionally has existed in the construction industry. The Quality Control Group was clearly placed in the role of verifying that the required quality had been achieved through inspection and testing.

Procedural Aspects

The following observations were made with regard to the procedural aspects related to design and document control, and nonconformance and corrective action situations:

1. With regard to the Review of Project Specifications, it was indicated that a number of parties are typically involved in the review process and that the level of involvement of these parties varies from project to project. It was noted that the QA/QC groups were usually in a unique review position because they could indicate the portions of

the specification which would give them the most difficulties during the construction phase.

2. With regard to The Use of General Construction Codes and Standards, it was noted that for the concrete situation, the codes and standards which are referenced on most projects were probably never intended for the "literal" interpretation stance which occurs on nuclear construction projects. In the nuclear situation, it was noted that the Design Engineer must assume a very active "interpretive" role if later quality related problems are to be avoided. It was felt that an industry-wide assessment of the current situation in the concrete area is needed to determine if the codes and standards should be revised to more closely reflect the nuclear construction situation.

3. Several different approaches to the Incorporation of Codes and Standards into Specifications were observed. The "Blanket Reference" of a concrete code was indicated as one which required less Engineering effort during the specification development stage, but can create later problems when a "literal" interpretation of certain sections of the cited code were made by the QA/QC groups. The "Selective Reference" method and the "Blanket Reference With Exceptions" methods were also observed. It is felt that the Engineer plays a crucial role with regard to the clear interpretation of the requirements for a project. It was felt that if the Engineer does not assume this interpretive role, then the QA/QC and Construction personnel must assume it be default.

4. With regard to the Quantification of Specification Requirements in the concrete area, it was noted that on some projects, too much emphasis is probably being placed on those process characteristics which can be quantified and not enough emphasis is being placed on the more qualitative aspects of the concrete process.

5. An area of concern on many of the projects in the research study dealt with the amount of the Engineering Changes which occurred. Some of the possible reasons for the extensive number of specification changes which occur on nuclear construction projects were felt to be: (a) The development of specifications with unrealistic requirements, (b) The lack of a feedback loop between the Construction and Engineering groups which could be used to make the Engineering documents more responsive to the construction situation and (c) The reuse of an Engineering Firm "Standard Specification" at the beginning of each new project. It was found that problems often arise because of the lag between the approval of Engineering Changes and a formal revision to the project concrete specifications.

6. One of the impacts of Engineering Changes is the need to revise supportive Construction and QA/QC procedures whenever a specification is revised. A second impact arises because of the need for strict compatibility between the PSAR and the specifications. The practice of "waiving" a specification requirement rather than attempting to revise either the specification or the related industry code and standard was also discussed. The disposition of Nonconformances by the "Use As Is" choice was also discussed as a possible example of an effective "waiver" of specification requirements.

CONCRETE Q.A. PRACTICES 595

7. The Nonconformance Control system which is adopted on a project has a great deal of impact on the "smooth progress towards construction completion" which is achieved. For instance, it appears that on most projects, the Nonconformance document must travel through a number of organizations (sometimes more than once) before a Nonconformance event is finally "closed". This practice not only consumes the resources of a large number of people, it also increases tremendously the time which is required between the identification of the Nonconformance, the receipt of an approved disposition and the completion of the corrective action. The length of time also appears to be influenced by the location of the organizations which are involved in the process. It would appear advantageous to have the personnel who are responsible for the dispositioning of the Nonconformance documents permanently assigned to the project site, for instance, since a routing of the document to a "Home Office Engineering Group" does not appear to be time-effective.

It would also appear to be advantageous to require a definable team such as the "Deviation Review Board" on Project B or the "Project H Review Board" to meet on a day-to-day basis if necessary, in order to obtain the immediate resolution of most of the Nonconformance documents which are written. It is felt that such meetings would reduce the decision-making time (which represents time during which construction is often essentially stopped) to a minimum.

8. A comparison of some of the Nonconformance documents which were presented indicates that there is a considerable amount of information that must be recorded every time one of these documents is initiated.

Some of the data input practices which were cited should be recommended as additions to the Nonconformance documents which are used on some of the projects in the research study. The practice implemented on Projects B and H, for instance, which requires the affected party to indicate the corrective action which will be taken to prevent the reoccurrence of the nonconformity, is one which is recommended. Another is the input of the Nonconformance Code and Cause Code information for each nonconformity which is required on Project H. This practice would appear to provide an excellent basis for later Trend Analysis activities.

It is also suggested, however, that the Nonconformance documents on a number of projects should be scrutinized in great detail to determine if all of the required information is really necessary (and in fact ever used). A reduction in the time required to fill out the forms could possibly be achieved on a number of projects if such a study were made.

9. A number of projects have adopted a set of forms for Nonconformance situations which consider the relative severity of the Nonconformance (e.g.: Major, Minor or Incidental). It is felt that the concept of several types of Nonconformance documents to record several levels of severity should be given serious consideration by the industry. Perhaps only the Major classified discrepancies would have to be fed into the formal nonconformance reporting and corrective action system. The Minor and Incidental classified discrepancies should be documented and dispositioned by the lowest competent level of

personnel.

10. From the amount of data which was made available on the projects in the research study, it appears that a large percentage of the Nonconformance Reports related to structural concrete on must projects are dispositioned "Use As Is". In light of the cost of a typical Nonconformance Report in terms of both manpower and project time, it seems that an industry wide study should be initiated in order to identify a number of general areas related to structural concrete which are consistently dispositioned "Use As Is". This might lead to a reassessment of the necessity for all of the requirements which currently exist in the concrete specifications, industry codes and standards, etc.

Procurement and Testing of Concrete Components

The requirements for high quality concrete for a two-unit nuclear power plant can exceed 500,000 cubic yards. Due to their nature, projects of this type require materials of the highest quality to insure the integrity of the structures. The components which are used in the production of concrete must be of acceptable quality if the finished product is to be acceptable. The research effort therefore included an analysis of the specification limits and testing requirements which were imposed upon the materials on the various projects. Specification limits were compared with the applicable codes and standards which govern the production of concrete for non-nuclear work, as well as on a project-by-project basis. The research report presents a number of tables which summarize this information. The following observations should be highlighted:

1. With regard to Cement it was noted that the cement requirements for the projects studied were quite similar because they all stipulated that the cement had to meet the requirements of ASTM C150. Several of the projects did however place additional requirements on cement which were not included in ASTM-C150. These requirements were related to the required strengths of the mortar cube specimens which were made from the cement. One of the projects had a statistically based mortar cube strength requirement which included a price adjustment provision. This was one of the few instances among the projects studied where statistical concepts were employed to allow for the natural variability which is associated with construction materials.

The procedure for receipt inspection of the cement at the project site was also studied. All of the projects required certified copies of mill test reports on the cement which was supplied. In addition to the mill test reports, the projects also required periodic in-process testing of cement. In addition to the normal testing frequencies, several of the projects required additional retesting for cement which had been in storage. Seven of the projects studied placed a maximum limit on the delivery temperature of the cement. The limits ranged from 130°F to 170°F. The need for an upper limit on cement delivery temperature may be questioned, particularly during cold weather operations and because aggregate and water temperature have a much greater effect on concrete temperature than does the temperature of the cement.

2. With regard to Aggregates it was noted that the primary source of the requirements for concrete aggregate is ASTM C33. All of the projects studied require conformance to this standard specification. It was indicated that there are two types of testing which must be performed. These include the initial qualification of the aggregate source and the in-process tests which are performed during operations. It was noted that the primary sources for guidance in the area of testing frequencies are ANSI N45.2.5 and the ASME Boiler and Pressure Vessel Code. It was found that the later projects in the study tended to follow the ANSI guidelines either exactly or very closely, whereas the earlier projects in the study, whose requirements were developed before ANSI was published, tended to have some testing frequencies that are quite a bit higher than those in ANSI.

It was again noted, with regard to the aggregate testing measurements that statistical procedures for describing the natural variability of materials were not applied.

3. With regard to mixing water, it was noted that there were some discrepancies among the codes in the test methods which are used to determine water acceptability. A broad range of in-process mixing water testing frequencies were encountered on the projects studied. There appeared to be a great deal of variation in the chemical tests which are performed and in the acceptance limits that were placed on these tests. This probably stems from the fact that the codes contain broad qualitative statements concerning mixing water requirements rather than specific testing requirements.

Concerning the use of crushed or flaked ice, it was noted that all of the projects allow ice to be used in the concrete mix to reduce the concrete placement temperature. The testing requirements for ice on all of the projects were generally the same as the ones specified for mixing water.

4. The advantages and disadvantages of the use of chemical admixtures were considered with project personnel. It is felt for instance that the use of entrained air for all nuclear concrete, regardless of exposure, might not be necessary and it is recommended that guidance be taken from ACI 211 concerning its use. ACI 211 includes a graded table which can be used to determine air content requirements. It was noted that all of the projects prohibited the use of set-accelerating water reducers. Some projects allowed the use of both water-reducing and water-reducing set-retarding admixtures, while others allowed only the use of water-reducing agents. The use of water-reducers in conjunction with air-entraining agents was also examined, and it was noted that water-reducing agents may contribute to the air content of the concrete.

It was also felt that superplasticizing mixtures should be given serious consideration for use in concrete which is placed in highly congested rebar areas. These admixtures have been used for a number of years in Japan and Germany. Their use allows the production of very high slump concrete at the low water-cement ratios necessary for high strengths.

It was also noted that fly ash was used on six of the projects in the study. The potential advantages and disadvantages of the use of fly ash were examined. It was pointed out that two of the projects experienced difficulties with the use of fly ash and discontinued its use.

5. As noted above in relation to the discussion of cement and aggregates, there was very little indication of the use of statistical concepts for considering the natural variability of materials on the projects in the study. It is recommended that statistical techniques, similar to those which have been employed for years in highway construction, be incorporated into the specification requirements on nuclear projects. Techniques such as random and multiple sampling, statistical control charts and price-adjustment tables (which are not currently being employed on the projects), could help to give a better representation of the properties of the materials which are being tested. These techniques could also possibly help to reduce some of the high testing frequencies which are currently employed on nuclear projects, while, at the same time, maintaining the level of quality of the concrete.

6. It is recommended that the acceptance practices which are currently being used for concrete be reevaluated. The practice of using in-process tests of concrete ingredients as partial acceptance criteria results in the generation of many nonconformance reports which are ultimately dispositioned "Use As Is" because final compressive strength test results are shown to be acceptable. Since ultimate acceptance is based on compressive strength, it is recommended that the in-process tests be conducted as Process Control Tests, with the final acceptance criteria being conrete compressive strength. This would result in the elimination of many nonconformance reports which will ultimately all be dispositioned "Use As Is".

Production and Evaluation of Fresh Concrete

The very large volumes of concrete which are required on major projects such as nuclear power plants have resulted in the development of very sophisticated equipment for the batching and mixing of concrete. A comparison of the specification requirements for the concrete batch plants on each of the projects was therefore made. The Quality Control practices and the testing requirements of the concrete production phase were also examined. Some of the observations which were made are:

1. The batch plant requirements on the projects were compared with each other and with those of ACI 301, ACI 304, and ASTM C94. It was noted that all of the batch plants were equipped for automated batching, automated recordation and were capable of automatically compensating for varying moisture content in the aggregate. It was noted that ACI 301 and ACI 304 have different requirements for batching tolerances and that five of the projects employ the simpler approach of ACI 301. Most of the projects studied required mixed uniformity tests in accordance with ASTM C94. There was a broad range, however, in the frequency with which this test was conducted on the various projects. On one of the projects the fairly simple uniformity requirements recommended by the Bureau of Reclamation, rather than the requirements of ASTM C94, were used.

2. All of the projects required a QC Inspector to be present at the batch plant at all times during concrete production. The level of responsibility of the batch plant inspector varies, however from project to project. On some projects the batch plant inspector had so much responsibility that it appeared that the batch plant operator was not really necessary. It was suggested that this point be seriously examined since the role of QC should be the inspection of construction operations rather than their direction.

3. On eight of the projects, the design mix water/cement ratio of the concrete was used as a control measure. Several of the projects use this water/cement ratio to establish a limit on the maximum amount of water which may be used in the batch. This is referred to as the "locked water meter" approach. A better approval perhaps would be to allow the water requirements to be field controlled by focusing on slump rather than water-cement ratio. This will help to account for batch to batch variation in moisture content of the aggregate. In this way the average moisture content of the aggregates and the average additional water requirements at the mixer could be used to determine whether the average mix conforms to the specifications regarding water-cement ratio.

4. It was found that sampling was generally performed at the point of placement when concrete was transported by pump, and at the truck discharge when the concrete was placed by other means. It was felt by a number of project personnel that sampling and testing was impaired when performed at the point of placement rather than under the controlled conditions of the batch plant. Several of the projects established correlation testing programs in an attempt to predict changes of material properties between the batch plant and the forms. It was noted that such correlation testing was extremely difficult for the case when concrete is placed by pumping.

5. With regard to compressive strength requirements, it was noted that the testing frequencies required by ACI 301, ANSI, and ASME were higher than those required by other codes and those normally employed in mass concreting operations. It was suggested that the use of the lower 150 cubic yard frequency of ACI 318, ACI 349, and ASTM C94 may be quite sufficient to provide adequate control of the concrete production operations.

6. On several of the projects it appeared that the concrete strengths resulting from the mix designs that were used were quite a bit higher than required. On most projects in fact it was extremely rare for a cylinder test to fail. The high cement contents necessary to achieve these high strengths can add to potential heat of hydration problems, particularly in the case of massive concrete placements. It is recommended therefore that mix designs be adjusted in accordance with the guidelines provided in ACI 318 and AC 349 when the average compressive strengths results are high and suitable control is indicated by a low standard deviation. Very little evidence that such adjustments were made could be found on the projects that were studied.

7. The required testing frequencies for in-process tests for slump, air content and temperature were also examined. It was noted that ANSI and ASME stipulate frequencies that are quite a bit higher

than those required by other codes and standards. Perhaps these frequency requirements are unrealistic. The purpose of sampling should be to verify that the process is in control and not to eliminate the need for judgement on the part of the inspector. Slump tests taken when cylinders are cast should be sufficient to verify that the process is in control.

8. With regard to slump requirements, an approach used on three of the projects should be noted. This approach stipulates <u>Working Limits</u> and <u>Reject Limits</u> and allows an "<u>Inadvertency Margin</u>" for batches which may occassionally exceed the working limit but are still below the reject limit.

9. Most of the projects in the study place a 90-minute time limit after which concrete cannot be placed. It is suggested that the condition of the plastic concrete should determine its placeability, and not some arbitrary time limit. This appears to be an example of the general trend on the projects of removing any judgement whatsoever from the inspectors and expressing everything in terms of absolute limits. Common sense and the judgement of a skilled inspector are assets that should not arbitrarily be eliminated simply because the concrete is being placed on nuclear power plant project.

Concrete Preplacement, Placement and Post Placement Practices

The major observations that were made with regard to the placement phases noted above are:

1. The Project Control System for each project was examined with the primary emphasis being placed on the involvement of the Quality Control group in the preplanning phase as applied to the preparation of detailed work plans or construction plans for individual concrete placements. On a number of projects (particularly the early vintage ones) it appeared that Quality Control was not involved in the review or approval of detailed work plans, and thus it was theorized that the preplanning phase fell outside of the scope of the Quality Assurance Program for these projects. Their major responsibility appeared to be at the point of final inspection. On several other projects, however, it was noted that the Quality Control group was very actively involved in preplanning. These activities included: reviewing documents for correctness and completeness and checking for open NCRs affecting the work etc. While one would anticipate improved quality with greater QC involvement, two examples which indicated that such systems had not functioned as intended were uncovered. The hypothesis is therefore presented that increased involvement of QC beyond a certain level of intensity probably does not improve quality and in fact may have a directly opposite effect. This is particularly true if such involvement overlaps the responsibilities of engineering and construction personnel or it places a greater emphasis on documentation and written communication rather than engineering judgement.

2. Mass concrete does not appear to be adequately described in the ACI Code. It was noted that ACI 301 defines mass concrete as any member whose least dimension exceeds 2 1/2 feet; however, many in the

industry feel that ACI 301 does not apply to nuclear construction since the provisions result from experiences gained on large gravity dam projects. Placements on such projects are usually orderly and members are not normally heavily reinforced. Since nuclear power plant construction is characterized by complex heavily reinforced placements, the applicability of the Code is questioned. Eight out of nine projects use a critical dimension to define mass placements although the critical size varies from 30 to 48 inches. Occasionally exceptions are made for fill concrete and isolated footings. One project evaluated each placement to determine if the heat of hydration will cause a problem, and therefore, if a reduced concrete temperature will be required. It was noted that this procedure is recommended by ACI 349.

3. The methods used to define hot and cold weather conditions on each project were also examined. A number of projects reported some confusion and dissatisfaction with the ACI 306 provisions; on a few projects studies have been conducted regarding when to initiate cold weather practices. Regarding hot weather practices, the specification revisions on one project were traced to illustrate the apparent confusion over the perceived requirements for hot weather concreting. It was felt that in that instance the project engineering staff was searching for a set of workable provisions which provided a sufficient degree of flexibility.

4. With regard to delivery systems the emphasis was placed on the effect which various practices had on delivering, conveying, and placing the concrete as rapidly as possible. In this respect, the addition of supplemental water, sampling at the point of placement, and the frequency of testing were all felt to be important. It appears that the elimination of routine testing at the point of placement first requires the completion of a correlation testing program. Most of the correlation testing programs which were studied did not appear to be well conceived and thus were very lengthy and costly. They all had their own unique characteristics, objectives, and in most cases, statistical deficiencies. In addition, the data are difficult to analyze, and no information was found to suggest that they were analyzed statistically. There did not appear to be an industry guideline which suggested how a correlation testing program should be conducted. In view of the time and expense involved in such programs, it appears that a research study which established the methodology to be used in a statistically based correlation testing program would be cost effective.

5. With regard to the requirements for formwork tolerances, reinforcement placement tolerances and reinforcement fabrication tolerances it is felt that they should be reevaluated with particular attention being given to the relationship between each aspect. It appears that nuclear structures should qualify as special structures in accordance with ACI 301 and 318 and that modified tolerances should be considered at least in certain portions of the plant. The number of nonconformances which are written because these tolerance limits are exceeded in no way indicates the difficulties and manhours expended in meeting the specification requirements.

The method of obtaining relief from formwork tolerances on a case by case basis by requesting an engineering design change was cited.

On several projects the frequent use of this approach appears to indicate that the project site organization was using this as a conveninet mechanism for modifying certain specification requirements without requesting a major change in either the project specification or the industry standards.

With regard to reinforcing bar tolerances it was noted that there is a relationship between the formwork tolerances, rebar fabrication tolerances, rebar placement tolerances, and the cover requirements. Several projects indicated that fabrication and placement tolerances took up most of the formwork tolerances, thus making the formwork erection process much more difficult.

6. Reinforcing steel was adressed with respect to user tests, rust, and control of nonconforming or rejected items. The problem of rust was particularly troublesome, and it was pointed out that previously published research reports indicated that the concerns over excessive rust as it applies to bond characteristics are generally unnecessary since the condition of the bar surface does not seem to be nearly as important as the height and number of deformations in the bar.

7. Cadwelding practices were examined by considering the data in terms of the pre-Reg. Guide 1.10 group and the post-Reg. Guide 1.10 group. It was noted that the requirements for these two groups differed significantly in the areas of crew qualification tests, acceptance criteria, provisions for substandard test results and preparation requirements. The later vintage plants generally complied with the Reg. Guide provisions. With respect to testing, it appeared that the testing frequencies were excessive since the visual inspection and tensile test data from several projects indicated that virtually all of the defective cadwelds were identified by visual inspection. In light of this information it is suggested that the industry should reevaluate cadweld requirements, particularly with regard to the testing of sister splices.

The cadweld preparation requirements were discussed and it was noted that the primary concerns appeared to be focused on insuring that the bar ends were properly cleaned and free of moisture. The various project specification requirements for cleaning and heating the bars were examined, and it was noted that there was a great deal of variation between specification requirements on the projects in the study. Some projects go to great lengths to describe the temperature and cleanliness requirements whereas others merely state that the bars should be cleaned and free of mositure. It was pointed out that there appeared to be no correlation between the generality of requirements and the rate of visual rejections.

8. Construction joints were examined, and it was noted that both starter mixes and grout were used on the projects in the study. These practices were compared with the Bureau of Reclamation practices which suggested that starter mixes provided superior joints and with less difficulties being encountered at the placement point. It was found that two methods of cleaning a joint were generally used. From an economic standpoint, sandblasting appears to be preferable over "green

cutting". A number of projects reflect this in their specifications. The requirement that aggregate be exposed during the cleaning operation was also examined. It was noted that data from previously performed research has shown that the exposure of aggregate to a specified amplitude offers no increased advantage in water-tightness or bonding characteristics.

9. Placement practices were also examined with a particular emphasis being placed on consolidation. The level of detail in several project specifications and QC procedures with respect to consolidation were analyzed. It was felt that in several cases too much emphasis was being placed on the details rather than the overall aspects of consolidation. Several projects make mention of the fact that forms and reinforcement should not be touched. It is felt that this type of a requirement forces the vibrator operators to become timid and more concerned about contract rather than good consolidation. No research data was available to support this hypothesis, however. On the other hand, three projects had provisions which stressed the importance of consolidating around embedments, reinforcement and in the corners. Revibration was also examined, and the projects on which this technique was not allowed or discouraged were noted. It appeared that revibration was contrary to accepted industry practice.

10. It is felt that the requirements for curing hot and cold weather should be reviewed since many projects have encountered difficulties in administering the ACI 305 and ACI 306 provisions. There also seems to be some confusion regarding the curing period during cold weather.

Conclusions and Recommendations

The introduction in the brochure for this Speciality Conference states:

> Over the next twenty-five years, if the critical energy needs of the country are to be met, utility companies throughout the United States must continue to make a large capital investment in the construction of both fossil fuel and nuclear electrical-power generation facilities. The industry is, however, at a critical crossroads as it attempts to balance factors such as complexity of design and construction, length of time required for completion and magnitude of financial investment required against the ever increasing industry codes and standards, and governmental regulatory and quality assurance requirements which are being imposed.
>
> *This speciality conference is designed to focus on this critical issue by not only documenting what has occurred but also be serving as a forum which will establish what improvements can be made in the future as a result of the lessons learned in the past.*

This paper has attempted to summarize the major observations of the nuclear portion of the U.S.D.O.E. sponsored research study entitled

"A Comparitive Analysis of Structural Concrete Quality Assurance Practices On Nine Nuclear and Three Fossil Fuel Power Plant Construction Projects". When that study is considered in conjunction with the information which has been provided by all of the speakers in the Opening Keynote Session and the Eight Concurrent Sessions in this conference it should be recognized that a satisfactory bridge between what has occurred in the past and the improvements that are needed in the future has been established.

This conference will be considered successful if this information is instrumental in causing fundamental changes to be made to industry codes and standards and governmental regulatory and quality assurance requirements where they are required. Those changes will certainly influence how effectively electric power generation facilities are built from both a time and a cost standpoint during and beyond the 1980's.

References

1. Willenbrock, J.H., Thomas, H. R. Jr., and Burati, J. L., "A Comparative Analysis of Structural Concrete Quality Assurance Practices On Nine Nuclear Power Plant Construction Projects," Report No. COO/4120-1 to the U.S. Department of Energy under Contract No. EY-76-S-02-4120.M002, June 1978.

2. Willenbrock, J.H., Thomas, H. R. Jr., and Burati, J.L., "A Comparative Analysis of Structural Concrete Quality Assurance Practices on Three Fossil Fuel Power Plant Construction Projects," Report No. COO/4120-2 to the U.S. Department of Energy under Contract No. EY-76-S-02-4120.M002, June 1978.

3. Willenbrock, J.H., Thomas, H. R. Jr., and Burati, J. L., "Final Summary Report: A Comparative Analysis of Structural Concrete Quality Assurance Practices on Nine Nuclear and Three Fossil Fuel Power Plant Construction Projects," Report No. COO/4120-3 to the U.S. Department of Energy under Contract No. EY-76-S-02-4120.A003.

4. Thomas, H. Randolph, Jr., "Quality Control of Cadweld (Mechanical) Splices," Journal of the Construction Division, ASCE, Vol. 105, No. CO3, Proc. Paper 14803, September 1979, pp. 201-216.

5. Thomas, H. Randolph, Willenbrock, Jack H., and Burati, James L., "Concrete Production on Nuclear Power Plant Projects," Journal of the Construction Division, ASCE, Vol. 106, No. CO3, Proc. Paper 15703, September 1980, pp. 327-339.

6. Thomas, H. Randolph, Jr., "Concrete Slump in Nuclear Power Plant Construction," Journal of the Construction Division, ASCE, Vol. 106, No. CO4, Proc. Paper 15881, December, 1980, pp. 567-584.

7. Burati, James L., Thomas, H. Randolph, Jr. and Willenbrock, Jack H. "Nuclear Construction Quality Assurance Systems," Journal of the Construction Division, ASCE, Vol. 107, No. CO2, Proc. Paper 16304, June, 1981, pp. 349-360.

SPECIALITY CONFERENCE
ON

EXPERIENCE WITH THE IMPLEMENTATION OF

CONSTRUCTION PRACTICES, CODES, STANDARDS, AND

REGULATIONS IN THE CONSTRUCTION OF POWER GENERATION FACILITIES

OPENING KEYNOTE SESSION - September 16, 8:45 a.m. - 12:00 noon

8:45 - 9:10	Welcome to The Pennsylvania State University • Jack H. Willenbrock, Professor of Civil Engineering • Edward H. Klevans, Associate Dean for Research Introductory Remarks • Aldo Palmeri, Supervisor, Civil Engineering, Ebasco Services, Chairman: ASCE Committee on Construction of Nuclear Facilities
9:10 - 9:30	Keynote Address: "Utility Perspective" William G. Counsil, Senior Vice President, Nuclear Engineering and Operations Group, Northeast Utilities Group
9:30 - 9:50	Keynote Address: "Architect-Engineer Perspective" Russell J. Christesen, Executive Vice President, Ebasco Services, Inc.
9:50 - 10:10	Keynote Address: "Constructor Perspective" Howard W. McCall, President, Power Group, Daniel Construction Company
10:10 - 10:40	Coffee Break
10:40 - 11:00	Keynote Address: "Government Perspective" James G. Keppler, Director, U.S.N.R.C. Office of Inspection and Enforcement, Region III
11:00 - 11:20	Keynote Address: "Concrete Industry Perspective" Robert E. Philleo, Chief, Structures Branch, Office of Chief of Engineers, U. S. Corps of Engineers; Past president of the American Concrete Institute
11:20 - 12:00	Panel-Audience Discussion Period
12:00 - 1:30	Lunch

POWER GENERATION FACILITIES

CONCURRENT SESSIONS I AND II

SESSION I – September 16, 1:30 – 5:00 pm

1:30 – 1:55	(1) Nuclear Standards Applicable to the Civil-Structural Design of Nuclear Power Plants Dr. John D. Stevenson, Vice President and General Manager, Structural Mechanics Associates
1:55 – 2:20	(2) The Nuclear Licensing Picture—Overview and Outlook Richard H. Vollmer, Director, Division of Engineering, Office of Nuclear Reactor Regulation, U.S.N.R.C.
2:20 – 2:50	Coffee Break
2:50 – 3:15	(3) The U.S.N.R.C. Enforcement Policy Charles E. Norelius, Acting Director, Division of Engineering and Technical Inspection Program, U.S.N.R.C, Region III
3:15 – 3:40	(4) The U.S.N.R.C. Construction Inspection Program of Reactors Under Construction Robert F. Heishman, Chief of the Performance Appraisal Section-Division of Program Development and Appraisal Office of Inspection and Enforcement U.S.N.R.C.
3:40 – 4:05	(5) Have We in the Construction Industry Effectively Adjusted to Nuclear Regulatory Requirements? Robert J. Washabaugh, Manager, Quality Assurance, Duquesne Light Company
4:05 – 5:00	Panel-Audience Discussion

SESSION II – September 16, 1:30 – 5:00 pm

1:30 – 2:00	(1) Aggregates: Gradation, Deleterious Substances, Potential Alkali Reactivity, Durability Ward R. Malisch, Director of Educational Activities, Concrete Construction Publications, Inc.
2:00 – 2:30	(2) Portland Cements: Requirements vs Performance William F. Perenchio, Consultant, Wiss, Janney, Elstner and Associates, Inc.
2:30 – 3:00	Coffee Break
3:00 – 3:25	(3) Fly Ash, Admixtures and Water Quality Rupert E. Bullock, Principal Civil Engineer, Civil Engineering and Design Branch, Tennessee Valley Authority
3:25 – 3:50	(4) Concrete Mix Design Requirements: Strength and Durability Gerald R. Murphy, Manager, C. L. Grover, Engineer, Concrete Technology, Power Group, Brown and Root, Inc.
3:50 – 4:15	(5) Heavy Weight Aggregates, Cement Grout, Cement Grout in Prestress Applications, Corrosion Prevention Systems Richard A. Bradshaw, Jr., Manager, Civil Technology, Technical Services Division, Daniel International
4:15 – 5:00	Panel-Audience Discussion

PROGRAM

CONCURRENT SESSIONS III AND IV

SESSION III – Sept. 17, 8:30 – 12:00 Noon

- 8:30 – 8:55   (1) Organizing for QA/QC Functions
  Clyde L. Hawn, Quality Assurance Manager, Ebasco Services, Inc.

- 8:55 – 9:20   (2) Qualification and Training of Inspectors
  Donald R. Johnson, Manager of Quality, Bechtel Power Corporation, WPPS Unit 2

  Dennis L. Vanderpol, Chief Construction Engineer, Bechtel National, Inc.

- 9:20 – 9:45   (3) QA/QC Indoctrination Requirements
  Kevin T. Kimmel, Indoctrination and Training Coordinator, Quality Assurance Division, Gilbert Commonwealth Companies

- 9:45 – 10:15   Coffee Break

- 10:15 – 10:40   (4) Civil Engineering Unit Documentation Practices
  James H. Olyniec

- 10:40 – 11:05   (5) Information Processing Methods and Applications
  Dennis D. Millican, Formerly: Manager, Information Services, Gilbert Commonwealth Companies
  Presently: President, Dennis D. Millican Assoc.

- 11:05 – 12:00   Panel-Audience Discussion

SESSION IV – Sept. 17, 8:30 – 12:00 Noon

- 8:30 – 8:50   (1) Freshly Mixed Concrete Control and Mix Uniformity Tests
  James R. Wells, Corporate Quality Assurance Manager, Duke Power Co.

- 8:50 – 9:10   (2) Concrete Sampling Procedures and Correlation Testing for Pumped Concrete
  Douglas J. Haavik, Engineering Specialist, Bechtel Power Corp.

- 9:10 – 9:30   (3) Evaluation of Compressive Strength and Special Properties of Hardened Concrete
  James P. Allen, III, Chief Structural Engineer, Stone and Webster Engineering Corporation

- 9:30 – 10:00   Coffee Break

- 10:00 – 10:20   (4) Experiences With Special Potential Volume Change Test Requirements of ASTM C-33 and ASTM C-342
  A. J. Hulshizer, Supervising Structural Engineer, United Engineers and Constructors, Inc.
  Ashok J. Desai, Structural Engineer, United Engineers and Constructors, Inc.

- 10:20 – 10:40   (5) Utilization of Inspection and Test Data to Obtain Greater Quality Control
  Reginald C. Coupland, Director, Site Quality Control, Beaver Valley Nuclear Power Plant, Energy Consultants, Inc.

(continued)

Session IV - (continued)

10:40 - 11:10 (6) Creep and Shrinkage Studies of Concrete Mixes for Prestressed Concrete Nuclear Containment Structures
Donald W. Pfeifer, Vice President Wiss, Janney, Elstner and Associates, Inc.
Mauro J. Scali, Petrographer, Wiss, Janney, Elstner and Associates, Inc.

11:10 - 12:00 Panel-Audience Discussion

12:00 - 1:30 - LUNCH

PROGRAM

CONCURRENT SESSIONS V AND VI

SESSION V - Sept. 17, 1:30 - 5:00 pm

1:30 - 1:55 (1) The Field Installation of Concrete Anchorage Systems
James A. Flaherty, Manager Engineering Design and Testing Teledyne Engineering Services
Louis J. DiLuna, Senior Eng., Teledyne

1:55 - 2:20 (2) Replacement of Steam Generators in a PWR Nuclear Station
Wallace G. Sanborn, Construction Manager, Stone and Webster Engineering Corporation

2:20 - 2:50 Coffee Break

2:50 - 3:15 (3) Millstone III Containment Dome Lift
James A. Galinsky, Manager of Engineering, Graver Energy Systems, Inc.

3:15 - 3:40 (4) Steel Fabrication in the Nuclear Power Industry
Denis Mason, Quality Assurance Manager, Thames Valley Steel Corporation

3:40 - 4:05 (5) Experiences with Concrete Anchors on Northeast Utilities' Construction Projects
Thomas W. Deshefy, Civil Engineer, Generation Construction Department, Northeast Utilities Service Company

4:05 - 5:00 Panel-Audience Discussion

SESSION VI - Sept. 17, 1:30 - 5:00 pm

1:30 - 2:00 (1) Standards for Concrete Plants and Production Operations
Clyde B. Tatum, Construction Superintendent, Ebasco Services, Inc., WPPSS Nuclear Projects No. 3 and 5.
Albert P. Demers, Project Manager, Associated Sand & Gravel Co., Inc.

2:00 - 2:30 (2) Impacts of Concrete Standards on Mix Verification and Adjustment in the Field
Bruce C. Bennett, Resident Engineer, Ebasco Services, Inc., WPPSS Nuclear Projects No. 3 & 5
Robert V. Potter, Assistant Construction Superintendent, Ebasco Services, Inc., WPPSS Nuclear Projects No. 3 and 5
John R. McCutchen - Construction Engineer-Civil, Ebasco Services, Inc., WPPSS Nuclear Projects No. 3 and 5.

2:30 - 3:00 Coffee Break

3:00 - 3:25 (3) From Batch to Form
William H. Brown, President: Sparks Equipment Company

3:25 - 3:50 (4) Intent of Criteria for Adjustment of Concrete in the Field
Timothy L. Moore, Structural Eng. Gilbert/Commonwealth Companies

3:50 - 4:15 (5) Is Criteria As Written Enough?
Robert M. Esbach, Staff Engineer, Gilbert/Commonwealth Companies

4:15 - 5:00 Panel-Audience Discussion

CONCURRENT SESSIONS VII AND VIII

SESSION VII – Sept. 18, 8:30 – 12:00 Noon

8:30 – 8:55	(1) ACI Standard 117, Tolerances for Concrete Construction and Materials J. Doug Sykes, Jr, Chief Civil Engineer, Ebasco Services, Inc.
8:55 – 9:20	(2) Specified Construction Tolerances: Are They Relevant to Design Requirements and the High Cost of Perfection? Allen J. Hulshizer, Supervising Structural Engineer, United Engineers and Constructors, Inc.
9:20 – 9:45	(3) Engineering Tolerances from a Construction Man's Point of View Frank J. Freiseis, Senior Construction Engineer, The Construction Management Division, Sargent & Lundy Engineers
9:45 – 10:15	Coffee Break
10:15 – 10:40	(4) Nuclear Construction Tolerance Requirements and Their Impact on the Supplier Robert A. Yockin, Assistant Chief Engineer, Reinforcing Bars, Piling, and Construction Specialty Sales Bethlehem Steel Corporation
10:40 – 11:05	(5) Engineering Modeling of Complex Reinforcing Steel Placements Alan J. Boos, Assistant Project Manager, Bechtel Power Corporation, Midland Plants Units 1 and 2
11:05 – 11:30	Panel-Audience Discussion

SESSION VIII – Sept. 18, 8:30 – 12:00 Noon

8:30 – 9:00	(1) Concrete Placement and Post Placement Practices at the Limerick Generating Station Alex C. McLean, Engineer, Construction Division – Limerick Generating Station Philadelphia Electric Company
9:00 – 9:30	(2) Hot Weather Temperature Placement of Mass Concrete James P. Lonergan, Sr. Construction Engineer, Bechtel Power Corp.
9:30 – 10:00	Coffee Break
10:00 – 10:30	(3) Cold Weather Concrete Placement and Post Placement Control Donald E. Dixon, Materials Consultant, Law Engineering Testing Co.
10:30 – 11:00	(4) Evaluation of Quality of Nuclear Concrete Structures John F. Artuso
11:00 – 11:30	Panel-Audience Discussion

FINAL WRAPUP SESSION

11:30 – 12:00   Results Obtained on the U.S.D.O.E. Sponsored Research Project: "Structural Concrete Quality Assurance Practices on Nuclear & Fossil Fuel Power Plant Projects"
Jack H. Willenbrock, Professor of Civil Engineering
H. Randolph Thomas, Associate Professor Civil Engineering
The Pennsylvania State University

SUBJECT INDEX

Page numbers refer to first page of paper

Accelerators, 554
ACI standard 117, 474
Admixtures, 162
Age-strength relation, 335
Aggregates, 12, 453
Anchor bolts, 361, 403
Anchorage systems, 361, 403

Bellefonte Nuclear Plant, 242
Building codes, 170, 435, 465

Cement grout, 185
Certification, 234
Chemical admixtures, 150
Civil penalties, 112
Code maker, 488
Codes, 40, 185
Codes and standards, 396, 417, 435
Cold weather construction, 554
Comprerssive strength testing, 299
Concrete, adjustment criteria, 453
    aggregate, 309
    anchorage systems, 361
    anchors, 403
    batching equipment, 417
    components, 588
    constituents sampling, 288
    construction, 465
    curing, 554
    industry, 58
    mix design, 170, 435
    mixes, 170, 335
    mixing, 279
    practices, 532
    production, 417
    quality, 299
    reinforcement, 522
    sampling, 288
    technology, 588
    testing, 299, 309
    tolerances, 474, 488
Construction, 41, 488
Construction engineer, 497
Construction industry, 128
Construction models, 522

Construction permits, 120
Construction procedures, 465
Construction quality, 465
Containment dome lift, 389
Containment structures, 58, 335
Cost-effectiveness, 25
Costs, 532
Creep tests, 335
Creep/shrinkage, 335

Deleterious substances, 12
Design standards, 71
Designer, 488
Destructive testing, 566
Document storage, 260
Drilled-in anchor installation, 403

Economic analysis, 25

Fabricating techniques, 508
Fabrication practices, 396
Field adjustment, 435, 453
Field erected subassemblies, 389
Field testing, 279, 532
Fly ash, 58, 162
Formwork construction, 497

High ambient air temperatures, 542
High density concrete, 185
Hot weather, 542

Impact studies, 25
Indexes, 260
Industry standards, 588
Information retrieval, 260
Inspection, 8, 234
Inspector traning, 10
Inspectors, 10

Licensing, 107
Low ambient temperatures, 554

Mass concrete, 542
Material requirements, 185
Materials testing, 58, 279, 288
Mix design requirements, 170

Non-destructive testing, 566

Non-standard bending, 508
Nondestructive testing, 299
Nuclear construction code requirements, 140
Nuclear electric power generation, 6, 40
Nuclear license, 107
Nuclear power plants, 170
    construction, 6, 10, 51, 112, 120, 185
    design, 6
    licensing, 6
    operation, 112
Nuclear regulation, 128
Nuclear Regulatory Commission, 107, 112
Nuclear Regulatory Commission inspection, 120
Nuclear tolerance requirement, 508

Operating licenses, 120

Performance tests, 335
Plant licensing, 242
Portland cement, 150, 162, 309
Pre-outage planning, 372

Quality assurance, 128, 205, 588
Quality assurance training, 239
Quality control, 205, 325
Quality control documentation, 242
Quality evaluation, 566

Radiation exposure, 372
Reactor containment dome, 389
Ready-mixed concrete, 447
Regulations, 8, 25, 41, 185
Reinforced concrete, 497
Reinforcing bars, 508
Reinforcing steel, 522
Restrictive tolerances, 508

Seismic restraint, 361
Shrinkage, 335
Slump loss, 150
Specifications, 40, 58, 453
Standard specifications, 309
Standards, 41, 58, 71, 185
Steam generator replacement, 372
Steel fabrication, 396
Structural integrity, 566

Temperature, 453
Tennessee Valley Authority, 242
Test data, 325
Test/inspection, 325
Three Mile Island, 107
Time limitations, 453
Tolerance associated cost, 488
Training, 10
Transportation of concrete, 447

Unit documentation practices, 242

Water quality, 162

AUTHOR INDEX

Page number refers to first page of paper

Allen, III, James P., 299
Archer, John C., 137
Artuso, Joseph J., 276, 566

Bennett, Bruce C., 435
Boos, Alan J., 522
Bradshaw, Richard A., Jr., 185
Brown, William H., 447
Bullock, Rupert E., 162

Chauvin, Glen A., 472
Christesen, Russell J., 25
Counsil, William G., 6
Coupland, Reginald, 325

Demers, Albert P., 417
Desai, Ashok J., 309
Deshefy, Thomas W., 403
DiLuna, Louis J., 361
Dixon, Donald E., 554

Eshbach, Robert M., 465

Fisher, John M., 415
Flaherty, James A., 361
Freiseis, Frank J., 497

Galinsky, James A., 389
Gallagher, Eugene J., 68
Gannon, Alice, 529
Grover, Chaman L., 170

Haavik, Douglas J., 288
Hawn, Clyde L., 205
Heishman, Charles F., 120
Hulshizer, Allen J., 309, 488

Johnson, Donald R., 225

Keppler, James G., 51
Kimmel, Kevin T., 234

Lonergan, James P., 542

Malisch, Ward R., 140
Mason, Denis, 396
McCall, Howard W., 41
McCutchen, John R., 435

McLean, Alex C., 532
Mercurio, William F., 200
Millican, Dennis D., 260
Moore, Timothy L., 453
Murphy, Gerald R., 170

Norelius, Charles E., 112

Olyniec, James H., 200, 242

Palmeri, Aldo, 1, 472
Perenchio, William F., 150
Pfeifer, Donald W., 335
Philleo, Robert E., 58
Potter, Robert V., 435

Sanborn, Wallace G., 372
Scali, Mauro J., 335
Stevenson, John D., 71
Sykes, J. Doug, 474

Tatum, Clyde B., 417
Tatum, C.B., 415
Thomas, H. Randolph, 588
Toth, Stephen R., 357

Vanderpol, Dennis L., 225
Vollmer, Richard H., 107

Washabaugh, Robert J., 128
Wells, J.R., 279
Willenbrock, Jack H., 1, 529, 587, 588

Yockin, Robert A., 508